The Lord of Uraniborg

Frontispiece from Tycho's *Astronomical Letters* of 1596. The inscription reads "Likeness of Tycho Brahe, son of Otte the Dane, Lord of Kundstrup and founder of the Castle of Uraniborg on the Island of Hven in the Danish Hellespont, and inventor and builder of the astronomical instruments used there. Done in 1586 at the age of 40."

THE LORD OF URANIBORG

A Biography of Tycho Brahe

VICTOR E. THOREN
Professor of History and Philosophy of Science
Indiana University

with contributions by
JOHN R. CHRISTIANSON

CAMBRIDGE UNIVERSITY PRESS

CAMBRIDGE
NEW YORK PORT CHESTER MELBOURNE SYDNEY

CAMBRIDGE UNIVERSITY PRESS
Cambridge, New York, Melbourne, Madrid, Cape Town, Singapore, São Paulo

Cambridge University Press
The Edinburgh Building, Cambridge CB2 2RU, UK

Published in the United States of America by Cambridge University Press, New York

www.cambridge.org
Information on this title: www.cambridge.org/9780521351584

First published 1990
This digitally printed first paperback version 2006

A catalogue record for this publication is available from the British Library

Library of Congress Cataloguing in Publication data
Thoren, Victor E.
The Lord of Uraniborg : a biography of Tycho Brahe / Victor E.
Thoren; with contributions by John R. Christianson.
p.   cm.
ISBN 0–521–35158–8
1. Brahe, Tycho, 1546–1601.   2. Astronomers – Denmark – Biography.
3. Astronomy – History – 16th century.   I. Christianson, J. R. (John
Robert)   II. Title.
QB36.B8T49   1990
520′.92–dc20
[B]   90–1477
                                                 CIP

ISBN-13 978-0-521-35158-4 hardback
ISBN-10 0-521-35158-8 hardback

ISBN-13 978-0-521-03307-7 paperback
ISBN-10 0-521-03307-1 paperback

Contents

Preface   vii

1   A Noble Humanist   1

2   The New Star   40

3   Becoming a Professional   74

4   The First Years on Hven: 1576–1579   105

5   Urania's Castle   144

6   The Flowering of Uraniborg   192

7   First Renovations: The Solar Theory   220

8   The Tychonic System of the World   236

9   High Tide: 1586–1591   265

10   The Theory of the Motion of the Moon   312

11   The Last Years at Uraniborg   334

12   Exile   376

13   A Home Away from Home?   416

Epilogue   471

Appendix 1: Abbreviations for Frequently Cited Sources   481

Appendix 2: Glossary of Technical Terms   483

Appendix 3: The Tychonic Lunar Theory   486

Appendix 4: Figures for Footnotes   497

Appendix 5: Tycho's Dwellings in Exile   500

Appendix 6: Letters, 1599–1601   502

Author Index   519

Subject Index   521

Preface

PERHAPS the best measure of the perception of Tycho's influence through the years is that this book is at least the fifth serious biography of him. Less than fifty years after Tycho's death, when the concept of biographical writing had barely begun to be extended from the lives of saints to the lives of kings, the French Catholic philosopher Pierre Gassendi had already conceived the notion of portraying the life and career of the Danish Lutheran astronomer Tycho Brahe. Because he did, and because he wrote to Denmark to get as much information about his subject as he could, we have details of Tycho's life that would probably not otherwise have been preserved. Gassendi used those details – along with Tycho's observations, letters, and published descriptions of his scientific work – to amplify the seven-thousand-word autobiographical sketch written by Tycho in his last years into an eighty-thousand-word biography published in 1654 as *Tychonis Brahei, Equitis Dani, Astronomorum coryphaei, Vita.* By the following year a second edition had been printed, and reprintings appeared in 1658 and 1717 as the fifth volume of Gassendi's own *Opera Omnia.*

As was the case for history generally, the nineteenth century was the great period of discovery in Tycho studies. Numerous documents touching on Tycho's life were found in repositories in Copenhagen, Prague, Vienna, and Basel. The most active excavator was the Danish historian F. R. Friis, who, unfortunately, did much of his work *after* he published the first modern biography of Tycho in 1871. The only serious shortcoming of Friis's work was that he had very little feel for Tycho's life's work.

This crucial facet was added by the distinguished astronomer and historian of astronomy, J. L. E. Dreyer. Utilizing the general commentaries on Tycho's work that had already begun to appear a century earlier in various histories of mathematics and/or astronomy, Dreyer produced an authoritative biography in English (1890; German edition, 1894), accurately subtitled "A Picture of Scientific Life and Work in the Sixteenth Century." (For a complete bibliographical citation, see Appendix 1.)

Unfortunately, Dreyer, like Friis, also did his work backwards, so to speak. After writing the biography (extended sections of which are virtual translations of either Gassendi or Friis) – and, no doubt, to a considerable degree because of the enthusiasm generated by his

extremely readable portrayal of Tycho – Dreyer embarked on what
turned out to be a fifteen-year task of publishing Tycho's collected
works. The resulting fifteen volumes of *Tychonis Brahe Dani Opera
Omnia* – published between 1913 and 1929 with the collaboration of
the classicist Hans Raeder – are widely recognized as models of the
enterprise and have been indispensable to the writing of my biography. The virtue of having quick and easy access to such a mass
of materials, all of which consist of either rare books or unique
manuscripts, will become obvious. The difference between puzzling
through convoluted Renaissance Latin *in print* and deciphering it
from the hand of someone writing in an observation log at night, or
entering additional notes in limited space between the observations at
some later time, will be less obvious but surely has been even more
important. Since the publication of the *Opera Omnia*, a few more
documents have been discovered. But with the possible exception of
one displaying an intermediate form of Tycho's lunar theory, thus
far they have been of only marginal interest.

   Among the spate of quick and worthless publications that appeared as a cruel jest to commemorate the fourth centenary of
Tycho's birth was a biography written in English by John A. Gade.
It was sufficiently readable to have been reprinted since then, but it is
not authoritative in any sense and, moreover, suffers from the
author's proclivity for embellishing historical fact without giving
any warning that he is doing so. How little Gade or most of the other
impromptu writers of 1946 had to contribute to Tycho scholarship
may be judged by the fact that in the following few years both
Dreyer's seventy-year-old biography and Gassendi's three-hundred-
year-old original (translated into Swedish) were republished. The
latter was accompanied by extensive annotations that effectively
doubled the total content of the original, and in 1971 the translator
and annotator of this work, Wilhelm Norlind, published his own
biography (in Swedish), which presented much new information
gleaned from the archives of Europe, primarily concerning the
history and preservation of Tycho's printed works.

   Historians have been described as people who carry bones from
one grave to another. Philosophers have been characterized, similarly unflatteringly, as people who make a living taking in one another's
wash. Inevitably, in the fifth biography of a subject, there will be
some of each enterprise. But the twentieth century has produced a
great deal of scholarship, and some of it has shed light even on the
basic data of Tycho's life. This light has been mostly indirect and has
been thrown mostly into such shadowy areas as Tycho's fostering by
his uncle and aunt, the duel that disfigured him for life, and the
background and implications of his morganatic marriage. Looming

behind these curious characteristics of his life and fundamentally responsible for all three – and for most of the rest of Tycho's aspirations and actions – is the fact that he was born into the highest ranks of the Danish nobility. Previous biographers of Tycho have paid little attention to this, either because they regarded the consequences as too obvious to require discussion or because they felt them to be an inappropriate subject for comment. For people raised in twentieth-century, more or less classless, society, however, the issue cannot be left to chance (no matter how much the notion of a noble class may offend our sensibilities), because the implications and ramifications of *privilege* are infinitely more subtle than are the technicalities of Tycho's astronomy and every bit as significant for his career: To try to understand Tycho the noble as Tycho the man is futile.

Fortunately, however, the basis of noble privilege was noble lineage. And because of this, noble families were collecting, already before the sixteenth century, records that preserved vital facts concerning the conditions of noble existence. At the end of the last century these genealogies were published in the "Annals of the Danish Nobility" (see *DAA* in Appendix 1). I have mined them extensively for facts and generalizations that reflect light on the noble life-style. I have also benefited considerably from the labors of other Danish scholars who compiled the extended sketches of prominent personages available in the "Danish Biographical Lexicon" (*DBL*).

It is in the description of Tycho's scientific work, however, that previous biographies are now truly obsolete. Because the history of science has become a considerably less lonely undertaking than it was for my predecessors, it is possible to say that virtually every aspect of Tycho's work in astronomy has now been elaborated well beyond the level at which it was presented by Dreyer, who alone among Tycho's previous biographers was able to research it, or even understand it. The fruits of the interests, talents, and industry of Yas Maeyama, Owen Gingerich, Robert Westman, and Edward Rosen will be clear from my references to them in Chapters 7, 8, and 13. I have not always used their ideas in precisely the form in which they advanced them, but (for that reason, among others) I have used their translations wherever they offered them. Most of the translations, however, are my own renditions.

The nuclei of Chapters 5 through 10 are my own publications and depict the life's work of an astronomer. Most of the discussion will be comprehensive to the lay reader who is willing to give it serious attention and to spend perhaps five minutes coming to grips with the concept of parallax. Chapters 7 and 10 will probably be heavy going for most readers, even though I have included a glossary

(Appendix 2) of technical terms in order to ease the burden. But the chapters are worth the perseverance required to understand them, because they present the most impressive (and so the least-known) aspects of Tycho's work and should at least convey an appreciation for the extent to which sixteenth-century astronomy had advanced beyond its sister sciences.

Those who contributed smaller pieces to my particular puzzles are, in general, too numerous to mention except in my footnotes. But I would like to thank Kristian-Peder Moesgaard, John North, Sam Westfall, and Curtis Wilson for having also provided the inspiration and comfort that comes from knowing that they have read and reacted to the various results I have presented in the scholarly literature over the years. My greatest debt of all is to John Christianson, to whose writings the reader will see numerous references. This biography was begun as his project and would surely have been a better one if it had been possible for him to contribute more to it. As it was, he drafted Chapters 1 to 4 and part of Chapter 11, translated the Danish letters I selected to illustrate Tycho's problems and attitudes in exile, and answered numerous queries concerning matters of fact, sources of information, and choices of interpretation. Those who differ with my views might find that John also objected to them, but that I was unwilling to be persuaded by his judgment.

The development and testing of ideas are not the only cooperative aspects of book writing. Funding for research time and travel is crucial, and I have been the recipient of a considerable amount of it from the National Science Foundation, the National Endowment for the Humanities, and the Research Grants Office (under a succession of names) of Indiana University. The encouragement of my colleagues and the material resources of the Department of History and Philosophy of Science at Indiana University have also been significant assets. Less direct but surely also even more indispensable to the scholar are the other institutional funds that support research libraries and subsidize the various practical aspects of generating ideas and getting them into print. Most important to me has been the Royal Library (Det Kongelige Bibliotek) in Copenhagen, which has preserved Tycho's letters and observations and made them available for the perusal of the few scholars in each century who might be interested in them. I am also indebted to professional journals with such unlikely titles as *Isis*, *Centaurus*, and *Journal for History of Astronomy* for having published my articles and given me permission to quote from them. Abaris Books, Inc., and Pergamon Press, Inc., respectively, have given me permission to use the extended translations presented in Chapter 13 and the graphs reproduced in Chapter

7. And finally, the Cambridge University Press has provided traditional editorial services that are rapidly becoming extinct in academic publishing.

The writing of a book takes a toll on anyone who gets close to the would-be author, but I take special pleasure in thanking my good friends Paula and Fritz Taggart for having tolerated many summers of books and manuscripts in their living room. And no one has encountered this burden more capriciously or assumed it more graciously than my cousin Hr. Simon Isaksson of Solna, Sweden, who chauffeured me around to most of Tycho's old haunts in Skaane. One of them was Knudstrup, where the family that has owned it since 1771 – specifically, Hr. Rutger Wachtmeister – graciously escorted us through the edifice built by Tycho's father in 1551 and showed us the very extensive collection of Tychoniana.

For the most part, however, the encouragement to assume and carry out this extended undertaking has come from the women of my life, and I hereby dedicate the result to them:

Helen (Ling) Thoren: 1913–41
Alice Thoren: 1883–1967
Janice (Thoren) Day
Shirley (Thoren) Knowlen
Zandra (Sloan) Thoren: 1935–86
Vikki Thoren
Krista Thoren
Lori (Thoren) Oh
Paula Thoren
Judith Mann Thoren.

Chapter 1
A Noble Humanist

A MONG the mass of detail that constitutes the personal, social, cultural, and intellectual background of Tycho Brahe's scientific achievement, the one indispensable fact is that he was born a Brahe, that is, born not merely into the Danish nobility but also into the small fraction of the noble class that had historically played significant roles in the administration, governance, and defense of the realm. The epitome of this special status was membership in the Rigsraad, or Council of the Realm. Nominally an advisory body for the king but actually an oligarchical institution devoted to defending the interests of the most powerful noble families, the Rigsraad consisted of twenty-odd members who declared war, concluded treaties of peace, appointed regents (among themselves, naturally), seated kings, and participated with kings in virtually every aspect of the daily affairs of state.[1]

All four of Tycho's great-grandfathers and both of his grandfathers had been councillors (see Fig 1.1).[2] His paternal grandfather and namesake, Tyge Brahe, had held that honor only briefly before being killed during the siege of Malmø in 1523, fighting in the cause that put Frederick I on the throne and brought the Reformation to Denmark. But Tyge's widow, Sophie Rud, was descended from the equally powerful Rosenkrantz and Gyldenstierne families and thus had her father and brother on the council to look after the interests of her young family.

In addition, Tyge's brother Axel was not only a *rigsraad* but long served as governor of the province of Skaane in which the Brahe heritage was seated. Axel was among the first Danish lords to convert to Lutheranism, and he supported the militant Lutheran King Christian III so effectively during the Danish phase of the Protestant Reformation that he won the honor of carrying the scepter at Christian's coronation in 1537. During these years of civil strife and religious upheaval, the sons of Tyge reached adulthood. Jørgen (George), the oldest, was brought to court in 1535, at the age of twenty, and Otte joined him shortly thereafter. In 1540 they

---

[1] See Chapter 11 for a summary of the council membership as of 1552 and 1590 and a discussion of the kinship of the members with Tycho.

[2] The genealogy of the Brahe family, with the vital statistics of all 177 members known to have existed from the fourteenth century until the line died out in the eighteenth century, is in *DAA* V, 97–115.

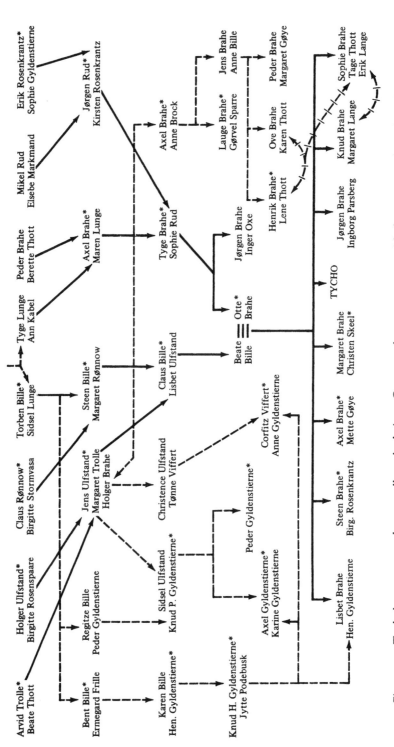

Figure 1.1. Tycho's ancestors and some collateral relatives. Compare the top two rows with the names on the arch (males) and supports (females) of the Frontispiece. Asterisks signify rigsraads.

received their first recognition for service to the realm, in the form of joint fiefdom of Storekøbing. By 1542, Jørgen had advanced to the command of Tranekaer Castle, from which, through the next fifteen years, he was to move upward to successively more important posts. Otte's career was even more distinguished and culminated in the governorship of crucial Helsingborg Castle and a seat among the elect in the Rigsraad.

In 1544, Otte Brahe married Beate Bille. Unlike the Brahe family, which had achieved and maintained its influence by prowess with the sword, the Bille family owed much of its ascendancy to persuasion with the word. From as far back as Archbishop Peder of Lund, who was primate of Denmark for eighteen years before his death in 1436, the Bille family had combined ecclesiastical influence with secular service to channel great wealth into the hands of those members who did not enter the Church.

When the Church was threatened by the Reformation in the third decade of the sixteenth century, the traditions of a family whose blood relations included seven of the eight current (Catholic) bishops of Denmark could permit only one response. And because no fewer than six Billes sat in the Council of the Realm, and most were warriors who commanded key strongholds throughout Denmark and Norway, that response was not restricted to words. But after engaging in a determined, if ultimately losing, struggle in the cathedral, in the Rigsraad, and on the battlefield to retard the advance of Lutheranism, each sought to repair his fortunes in his own way.

One of the means chosen by the most powerful of them, Claus Bille, was alliance with the Brahes, by the time-honored practice of intermarriage. He gave the hand of his eighteen-year-old daughter, Beate, in marriage to the still-unproven twenty-six-year-old Otte Brahe. In so doing, this hard-bitten veteran of the Stockholm bloodbath and second cousin of the reigning (1523–60) king of Sweden, Gustavus Vasa, created the conditions under which he would become the grandfather of Tycho Brahe.[3]

For the bride, the aristocratic splendor of the wedding was followed by the typical travail of repeated pregnancy and childbirth that was the common fate of women of all stations in that era. In Beate's case, it was twelve children in twelve years, of whom eight survived infancy. And Beate Bille – aristocratic women in sixteenth-century Denmark retained their maiden names after marriage – was one of the hardy and fortunate few who survived to live to the ripe old age of seventy-eight. Her first child, a daughter, Lisbet, was born within

---

[3] Claus Bille's maternal grandmother, Birgitta Kristiensdatter (Vasa), was a sister of King Gustavus's grandfather. The genealogy of the Bille family is in *DAA* VII, 58–94.

ten months of the wedding. A second child, who died very young, must have been born prematurely, because on 14 December 1546, fifteen months after Lisbet's birth, Beate gave birth to twins. The firstborn lived to be baptized with the name of his paternal grandfather, Tyge.[4]

The conditions of baby Tyge's birth virtually ensured him the opportunity to live as a veritable prince. But they did not dictate his destiny completely. For Tyge would not be raised by his parents. As he understood the situation in his mature years, his uncle Jørgen "without the knowledge of my parents (took) me away with him while I was in my earliest youth [and] brought me up and thereafter supported me generously during his lifetime . . . and always treated me as his own son."[5] Pierre Gassendi, writing a hundred years after the event and after having talked to two grandsons of Tyge's brother Steen, amplified this account slightly. According to him, Jørgen justified his action to Otte and Beate by pointing out that they had a second son, Steen, to raise and claiming that it was only fitting and proper for them to share their wealth, so to speak.[6]

There was nothing particularly unusual about taking in foster children. Indeed, when epidemic disease was rampant, warfare endemic, and child bed an ever-present mortal danger, children frequently lost one or both parents. In the Danish colony of medieval Iceland, blood feuds had rendered fostering the resort of choice, to reduce the likelihood that an entire family would be wiped out in a raid. Among the sixteenth-century Danish nobility, extended missions abroad for warfare or diplomacy often meant that children were left behind to be raised by grandparents or other near kin. In Tyge's case, however, it seems not to have been the interest of either the child or the parents that was being accommodated but that of the foster parents.

The sole basis for Tyge's fostering was that Jørgen Brahe and his wife were childless. And even this characterization was clearly an understatement, as at the time of Tyge's birth his foster mother (and aunt), Inger Oxe, was only about twenty.[7] Because Steen was born a

---

4   Sometime around 1556, Otte recorded the birth dates of his family. The list was published as the first of 301 documents concerning the life of Tycho (exclusive of three volumes of letters) in vol. XIV of his *Opera Ømnia*.
5   V, 106, as translated in *Raeder*, 106. For a list of abbreviations of commonly used sources, see Appendix 1.
6   *Gassendi*, 3–4.
7   Inger's birth and marriage dates are unknown. However, if the birth dates and order of birth of various siblings are accurate, the earliest that Inger could have been born was late 1526. See *DAA* XXIV, 343 And unless she was married at a much younger age than her sisters were, she was not married very long before Tycho's birth. The earliest date documented for her status as Jørgen's wife is 10 February 1548. The closest thing to a justification for what

year and a week after Tyge, it seems reasonable to speculate that the "transfer" occurred in the middle of Tyge's second year, after baby Steen was past the greatest uncertainties of infancy, presumably after Jørgen had unsuccessfully attempted to obtain Tyge by negotiation, and during the months of pleasant weather when Otte and Beate would be away from home socializing instead of at home protecting their little sons against such an unlikely event as kidnapping. When the parents finally accepted the situation (consoled, probably, by the knowledge that another little one was already on the way), Tyge settled into a household very much like the one he had left.

Although Tyge was raised more like a cousin than a brother of his siblings, and therefore undoubtedly spent more time by himself than he would otherwise have done, in many ways he reaped the benefits of both possibilities. He was to grow up as the only charge of a doting aunt and uncle but would later compete in the adult world of power politics as the oldest of five brothers.

Tyge's stepfather and uncle, Jørgen Brahe, was a man cut from the ancient warrior pattern. He was a man of action rather than a statesman, but he was a convivial person who could hold his own in the hard drinking circles at court and function efficiently as an administrator of fiefs. Like his brother, he married into a family whose traditions were somewhat broader than the exclusively martial ones of the Brahes.

The Oxes were relatively recent arrivals in Denmark, having come from France at the end of the fourteenth century.[8] Although much less prolific than the Billes, they had produced four *rigsraads* during the preceding hundred years before losing their influence as a result of the civil upheavals of the 1530s. By the late 1540s, however, the Oxes were on the rise again, largely through the drive and ingenuity of Inger's oldest brother, Peder Oxe. As early as 1548, at the age of twenty-eight, Peder led Princess Anne's entourage to her wedding with Duke Augustus of Saxony: A few years later he was to mediate the peace that made Augustus the elector of Saxony. By 1552, Peder's diplomatic talents had propelled him into the select group of older men that constituted the Council of the Realm.

Peder's sister Inger seems to have shared many of his intellectual interests and capacities. Her letters reflect a woman of charm and highly developed social grace, and she is known to have remained very close to Princess Anne of Denmark/Saxony, who later gained

may have been arranged as a platonic marriage is S. C. Bech's statement: "Although it was very abnormal not to be married, because holy marriage was almost a Christian duty, prominent personalities ... could get permission to live and die as batchelors" *Danmarks historie* 6 (Copenhagen Politikens: Forlag, 1963), p 427.

[8] *DAA* XXIV, 339–45.

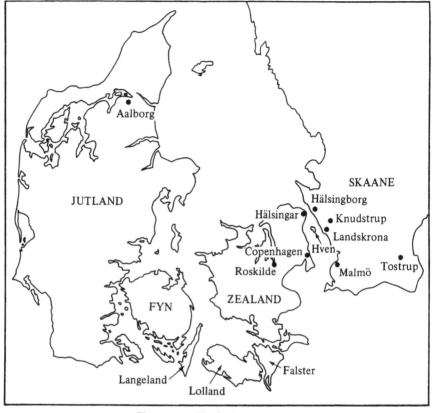

Figure 1.2. Tycho's Denmark.

some fame as an alchemist. Most significantly – for we know little of Inger's actual attainments – she came from a learned and cultivated family. Her mother was a Gøye, a daughter of the kingmaker of the Reformation era, councillor Mogens Gøye of Krenkerup. And unlike Beate Bille, Inger used her family traditions to impel Tyge toward an education that was markedly different from what his natural father regarded as appropriate to a Brahe.

We know little about Tyge's early youth, not even where he was raised. "Home" was his uncle's ancestral seat of Tostrup, in the eastern portion of the province of Skaane, on what is now the Swedish side of the shipping channel (Øresund) which then constituted the heart of the sea kingdom of Denmark. But there must have been at least occasional visits to Otte's somewhat distant seat at Knudstrup (Figure 1.2) to see his natural parents and their ever-increasing brood. When Tyge was five years old, the manor house in which he had been born was torn down and replaced by a great fortress of red brick, part of which still stands. This new building, at that time surrounded by the waters of a broad moat, would one day belong to him.

Otte's and Jørgen's status as members of one of the conciliar families entitled them to "employment" in the profitable task of administering a royal fief. Aside from the element of preparedness for defense, traditionally implicit in most of these fiefdoms, conscientious administration demanded a certain amount of time in residence at the fief, which might be located anywhere in the realm. Moreover, because the distribution of these political plums was subject to the vicissitudes of power politics and the vagaries of royal whim, one had to maintain enough presence at court to protect one's interests. Any vassal, therefore, who was at all ambitious, found himself frequently moving. Jørgen was promoted from Tranekaer Castle on Langeland to Naesbyhoved Castle on Fyn in 1549 and then to Vordingborg Castle in 1552. By this time, Tyge would probably have been old enough to remember the move and the ceremonies by which the command of this immense medieval stronghold was transferred to his foster father.

Vordingborg was on the south coast of Sjaelland, on the main travel route between Copenhagen and the continent and it therefore, attracted many visitors besides the various members of the Brahe, Oxe, Rud, and other related families who came regularly. Duke Ulrich of Mecklenburg arrived with his court in 1556, and the party of young Princess Elizabeth of Saxony passed through in 1557, accompanied by sixty knights, on the way to visit her Danish grandparents. The peripatetic court of King Christian III also stayed in Vordingborg from time to time. Vordingborg was near the estates of Peder Oxe and across the waters of the Smaaland passage from the fertile islands of Lolland and Falster, which were part of the widow's jointure of Queen Mother Sophie, King Christian's stepmother. Their administrative center was Nykøbing Castle, another medieval fortress that stood on an islet in the straits between the two islands. In 1555, Queen Sophie invested Jørgen Brahe with the fief and command of Nykøbing, giving him one of the greatest assemblages of fiefs in Denmark.

Jørgen's perhaps ill-advised attempt to serve simultaneously two masters who had never been on good terms with each other came to grief rather quickly. In the opening rounds of a power struggle that was to culminate in the spectacular fall and exile of Peder Oxe in mid-1558, Jørgen relinquished his fief from the king and transferred his seat to Nykøbing.[9] Presumably, Inger and Tyge followed Jørgen on his periodic moves from one fief to another and on his frequent

---

[9] For a brief sketch of Jørgen's life, see *DBL* III, 566–7.

trips back to Tostrup in Skaane. When they did not accompany him to court, they may well have gone to check on the property held by Inger as her share of the Oxe family domain. This all was part of the life of a lord, and it was the only life Tyge would know until he started school at about the age of seven.[10]

As was the case throughout Europe for at least two hundred years both before and after Tyge's day, grammar school in Denmark was the place where one learned Latin grammar. Not surprisingly, such institutions were almost invariably associated with the church. Most were monastic schools. If a nobleman's son attended one of them, it was probably because the school was under the administration of some noble (Lutheran) governor who was related to him.[11] At such schools, the noblemen's sons chanted and studied with the common schoolboys but served as pages in their kinsman's household to develop the aristocratic graces appropriate to their station.

Most of the noble children, however, went to cathedral schools in the episcopal towns. In order to ensure suitable accommodations and proper supervision in cultivating the habits of a gentleman, they were generally lodged in the household of the bishop or some other substantial clergyman of the city. Twenty years earlier, such a household would have been the establishment of a celibate Catholic aristocrat like Tyge's relatives on the Bille side.

Since the Reformation, however, the bishops had been Lutheran theologians of middle-class origins, whose livings were prosperous but not princely and whose households centered on their families. Virtually all of them had studied at Wittenberg, and they modeled their households on those of professors like Martin Luther and Philipp Melanchthon, with whom they had lived as student boarders. Around the long tables in their paneled chambers, family, pupils, guests, and cathedral colleagues all gathered for meals, just as their various counterparts had done at the castles in which Tyge had been raised until then. As the focal point of a center of learning, however, this table featured conversation very different from that at a castle table. Because of this innovation of the Lutheran Reformation, Tyge was able to participate in table talk of a kind to which he would not have had access just a generation earlier, when all learned communities in Europe had still been organized as celibate colleges, monasteries, and cathedral chapters.

Tyge's next younger brothers, Steen and Axel, are known to have gone to the cathedral school at Aalborg, where Otte Brahe was in

[10]   V, 106.
[11]   On the Danish grammar schools of this period, see Birte Andersen, *Adelig opfostring: Adelsbørns opdragelse i Danmark 1536–1660* (Copenhagen: G, E, C, GAD, 1971), pp. 62–80.

command of castle and county. If Tyge's experience can be inferred from theirs, he was sent to a cathedral school near Vordingborg. All we actually know is that he "was sent to grammar school in my seventh year" and continued his elementary studies until he was about twelve.

All the schools had a decidedly clerical stamp, one that had not changed greatly through the Reformation. Except for an occasional rector, the teachers were unmarried men. The students were nominally in the Church too and joined their teachers in wearing clerical garb. Schooling began at seven o'clock and lasted until late in the afternoon, except for Wednesday and Saturday afternoons, which usually were free. Latin grammar and Lutheran religion dominated the curriculum, but some schools taught Greek and even elementary mathematics. Music and theater were also regarded as essential to a basic education. Students sang at church services, weddings, funerals, and festival processions. They memorized whole plays of Terence and Plautus as part of their study of Latin, and they frequently gave performances of them. By the time they reached their early teens, they were supposed to be sufficiently grounded in the teachings of Lutheranism and the essentials of Latin grammar to proceed to the university.

The transition from Latin school to university was primarily the progression from studying the Latin language to studying the classical works written in it. Personally, it was – for a student of Tyge's class – only the move from the household of a bishop to the intellectual milieu of a university professor's home, where the young students began by sitting in on discussions at the table, read and attended lectures under the supervision of their learned host, and often were tutored by one of the older students living in the same household. For many students, the biggest change was probably the transition from a rural or provincial town to metropolitan Copenhagen – the capital, largest city, and seat of the only university in the realms of Norway, Denmark, and Schleswig-Holstein.

The University of Copenhagen occupied the old palace and grounds of the Catholic bishops. Its portal was emblazoned with an eagle, under which was the Latin inscription translated as "He looks up to the light of heaven."[12] No one who ever entered the university would fit that description better. Tyge's three years there were so uneventful that we have virtually no information about them. Not even the place of his lodging is known, but the relatively early age at which Tyge's mathematical interests began suggests that his aunt and uncle may have placed him in the household of Nicolaus Scavenius,

[12] *GADE*, 15.

professor of mathematics and client of the Oxe family. There are
other reasons to think that he may have lived with the famed professor of theology, Niels Hemmingsen, as his future brother-in-law,
Christen Skeel, and as his preceptor of later years, Anders Sørensen
Vedel, did. The fact that we have a date for the beginning of his
tenure in Copenhagen, 19 April 1559,[13] suggests that Tyge may
actually have matriculated at the university. In general, however, noblemen's sons did not register for university degrees but
merely attended selected courses of lectures and various other exercises as part of a more widely diversified program of study.

Although the sons of middle-class merchants, urban patricians, or
Lutheran clergymen could benefit from university degrees in their
pursuit of clerical or academic careers, boys of Tyge's rank were
born with all the credentials that they would need. However, Tyge's
later career is evidence that he acquired as good a classical education
as any of his contemporaries did. Basically, this meant expanding
his control of Latin grammar to acquire skill in logic and rhetoric,
the other two arts of action comprising the medieval *trivium*. Such
studies in the formal techniques of debate and public speaking were
regarded as relevant to a career in power politics. We also know that
Tyge learned some Greek and possibly even a bit of Hebrew.

Already in these early years at Copenhagen, Tyge was apparently
developing an interest in the four mathematical sciences of the
quadrivium: arithmetic, geometry, astronomy, and music. During
1560, Tyge acquired the great elementary astronomy text of the
Middle Ages, Sacrobosco's *On the Spheres*, which Professor Scavenius used in his lectures. In the following year Tyge purchased the
much more advanced *Cosmography* of Peter Apian and the *Trigonometry* of Regiomontanus. In his inscription in these works, "Tycho
Brahe, Anno 1561," we have the first appearance of the Latin form of
his first name (pronounced Teeko),[14] under which Tyge (pronounced
Teegeh) was to make his way in the learned world and by which he is
remembered today.

Even though Tycho undoubtedly pursued astronomy further and
more successfully than his schoolmates did – to the point of even
purchasing an ephemeris of planetary motions[15] during this period –

---

[13] Tycho recorded the dates of his various moves in a horoscope of his life, some of whose
details were preserved by one of his students. See John Christianson, "Tycho Brahe's Facts
of Life," *Fund og Forskning* 13 (1970); 20–25.

[14] Tycho always signed his name with either an *ij* or the equivalent, frequently used but
somewhat nonsensical, *ÿ*. In the handwriting of Tycho's Denmark, *ij* stood for the sound
pronounced like the *ee* in the English "sheet." If Tycho had wanted his name to be
pronounced to rhyme with the English "high," he would presumably have spelled it
"Taecho," which he never did.

[15] V, 107. The ephemeris was that of Stadius (*Gassendi*, 6).

his interest was by no means unusual. Indeed, a general concern for astronomy permeated the whole intellectual atmosphere at Copenhagen.

The basis of this orientation was a movement called Philippism. Although it was primarily a theological doctrine developed by Luther's chief lieutenant, Philipp Melanchthon, Philippism articulated a conception of the church that emphasized education. To Melanchthon, the church was essentially an educational institution, whose great and vital purpose was to teach the true path to salvation. In order to succeed at this mission, Melanchthon believed that the church had to be staffed by a clergy of scholars and teachers, men whose theological training was firmly grounded in a mastery of the seven liberal arts. Fluency in Latin, Greek, and Hebrew was prerequisite to a true comprehension of the Holy Scriptures and the writings of the church fathers. Competence in rhetoric and dialectic as well as broad familiarity with literature and history were essential to effectiveness in the pulpit. And knowledge of the four mathematical disciplines paved the way to understanding the secular and spiritual worlds. Music required no justification. Arithmetic and geometry were subjects of great practical use, as well as the path to knowledge of the heavenly science of astronomy. Astronomy, finally, not only established the calendar of church ceremonies but also led to contemplation of the Creator and revealed the cosmic (astrological) influences that affected people's lives.[16]

By Tycho's day, Melanchthon's ideas had won sufficiently broad acceptance to have been institutionalized in many of the leading Lutheran universities. This development was important to both astronomy in general and Tycho Brahe in particular. It created the environment in which Tycho acquired his interest in the heavens and provided the resources that were to allow him to develop it. It meant that the Lutheran universities had at least one professional chair in mathematics (which consisted largely of astronomy) and frequently had two, as did Leipzig, Wittenberg, and Rostock, where Tycho was to study subsequently. In England, by contrast, where Melanchthon's theological ideas did not prevail, there was not a single university chair in mathematics throughout the whole of the sixteenth century.

It was not merely a matter of having formal instruction: In fact, there is no indication that Tycho attended many lectures. But it cannot be irrelevant that two of his earliest texts, those by Sacrobosco and Apian, were the ones on which the professor at Copenhagen based his lectures in astronomy. And if the eclipse of the sun

[16] Karl Hartfelder, *Philipp Melanchthon als Preceptor Germaniae* (Berlin, 1899), pp. 183–97.

that occurred on 21 August 1560 was indeed the source of Tycho's serious interest in astronomy, as his first biographer asserted,[17] he was almost surely drawn to the phenomenon by his university contacts, for he is unlikely either to have found a reference to it in the literature by himself or to have noticed on his own the obscuring of less than half the sun.

It is not inconsistent with Gassendi's story that Tycho might have learned about the eclipse only after the event, from a tract published at the end of the year by one of his professors. Written in Danish by the professor of rhetoric, this piece interpreted the eclipse as a sign that doomsday was near. At that time such apocalyptic ideas were not considered farfetched. Moreover, Melanchthon had been convinced that humanity and the stars were closely linked. And although he had been unable to carry the day with his views on this issue (Luther had scoffed at all ideas of astral influences or astrological portents), astrology remained a subject of great general interest throughout the sixteenth century. In any case, the incorporation of mathematics into the standard curriculum provided a stimulus for astronomical interests. The result was the creation of a reading public for astronomical literature that included students, professors, and university graduates in all walks of life (but primarily in the clergy) and called forth a large number and variety of publications from the German presses of the era.

During his years at Copenhagen, Tycho joined that public and began to explore the astronomical and astrological literature. He learned that the motions of the heavens were not works of caprice but were subject to calculations that allowed them to be predicted well in advance. Tycho was developing other academic interests too and was forming friendships with fellow students, older tutors, and perhaps even professors, such as the young Dr. Johannes Franciscus, who came to the university in 1561 as professor of medicine. Certainly, as the first Brahe to enter a university, living in his professor's household and frequenting the streets and halls of the Latin quarter, Tycho was being drawn into a way of life very different from that of his uncle's or his father's noble households.

At the end of 1561, when Tycho turned fifteen, it was time to move on. The sixteenth-century nobleman who contemplated entering public life needed to know the languages of foreign lands; the customs and personalities of foreign courts; the polity and policies of foreign kingdoms; and the history, political theory, music, literature, art, architecture, and military science that made up the common European heritage. The traditional way of acquiring this

---

[17] *Gassendi*, 5. This notion is refuted convincingly in *Norlind*, pp. 14–15.

knowledge was by attendance at foreign courts. It remained the form of education chosen for Tycho's four younger brothers. After apprenticing as pages in the households of some noble kinsman, they would proceed to training as squires under some prominent foreign lord. They would win their spurs around the age of twenty-one, broaden their experience with further service as courtiers or armed knights, and eventually return home to serve the Danish court. By the time they were ready for marriage, they could expect to have the credentials necessary for governance of a major fief. Informal as it was, it was an education that could still in Tycho's day equip men for careers in the highest echelons of government. Accordingly, his brothers all matured into men of culture and social grace, who inspired confidence in their ability to plan and administer competently both civil and military matters. Two of them were to become councillors of the realm. But their education could never have produced an astronomer.

Tycho Brahe escaped from it by a hair's breadth – by the quirk of fate that took him out of Otte's and Beate's care and into Jørgen's and Inger's. With that turn of fortune, he was brought into the tradition of the Oxe family, under which the great Peder himself had spent five years traveling with a tutor among the universities of Europe. Whether Jørgen simply yielded to the family tradition of his charismatic brother-in-law or whether he perceived that the hoary exercise of arms in a courtly atmosphere was no longer sufficient in itself as training for the life of a great aristocrat is not known. Either way, he probably had to debate the issue with Otte. By the time Tycho had spent three years in city and university, however, there was probably no returning to the feudal pattern of castle and court education, and in the end, Jørgen and Inger prevailed. When the time came to take the Grand Tour that had become a standard feature of the education of Danish aristocrats, Tycho followed the path of the Billes and the Oxes to foreign universities, rather than the path of the Brahes to foreign wars.

The obvious place to start was Saxony. It was a land where the purest form of High German – still a language of the Danish court – was spoken and where the holy places of Lutheranism could provide a source of inspiration. It was also a land where Jørgen and Inger had close ties to the court, because they had accompanied Electress Anne to her wedding ceremonies there when Tycho was a baby, and Inger had continued to correspond with her since.

Tycho left Denmark on 14 February 1562. He was not alone. Not only did he travel most of the way in some kind of ad hoc caravan, but he also was being looked after by a preceptor who had been chosen with great care by Jørgen and Inger. This companion-guide-

tutor was Anders Sørensen Vedel, a twenty-year-old Dane of
middle-class background and advanced standing at the University of
Copenhagen.[18] In return for his expenses on the trip, he was expected to act generally in loco parentis – to direct his fifteen-year-old
charge's university studies; arrange for private extracurricular instruction in modern languages, fencing, horsemanship, and dancing;
provide moral and spiritual guidance; and administer the money and
carry out the instructions sent to him from Denmark.[19] It was a
demanding position, but it provided valuable contacts and paid the
expenses of an education abroad. And not surprisingly, it also frequently led to lasting friendships.

The journey by ship across the Baltic and on horseback along the
roads that followed the Elbe and the Saale rivers took five weeks.
The path was well trodden both literally and figuratively, because
large numbers of Danes went to the University of Wittenberg.
Tycho and Vedel, however, passed through Wittenberg and then
rode for about two more days to reach Leipzig (Figure 1.3). The
university there, after having been all but destroyed in the aftermath
of the Reformation, was once again one of Germany's largest. And
because it was dominated by Philippists, just as Copenhagen was, and
because its courses were conducted in Latin, its general ambience
would have been quite familiar to the two young men, even though
they had few compatriots there. Indeed, there were only two other
Danes matriculated when "Andreas Seuerinus Cimber" and "Ticho
Brade ex Scandia" registered on 24 March 1562.[20] Perhaps it was the
fact that one of the two was a brother of a classmate of Vedel, Peder
Hegelund, that drew them there. During the next two years, only
Peder Hegelund, yet another of his brothers, and the noble youth
Knud Skram and his preceptor joined the Danish-speaking contin-
gent. Knud Skram's mother, like Tycho's foster mother, was also a
close friend of Electress Anne.

At Leipzig, Tycho seems to have followed the normal curriculum,
studying what he later described as *humaniora*, primarily classical
languages and classical culture. In addition, he continued with the
sciences, particularly astronomy. The serious study of astronomy
was not part of the program envisioned even by his foster parents, let
alone his father, but Tycho, by his own testimony, "bought astronomical books secretly, and read them in secret." He acquired a little
celestial globe "no bigger than a fist" and the pioneering celestial

---

[18] For a brief biographical sketch of Vedel, see *DBL* XXV, 183–92.
[19] Henny Glarbo, "Studier over danske adelsmaends udenlandsrejser i tiden 1560–1660,"
Historisk Tidsskrift, 9th series, Vol. IV (Copenhagen, 1926), pp. 221–74.
[20] *Dreyer*, 16. See also Christianson, "Tycho Brahe's Facts of Life," pp. 24–5.

Figure 1.3. Renaissance Europe.

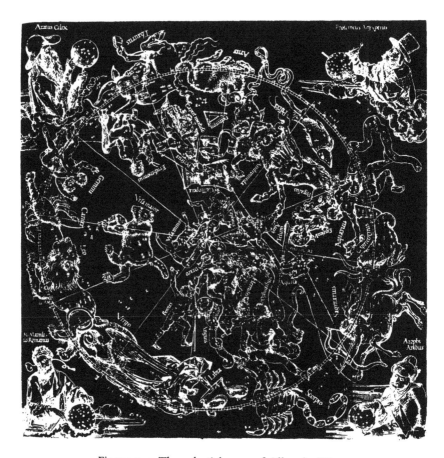

Figure 1.4. The celestial map of Albrecht Dürer.

maps of Albrecht Dürer (Figure 1.4) and "by and by, got accustomed to distinguishing all the constellations of the sky."[21] He
purchased ephemerides and began to keep track of the planets.
Lacking a proper instrument, Tycho could only check the predictions of the ephemerides by lining up a planet and two stars by means
of a taut string and estimating the positions of the planet from the
positions of the two stars on his little globe. But, rough as this
method was, he soon satisfied himself that both the Alfonsine Tables
constructed from Ptolemy and the Prussian Tables done from
Copernicus left a great deal to be desired in their predictions.[22]

As part of his studies, Tycho also pursued astrology. He bought an

---

[21] V, 106–7, as translated in *Raedar*, 107.   [22] Ibid.

astrological work by Johannes Garcaeus and began a special notebook devoted to predicting the various terrestrial effects of celestial
influences.[23] Following Garcaeus's example, Tycho began to cast the
horoscopes of famous men, beginning with Caspar Peucer, who was
Garcaeus's master, Melanchthon's son-in-law, and a notable professor of mathematics in his own right.

In August 1563, Tycho started another notebook, the first of his
observation logs. The occasion was his opportunity to witness a
conjunction of Jupiter and Saturn, a phenomenon that occurred only
at twenty-year intervals and was supposed to have considerable
astrological significance. By this time, he had abandoned his string in
favor of a pair of large compasses, which he used by sighting from
the vertex along each leg to the two objects being observed. The new
method produced the same results he had been getting with the old:
slightly superior performance by the Copernican predictions, but
still errors of sufficient magnitude to be detected by a sixteen-year-
old with no formal tutelage in the subject and no instrument. The
Ptolemaic ephemerides, in fact, were off by a whole month.[24]
According to his memory many years later, Tycho already had
cherished ambitions of rectifying this sorry state of affairs and
realized that he could do this only by means of a large body of
observational data. With the recording of his observations of the
great conjunction, he thus took the first step toward building such a
collection.

Hand in hand with the decision to start a journal must have come
the ambition to own an appropriate instrument. Although Vedel
tried to keep Tycho's mind on the studies intended for him, Tycho
was eighteen now and harder to control. Moreover, he had fallen
into bad company and had some astronomically inclined friends to
reinforce him in his unsanctioned activity. He had become acquainted with the elector's court astrologer, Valentine Thau,
through Vedel[25] and was receiving instruction from Bartholomew
Schultz at the university.

Scultetus – as he latinized his name for his later publications – was
a thirty-year-old German from a prosperous family, who had studied with Leipzig's noted professor of mathematics, Johannes Homilius until the latter's death in 1562. Schultz was considerably more
advanced than Tycho and was in the process of completing a text
on sundials from Homilius's notes and following up on the task

---

[23] Dreyer, 21. See the catalogue of Tycho's library reconstructed in Norlind, p. 347.
[24] V, 107.
[25] C. F. Wegener, Historiske Efterretningar om Anders Sørensen Vedel (Copenhagen, 1851), pp.
28–32.

of surveying and mapping Saxony that Homilius had begun.[26] Through informal contacts with Schultz, Tycho became familiar not only with technical astronomy but also with the associated mathematical disciplines of geography and cartography and the practical arts of navigation, surveying, and instrument making. In Tycho's day, the most famous practitioner of these arts was a disciple of Gemma Frisius, the great cartographer Gerhard Mercator.

Homilius and Schultz had used a cross staff (or a *radius*, as it was called by astronomers) for their observations, and so it is not surprising that it was what Tycho chose as his first real instrument. Tycho's radius appears to have been made commercially by a nephew of Gemma Frisius and, if it was, would have resembled the one shown in Figure 1.5.[27] In Gemma's exhaustive treatise on the astronomical radius, the crossbar was moved along the staff until the angle to be measured was exactly covered by the length of the bar. In Tycho's (and the depicted) model, the crossbar was equipped with one movable sight. When the observer had adjusted the bar on the staff so that he could see his two stars along the lines of sight running from the end of the staff through each sight on the crossbar, he could obtain the tangent of their angle of separation by reading the scales etched on the bar and the staff.

Characteristically, Tycho soon became dissatisfied with the coarseness of the scale on the staff. Schultz came to his rescue by showing him a trick that Homilius had taught him for obtaining finer divisions. Thus came into Tycho's life the transversal points that eventually became (see Chapter 5) virtually a trademark of his instruments. Once he got his refined divisions, Tycho began to notice systematic errors in his data, which he soon traced to faulty logic in the construction of the instrument and rectified with a table of corrections. After 1 May 1564, when he entered the first observations from the cross staff into his log,[28] Tycho was finally in a position to do something more than just contemplate the heavens.

It seems reasonable to view the cross staff as a symbol of Tycho's coming of age. Until that time he had pursued astronomy furtively.

[26] *Zinner*, 388–9, 532–4.
[27] Tycho says (V, 107–8) only that "he secretly had a wooden astronomical radius made according to the direction of Gemma Frisius." In his technical discussion of the merits of the radius, or cross staff (V, 97), Tycho mentions owning a radius "constructed not by myself, but by Walter Arsenius, a nephew of . . . Gemma Frisius . . . and one made" later by his craftsmen. It is not impossible that the two references to Gemma Frisius are to different instruments and that the former was merely thrown together for Tycho at a shop in Leipzig. But if Tycho actually owned three cross staffs, it would be very uncharacteristic of him to have failed to mention the one to which he would have had the greatest emotional attachment.
[28] X, 5.

Figure 1.5. An Astronomical Radius, or cross-staff, made by Gualter Arscenius, maker of Tycho's first professional instrument. (Photograph by Kushber 86, 6–30, 1; courtesy of the British Library.)

But he could scarcely have· ordered the expensive brass-bound instrument all the way from Louvain without the knowledge and acquiescence of his long-suffering preceptor. As Tycho progressed through his eighteenth year, moreover, he seems to have become more assertive about his current activity and also more conscious of his future. He had struggled conscientiously to pursue the legal studies that had been the rationale for his university education but had come to find both them and the whole idea of being a courtier increasingly distasteful.

On the occasion of his eighteenth birthday, therefore, when one of the professors at dinner that evening told of having known an illiterate craftsman who was "an astronomer by nature," it struck such a responsive chord in Tycho that he recorded the anecdote in his notebook.[29] A few days later, he embarked on his first research project. Recording every imaginable nightly feature of the heavens during the twelve days of Christmas, he prepared to check faithfully through the coming year to see whether there was any validity at all in the notion that these day-by-day appearances presaged the month-by-month patterns of the weather.[30]

In the spring of 1565, Tycho and Vedel, after three years in Leipzig, set out for home. Peder Hegelund recorded the day of their departure as 17 May. They passed through Wittenberg to see Schultz, who had moved there in the previous fall to study for his master's degree and reached Rostock on 25 May. Swedish and Danish fleets were grouping in the sound for another encounter in the two-year-old war between Scandinavian cousins, but the two scholars would have had no difficulty finding a ship that was going to thread its way between the fleets to Copenhagen, for war in the

[29] *Norlind*, pp. 16–17.   [30] *Dreyer*, 21.

sixteenth century was far from being the total enterprise that it has become in modern times. Tycho had doubtless already heard that Jørgen Brahe had assumed command of a warship and had distinguished himself in 1564 by capturing the admiral and flagship of the Swedish fleet. And if he did not already know, he learned at Copenhagen that Jørgen had been commissioned as vice admiral of the Danish fleet and was deeply involved in preparations for the imminent engagement between the fleets. Accordingly, Tycho proceeded to Helsingborg, Malmø, and finally Knudstrup Manor, making determinations of the local latitude at each stop.[31]

The war was in evidence at Knudstrup, too. Otte Brahe, who had been elevated to the Rigsraad in 1563, was still governor of Aalborg Castle and had been spending most of his time there. But Swedish border raiders had been making themselves sufficiently obnoxious around Knudstrup that his wife's brother, Steen Bille, had organized a defense of the area by fortifying the old Cistercian Abbey of Herrevad, a few kilometers away.

By April 1566 the war was to come close enough to claim Steen's parents (Tycho's grandparents) who died commanding the defense of Baahus Castle. When Tycho first joined his family, he was visiting in his curious role of nephew/cousin of his parents and siblings. In less than a month, however, his status changed. Following an indecisive battle on 4 June 1565, Jørgen Brahe had sailed with his fleet back to Copenhagen to regroup. As Admiral Trolle lay dying of his wounds, and the ships were being repaired and reprovisioned, there was an accident. Both King Frederick II and Jørgen Brahe fell into the water under Amager Bridge near the royal castle at Copenhagen. Some of the contemporary sources suggest that they had been drinking, that the king fell in first and Jørgen fell in while trying to fish him out from a boat. Whether from the carousing or the ducking, Jørgen never recovered, and when the fleet put back to sea, Vice Admiral Brahe was dead and his cousin Otte Rud was admiral.

About a year before his death, Jørgen had been seriously sick and had gone home to rest for the summer. Then, during much of the winter, Inger Oxe had been sick. These misfortunes – and the war, no doubt – had led fifty-year-old Jørgen to initiate a process he should probably have been contemplating before then: that of making Tycho his legal heir. With Jørgen's death, however, the plans for transmitting his estates to Tycho were aborted. Instead, the castle of Tostrup and the income from the hundreds of tenants who made up its rent rolls were to revert to general distribution among the Brahe family, but only after Inger Oxe's death.

---

[31] *Norlind*, 20. The observations were not printed by Dreyer.

Inger continued to live in great wealth and dignity as a widow until 1591. In addition to obtaining a life tenancy for the whole Tostrup inheritance as a widow's jointure (although it was probably encumbered by the obligation to make long-term payments to Jørgen's two sisters and two half sisters), she owned Sollested Manor and assumed possession of her husband's fiefs as a vassal in her own right. The only thing she lost was the supervision of her foster son. From mid-1565 until he reached his majority, Tycho's affairs were to be directed by his natural parents, Otte Brahe and Beate Bille.

Tycho stayed in Denmark for nearly a year. It was probably the first extended period he had spent with his family. Even then, the members he probably knew best were gone: brothers Steen and Axel (eighteen and fifteen) were off serving as squires in Germany, and older sister Lisbet had married and then died in childbirth at the age of eighteen, in 1563. But there were several younger members with whom to get acquainted. Sisters Kirsten (who died the next year at thirteen and a half) and Sophie were at home with Beate and Otte in Aalborg Castle. So was little granddaughter Lisbeth Gyldenstierne. Brothers Jørgen and Knud, eleven and ten, were in school at nearby Vittskøl Abbey, and sister Margrethe, fourteen, was schooling with an aunt in the Lutheran nunnery of Gudum Cloister near Aalborg.

Tycho probably spent much of his time with his father. The two of them may have given some thought to trying to make a special claim on Jørgen's estate for Tycho and undoubtedly began the complex task of settling the estate through the regular legal processes. It would have been especially good practical exprerience but was probably only one of a number of administrative enterprises to which Otte wanted to introduce his reluctant firstborn son. Certainly there was not much else for Tycho to do, because he simply had not had the martial education required for the kind of contribution to the war effort being made by all of his older relatives.

Although his father doubtless regarded Tycho's ignorance as vindication of his objections to Tycho's literary education, he must have pointed out pragmatically and patiently that the war presented other opportunities as well and that Tycho could still take advantage of them by embarking on a career of civil service to the king. But that did not appeal to Tycho. Whether he actively disliked the notion of going to work as a courtier, as later comments suggest, or whether he was simply so consumed by his passion for astronomy that he was unwilling to settle for any lesser interest is not clear. Nor is there any record of how he defended his desire to continue his schooling. Certainly an argument based on his alleged legal studies would have had a hollow ring, for the furtiveness of his early pursuit of astronomy had gradually given way to a determination to pursue his own interests (although how much Vedel had perceived and reported to Jørgen or Otte is not known). In any case, by the spring of 1566, Tycho had either won some kind of stalemate in the discussions or simply asserted himself. When he left Denmark this time, his destination was the birthplace of Lutheranism, Wittenberg.

Tycho arrived in Wittenberg on 15 April 1566, traveling in the company of at least two fellow Danes. One of them, a twenty-four-year-old master of arts Hans (from) Aalborg, may have been recruited by Otte to function as Tycho's preceptor, for Vedel was already in Wittenberg completing the requirements for his master of arts degree. Many other Danish students were there too. They habitually came in such numbers, in fact, that Danish kings paid regular pensions to Wittenberg professors to ensure that their subjects would be well received.[32] One modern scholar described the university as the postgraduate school of the University of Copenhagen in that era.

Tycho did not matriculate, however, and seems not to have lodged with any of the professors. Perhaps he stayed with Schultz, who seems still to have been there at that time. Whatever he was doing, he did it for only five months before a serious epidemic struck the congested little town and caused a general exodus of the students. Leaving on 14 September 1566, he and Aalborg arrived in Rostock (where Aalborg had obtained his M.A.) ten days later.[33] The following month Tycho matriculated at the university.

Although Rostock had long been a favorite retreat of Danish students, and Tycho probably found it the most congenial of the universities he attended, his first months there were traumatic. First, there was a lunar eclipse on 28 October 1566. As a result of analyzing it, Tycho concluded that it foretold the death of the Turkish sultan, Suleiman the Great. After he had written and posted a little poem of Latin hexameters announcing his conclusion, the news came that the sultan had indeed died, but almost six weeks before the eclipse.[34] This misfortune, however, was insignificant compared with what happened during the Yuletide festivities of that year.

Tycho was staying in the household of Lucas Bacmeister, a professor of theology. On 10 December 1566 there was a celebration of a betrothal. During the dancing, Tycho fell out with another aristocratic Danish student named Manderup Parsberg (actually a third cousin of his), who seems to have left Wittenberg about the same time Tycho did. It is not impossible that Tycho's ill-fated

[32] Bech, *Danmarks historie* VI (Copenhagen, 1963), p. 265.
[33] See Christianson, "Tycho Brahe's Facts of Life," pp. 24–5.
[34] Tycho told this story on himself twenty years after the event: I, 135–6.

prediction of the sultan's death was part of the dispute, as it certainly provided an opportunity for some sport at Tycho's expense.[35] They renewed their quarrel at a Christmas celebration on 27 December and met for a third time two days after that. Young noblemen always wore swords, or had them close at hand, and their fencing masters had taught them how to use them. Fortunately, the broadsword had not yet given way to the rapier, so the hack and slash were more common than the quick thrust and kill. A remarkable account of what happened on that dark night was preserved orally for almost a century and then was put into writing by a Lutheran clergyman of Lübeck named Master Jacob Stolterfoht.

His account begins with the assertion that Tycho's astrological art allowed him to foresee a contingent or accidental occurrence and, consequently, that he had kept to his room all that day. In the evening, however, he came down to supper

> but unexpectedly got into an argument with one of the table companions, and soon they were so wrought up, speaking in the Danish language, that they demanded swordplay of each other, stood up forthwith, and went out. My late grandmother, who knew the Danish language and was eating in that same room, admonished the other table companions to follow them straightaway and to try to hinder any misfortune, which they indeed did do. But when they came out into the churchyard, the others were in full brawl, and Tycho had received a stroke that had hacked away his nose.[36]

Beyond the fact that it definitely disfigured him for life, the details of Tycho's wound are vague. Portraits show a long diagonal scar across his forehead and a rounded line across the bridge of his nose, suggesting a near miss that clawed its way down to his nose and hacked out much of the bridge. If it was a much nastier bite than a slice off the end of the nose, it was not the mortal blow that another centimeter or two the other way would have been.

Dueling deaths were common among the Danish nobility at that time. Tycho's second cousin, likewise named Tyge Brahe, was killed by his uncle Eiler Krafse in a duel in 1581. A third cousin, Anders Bille, killed a man in 1568, and his brother, Erik Bille, killed his first cousin (and Tycho's second cousin) Jørgen Rud, in 1584. Jørgen's brother, Peder Rud, was killed in 1592 by Christen Baden. Perhaps the best indication of the general violence of that age is the fact that a

---

[35] *Gassendi*, 10. The origin of the apocryphal tale that they quarreled over who was the better mathematician is decisively exposed by Norlind in *Gassendi*, p. 209.

[36] Jacob Stolterfoht, *Consideratio Visionum Apologetica, Das ist, Schrifftmässiges Bedencken, Was von Besichtern heutiges Tages zu halten sey* ... (Lübeck, 1645), p. 306.

law was passed in 1576 stating explicitly that a nobleman who killed his brother could not inherit any part of his brother's estate.[37] There must have been some anxious moments during the early days, as the attending physician (or surgeon, more likely) prayed that the dreaded, unpredictable infection would not set in, and then a long period of convalescence, occupied primarily by anguished reflection on his fate while the scar tissue formed. As is almost inevitable during such times of recuperation and rehabilitation, Tycho's attention was drawn to the treatment he was receiving. If he had not already developed the interest in medicine that he retained for the rest of his life, he surely did so at this time.

Like astronomy, medicine was a Greek science, transmitted to Europe by the Arabs. Likewise, it had been the subject of attempts both to purify and to improve it. No sooner had the medieval Arabic authorities been replaced by texts translated from the source of their knowledge, Galen, than Galen himself came under fire from empirically minded anatomists who found that even Galen had been less than perfect, and they argued that Nature, itself, should be the only authority.

At about the same time, the views of Paracelsus began to undermine the basic tenets of Greek medicine. By 1567, each of the new schools of thought was represented at Rostock, owing to the deaths (from the plague) of two traditionalists in 1565 and 1566. The foremost of the new appointees was Henrich Brucaeus (van den Brock), a native of Ghent, with a medical degree from Bologna, and former physician to the house of Braganza. Brucaeus was an empiricist, who introduced postmortem examinations to Rostock and became famous for his meticulous study of scurvy. Although he was an implacable foe of astrology, he was sufficiently interested in astronomy to teach the subject at the university and even to publish an elementary text.[38]

Levinus Battus was a younger man, likewise a native of Ghent, but with a medical degree from Padua. He had been influenced by Paracelsus, whose posthumous manuscripts were just being published. He was greatly intrigued by Paracelsus's belief that the spirit of God permeated and unified all of Nature and that there must therefore exist in the bowels of the macrocosm (the earth) mineral remedies that could alleviate any of the ills that could inflame the bowels of a microcosm (person). As a Paracelsan, Battus scorned

[37] V. Møllerup, *Danmarks Riges Historie: 1536–1588* (Copenhagen: Det Nordisk Forlag, 1933), p. 264.

[38] Brucaeus was actually appointed professor of medicine and higher mathematics (astronomy). See *Allgemeine Deutsche Biographie* III (1876), pp. 374–5.

anatomical study for dealing with superficialities instead of probing into the true nature of health and disease and so looked to alchemy and astrology as more promising lines of research.[39] In connection with the latter, he had pursued astronomy vigorously. In his capacity as a teacher of mathematics at Rostock since 1560, he is known to have lectured on such highly specialized topics as eclipses. Tycho may, therefore, have made his acquaintance even before his accident.

The ideas of Paracelsus found fertile ground in Tycho's mind. Melanchthon's conception of a universal learning, which had set the pattern of education in all the Lutheran universities that Tycho had attended, made him receptive to the proposition that medicine, alchemy, astrology, and astronomy were linked in a great cosmic harmony. At the same time, an emphatic doctrine of empiricism ran through Paracelsan alchemy, as it also did through Humanistic anatomy. And Tycho's early astronomical observations reinforced the lesson that truth is to be found in nature, rightly observed. How specifically his new interests were connected to the aftermath of his violent accident can only be a matter of speculation, but they do seem to be connected with his stay in Rostock and certainly retained a lasting appeal for him.

Tycho returned to Denmark some time after observing the solar eclipse of 9 April 1567.[40] If his scars earned him some sympathy from Inger and Beate, they probably provoked some other kind of moralizing from Otte. At the same time, disfiguring wounds were far from uncommon in that martial society. Otte's cousin, Lauge Brahe, who had just died that spring, had borne through his adult life as a *rigsraad* the mark of a spear that had struck his face in battle. Yet this wound had not prevented his marriage to one of the richest (Swedish) heiresses in Scandinavia. And Tycho was beginning to experiment with a prosthesis to try to disguise his wounds a bit.

According to a later account, the mature Tycho wore a nosepiece made of gold and silver blended to a flesh color.[41] Exhumation of Tycho's body in 1901, however, revealed stains of copper in the

---

[39] On Levinus Battus, see Sten Lindroth, *Paracelsismen i Sverige til 1600–talets mitt* (Upsala. Universitatsforlag, 1943).

[40] X, 13.

[41] *Gassendi*, 10, 209. If Tycho had lived in India he could have had a skin graft done from his forehead. Because the penalty for adultery there was amputation of the nose, the technique of grafting (done by members of the potters guild) was highly developed and routinely successful. Tycho visited Venice in 1575. At that time Gaspare Tagliacozzi (1545–99), the creator of the specialty of plastic surgery, was just starting his practice in Bologna. Within a few years he was routinely correcting facial mutilations by means of skin grafts. In 1597 he published a book (*De curtorum chirurgia per insitionem*) describing and depicting procedures for ameliorating problems such as Tycho's missing nose. See Leo M. Zimmerman and Ilza Vieth, *Great Ideas in the History of Surgery* (New York: Dover, 1967), pp. 261–7.

nasal area.[42] One might assume, therefore, that Tycho had a piece made of precious metals that he used for "dress" occasions and a lighter piece for everyday activity (or, at least, for burial) which would have been less irritating and more easily kept in place by the adhesive salve that he always carried around with him in a little box.

Tycho's journals give no indication of his activities in the second half of 1567. He surely spent some time with his family, who had just moved to new quarters at Helsingborg Castle.[43] But the new command near the front line and a difficult campaign into Sweden in November probably kept his father too busy for much discussion. Tycho doubtless went to the university to renew old acquaintanceships, too. Remarkably enough, he may also have appeared at court, for some profound changes had occurred there while he was abroad, which had great significance for him: Peder Oxe had returned from exile.

Tycho had been twelve when his Uncle Peder had fallen from power. If the details of the fall have thus far eluded historical analysis, they must have become quite familiar to Tycho over the years. Jørgen's loss of Vordingborg Castle in 1557 had almost surely been part of the power struggle that preceded Peder's fall, and he had remained loyal throughout Tycho's youth to the cause of getting his brother-in-law reinstated.

Suffice to say that both Peder and his brother Eskild, master of the exchequer, had trampled on the interests of many powerful men during their meteoric rise to power and that, one way or another, their enemies finally managed to undermine old King Christian III's confidence in them. In 1558, Peder had fled the realm under cover of night, accompanied by only two men. From exile, he petitioned the king and then, after 1559, his successor, Frederick II, for mercy. But even though he managed to enlist both the French king and the Holy Roman emperor in pleading his cause and was aided from within the country by such staunch supporters as Jørgen Brahe, his pleas had been futile. When humility failed, Peder turned in his rage to the enemies of the crown. He went to Lorraine, where Christine of Denmark, daughter of the deposed Catholic king of Denmark, Christian II, was ensconced. From there he spun a web of anti-Danish intrigue that actually raised fears of invasion in Denmark in 1560. War did not come until 1563, however, and then it had been with the Swedes, not with Danish exiles. But as the war dragged on, taxes soared (even though the royal debt was rising to dizzying

---

[42] Heinrich Matiegka, *Bericht Über die Untersuchung der Gebeine Tycho Brahes* (Prague, 1901), pp 10–12.

[43] *DBL* III, 574.

heights), and the Danish forces were decimated by mutiny and desertion. The conviction mounted that Peder Oxe was the only one who could retrieve Denmark's fortunes. Through machinations that can only be explained by the acknowledged fact that the suave Peder had many powerful kinsmen and friends at home, the anti-Oxe bloc was overcome. By a series of delicate, face-saving negotiations, the king and Oxe were gradually and cautiously drawn into a wary reconciliation.

Peder returned in 1566, a month after Tycho left for Wittenberg. Almost immediately he became the leader of a strong-willed Rigsraad that strove for a larger share in the government of the realm. In February 1567, Peder was made governor of Copenhagen and, as such, head of the naval wharves at Holmen, with the whole fleet under his supervision. One by one, his supporters were moved into key positions in the realm: Otte Brahe's promotion to the strategic fortress of Helsingborg at the northern entrance to the sound may have been one of those moves. By August 1567, Oxe was lord high steward and remained until his death as the head of the government in all but name. As the queen mother summed it up in a letter to her daughter Anne in Saxony, "Peder Oxe is now high in the saddle, higher than he has ever been; he goes and does with our son the king as he will."[44]

During the years between his return from exile and his death in 1575, Peder Oxe presided over a veritable revolution. On the political scene he ended the futile Northern Seven Years' War, commandeered a third of the property of all nobles to pay off the immense war debt, tripled the toll on merchant shipping through the Øresund so that it became a lucrative and stable source of royal revenue, and reorganized the Danish-Norwegian defense so that it rested on the popular national navy instead of an expensive and unreliable army of foreign mercenaries. Through these moves, brilliantly attuned to the economic trends of the time and the geopolitical realities of a far-flung sea-linked realm, Oxe laid the foundations for a half-century of peace and prosperity (and earned Frederick's assessment, by 1572, that he "now has the realm's money in hand").[45]

This was a period during which the pastimes of peace flourished as never before in Denmark, and Oxe, himself, influenced those pastimes greatly. At Gisselfeld, he built a great Renaissance palace as his seat, surrounded it with imported fruit trees and exotic plants, and transformed the swampy pools into an intricate complex of carp ponds. He became the patron of architects, painters, musicians, a

[44] *DBL* XVII, 547–62. See also Bech, *Danmarks Historie*, pp. 269–79.
[45] Ibid., pp. 427–8, 456.

state historian, and even an astronomer. The historian was Anders
Vedel. The astronomer, of course, was Oxe's nephew by adoption,
Tycho Brahe.

When Tycho returned from Rostock, Peder Oxe was just consoli-
dating his position. Otte Brahe knew as well as anyone how the
games of power politics were played in Denmark, and he knew
better than most how the power was shifting. In the past, when
political success for a Brahe had been a matter merely of probability,
he must have become progressively more disappointed by his eldest
son's disinclination to strive for power and influence. Now that
Tycho's relationship to Oxe made success a virtual certainty, Otte
must have been frustrated beyond description when Tycho refused
to grasp the opportunity. Yet that is exactly what happened.

After staying in Denmark through his twenty-first birthday and
the Christmas holidays, Tycho embarked on his third trip abroad.
Not surprisingly, there is evidence of friction between father and
son. Two weeks after arriving at Rostock on New Year's Day, 1568,
Tycho alluded to his problems in a letter to Hans Aalborg, the
earliest surviving piece of his correspondence:

> I have decided to stay here the winter over, if it please God. What will
> happen then, time will tell, and I commit that to God's custody....
> But you, my dear Hans, must say nothing about the reasons for my
> departure, which I have told you in confidence, so that nobody will
> suspect or see that I am complaining about anything or that there was
> anything in my native land that would have driven me to leave it. For
> it is of the utmost importance that nobody should hear my complaints
> about anything, nor in truth do I have much cause to complain. For I
> was better received in my native land by family and friends than I
> deserved; the only thing lacking was that everybody be pleased with
> my studies, which can certainly be forgiven....[46]

It is unfortunate that Tycho was not more specific. Who was not
pleased with his studies? Anyone besides Otte? Probably, for Tycho
continued to allude to problems on this score even after his father
died in 1571. Why was Tycho so concerned that his complaints not
reach his family? Was it anything more than a desire not to be found
out in an airing of the family's dirty linen? Again, probably. Tycho
was, after all, now dependent on his father for support in his studies.
Moreover, friends and perhaps even a relative or two at court were
working to find a way to reconcile Tycho's noble station with his
desire to pursue a career in science. This was no easy task because
noblemen were barred by custom from virtually all scholarly posi-

[46] VII, 3.

tions in schools, the university, and the church. They could enter the royal chancery and become diplomats, but that did not appeal to Tycho. The crown granted a few pensions to men of learning, but only for the duration of the royal pleasure, and so a pension could not be the foundation of a career. This left only one alternative (now that he was not going to inherit his uncle's wealth): a canonry.

The cathedral chapters of Denmark and Norway had survived the Reformation with their landed endowments intact. In Tycho's day, the canonries of the Lutheran cathedral chapters were awarded by the royal administration to government servants and men of learning. They were the only such offices that were granted to noblemen as well as commoners. If Tycho could obtain one, his way would be clear to pursue a lifetime career as an astronomer and scholar, without offending social conventions or neglecting his obligations as a nobleman. This was important to him because he could never have imagined giving up the privileges and status of his nobility, whatever his grumblings over the burdens it imposed. But on the substantial incomes of a canonry, together with whatever wealth he might inherit, he could live as befit the dignity of a nobleman.

Moreover, the office of Lutheran canon would not obligate him to enter holy orders or live otherwise than as a secular lord. At least one Danish canon, Morten Pedersen of Roskilde, was already pursuing studies in astronomy, so Tycho would not even be embarking on an unprecedented activity. Tycho would certainly have remembered that the greatest astronomer of his century, Nicolaus Copernicus, had spent his life as an unordained administrator of a cathedral chapter. Tycho's nomination went forward with a speed that only Peder Oxe could have inspired. On 14 May 1568, royal letters patent were issued at Copenhagen providing that the next vacant canonry at Roskilde cathedral be reserved for Tycho.[47]

Tycho spent his first two weeks at Rostock in the house of Professor Levinus Battus. He then found his own quarters in the law college, which happened, Tycho mentioned to Aalborg, to have a very suitable place for observing. Actually, to judge by the relationships Tycho formed with several of the professors at Rostock – none of whom were members of the law faculty – he was probably spending most of his time doing medical alchemy.

For the second time, however, Tycho's stay was destined to be short, and for the same reason. Only five months after his return, the university authorities levied against him a stiff fine of twenty Joachimsthalers, probably because of his duel with Parsberg, and Tycho decided he would rather leave the university than post the money pending his appeal. When the university council heard his appeal in October, Peder Oxe's influence was felt in Tycho's favor, and Brucaeus testified that Tycho had presented a celestial globe to the university. But the best these efforts could produce was a resolution that Tycho's delay in paying the fine should not be held against him. If there were further appeals, they also came to naught, for Tycho apparently paid the fine about two years later.[48]

During the summer of 1568, Tycho traveled southward across Germany. Perhaps he met his brother Steen at Arnstadt to witness the ceremony at which Count Gunther of Schwarzburg-Rudolstadt awarded Steen Brahe his spurs and presented him with a warhorse and harness of armor. By September Tycho was in Basel, where Peder Oxe had spent some of his youth and some of his exile and where Tycho and two of his cousins and several of his friends matriculated at the university. Apparently Tycho felt Basel had possibilities, for he was to return to the city several years later and even decide to spend his life there.[49] But at this time, it did not provide what he wanted. Perhaps Tycho did not quite realize that he had now learned about all he could from teachers and texts and that after nine years of university work, it was time to start participating more actively in scholarly life. Tycho actually did start working on his own in Basel, experimenting with the construction of quadrants with the assistance of a young Belgian named Hugo Blotius.[50] After no more than a few months, however, Tycho left Basel. In Freiburg he noted some celestial models designed by Schreckenfuchs that demonstrated planetary motions according to the theories of Ptolemy and Copernicus.[51] In Lauingen he was invited home and entertained in conversation late into the night by Cyprianus Leovitius, a mathematician and astrologer whose ephemerides he had used.[52] In Ingolstadt he met Phillip Apian, son of the famous mathematician whose textbook Tycho had used to teach himself astronomy.[53] Only in the spring of 1569 when he reached the old imperial city of Augsburg, however, did he find circumstances sufficiently appealing to warrant his settling there for a while.

It is hard to say precisely what Tycho found so attractive about Augsburg. To be sure, it was graced by a circle of wealthy humanists who dined together and deliberated on learned matters, but numerous cities boasted such informal academies in that era. And if one of its artisans, an instrument maker by the name of Christoph Schissler,

---

[48] XIV, 3.   [49] V, 108.   [50] II, 343; VII, 328.
[51] VII, 15, 18, 79. See also Zinner, 530.
[52] III, 221–2, 400. Leovitius (1514–74) was the son of Johannes Karasek, mayor of Königgrätz.
[53] III, 157, 397. See also Norlind, 24.

impressed Tycho sufficiently to be singled out by him in later years as "a clever craftsman ... for whom I had long searched in vain."[54] it is, nevertheless, hard to see how even he could have held Tycho there for very long. Most likely it was the combination of these attributes, together with the fact that Tycho personally found several of the people there extremely congenial, that held him in Augsburg. Clearly, the attraction was mutual.

Long after he left Augsburg, Tycho received letters from people in the circle there. One of his correspondents was Hieronymus (Jerome) Wolf, a man older than Tycho's father, who had been a friend of Melanchthon and a secretary to Jacob Fugger, and was said to be so erudite that he could speak Greek even more fluently than Latin. Another was Paul Hainzel, alderman of Augsburg and kinsman to the "strange lord of Elgg," J. H. Hainzel, who later became the patron of Giordano Bruno. Paul's brother, J. B., had been a fellow student of Peder Oxe in Basel. A third was Johannes Major, a teacher at the Augsburg Gymnasium. Aristocratic, learned, and highly sophisticated, this circle of men pursued a life-style that helped give focus to Tycho's ideals. In addition to talking about the novelties of European intellectual life, they collaborated in ventures that combined art and technology in the service of empirical science. The most noteworthy instance of this collaboration was precipitated by Tycho himself.

Of all the things that might be said about Tycho's sojourn in Augsburg, one feature of his activity there dwarfs the rest in significance: the fourteen months he spent establishing his interest in instrument making. That interest had begun with his discovery in Leipzig that the cross staff he had purchased from the artisan with the best credentials in the trade simply would not perform to the expectations of a seventeen-year-old.

It was not merely a matter of the divisions being inadequate, for after Tycho had had them modified, he still found that the piece was subject not only to systematic error but also to considerable random error. For the one he could compensate with a theoretically derived table of corrections, but for the other there was no cure short of designing a new instrument. It was thus in Augsburg that Tycho finally put together the combination of time, money, inspiration, and craft skill requisite to the task. What he produced was essentially a pair of giant compasses, consisting of a graduated brass arc (of 30°) and wooden legs about one and a half meters long. Large enough to yield reasonably fine measurements and light enough to be transported and used without difficulty, the instrument proved sufficient-

54 V, 103. *Zinner* (503–20) lists numerous extant articles crafted by him.

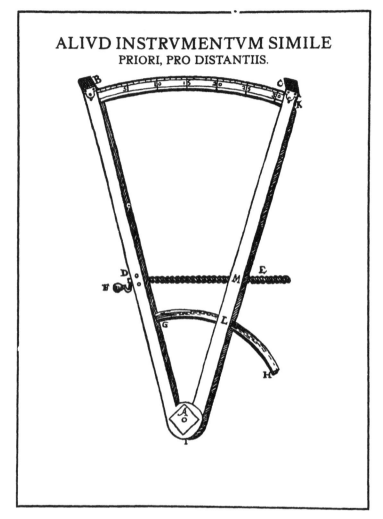

Figure 1.6. Tycho's first self-designed instrument (1569).

ly serviceable to merit description and depiction (Figure 1.6) in the catalogue of Tycho's instrumental achievements that he was to publish more than twenty-five years later. But it was far from ideal. It had a systematic sighting error that Tycho did not manage to circumvent for another decade and that, in the meantime, could be rectified only by another table of corrections.[55] In addition, the

---

[55] V, 80–3, The table is in X, 19. The instrument seems to have become available for use in November 1569.

instrument was liable to random uncertainties arising from the hazards of reading an instrument that had neither a mounting to facilitate steady sighting nor a scale graduated to the precision Tycho hoped to achieve.

Tycho's dream was minute-of-arc accuracy. To achieve it, he realized that he would have to have an instrument large enough to permit estimation of fractions of a minute of arc. While he was standing outside a shop one day, discussing the problems of building an instrument of such size, Paul Hainzel came down the street. Joining the conversation, he found his imagination fired by young Tycho's vision and offered to underwrite the cost of whatever Tycho could design and the craftsmen of Augsburg could fabricate. The result of this collaboration was the largest instrument Tycho would ever see, constructed at Hainzel's country estate just outside Augsburg.

As Tycho's picture of it shows (Figure 1.7), it consisted of an arc of 90°, made entirely of well-seasoned oak, except for the brass graduation strip on the arc and the plumb bob suspended from the axis. An astonishing five and a half meters in radius and so heavy and bulky that forty men were needed to put it in place when it was completed, it was nevertheless built in a month, apparently during March 1570.[56]

Tycho's logs contain observations made with it at approximately weekly intervals from 1 April to 16 May.[57] How satisfactory Tycho thought they were is hard to estimate. Given the enormous combined weight of the mast and quadrant, it must have required considerable effort just to rotate the piece into the desired azimuth plane, let alone swing the arc up to the appropriate elevation and hold it there until further adjustments in azimuth and elevation brought the object into the sights (DE) so the elevation could be read off at the plumb line (AF). Perhaps the fact that each night's observation consists of only one entry is implicit evidence of the strain on Hainzel's servants. If there were any serious difficulties, however, Tycho glossed over them and insisted even in later years that the Quadrans maximus rendered possible an accuracy "hardly ever attained by our predecessors." Only in respect to that sincerest form of flattery – imitation – was he silent: He would never attempt to duplicate it for himself.

Whatever else the great quadrant may have accomplished for Tycho, it provided his first brush with fame. In April 1570, Petrus Ramus arrived in Augsburg. The most celebrated philosopher of his age, Ramus had made himself notorious by attacking the meta-

---

[56] V, 88–91; II, 342–7. [57] X, 36–7.

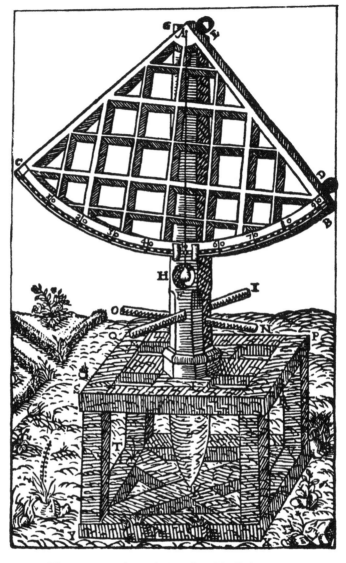

Figure 1.7. The great quadrant designed and built by Tycho at Augsborg in 1570.

physics, epistemology, and methodology of sixteenth-century Aristotelianism. Starting already in 1536, by defending for his M.A. degree the thesis that everything Aristotle had said was wrong, he espoused a radical empirical outlook that extolled the virtues of observation and induction and advocated a replacement of the deductive Aristotelian physics with mathematical laws based on empirical evidence. It was a philosophy that resembled in many ways the Philippism of Tycho's university instruction – not least that its author had become so enthusiastic in applying it as to decide that he had to convert to Protestantism.

This philosophy was not one, however, that Tycho could swallow whole, particularly those parts that concerned astronomy. Just a few months before his appearance in Augsburg, Ramus had published his ideas for reform in the field of astronomy, calling for nothing less than the complete eradication of all "hypotheses" from astronomy, and the building of a new science by observation and induction. If Tycho were not acquainted with those specifics at the time, he soon received instruction in them from Ramus himself. Meeting through Hieronymus Wolf, the two soon found themselves discussing the reform of astronomy. Tycho was willing to concede that the axioms of physics could not be considered immutable and valid for all time, but he did not see how deductive axioms could be eschewed entirely. He pointed out that the axioms of geometry were deductive and argued that an astronomy without hypotheses of some kind – circles, epicycles, uniformity, and so forth – was impossible. Equally important, he thought, were the humanist and Philippist conceptions of cosmic order and harmony, at least until empirical evidence demonstrated that, in a particular context, they were untenable.

Exactly what impact Ramus had on Tycho can only be conjectured. Certainly Tycho, when he described the encounter in a letter some sixteen years later,[58] could not discern any effect. He was still unable to subscribe to much of Ramus's philosophy and definitely felt that the French philosopher did not completely grasp the inner workings of astronomy. And if the two did agree on the necessity for renovating astronomy by means of numerous and exact observations, it was Tycho who had begun to translate his thoughts into actions. But Tycho can scarcely have failed to be inspired by the mere chance to converse with the famed iconoclast, and he must have been exhilarated by Ramus's admiration for the newly built *Quadrans maximus*. Only a year later, he was to find the existence of a young Danish nobleman named Bracheus, and the great instrument he had designed, mentioned briefly in a book[59] describing Ramus's travels in exile.

Not long after Ramus's departure from Augsburg, Tycho left too. It seems likely that he was called home, as his brother Steen is known to have been, by concern that his father's health was failing. But

[58]   VI, 88. On Ramus's work, see *DSB* XI, 286–90.
[59]   *Defensio pro Aristotele adversium J. Scheccium*, 1571.

although Tycho left Paul Hainzel with the task of overseeing completion of a large wooden celestial globe he had commissioned from
the instrument maker Schissler, indicating a somewhat impromptu
departure, the trip itself seems to have had little urgency to it.

Tycho did not reach Wittenberg until September[60] and probably
visited with his former professors in Rostock before crossing the
Baltic to Denmark. While he was traveling homeward, the long war
was grinding to a halt. Negotiations were taking place in Stettin,
mediated by the Holy Roman emperor and other great sovereigns.
One of the mediators was Charles de Danzey, France's ambassador
to Denmark, who was part of the Copenhagen circle of intellectuals
and who was probably already acquainted with Tycho. By the time
the negotiators pronounced themselves satisfied on 13 December
1570, Tycho was probably home. He certainly was by Christmas,
when he climbed the great tower of Helsingborg Castle to record
some observations of the moon.

As those around him probably suspected, this was to be Otte's last
Yule season. In addition to calling his sons back from abroad, the old
warrior had begun putting his affairs in order by compiling a legal
listing of his landed estate. Otte was not actually old, but fifty-three
years of aristocratic life had taken a heavy toll. He breathed with
difficulty and coughed violently. His feet swelled while his legs
withered. By spring he was bedridden, suffering from a catarrh of
the bladder according to some, but from fluid in the lungs according
to the diagnosis of his concerned eldest son. On 9 May, finally, after
a severe three-day ague, he died. An exhausted Tycho poured out his
emotions in a moving letter to Vedel, dated 18 May 1571.

> I have received and read your letter filled with sweet words of con-
> solation, which raised my spirits no small amount from my distress
> and great sorrow over my father's departure from this life.... There
> are certainly many theological grounds for consolation that can be
> extracted from the divine expressions of the Holy Scriptures, and
> philosophical ones that can be derived from the common fate of
> humanity and from the inconsistency and unceasing alteration of
> everything earthly, on which I draw for support in my distress and
> sorrow. But I also take no slight consolation from the fact that my
> father passed from this wretched life of mortality so peacefully and
> quietly to the heavenly and eternal realm of our father, where we,
> according to Paul's testimony, have a secure home and city.... When
> he at last opened his eyes and saw all of us standing around the bed
> grieving and crying, he asked us if there seemed to be any signs of

[60] Tycho's last observation with the *Quadrans maximus* was on 16 May (X, 37). He autographed a student's travel diary in Wittenberg in September (*Norlind*, 26–7).

death on him (for he said he did not feel any stronger pain than he had suffered earlier). Some of us then answered that he was very weak and that we feared he would not be able to make it through. He thereupon said that he would like to talk to his wife. And when she was brought forward half-unconscious and he thanked her for having lived so well and virtuously with him in his life, he said farewell to her and entrusted her to us children and his friends. Then he shut his eyes again and listened to the priest's words of consolation. . . . Therewith he closed his eyes and emitted a sigh so soft that those present did not know if he was dead. For he habitually slept almost deeper than death, and during his actual death throes he did not change even once the position of his body, feet, arms, hands, or head but slept so quietly and peacefully that even the priests stated that they had never witnessed a more sublime departure by anyone.[61]

As befit the professional administrator he had been, Otte left his affairs in good order and his heirs well provided for. The cadastre summing up his worldly accumulation is still extant and is an imposing tribute to Otte's organizational and entrepreneurial instincts. Besides Jørgen Brahe's seat of Tostrup, which would remain at Inger Oxe's disposal for the duration of her life, Otte's estate consisted of the manors of Knudstrup, Elvedgaard, and Braheslykke; some five hundred farms, sixty cottages, and fourteen mills scattered through Denmark and Norway; four residences and forty-odd pieces of rental property in Copenhagen and various other cities of the realm; and an undivided share of an extensive forest in Jutland, which annually yielded income from timbering and the rooting of sixteen hundred swine.[62]

The extent of Otte's share in this forest is known and provides a nice illustration of the complexities of Danish inheritance. Originally the property of Otte's maternal grandfather, Jørgen Rud, the forest had been left intact at his death in 1505 but passed to his three children as shares of income, two parts to his son and one part to each daughter according to the general law of Denmark. Otte's mother, Sophie Rud, was twice married and was survived by four daughters as well as by Jørgen and Otte Brahe when she died in 1555. Otte therefore inherited one-quarter of his mother's quarter share and acquired another quarter in his share of Jørgen's estate. In addition, he had purchased the one-eighth shares of two of his sisters over the years. When he died, then, his various heirs became joint owners of the forest land, along with a dozen or more kinsmen among the descendants of Jørgen Rud. Such split ownership was generally

---

[61] VII, 3-4.
[62] "Otte Brahes Jordebog, 17 April 1570," Rigsarkivet, Copenhagen.

recognized as being inefficient at best and as problematical at worst. One of the goals in settling estates, therefore, was to consolidate individual holdings by the judicious trade-off of jointly inherited properties.

The customary procedure in estate settlements was for all the heirs to meet in the presence of at least two noblemen who held the crown's commission to mediate the settlement. Otte's heirs were Beate, her seven surviving children, and her one grandchild. But many other people had an interest in the settlement, especially Peder Oxe, who had owned three manors jointly with Otte, and Inger Oxe, who still held a substantial portion of the Brahe inheritance in life tenancy. Beate could claim whatever inheritance had come into the marriage from her Bille relatives over the years, half of the property purchased during their marriage, and other property reserved to her in writing, as her widow's jointure. In addition, because Otte held a number of small fiefs as security (and interest) for loans to the crown, she would assume the administration of and collect the income from those fiefs as a vassal in her own right. Denmark had no custom of primogeniture (as in England), whereby the oldest son inherited almost everything. Tycho's immediate inheritance from the remainder of the estate thus would be only a share more or less equal to that of the rest of the heirs. As a son he got two shares, whereas his sisters got only one. As a son he was also entitled by custom to include a manor house in his division, and as the eldest son he had a special right to the patrimonial manor of Knudstrup, given that Otte owned more than one manor. But in order to do everything fairly, the entire estate had first to be appraised item by item, then liquidated as far as necessary to pay off heirs who would not get appropriate shares of indivisible items, and finally distributed according to the priorities of individual heirs.

For Otte's estate, all of this seems to have taken around three and a half years, until late 1574. At least, that is when Tycho made final plans for his next trip abroad and two of his brothers made arrangements for marriage. The details of the settlement have not been established, but the outline is clear. Because of the sheer size of Knudstrup and its associated real estate and possibly also because Steen had been effectively the "oldest" son for so many years, Tycho and Steen shared their inheritance of the patrimonial estate. Both were to style themselves "of Knudstrup" for the rest of their lives, just as Otte Brahe had. Axel Brahe inherited the ancient Rud seat of Elvedgaard. Jørgen Brahe (the younger) and Knud Brahe inherited their uncle Jørgen's estate of Tostrup, so the settlement must have included compensation to Inger Oxe for her expected lifetime income from that estate. The three sisters' shares were undoubtedly settled in cash, scattered farms and city properties, and such valuables as jewels, furs, and plate.

The register of Otte's estate made for him in 1570 listed Knudstrup as consisting of 322 farms, 29 cottages, and 7 mills, of which 122 had been purchased during the marriage and were thus half Beate's. Thirty-three had been an inheritance of Beate's and another 34 were held, and would continue to be administered by Beate, as surety for loans.[63] What remained to be split by Tycho and Steen, therefore, were the annual incomes from 200 farms, 25 cottages, and $5\frac{1}{2}$ mills, and the manorial production and seigneurial rights of Knudstrup. It was a rather small share for an heir of one of the Danish conciliar families, particularly considering how close Tycho came to inheriting all of his uncle Jørgen's estate. But it would have represented wealth of staggering proportions to a middle-class student such as Vedel. Certainly it was sufficient to allow Tycho to live in financial independence and do as he pleased for the first time in his life.

[63] Kristen Erslev, *Konge og lensmand i det sextende orhundrede* (Copenhagen, 1879), pp. 7, 68, 70, 72.

Chapter 2
The New Star

U NTIL the Danish Reformation of 1536, learned Danish noblemen had inevitably been clergymen. During the Reformation,
however, the episcopal offices and episcopal estates that they had just
as inevitably monopolized were seized by the crown and turned over
to suitably accredited Lutheran "superintendents" of middle-class
background who posed no problems in respect to either loyalty or
competence. The lower-level cathedral chapters, not offering the
wealth or power of the episcopal offices, were still accessible to the
nobility. It had been common before the Reformation to reward
royal secretaries and various other crown servants with canonical
prebends, and it remained so afterward. And because the noblemen
who were now excluded from high office in the church began to
dominate crown and chancery offices more than ever before, canonry incomes were increasingly diverted to the nobility. After the
radical changes of 1536, however, these royal servants were no
longer clerics but, instead, Lutheran laymen. From 14 May 1568,[1]
Tycho Brahe knew that he would one day be one of them.

Before the Reformation, formal education was associated with
clerical celibacy. And because clerical celibacy was a threat to familial
continuity, no aristocratic parents could send more than one or two
of their sons to university. But after the Reformation they could, and
some of them did, not so their sons could become universal men or
achieve satisfaction and even glory, in the terms of the Italian
Renaissance, but, rather, so they could be better trained to fulfill the
traditional role of their class: service to the realm. For the first
generation – those who came of university age between 1537 and
1571 – the fraction (of a restricted sample)[2] who received a scholarly
education was 29 percent. For those in the generation following
Tycho it was 66 percent, and by the middle of the seventeenth
century it was 90 percent. Tycho was thus in the vanguard of a new
order. But even if he were in a minority, it was not a particularly
small one. And although he was unusual in wishing to direct his
learning toward the pursuit of new knowledge, he was not isolated
in his interest. Both at the university and at court, among friends
who were commoners and patrons who were nobles, he found

---

[1] XIV, 3. For a list of abbreviations of frequently cited sources, see Appendix 1.
[2] Birte Andersen, *Adelig oppostring: Adelsbörns opdragelse: i Denmark 1536–1660* (Copenhagen: 1971), pp. 62, 149.

40

people who shared his ideals, if not his particular combination of resources and motivation.

Shortly before his death in 1560, Philipp Melanchthon had remarked that no realm in Christendom outshone Denmark in learned men. His statement was doubtless colored by the fact that all the men to whom he was referring were learned in theology and were zealous followers of Melanchthon himself. But Melanchthon preached the ideal of a well-rounded education and could neither have produced nor admired adherents who were strictly theological specialists. Throughout the bishoprics and cathedral chapters of the realm and even at court, there were alert intellectuals interested in all aspects of learning.

Thus it is not surprising that the true seat of learning was the University of Copenhagen and that Copenhagen's intellectual circle therefore offered as much stimulation as could be found anywhere in Europe. Its patrons were influential aristocrats, and its participating members were learned humanists, scholars, and physicians.

Dean of the circle until his death in 1570 at the age of seventy-six was Johann Friis, who had an M.A. from Cologne and had lived in Paris and Rome before the Reformation, known Luther and Melanchthon during the years of transition, and met every important person in Danish history since that time.[3] As chancellor of the University of Copenhagen since its reendowment in 1537, Friis had been a constant patron of scholarship. Among the last of his clients was Tycho's former preceptor, Vedel, who had come home from Wittenberg in 1567 with an M.A. and received an appointment as chaplain of Copenhagen Castle. It was probably Friis, who had seen – and made – so much of recent Danish history, who inspired Vedel to write an official history of Denmark. It was certainly he who bequeathed to Vedel the unfinished translation of Saxo Grammaticus that eventually emerged complete, as Vedel's contribution to Danish history and culture.[4]

The death of Johann Friis left Peder Oxe as the foremost patron of learning in Denmark. Lord treasurer of the realm and head of the Rigsraad, Oxe governed Denmark while King Frederick moved from one hunting lodge and castle to another throughout the realm. Oxe had an urbane and brilliant personality and an intellect developed by five years of study in foreign universities during his youth. He spoke several languages with ease and maintained a life-style – including indulgences in rare delicacies such as oysters and pheasant – that was splendid even among lords.[5] With Oxe's encouragement, King Frederick decided to emulate his brother-in-law's patronage of

[3] *DBL* VII, 418–25.   [4] *DBL* XXV, 183–92.   [5] *DBL* XVII, 560–1.

the Saxon universities of Leipzig and Wittenberg, by raising the
emoluments of the Copenhagen professors, endowing a fund to
provide free board for a hundred undergraduates a year, and establishing royal scholarships to allow each year four advanced students
to pursue their studies abroad.

Another member of the older generation of Copenhagen aristocrats, and one who almost surely interacted more with Tycho than
did both of the other two combined, was Charles de Dançey, the
French ambassador to Denmark. More than thirty years older than
Tycho, he had been raised on the estate of his ancestors near Poitiers
but had been sent to Strasbourg in his youth to learn German. As it
turned out, Dançey absorbed more German culture than his parents
could have wished, for he returned a convert to Protestantism and an
ardent believer in astrology. The latter he was induced to abandon
by discussions with the venerable John Calvin himself, despite the
lingering feeling – right up to the time he reported it to Tycho in a
letter – that astrology had proven to be an uncanny predictor of
unfaithfulness in women.[6] But the Protestantism held and made
Dançey an ideal envoy to a Protestant country, particularly for such
nasty tasks as informing Denmark in 1572 that over two thousand
Protestants, including Petrus Ramus, had been slaughtered in Paris
on St. Bartholomew's Day. By that time, however, he had been in
Denmark more or less continuously for nearly twenty-five years,
seeking consideration for French interests in the policy decisions of
Danish statecraft. Well liked and highly regarded, Dançey played an
important role in mediating between Denmark and Sweden during
the Northern Seven Years' War and was later to allow Vedel to use
his correspondence to write a manuscript history of the war.

Most of the rest of this intellectual circle was associated with the
university. The intellectual patriarch of that group was Niels Hemmingsen, who was at the zenith of his influence as the foremost
theologian of Denmark and whose reputation extended far beyond
the borders of Denmark. His colleague on the theology faculty was
Dr. Erasmus Laetus, whose writings included a long religious poem
written in 1557 in honor of Dançey and his *Bucolica* of 1560 featuring a dedication to King Frederick written by Melanchthon. After
Laetus's ennoblement in 1569,[7] a rare honor in sixteenth-century
Denmark, he followed with enough other humanist compositions to
allow him to salute in dedication most of the monarchs of Europe.
In 1573, Hemmingsen's influence was extended by the appoint-

[6] VII, 41. For a biographical sketch, see H. F. Rørdam, "Charles de Danzay, fransk resident
ved det danske hof," *Historiske Samlinger og studier* II (Copenhagen: Gad, 1898).
[7] *DBL* XV, 84–6.

ment of a former student of his, Niels Kaas, as royal chancellor and thus head of the University of Copenhagen. A veteran of four and a half years in Hemmingsen's household, and study abroad under Melanchthon, Kaas had been the rising star in the Danish central administration since 1560 and later became an important patron of Tycho's interests.[8] Like the rest of these men, he was a humanist, with interests ranging across the literary world but with no special concern for science. The circle also included, however, three people with a special concern for science in the particular form of medicine.

The oldest of these physicians was Hans Frandsen of Ribe (Johannes Franciscus Ripensis), in Jutland. Professor of medicine, with a wide practice in Copenhagen, he had been known to Tycho since his student days.[9] Although he was a traditional Galenist in his medical outlook, he remained a source of encouragement throughout Tycho's young manhood, sufficient to inspire Tycho to compose, and apparently print, a Latin ode to his memory in the 1580s.

Close to Tycho in both age and rapport was Ripensis's newly appointed colleague on the faculty of medicine, Johannes Pratensis. His father, Philip du Pré, had come to Denmark in the entourage of Isabella of Hapsburg when she married Christian II. With the coming of the Reformation to Denmark Philip had converted to Lutheranism and had eventually been appointed canon of Aarhus Cathedral. The son had come to Copenhagen in 1560, just a year after Tycho, and had earned his M.A. by 1564.[10] In 1565, he and another student had been granted jointly the income of a vacant professorship in medicine to support their medical studies abroad. After six years in foreign universities, including some time at that famed bastion of empirical science, the University of Padua, they had returned home with medical degrees – Pratensis to be elected to the professorship whose proceeds they had been enjoying and his companion, Peder Sørensen to be installed as royal physician to King Frederick.

An alderman's son from Ribe, Sørensen had distinguished himself sufficiently in his studies at the University of Copenhagen to be invited at the age of twenty to lecture on Latin poetry. Sustained brilliance in his M.A. studies had led to the extraordinary award that allowed him and Pratensis to attend various universities in France, Germany, and Italy. While abroad, Sørenson had fulfilled the promise of his early career by writing a book that brought him immediate recognition throughout Europe.[11] Entitled *Idea medicinae philosophiae* (Basel, 1571) and published under the Latin form of his name, Petrus

[8] *DBL* XII, 292–8.   [9] *DBL* VII, 217–18.   [10] *DBL* XVIII, 575–6.
[11] *DBL* XIII, 306–9. See also *DSB* XII, 334–6.

Severinus, the book systematized the general doctrines of Paracelsus and defended those aspects of his thought that conflicted with traditional medicine. It was not a first such attempt to rationalize the often eccentric and generally chaotic utterances of Paracelsus, but it was probably the most effective in making his ideas respectable to men of learning and thus bringing them into the mainstream of sixteenth-century thought.

As we have seen, Tycho was certainly very sympathetic to the Paracelsan ideal of creating new knowledge based on new observations, even if he probably would not have endorsed Paracelsus's notorious public burning of the works of Galen. Between the new presentation of the concepts by Severinus and the almost firsthand elaboration of his thought that Pratensis, also a Paracelsan, must have been able to provide, it is not surprising that Tycho, during the months following their return in mid-1571, found his own thoughts occupied by Paracelsan medicine.

It should be clear from the foregoing discussion that Tycho was by no means an isolated intellectual giant in a cultural desert. In the environs of Copenhagen, he could move from castle to manor house to embassy to Latin Quarter among noblemen and scholars with broad interests, cosmopolitan backgrounds, and imposing capacities. For a variety of reasons, however, Tycho seems not to have spent much time at Copenhagen. Tension between him and most of his peers, it seems obvious, was one factor.[12] The basis of this tension is harder to discern, but it seems to have been simply that Tycho was different. Whether it was the fact that he preferred the "clerkish" pursuits of academia to the noble exercises in military, civil, and social domination, or whether it was a matter of his being too outspoken about his preference, it is impossible to say. It is clear, however, that Tycho actively disliked the gamesmanship of court life. If being at the capital to participate in one set of activities obligated him to participate in the other, he may simply have preferred to forgo both.

There were also, of course, positive reasons for Tycho's staying near home. Brothers Steen and Axel had returned abroad after their father's death, so whatever help their bereaved mother needed with the final rendering of the crown's accounts at Helsingborg, the

---

[12] Friis (39) reported an exchange at court between Tycho and Tage Krabbe, a second cousin of his. Tage approached Tycho with the greeting "Why, here we have the cynical Diogenes!" In response to Tycho's confusion, Tage explained, "You are buried in your nonsense just as Diogenes was in his tub." Tycho then retorted that he would not compare Tage with such a little fellow but that he reminded him of Julius Caesar, in that Caesar preferred to be first in any other city rather than second in Rome, and Tage would rather be first among fools than second or third among the learned.

removal of the family back to Knudstrup, and the negotiations in settling the estate probably fell to Tycho.

This occupation with family affairs, and perhaps the first period of real intimacy he had ever had with his mother, turned Tycho's thoughts to – or even revealed to him for the first time – the circumstances of his own birth. He was moved to compose a Latin poem, humanist in spirit and expressed in the words of his stillborn twin brother.[13] "He dwells on earth," the brother was made to say of Tycho, "but I live on Olympus." To the humanists of Tycho's circle, poetry was not simply a felicitous expression of feelings but something possessing spiritual, almost magical power, an ability to infuse matter with life. Petrus Severinus once explained how this worked:

> Plato asserts in one of his dialogues that there is a divine power in the words of poets.... The very bonds which hold body and soul together seem to be loosened when the senses are overwhelmed by a beauty and delight which they have never before known.... These invisible ideas possess such great power that their force can color, sustain and enliven the physical bodies of nature and endow them with a multitude of allurements.[14]

Having completed his effort to infuse new life into the memory of his twin brother, Tycho took the poem to Copenhagen and had it printed in 1572 as a broadside from the press of Mads Vingaard, a humanist printer who had previously published some of Vedel's writings.

It is not to be expected that a man of twenty-five would live with his parents for very long, and Tycho seems not to have done so. Probably sometime during the last half of 1571, he moved a few kilometers away to quarters at or near his Uncle Steen Bille's residence of Herrevad Abbey. And, probably not coincidentally, Tycho also at this time met, wooed, and won the woman who would be his life's mate.

Out of the contemporary rumors of scandal, two solid facts emerge. Tycho's spouse was named Kirsten Jørgensdatter,[15] and she was a commoner. One of Tycho's last students described her as "a woman of the people from Knudstrup's village."[16] But twentieth-

---

[13] IX, 173. An English rendition is provided by *Gade* (12).
[14] E. Bastholm, ed., and Hans Skov, trans. *Petrus Severinus og hans Idea medicinae philosophicae: En dansk paracelsist* (Odense: Odense Universitäts forlag, 1979), p. 49.
[15] The patronymic, indicating that her father's name was Jørgen, is known only through a document first published in 1935. It was reprinted by *Norlind* (368–9) as a "supplement" to the *Opera Omnia*.
[16] *Gassendi*, 23.

46   The Lord of Uraniborg

century scolarship has provided strong reasons for qualifying that statement. Already in the seventeenth century there was a conflicting tradition that she was a clergyman's daughter.[17] As it turns out, a Jørgen Hansen was the pastor of Kaagerod, the parish church some five kilometers from Knudstrup, from 1546 to around 1569. He may therefore have baptized Tycho. Following a succession of two other pastors, a Hans Jørgensen was called to the parish in 1591 by the lords of the parish, Tycho and Steen Brahe. That Hans Jørgensen visited Tycho many kilometers away at his island home observatory in 1591.[18] It is possible that he was there simply to be interviewed by Tycho, either before or after his appointment. But it also is possible that both he and Kirsten Jørgensdatter were the children of Jørgen Hansen.[19] Given the much greater likelihood of Tycho's finding sufficient common ground for lifelong compatibility with a girl from the middle class than with one from the lowest ranks of the Danish population, this seems even probable.

Such a social distinction would have been meaningless in Tycho's day, however; the one that counted in that era was that between noble and non-noble. Because of this distinction, Tycho and Kirsten could not be formally married. This did not mean that there was necessarily anything scandalous about the liaison. Indeed, casual dalliances between nobles and commoners were so frequent as to be deemed unworthy of comment.[20] Even longer-term relationships were so common that they had been institutionalized in the oldest Danish law code, and their legal status had been perpetuated to Tycho's day.

According to the Jutish law, the woman who for three winters lived openly as wife in a house, eating and drinking and sleeping with the man of the house and possessing the keys to the household, should be his true wife. An ancient Jutish word had even been retained to describe the wife and children of such an alliance: *slegfred*. In the polygamous Viking days, *slegfred* referred to wives of secondary status, but the concept was still legally valid in Tycho's day. Indeed, as recently as 1568, another man named Jørgen Hansen had

---

[17]   *Dreyer*, 70.   [18]   IX, 101.

[19]   The cadastre of Otte Brahe's estate compiled in 1570 contains the names of some three hundred peasants who were tenants of Otte, including all of those associated with Knudstrup. Only two were named Jørgen, and one bore the patronymic Jørgensen, but none of the three lived near Knudstrup. Because Tycho's student described a wife named Kirsten who came from Knudstrup, and the only Jørgen in the area was the pastor, Jørgen Hansen, the argument is better than that often used in historical issues.

[20]   As long as it was noble*men* and non-noble women. The opposite situation was punishable by death for the offending male. See S. C. Bech, *Danmarks historie* VI (Copenhagen: Politikens Forlag, 1963), p. 356.

brought suit against a person for calling him a bastard. It seems that Hansen's noble father, Hans Skeel, had been twice married to noble wives but that, after his second wife died, he lived with a commoner who bore him eleven children. Hans's son argued that his parents had lived together honorably and openly for twenty years, "although because she was a commoner, they could not be married." A half brother and half sister, who, of course, were nobles, testified that they respected Jørgen Hansen's mother as an honorable woman and considered her children to be *slegfred* children and not bastards. The nobles on the provincial court thus ruled that Jørgen was legitimate and not a bastard.[21]

Tycho's biographers have unanimously seen his morganatic marriage as an essentially negative act. Unable to accept it at face value as a love match that Tycho, because of his increasing disenchantment with the mores of the noble class, was unwilling to sacrifice on the altar of conformity, they have interpreted it as evidence that Tycho regarded himself as unsuited to a normal aristocratic marriage. It is possible that Tycho, because of his reluctance to participate in the social politics of court life, had already begun to project the image of a dropout from noble society. It was surely this alienation – much more than the disfigurement from his misadventure at Rostock – that would have detracted from the eligibility conferred by the Brahe name.

But Tycho also should have been able to recognize that he was unlikely to be able to enter into an aristocratic marriage without making serious compromises with respect to both his interest in scholarly pursuits and his abhorrence of courtly life. On the other hand, there were certainly less drastic alternatives than an open and permanent alliance with a commoner. Many, perhaps even most, of the scholars whom Tycho knew were unmarried. And many, if not most, of the nobles he knew had "arrangements" with a commoner at some time in their lives. That Tycho did not choose either of these alternatives says something about the strength of his character, even if it leaves unanswered the question as to whether that strength derived from his feelings for Kirsten or his contempt for the values of his peers.

The flowering of Tycho's relationship with Kirsten and his move from Knudstrup to Herrevad Abbey are the only two incidents documenting Tycho's activities in 1571 and 1572. There are two reasons for believing that the former occurred in 1571. One is a

[21] C. Klitgaard, "Hans Skeel til Nygaard," *Vejle Amts Årböger* (Copenhagen, 1912), pp. 21–2.

statement written by Tycho's youngest sister sixty years after the
fact, attesting that he and Kirsten had lived together in Denmark for
twenty-six years before leaving in 1597.[22] The second reason is that
Tycho seems to have mentioned to Wolf, in a letter written no later
than the fall of 1571, that he was having some problems in connection with marriage and emigration, for in November 1571, Wolf
wrote suggesting that philosophy might substitute for a wife.[23]
Whatever underground ceremony might have been celebrated,
however, Tycho must have spent most of his time at Herrevad in
matrimonial quarters.

Herrevad Abbey – like Knudstrup, still standing today in essentially the same form and environs – lies to the northeast of Knudstrup,
only a few kilometers away as the crow flies, but on the other side of
a high, thickly forested ridge. Tycho must have ridden many times
through the oak forests around the eastern end of the ridge, past
villages of half-timbered farmhouses, and into the valley of the
Rønne River which runs at the northern base of the ridge. The
villages on the northern flank of the ridge belonged to the abbey.
Cavernous oaken barns marked the boundary of the abbey, and the
Romanesque church loomed above the buildings of the abbey itself.

Herrevad had been founded in 1144 by Cistercian monks, who had
come to the wooded vale of the Rønne River directly from their
home base of Cîteaux in France. In the ensuing four hundred years, it
had become a vigorous institution, thanks in no small part to the
contributions of wealthy and influential patrons.[24] Among them had
been Peder Brahe, known as the "knight from Halland" (a province
in Sweden), who had been buried at the abbey in 1390 under an
immense slab of stone engraved with the shield that still, in Tycho's
day, constituted the herald of the Brahes.

The principal weakness of the Scandinavian monasteries had been
their distance from Cistercian headquarters in France. Over the
years, the problems of communication had been solved by creation
of a regional vicar general. In the sixteenth century, however, this
solution became a problem in itself, for in the power struggle
between church and state that characterized the development of
modern Europe, the kings of Denmark gained direct access to the
administrators of the monasteries. Since 1523 the kings had
appointed the vicars general for Denmark, and in 1537 King Christian III incorporated all monasteries into the new Lutheran state
church. Spiritual authority, of course, was immediately transferred

[22] XIV, 299–300.   [23] VII, 5–6.
[24] For an extended discusson of Herrevad Abbey, see John Christianson, "Cloister and
    Observatory: Herrevad Abbey and Tycho Brahe's Uraniborg" (Ph.D. diss., University of
    Minnesota, 1964).

to Lutheran theologians, who ran the monastic school and tried to convert the incumbent monks to Lutheranism, in the hope that they might choose to marry and accept parish calls.

The monasteries' administrative authority was initially vested in the abbots, who functioned as vassals of the crown. But as conversion and attrition took their toll of the monastic population over the years, the abbeys were gradually secularized under aristocratic lay administrators. By 1560, only three were outside royal control, and two of them were quickly converted, in order to help finance the war with Sweden: Vittskøl went to the husband of Tycho's older sister, Henrik Gyldenstierne, and Herrevad went to Tycho's mother's brother, Steen Bille.

Steen Bille was a man who had used the path of learning to prepare for service in the royal chancery. It was a path that the Billes had trod for generations before him, and with such success that on the eve of the Reformation they were the preeminent clerical family in the realm. Even after the upheaval that transformed Denmark into a Protestant country, Steen had followed the steps of his ancestors with remarkably little deviation: introductory tutelage at home, primary schooling at a monastery, and then study under the supervision of his uncle, Torbern Bille, the recently deposed archbishop of Denmark, who was serving out his days in dignified retirement as lifelong fiefholder of the Benedictine convent of Bosjø. When he was ready for university, Steen and his brother Jens had taken the old roads to the University of Paris instead of the new one to Wittenberg. Their only departure from tradition, and concession to the realities of post-Reformation Denmark, was a Lutheran tutor to guide the young brothers into the new faith.

Steen was abroad for five years, in France, England, Scotland, and Germany. Unlike many of the Billes of past generations, he seems not to have taken a degree. But when he returned home he nonetheless had no trouble obtaining a post as secretary in the royal chancery, which had been reorganized after the Reformation by Master Ove Bille, late chancellor and bishop of Aarhus. Steen served in the chancery for five years, administering the foreign and domestic affairs of the realm, before leaving the government for private life. In the next ten years he parlayed inheritance, marriage, sound business instincts, and the king's confidence into an imposing estate that he commanded from the great gate house of Herrevad Abbey.

One of Steen's contemporaries characterized him as a lively and sociable person who got along well with people of all social classes but especially enjoyed the company of learned men.[25] Many of these

---

[25] *DBL* III, 51–2.

people were his close relatives, like Tycho. They engaged in the typical broad range of humanist activities, including writing poetry, collecting antique articles and manuscripts, and compiling genealogies. Brother Jens Bille, of the nearby manor of Gunnerstrup, was a student of runic writing and practiced his command of the ancient Nordic alphabet by exchanging letters with two cousins, Bent Bille and Mogens Gyldenstierne. Steen's most serious intellectual interest, however, seems to have been alchemy, the area in which Tycho's interests were concentrated.[26]

The Herrevad Abbey at which Tycho lived and worked in the early 1570s was the product of a four-hundred-year attempt to achieve economic self-sufficiency. In the belief that the monastic ideal of spiritual isolation from the world could be no more complete than a community's material isolation, the abbey had not only developed its own extensive buildings and grounds but also channeled the resources of the surrounding area to its use: Nearly three hundred farmers paid annual rents in grain, livestock, dairy products, and even honey. Thirteen parishes paid tithes. The abbey owned six mills on the Rønne River, salmon fisheries at the mouth of the river, and cod-fishing rights along the coast. By Tycho's day, its forests were being exploited to fuel iron works and brick works and to provide timbers, wheelbarrows, and even oak pipe for construction projects. The grounds themselves contained orchards, gardens, fish ponds, and a mill, in addition to the usual facilities for tending animals and working wood and iron.

If the variety and extent of the activity at Herrevad was not unusual among such institutions, it was, nevertheless, inspiring to a young man who had grown up in the artificial world of cities, courts, and universities. Motivated by his own reasons for seclusion and contemplation, Tycho began to conceive of and develop, in Herrevad Abbey, the institution that was to achieve widespread and lasting fame as the prototype of the modern research institute.

In the long term, the facilities at Herrevad that were to have the greatest impact on Uraniborg were those for fashioning wood and metal. Tycho was probably not at Herrevad very long before he decided to replace the half-sextant he had produced at Augsburg (See Figure 1.6) but had given to Paul Hainzel as a parting gift. The project went well. The half-sextant's straight walnut legs were relatively simple, and even the curved brass arc seems to have turned out well enough to encourage Tycho to have a larger, interchange-

---

[26] Tycho mentioned twenty-five years later that "at the time I had . . . only recently returned from my journey in Germany and was more occupied in chemical than in astronomical studies." (V, 83). See also V, 108.

able, 60° arc made, which would in fact provide the name *sextant* for the instrument.[27] At the same time, Tycho recognized the limits of what nonspecialists could do, even when he was at hand to provide inspiration and guidance. Thus, when he wanted more difficult or more decorative instruments made, he had to order them from an artisan in Copenhagen. What he would eventually look for in creating his own instruments would be the best of both worlds – skilled smiths with the best facilities available, working where he could easily supervise their efforts.

In the short term, the facilities most useful to Tycho were those he used for alchemical experiments. Some of them were probably already in place, for monasteries had traditionally produced cordials and medicines by the same methods of distillation that were used in alchemy. Moreover, Steen Bille is known to have been interested in alchemy and may already have done, or sponsored, some work of his own. But Tycho did his work in a building especially equipped for his needs, probably situated outside the walls of the abbey so as to limit the effect of noxious fumes.[28]

When Tycho had difficulty obtaining the glassware he needed for his distillations, he was able to induce a party of Venetian glassmakers to set up shop on the grounds of the abbey. From the beginning, of 1572 the glassmakers produced wares with a popular appeal. And by 1576, goblets, bowls, windowpanes – even red-tinted ones for the new royal castle at Kronborg – were being shipped from Herrevad. The demand for such items in Denmark was apparently not satisfied until 1583, at which time the artisans moved on to employment with an astronomical friend of Tycho, the *landgrave* of Hesse.[29] By then, Tycho had fitted out his own estate and was presumably well stocked with alchemical glassware, for he never set up a glassworks of his own.

Less successful – no doubt because the market for its product was considerably more limited – was Tycho's venture in papermaking. Again, this can be seen as the continuation of an ancient monastic tradition, that of producing vellum for the manuscripts that were copied, bound, and preserved in the old library of Herrevad. The impetus for this enterprise was Vedel's anticipated completion in 1575 of his translation of Saxo Grammaticus. In an effort to solve the problem of obtaining paper for the printing, Tycho prevailed on his

---

[27] V, 81, 85.
[28] In II, 307, Tycho implied that his laboratory was outside the walls of Herrevad. But there was a folk tradition that Steen had a laboratory inside the walls, and an excavation in 1939 did, indeed, reveal a cellarlike room: *Norlind*, 29.
[29] Christianson, "Cloister and Observatory."

uncle to establish a paper mill and also composed a poem calling on
the women of Denmark to sacrifice some of their linen to help
publicize the deeds of their ancestors, as recorded in Saxo's chronicles.[30] Apparently the paper for Vedel's edition was acquired elsewhere, but a mill did go into operation at Herrevad in 1576,
producing paper watermarked with Steen Bille's initials. Unfortunately, this first paper mill in Denmark did not long stay in production. Thus, ten years later when Tycho had things ready to print, the
acquisition of paper in Denmark was again enough of a problem to
persuade him to establish his own mill.

The abbot from whom Steen had taken control in 1565 lived at
Herrevad until his death in 1572, and there may have been a few
other old monks in residence, as well. There was a Lutheran lector
and probably a few lesser instructors, for the cloister school was still
in operation. Indeed, one of the students was Tycho's youngest
brother, Knud, then sixteen years old and serving Steen as a page. In
addition, young scholars and clergymen came to seek the patronage
of Steen Bille, who controlled the appointments to five parish
pulpits, maintained pupils in a number of schools besides Herrevad,
and was known for his influence in promoting the careers of capable
young men. Thus, something approaching a learned academy came
into existence at Herrevad. Tycho was never to regard himself as
simply an astronomer, and he especially did not at this time. He
would express himself poetically as worshiping all the "most de-
lightful Muses surrounding Apollo," and there is every reason to
believe that the statement was particularly descriptive of his work at
Herrevad. What had captured his imagination in his youth were the
putative astrological connections between the heavens and the earth.
What had expanded his ambitions at Rostock was the Paracelsan
view of the place of humanity in the whole scheme of nature. Plato
had rather vaguely described the universe as a cosmos in which all
parts were related to the whole, but Aristotle had later drawn sharp
distinctions between the celestial and the terrestrial regions. For
centuries, Aristotle's physics had prevailed, but this had begun to
change with the fifteenth-century revival of Platonic and Hermetic
thought. Early sixteenth-century reformers like Copernicus, Paracelsus, and Luther had challenged accepted worldviews, setting in
motion the attempts of individuals like Melanchthon, Ramus, Niels
Hemmingsen, and Petrus Severinus to build new syntheses. The aim
of much of this activity was to build a unified view of nature in the
spirit of Plato, to replace that of Aristotle. Tycho's friend Severinus
rationalized the ideas of Paracelsus by placing them in a Neoplatonic

[30] IX, 174.

context. Tycho likewise worked within the same context, of reforming humanism in the course of investigating aspects of the universe as cosmos.

In the view of Tycho and his friends, the medical theories of Galen, which Severinus linked to the philosophy of Aristotle, were no longer acceptable. Galen had described disease as an imbalance of the body's four humors. The cure was supposedly to restore the balance, by introducing the opposite humor, typically by means of hot and cold packs or herbal medicines. Medicine thus became essentially a study of the human anatomy and the four elements – blood, bile, black bile, and phlegm – that regulated its humors.

Paracelsus had rejected these theories, asserting that each human being is a microcosm of the universal macrocosm. This led him to see disease as something from outside the body that penetrated into it so that harmful macrocosmic forces began to work within the microcosm. But there was also a "spirit" in nature that drove out disease. Severinus called it (among many other names) *balsam* and described it as a celestial substance found in plants, animals, and minerals. According to him, balsam filled the whole body and gave it color and vitality, "calling forth movement when it fills movable parts of the body, and feelings in the sensitive parts. Elsewhere it will call forth digestion, elimination of waste, distribution of food and nourishment. Here again it thinks of fertility and brings about the colors and processes that are essential to reproduction."[31] Balsam also drove out the enemies that impinged on its place in the body.

The physician's task, as Petrus Severinus described it, was to study, understand, and refine alchemically this curative spirit in nature. The physician, he asserted, studied illnesses and balsams, not merely the human body and its humors. Real true anatomy was vital and alive, not the mere dissection of cadavers. Indeed, philosophical medicine studied the universal harmony of all creation and drew on many other disciplines, including astronomy, astrology, meteorology, physics, botany, and agriculture, applying all of this knowledge to the microcosm.

The Paracelsan ideal of the world, interconnected in every aspect and governed by agents whose operations were probably strongly analogous, had found a strong response in Tycho when he was at Rostock and remained an article of faith for him as long as he lived. Because this idea was much less well developed than astronomy was and therefore required little of the disciplined, sequenced study peculiar to astronomy alone among the sixteenth-century sciences, Tycho was already experimenting in Augsburg with the "spagyrical arts." Even without Peder Sørensen's new interpretations of Paracelsism, therefore, alchemy might well have preoccupied Tycho at Herrevad. But with them and the reinforcement of Pratensis and the facilities of Steen Bille, there can be little wonder that Tycho would subsequently describe himself as being occupied with chemical investigations before November 1572 and leave behind from this period not a single astronomical observation.[32]

Life was not all business. In the spring of 1572 Tycho received a letter[33] from Pratensis whose apparent purpose was to recommend a young Danish scholar recently returned from Cologne. But whether Pratensis doubted the man's competence, was reluctant to impose on Tycho, or was simply feeling frivolous, he couched the entire letter in Paracelsan allegory and generally nonsensical references to such things as the danger of wood nymphs. His candidate, he said, was skilled in Greek and Arabic and was a "Galenist, Paracelsan, theologaster, philosopher, physician, syrupmaker, chymist, or anything else you want; yet despite all this – nothing."[34] Tycho replied more or less in kind,[35] extemporaneously, with a model of humanist elegance, and in Latin verse, asking Pratensis to send over the young scholar to help in the labors of Vulcan. He also asked Pratensis to send some cucurbit flasks for his alchemical distillations and reminded him that the nymphs of the woodlands were a gift of Ceres.

A couple of months later there was another interruption. King Frederick married his young niece, Princess Sophie of Mecklenburg. Foreign rulers who were relatives of the bride and groom streamed into Denmark from such places as Saxony, Mecklenburg, and Schleswig-Holstein. Each ruler's entourage consisted of a long train of courtiers, councillors, lords and ladies, diplomats, theologians, heralds, and servants. In order to meet them in appropriate splendor, King Frederick called on all of the Danish nobility to come to the celebrations, dressed in new court attire, mounted on their best steeds, and each accompanied by two squires and a page. From Herrevad, Steen and his wife, Christine Lindenow, undoubtedly went, probably with young Knud Brahe riding as one of his uncle's squires. Tycho probably borrowed attendants for the occasion and went too. The wedding was held on 20 July 1572, in Copenhagen Castle.

---

[32] V, 83, 108.   [33] VII, 7–8.

[34] The young man was Niels Mikkelsen, the son of an influential clergyman in Viborg. He had studied in the Paracelsan center of Cologne from 1567 to 1569 and had flirted briefly with the Jesuits but was now home in Denmark, looking for a canonry or some other call. Whether he ever worked with Tycho is not known. See Vello Helk, *Laurentius Nicolai Norvegus S.J.* (Copenhagen: Gad, 1966), pp. 63, 431–2.

[35] VII, 9–10.

The next day, the young queen rode to her coronation in a silver carriage drawn by eight horses, accompanied by a great parade of princes and nobles. Peder Oxe carried her crown before her to the high altar of Our Lady's Cathedral. Such an occasion, of course, was celebrated with banquets, parades, tourneys, plays, masquerades, and other festivities. Scholars from the University of Copenhagen even entertained their colleagues from Saxony – who probably included some of Tycho's old professors from Leipzig and Wittenberg – with banquets, student comedies, and long theological discussions.

On 11 November 1572, there was an event that signaled the formal beginning of Tycho's career as an astronomer. As Tycho was returning from his alchemical laboratory that evening for supper, he noticed an unfamiliar starlike object in the sky, one not only clearly alien to the constellation in which it appeared but also brighter than any star or planet he had ever seen. If we can believe Tycho's description of his discovery, he did not feel he could trust his own vision but had to appeal first to his own servants and then to some passing peasants, for corroboration.[36] How much satisfaction he could have obtained from them is doubtful, for although Tycho claims to have been familiar with all the stars since boyhood, it is unlikely that any of his impromptu consultants had taken the same pains. But mere confirmation of the existence of something brighter than Sirius or Vega, or even Venus, must have greatly excited Tycho, because its location outside the zodiac ruled out the possibility that it was a planet and its appearance seemed to deny that it could be a comet. Even though Tycho had not yet seen a comet himself, he was quite familiar with the accounts that described them variously as being tailed, fuzzy, or hairy in appearance.[37] The crucial test, of course, was motion, and a few nights of observation quickly established that this new object likewise lacked the progressive movement relative to the stars that characterized comets.

It would be interesting to be able to examine Tycho's work on this, his first serious investigation. But unfortunately, these observations are the only ones of his entire career that are known to have been lost. They alone (as far as can be ascertained) were not copied when one of Tycho's assistants later compiled a notebook that constitutes the sole record of his observations from his first efforts in 1563 up to those in December 1577.[38] But it is not unreasonable to presume that Tycho recorded the object's position in some detail on the very first night, for even if he had already formed the opinion that it was not a comet, he had to be prepared for the possibility that

[36] I, 6: II, 330–1: III, 93.   [37] III, 107: I, 28.   [38] X, xii.

it was or, whatever else it might be, might be something that moved. Because the interloper appeared near the three stars that comprise the right-hand half of the familiar W of the constellation Cassiopeia, and Tycho later described it as forming with them a fairly exact parallelogram, he recorded its distances from the stars. Such observations were the kind for which Tycho's cross staff had been designed and were those that Tycho actually used in his subsequent discussion. It cannot have taken him long to determine that the object was too stationary to be a comet and even too stationary to be associated with the Aristotelian sublunary sphere.

By Tycho's day, the Aristotelian dichotomization between the earthly sphere of change and the heavenly spheres beyond the moon where things were assumed to retain forever a perfection unimaginable to anyone living on earth had withstood the scrutiny of philosophers of the Greco-Roman, Arabic-Islamic, and Latin-Christian cultures, for nineteen hundred years. It is thus not surprising that for most of Tycho's contemporaries, change was both theoretically and (through the gradually accumulating weight of scientific terminology) almost logically restricted to the sublunary world.

The primary reason for the longevity of Aristotle's cosmology was that in regard to prephotometric and prespectroscopic astronomy, it was essentially correct. Only with respect to comets was it seriously in error, and even there it was considerably less vulnerable than might be expected. First, the assumption that comets lay outside the domain of astronomy helped protect them from careful examination by astronomers. Most significant, however, was the fact that astronomical distances could not be determined routinely. Astronomers had known since antiquity that the moon was sufficiently close to the earth to show the effects of perspective when it was not observed precisely in the zenith, that is, when the observer was not in a direct line between the moon and the center of the earth. (For a glossary of technical terms, see Appendix 2.) From at least Hipparchus's time, it had been understood that under extreme conditions this apparent displacement of the moon from its true (*average*, as seen from all points on the earth) position could amount to more than a whole degree. Generally it was less, as the parallax changed continually and rapidly with the daily rotation of the earth (or, as Tycho and virtually everyone before him saw it, with the diurnal motion of the celestial sphere in the opposite direction). These smaller effects were still large enough to be detectable for the moon and would be proportionally larger for anything nearer the earth. But rationalizing them was a rather involved mathematical task, which presupposed a complete accounting for the intrinsic motion of the body in question. The combined problem was sufficiently difficult that it had never been solved satisfactorily, even for the moon as of Tycho's day. So it is no wonder that people had not worked seriously enough at resolving it for other supposedly sublunary objects to discover that there simply were no such objects for which it could be done.

Although this theoretical matrix may be unconvincing to modern readers, it is nevertheless important to appreciate its hold on the sixteenth century. Most of Tycho's contemporaries were so compelled by its logic that they could not even think of doubting that there would be parallax in a body so manifestly "generated" and thus so obviously associated with the sphere of change. Many others actually obtained observational results, through which they convinced themselves that the new star was below the moon where it was theoretically supposed to be, whereas others insisted on calling it a comet, even though they could find no parallax for it. Even people who were willing to accept observations that placed the star above the moon could remain mired in the rest of Aristotle's doctrine: The *landgrave* of Hesse, John Dee, and Thomas Digges all assumed that any change of brightness of the star must be purely apparent (rather than intrinsic) and therefore the reflection of a change in the star's distance.[39]

Tycho was neither an uncritical follower of, nor even particularly well disposed to, Aristotelian theory, but he could not fail to have been influenced by it to some degree in his initial reaction to the remarkable intruder. Yet, however much such things as the very terms available for discourse on the subject were commitments to the Aristotelian worldview, Tycho succeeded in detaching himself sufficiently from his preconceptions not only to test the object for parallax but also to believe the negative results he obtained.

Although determining distance by measuring parallax was generally rather complicated, there was one case for which it was very simple, the one in which there was no change of position at all. It was precisely this case that obtained in 1572. The first comparison of angular distances from a given star taken at two different times of night should have alerted any competent observer to the absence of change. There is no reason that Tycho could not have discovered this on the very first night, if he took the stellar form of the object

---

[39] The *landgrave* assumed that the star was moving straight up (VI, 49), without thinking that such a rectilinear motion in the supralunary sphere would be as embarrassing for Aristotle as would a change in brightness. John Dee suggested the same thing, in an unpublished (but lost) manuscript entitled "De stella admiranda in Cassiopeiae Asterismo, coelitus demissa ad orbem usque Veneris, iterumgue in Coeli penetralia perpendiculariter retracta Lib. 3.A.1573," as cited by F. R. Johnson, *Astronomical Thought in Renaissance England* (Baltimore, 1937), p. 156. Digges (III, 172) assumed that the variation in brightness might be due to the earth's moving away from the star through its Copernican annual motion.

seriously enough to remeasure its position after an interval of several hours. Only after a few nights of subsequent observation, however, could he have ruled out the possibility that there actually was some parallactic motion that was fortuitously canceled out by an equal and opposite intrinsic motion of the body itself. Then, once he knew that the gross parallax that would save Aristotelian theory was lacking, he would have examined his figures more carefully for evidence that the starlike object was, indeed, at a stellar distance.

At some point Tycho observed the distance between the new star and Schedir, the brightest star in Cassiopeia, and found that it was invariably 7°55′, whether he measured at upper culmination 6° from the zenith, lower culmination 28° above the horizon, or any place in between. Tests with other stars produced the same null result. Eventually he devised a third kind of check, which foreshadowed the elegance and ingenuity he was to show later in his design and handling of instruments. What he wanted was a direct measurement of the nova's declination, which would enable him both to certify that it remained constant from night to night and to compare the result with the declination he derived trigonometrically from other observations. Such measurements were universally obtained from a vertically mounted quadrant.

All Tycho had was a cross staff and a sextant.[40] And although he could mount his sextant perpendicularly so as to measure altitudes, he could obviously measure only up to 60°, whereas the new star crossed the meridian 6° from the zenith. Undaunted, Tycho turned the instrument around and set it in a north window to record lower culmination. How closely his observed declination agreed with his other figures we do not know, for Tycho did not mention the results in his write-up. Twenty years later, however, he still remembered the achievement with pride. When he published his descriptions of the instruments on which his reputation rested, one instrument alone – the oldest, most primitive member of his collection – received double coverage, so that he could depict it in both its normal usage and the extraordinary role into which he had pressed it for observations of the new star of 1572.[41]

Within three or four weeks, Tycho formulated and resolved all of his questions concerning parallax. An unsigned letter sent to him on 16 December asking his opinion on the new star suggests that he had already communicated his findings to a few friends.[42] He did not, however, send word as far as Copenhagen, perhaps because he was

---

[40] The sextant was really two instruments, in that it consisted of one pair of hinged legs and interchangeable arcs of 30° and 60°: II, 331, 333.
[41] Ibid. See also V, 80–7.   [42] I, 141.

not yet willing to volunteer his opinion in learned circles, but more likely because he did not want to expose his results prematurely. Whether he even composed a formal statement of his findings at this time is not clear, although it appears that he at least wrote up a Latin report of his observations.

It is unlikely that Tycho was able to spend much time on the report, because with his interest in astronomy reawakened by the new star, he had decided to compile an astrological meteorological almanac for the coming year.[43] If he were truly going to test the accuracy of his predictions, he would have to commit them to paper before year's end. As part of his almanac, he included a detailed prediction of the lunar eclipse that would appear on 8 December 1573. And because it would be an astrological almanac, he had to investigate the significance of the eclipse. When he did, he found that it seemed to foretell the death of King Frederick.[44] This was obviously not a prognostication that could be discussed openly, and in any case, Tycho was not confident of his data for either the moon or Saturn. At the same time, however, Tycho was unwilling to suppress a prediction that would be sensational if it were to turn out to be accurate. So he veiled it in impenetrable allegory about a Phoebic hero's triple hospitality in receiving an Olympian host, and some vague imagery of Callisto, Diana, and Cynthia.

When Tycho completed his almanac, he apparently liked the result well enough to consider publishing it, because he then composed an introduction to the almanac, in the humanist genre of an oration.[45] Tycho's oration drew on the themes of earlier humanist orators. Like Ficino, he emphasized the contemplative role of human beings, although under the influence of Philippist rationality, his focus was on the empirical study of perceptible phenomena in the universe, rather than on mysticism.

In his salutation, Tycho invoked the classical muse of astronomy, Urania, but the spirit that permeates the oration is that of Paracelsus, beginning with a description of the universe as comprising heavens, earth, seas, sun, moon, stars, animals, vegetables, and minerals. After describing the Creator in thoroughly Lutheran terms as incorporeal, immense, eternal, incomprehensible, and omnipresent but located in no single place, Tycho expressed the traditional Renaissance conviction that God created humans in his image and placed them on earth in the center of the universe to contemplate, as in a mirror, the nature and constitution of the whole of creation, so that

---

[43] III, 94.   [44] I, 132–3.

[45] The oration is printed in I, 35–44. It is signed "At our Museum of Herrevad, December 1572."

during their mortal life they might learn to know the majesty and wisdom of the invisible, incorporeal God through the visible objects of his creation.

The universe itself, Tycho thus concluded, is the best book of theology. This is true even for the vilest animal, the tiniest herb, and the most common metal or mineral, but evidence of divine wisdom is nowhere so clear as in the celestial regions, with their bodies of immense size, radiance, and perpetual regularity. In contrast, everything on earth, except the human soul, is in a constant state of dissolution and alteration. Although there is a hierarchy of intelligence among humans, most people are ignorant or even scornful of knowledge concerning the heavens. And among those who are not, almost none has proceeded further than the threshold of celestial knowledge. Yet they trumpet themselves as great authorities; do great harm to the disciplines of astronomy, astrology, and meteorology; and bring much ridicule on the compilers of almanacs.

After mentioning that he had written a book on the two astrologies of meteorology and horoscopy, entitled *Against Astrologers for Astrology*, Tycho turned to a discussion of his theory of meteorology by observing that manifest alterations occur in the atmosphere below the moon. He ascribed these changes in weather to celestial influences and especially to the moon, which exerts a strong influence because it is immediately next to the atmosphere. The lunar influences are strongest at the new and full moon and at the intervening quarters, but they also have an effect at the points halfway between each quarter. Tycho followed Ptolemy in associating the changes of these lunar octads with storms, as well as with crises in human illness, but he attached particular meteorological significance to the longitudes of the sun and moon at each octad.

For the task of correlating longitudes with these effects, Tycho rejected the Alphonsine and Prussian tables in favor of his own, more accurate observations,[46] and he ridiculed armchair astronomers who remain shamefully ignorant of the stars in the sky while they sit by the stove and study astronomy in books and papers. He also objected to the method of delineating the twelve houses of the zodiac normally used by astrologers and alluded to his own scheme based on the great circle coordinates of the horizon and meridian of a single place and divided into the eight houses of the lunar octads as well as the traditional twelve houses. This innovation he defended as more consonant with mathematical harmony. Another one, choosing, Copenhagen, the capital of Denmark, as his place of reference, he justified with a comment on the great debt that all persons owe to their native land.

46 I, 37.

In conclusion, Tycho described the manuscripts he had written on related subjects[47] before turning to the contents of his almanac for 1573: tables of the risings and settings of the sun, moon, and planets; their configurations for each day and (especially) each lunar octad; and the predicted weather for each day of the year. He warned against placing too much confidence in his weather prognostications because celestial influences on the atmosphere are indirect, working through many local conditions that vary from place to place. Saying that his aim was really to study the discrepancies between predicted and actual weather in order to learn more about the correspondences between the heavens and earth, he emphasized the need for systematic meteorological observations. Tycho ended his elegant oration with a quotation from Ovid on the joys of astronomy and a verse of his own in the form of a Platonic lament that the cares of the world – the demands of courtly life, the raw cold of the North – disturb the peace and calm required for sidereal contemplation.

Soon after the new year began, at the first break in the "raw cold of the North," Tycho went to Copenhagen, taking his several manuscripts with him. Upon his arrival, he found that no one in the city had yet noticed the new star.

In view of the awesome nature of the supernova – an indescribably massive explosion that causes the brightness of a star to increase hundreds of millions of times[48] – it is easy to assume that such phenomena will be considerably more conspicuous than they have actually turned out to be. In fact, the very rarity of the occurrence (which current astrophysical theory regards as a unique change of state peculiar to some of the larger stars) means statistically that supernovae will be exceedingly remote and hence relatively unspectacular even in their temporarily enhanced state. The result is that even the most prominent supernovae have magnitudes that distinguish them from their stellar background merely in degree rather than in kind.

Thus, although both Tycho's star and the supernova of 1054 were

[47] I, 38–9: *De variis astrologorum in coelestium domorum divisione, opinionibus, earumque insufficientia*; and *De horis zodiaci inaequalibus, quas planetarias vocant*.

[48] In 1945 Walter Baade produced a good light curve from Tycho's observations and concluded that it showed a Type I supernova; see *Astrophysical Journal* 102(1945): 309. The random uncertainty of each observation was only ±23″. But with the advent of radio astronomy, J. E. Baldwin and D. O. Edge, "Radio Emission from the Remnants of the Supernovae of 1572 and 1604," *The Observatory* 77 (1957): 139–43 found a 10′ difference between Tycho's coordinates and the associated radio source. More recently F. R. Stephenson and D. H. Clark, "The Location of the Supernova of A.D. 1572," *Quarterly Journal of the Royal Astronomical Society* 18 (1977): 340–50, found that the difference was only 4′ and that one-third of that error was due to a faulty position for one of Tycho's reference stars. Thomas Digges placed the star right on the radio source, by observations made with a cross staff, although those observations showed more scatter than Tycho's do.

visible in daylight to anyone who knew they existed, neither was so
bright as to arrest the attention of the untrained eye. In fact, as far as
is now known, no one in all of Europe ever noted the supernova of
1054; rather, it was documented by contemporary Chinese reports,
notations by some Islamic physicians, and modern assumptions
associating these reports with the Crab nebula.[49] In general, Europe
did much better with the new star of 1572. Several Scandinavians are
known to have noted it independently of published reports,[50] but as
of Tycho's first visit to Copenhagen in 1573, no one in its intellectual
community was aware of its existence.

The reaction of Tycho's friends was as gratifying as he could have
hoped. Dançey, in whose house they were dining when Tycho
mentioned the star, thought Tycho must be pulling their legs, perhaps to teach them an object lesson on the necessity of glancing at
the sky occasionally. Pratensis, in daily communication with the
other professors at the university, could not be persuaded that so
many had overlooked something so significant for so long. Tycho
did not press the issue but merely smiled and hoped for a clear
evening. When the star did appear and showed itself to be as unlike a
comet as he had claimed, it provoked an excited discussion. Sometime during the conversation, Tycho's manuscript emerged, and
Pratensis urged him to complete and publish it.

That an activity that modern society holds in esteem could be
viewed by Tycho's social peers as a questionable enterprise is a
measure of the cultural gulf separating us from the sixteenth century,
a gulf not readily bridged. Scientific ideas, however strange those of
another era may seem, are by definition capable of rational explanation. And because they tend to be taught consciously by one generation to the next, they usually are explained, in books that often
remain available for later generations as well.

On the other hand social mores, however logical they might once
have been, can evolve in mysterious ways and often outlive so much
of their original cultural matrix as to render impossible any reasonable reconstruction of their rationale. Moreover, people seldom feel

---

[49] But for a new examination of the evidence and the conclusion "that there is considerable
doubt whether the object of A.D. 1054 and the Crab Nebula are connected at all," see Ho
Peng-yoke, E. W. Parr, and P. W. Parsons, "The Chinese Guest Star of A.D. 1054 and the
Crab Nebula," in *Vistas*, 13, pp. 1–13. For evidence that some Europeans may have seen a
supernova in 1006, see Bernard R. Goldstein, "Evidence for a Supernova of A.D. 1006,"
Astronomical Journal 70 (1965):105–14. For a discussion of novae recorded in the pre-
telescopic era, see F. R. Stephenson and D. H. Clark, "Historical Supernovas," *Scientific
American* 234 (1976):100–7; and F. R. Stephenson, "A Revised Catalogue of Pre-Telescopic
Galactic Novae and Supernovae," *Quarterly Journal of the Royal Astronomical Society* 17
(1976): 121–38.

[50] For a listing of other observers, see *Norlind*, 46–7.

the need to attempt the feat, for custom tends to be transfused rather than explained, and departures from it are ignored, ridiculed, or punished rather than rebutted. The result is that we have very little insight into either the nature or the strength of the tide against which Tycho decided to swim. Basically, it was an expression of the almost genetically derived contempt with which the warrior ethic has always viewed all other occupations.

In Europe, however, this prejudice was institutionally reinforced by the traditional feudal obligation of the knight to devote his time exclusively to preparing for the defense of lord, vassal, and the faith. If the passage of time had made this rationale obsolete and had even removed enough of the stigma from the passive life of the clerk or cleric to permit the nobility a modicum of book learning, it had not, apparently, progressed so far as to produce a toleration for book writing. Even fifty years later, René Descartes was to experience the same difficulty in France. Thus, although Tycho's entrance on the literary stage marked a major turning point in his life, the information on the event is sufficiently contradictory to leave considerable doubt about the circumstances, and even the nature, of his decision to publish.

According to an account given by Tycho long after the events,[51] in about 1590, the original impetus toward publication came from Pratensis: Tycho himself had never considered publication because he still subscribed to the general prejudice that pursuit of the scholarly arts was unworthy of his rank. Initially, therefore, he dismissed the suggestion, pleading the imperfection of his manuscript and excusing its lack of polish by reminding Pratensis that he had not composed it for publication. Pratensis, however, persevered. While Tycho was at Knudstrup making plans for another trip abroad (apparently to make the final choice of a place to which to emigrate), his friend forwarded some documents on the nova that had begun to reach Copenhagen with the thawing of the sea-lanes. Tycho read the documents, he said, primarily because he was too ill to start on his travels anyway. But even though he was appalled by the general incompetence of the accounts and distressed by assertions that the star was a comet as close as twelve to fifteen earth radii away, when he went again to Copenhagen, it was only to check on some work that was being done for him at an instrument shop there. Pratensis naturally pressed him further, and when he uncovered the real problem (presumably Tycho wanted to avoid making an issue of the difference in their stations?), he urged consultation with Peder Oxe. But even though the most powerful Danish statesman of the six-

[51] III, 93–6.

teenth century was quite receptive to the project and even suggested that Tycho remain anonymous if that would ease the situation, he was still not convinced and again returned home with his manuscript. Only after a particularly persuasive letter from Pratensis did he finally agree to publication.

It is safe to say that this story has a few improbable features. First, in order to recognize the image of Tycho conveyed in it, we have to forget the resolute young man who since his teens had been challenging convention with his choice of a life work, and was in the process of defying it completely with his choice of a life partner. We have to dismiss as a temporary aberration the ambition that had prompted him to circulate six years earlier his astrological prediction of the death of Suleiman, and simply refuse to ask why he wrote his tract on the new star. To justify it, as Tycho did, in terms of the general astrological meteorological calendar he was preparing merely raises the larger question of the purpose of the calendar. We can then only wonder at the sudden conversion that resulted in his not only publishing but also attaching his name to the tract, and barely a month after the suggestion could rouse only a shocked refusal. Finally, we must be thankful that we do not yet have to compare this timid, reluctant individual with the forceful, not to say overbearing, Tycho of later years who was to engage in a virtual shouting match with a king when he felt that his personal and professional status were being threatened.

Nor are these the only problems with Tycho's curious tale. For although the letter from Pratensis begging Tycho to publish actually exists and was, in fact, printed as a preface to his book,[52] another one written by Pratensis three weeks earlier (but not printed by Tycho)[53] shows that the work was already in press at that time. It even obligingly mentions that Pratensis would immediately begin composing his letter for the preface.

If Tycho had never volunteered his account of the events, the verdict would be obvious: He went through the motions of covering himself and then took the plunge. If Tycho had even recorded his story in 1573, dismissing it would present few problems. But because he issued it so long after the facts that it could have had no conceivable propaganda value, it is as difficult to reject it out of hand as it is to swallow it whole. For the fact is that in the early 1590s, when Tycho was reprinting his own efforts as part of his larger analysis of all writings on the new star, he was under no obligation to comment on the circumstances of the original publica-

---

[52] I, 6–8, dated 3 May 1573.   [53] VII, 10–13, dated 16 April 1573.

tion. If he felt that his unusual prefatory letters required some explanation, he only had to corroborate the role of Pratensis's letter that they conveyed (for as he said in criticizing another writer at about that time, liars should have good memories).[54] But except for this aspect, which is the only known fabrication, Tycho's tale was completely gratuitous. In embellishing so elaborately the image of reluctance he had displayed in 1573, Tycho was therefore either inventing details for no purpose whatsoever or else presenting the truth – at least as he perceived it in 1573 and remembered it two decades later.

As of late 1572, Tycho had grappled for some ten years with what he and his contemporaries saw as a fundamental incompatibility between the serious pursuit of his scholarly interests and the behavior expected from a person of his class. If his refusal to start working up the totem pole of government service had tended to identify him as an impractical young man who had not yet settled down, at least his academic strivings had been sufficiently restrained to exclude any explicit action that would brand him a black sheep – except, of course, insofar as his relationship with Kirsten may have combined with his studies to place his life-style beyond the limits of mere unorthodoxy. It is doubtful that Tycho, now entering his twenty-seventh year, could have straddled the fence much longer. Unfortunately, the only way he could see to resolve the situation was to emigrate. On the assumption that even though he may not have yet been certain that he wanted to take such a drastic step, he would have been reluctant to take any other action that would commit him to it.

Whatever the degree of Tycho's initial reluctance to publish, he responded to Pratensis's encouragement by composing a statement of his conclusions on the new star. But because of his isolation at Herrevad Abbey and perhaps because Dançey and Pratensis constituted a very friendly audience, Tycho had no way of knowing how the scientific world was reacting to the new star. Although he did know that people would be predisposed to regard as sublunary an object so manifestly subject to generation and corruption (by December it had already begun to fade visibly), he seems to have had no conception of the reluctance with which they would greet what he regarded as a clear-cut geometrical demonstration to the contrary.

The result was a dispassionate manuscript with little urgency in its argument. Aside from a title and opening lines that proclaimed the appearance of a new star, and a passing allusion a few pages

[54] III, 236.

later to the fact that he would later prove it was not some kind of meteor in the elementary world, Tycho devoted his opening contentions to disputing the cause or origin of the new star. According to Tycho's interpretation of the nature and location of the star, it was a truly extraordinary pheomenon – as remarkable and even miraculous, Tycho asserted, as Joshua's stopping of the sun in the heavens or the darkening at noon during the Crucifixion. All the philosophers have agreed that the celestial world cannot change, and all the scientists through thousands of years have found that the stars remain unchanged in number, position, order, and so forth. Still, it is not completely unprecedented, for according to Pliny, Hipparchus also had seen a new star.[55] The philosophers have tried to explain away this troublesome counterexample, but Pliny's words are so unambiguous and Hipparchus's abilities as an astronomer so thoroughly documented that it simply is not reasonable to argue that the object in question might have been instead of a star a comet or meteor from the sublunary world.

Only at this point, four pages into his discussion, did Tycho pause to acknowledge the existence of an alternative to the proposition that the recent phenomenon was indeed a star, and to state that he would present observations that ruled out all other possibilities. Then, unaware that learned opinion would short-circuit his arguments with precisely such assumptions, he resumed his attack on what he thought would be the focus of debate, the origin and meaning of the star: Just as the philosophers are confounded by this rare object, the theologians are equally embarrassed, as they account for everything in terms of an original and unique creation. As for the Paracelsans, who suggest that this star until now has hidden itself in the ancient womb (to use their expression) and only recently has matured and revealed itself so mortals,[56] their explanation is inadequate too. For if the new star really is something that has just erupted after a long period of dormancy or has finally matured after an extended growth by more or less natural causes, why has this dynamic universe not produced other kindred appearances in such a vast stretch of space and time? None of these explanations will work: The new star can be regarded only as a special creation of God. But although it is of divine origin, it is not an analogue to the star of the Magi, which appeared at the birth of Christ, for nothing at a truly celestial distance could lead anyone anywhere because it would, ipso facto, appear to be (stay) at the same place in the heavens, wherever the wanderers moved on earth. Thus, because no one can explain the star satisfactorily, God clearly created it for his own purposes. Tycho

[55] I, 16–47. [56] I, 18.

could presume only to describe it and add some astrological comments on its effect.

With the nature and origin of the nova argued, Tycho turned to the empirical details, which he had probably already sketched out in December: Frequent observation with a very good instrument has shown that the star is stationary. Using the distances from two of the brighter nearby stars and the coordinates of the respective stars, the longitude and latitude and the right ascension and declination of the new star can be derived trigonometrically. Insofar as the given stellar positions are not completely accurate (for neither Ptolemy nor Copernicus can be trusted in this regard), the result may be slightly off. But the location will be close enough and is the best that can be achieved until all the stars are recharted, which Tycho himself, God willing, will do someday.

Having established that there is no proper motion to cloud the issue of parallax, Tycho considered the question of the "place" of the star – its distance from the earth. Although he had been (and was to remain throughout his life) extremely thorough in presenting his trigonometrical calculations, he treated parallax as though it would pose no problem at all for the reader, saying simply that if the star is seen in the same point of the heavens, both near the horizon and at the zenith, it is located either in the eighth sphere or not far below. And that, he asserted, is exactly the situation; he had checked the results on numerous stars but had never found a variation of even a minute of arc. There is, therefore, no doubt whatsoever that the star is far above the moon, rather than in the region of the elements below it.

It is probable that this conclusion marked the end of the astronomical portion of Tycho's original tract. The remainder of the (third) section contains some arguments that look very much as if they were added when the work went to press. The fourth section, describing the decay of the star up to May, was almost surely inserted de novo at the same time.

To point out the significance of the lack of parallax, Tycho continued with trigonometrical calculations (comprising nearly half the section) that showed that the star would have a parallax of $58\frac{1}{2}'$ at lower culmination (altitude = 28°) if it were as close as the moon. This proves, he reiterated, that the nova is in a very remote sphere, either the eighth one where the fixed stars are or one of the planetary spheres just below it. But if it were in one of the planetary spheres, it would surely have been affected in some way by the progressive motion peculiar to those spheres; yet it has not moved so much as a minute of arc in six months now. So it is clearly in the eighth sphere and therefore cannot be any kind of comet or any other fiery meteor, for all these are generated below the moon. At least, that is the opinion of everybody except Albattani,[57] who claimed that they originate in the heavens, in the sphere of Venus. Whether this is true, Tycho cannot say, but if he has the chance to see a comet in his lifetime, he will check this. At any rate, because the new star has neither the form of any recognized kind of comet nor the progressive motion common to all comets, it cannot be a comet.

Because this object is a star, and a very bright one at that, it must be very large, for all stars are many times the size of the earth, and first-magnitude stars are 105 times as large as the earth. Over time, the nova has declined considerably from its initial splendor. When it first appeared, it surpassed the maximum brightness of both Jupiter and Venus and was even visible at noon when not obscured by clouds. But by December, its brightness had declined to about the equal of Jupiter's, and by February or March it had decayed to first magnitude. Its color, too, has "degenerated" by stages: from a clear whiteness akin to Jupiter's to a reddish tinge resembling Mars or Aldebaran (but not, according to Tycho, as red as Betelgeuse), to a grayish "leaden" color in May similar to Saturn or Venus.

As he did in the third section, Tycho injected into the fourth section a modest rebuttal of the Aristotelian position, which he had found, since December, to be much more tenaciously held than he originally anticipated. The alterations, he argued, do not prove that the star is either sublunary or something in the nature of a comet or other earthly exhalation. To retreat to such a conclusion in the face of the lack of parallax would be more absurd than to admit change into the heavens. Because the star is an unnatural object, existing outside the laws of nature, it is not at all strange that it should have appeared suddenly. And if it should one day disappear, that would be no more remarkable for such a phenomenon.

Although these additions extended the content of Tycho's tract considerably they did little to change its fundamental character. The tract still is very short and remarkably unpolemical, especially in comparison with Tycho's later writings. The four sections describing the nova occupy only twenty-eight pages in the printed book (fewer than the astrological commentary that follows them), and those pages are so small that they were able to fit into ten when

---

[57] I, 27. The name Tycho was searching for here was Albumassar: He changed it (without note) when he reprinted his pamphlet in the *Progymnasmata* (III, 105), probably after noticing Maestlin cite the same passage. See Robert S. Westman, "The Comet and the Cosmos: Kepler, Maestlin, and the Copernican Hypothesis," in Jerzy Dobrzycki, ed., *The Reception of Copernicus' Heliocentric Theory* (Dordrecht, Netherlands: D. Reidel, 1973), pp. 7–30.

Tycho later reprinted them in his voluminous *Progymnasmata*.[58] It would appear that by May, after he had seen some other writings on the star, Tycho realized that his account was too short and too unpolemical. His additions seem to have been designed to rectify both defects, for they contain sharper arguments as well as supplementary information. But for the most part, Tycho relied on a new special preface to put teeth into his account.

The idea behind Pratensis's (second) letter was that it would be printed as a preface to the pamphlet, urging an unnamed Tycho to publish his findings. Tycho would then play out his part by writing a response – which likewise would be printed as a preface – giving his reluctant permission to publish. This much was just humanist convention. Even a century later, few authors would admit to publishing on their own initiative; they almost invariably excused themselves for burdening the literature with another publication by professing to have been persuaded to do it by enthusiastic supporters.

Thus, Tycho did indeed compose a letter giving Pratensis reluctant permission to publish his material as a feeble attempt to stem the tide of incompetence in the existing literature. He asserted that his instrument was really much better than the commercial ones used by other writers and pointed out that the competing writers had almost invariably been victimized by esoteric methods of observing that led them to calculate spurious parallaxes. But most of the problem lay with the "dull wits" and "blind observers" who insisted on regarding the star as a comet, despite all the mathematical and physical indications to the contrary.[59]

Tycho also expressed the hope that Pratensis would either keep him anonymous or shield him with an anagram, for many people did not think such activity appropriate to nobles. But Tycho was willing to leave the matter to Pratensis's discretion, because Pratensis knew that Tycho neither wrote the tract for publication nor had time to revise it, so that if any names were mentioned, Pratensis would have to share the blame for anything that might be incomplete or erroneous.

From the standpoint of the average reader, Tycho's interest in questions of the place and nature of the star was entirely academic. What concerned most people was the purpose of such an extraordinary occurrence, and Tycho shared this concern. In common with

---

[58] III, 97–107. C. D. Hellman, "Was Tycho Brahe As Influential As He Thought?" *British Journal for the History of Science* I (1963): 295–324, also noted the unpolemical nature of Tycho's tract.

[59] I, 12.

most of his contemporaries, he found it impossible to regard a spectacle like this as irrelevant to humankind. The fact that he was also
greatly impressed by its philosophical implications did not diminish in the least his interest in determining its practical significance.

But although the astronomical aspects of the problem had been
relatively straightforward, the astrological ones looked to be hideous,
for only one other comparable phenomenon had been recorded
– the new star observed by Hipparchus some 125 years before Christ,
and about 3,840 years after the Creation. That star had been a herald
of cataclysmic change in the civilized world, involving the spiritual
and political decline of the Jews, and the transfer of Mediterranean
hegemony from Greece to Rome. Given this kind of precedent, the
possibilities appeared to be unbounded, and the astrological configurations seemed to agree.

To wit: the new star of 1572 precedes by ten years the great
conjunction that will occur at the ending of the watery trigon and the
beginning of the fiery trigon, and the effects of these two major
events will probably coalesce. This in turn will cause great alterations
of empires and kingdoms, serious tumults, and changes in all things
in the world, including a new condition of kingdoms, different from
that of the past, and a new arrangement of religion and law. The new
star resembled Jupiter at first, and so it brought good fortune and
health in the beginning. But it later took on the color of Mars and
therefore will bring warfare, plague, rebellion, captivity of princes,
and the like. The regions that will be especially affected are those of
the north, including Russia, Livonia, Finland, Sweden, and southern
Norway, but the effects will spread to almost all of Europe. Tycho
would have liked to be more exact, but because nobody knew
precisely when the star appeared, it was impossible to say how the
heavens were situated. He decided therefore, that it was generated
at the new moon preceding everyone's sighting, on 5 November
1572,[60] referred to the problems of achieving accurate astrological
predictions, and finished with an allusion to the "more true and
secret sources of another astrology," known only to a few and of
such a nature that its mysteries must not be profaned.

If Tycho had been able to stick to his plan and merely had his name
"listed" on the title page, by Pratensis, he might have fooled a few
people. He might even have gotten away with being cited in the
poems by Johannes Franciscus and Vedel that grace the front and
back of the book, as Pratensis could have been supposed to have

[60]   As his authority for this notion, which Tycho retained for the origin of several later comets,
Tycho cited Halus (I, 32), an eleventh-century muslim (Ali b. Ridwan) commentator on
Ptolemy's astrology.

solicited them. But at some point, Tycho apparently decided that he would as soon be hanged for a sheep as for a lamb. Perhaps it was mounting frustration over the fact that the Roskilde incomes for which he had supposedly been first in line had been diverted to the University of Copenhagen in 1571, in order to increase professorial salaries. Perhaps it was questions about his future with Kirsten, now that she was manifestly pregnant. Perhaps it was his accession to Knudstrup or his final decision to emigrate.

In any case, in both of his last two additions to the manuscript, Tycho expressed his alienation by describing the ethic of the nobility in such pejorative terms as to leave himself thoroughly compromised if Pratensis should "decide" to use his name. Already in his prefatory letter, he had complained to Pratensis that the judgment of men had degenerated to perversity in their disdain for scholarly studies and their exaltation of riotous, slothful, and wasteful living, and he alluded to his plans to leave friends and homeland to go abroad. Then, in a 230-line poetic epilogue, *Elegy to Urania by the Author,*[61] Tycho deprecated the glories esteemed by others of his class: feats of arms, concourse with kings and princes, and the pursuit of wine, women, and song. What he hoped to achieve was the eternal glory of having successfully cultivated astronomy, and he would not be deterred by the opinions of others. Although his blood was as noble as anyone's, he took no pride in mere lineage: What he did not accomplish by himself he would not call his own.

The rest of his *Elegy* is allegorical, composed with the intent of capturing in verse the power of the star and the muse. Tycho reported having been wandering along a brook in the forest of Herrevad as the sun was setting and the great chariot of Luna was mounting the sky. Suddenly from the cloudless skies a goddess appeared, striking fear in him. She allayed his fear by saying that she was Urania, once worshiped by kings but now neglected by humanity. Apollo, recognizing her sadness, sent her to claim Tycho as her own. She then realized that when Tycho had seemed to be worshiping Vulcan in the laboratory, he had really been seeking the sidereal forces within the earth, for both heaven and earth are one nature. And although the alchemical labor of Vulcan could reveal the power of the earthly stars through balsams, the power of the heavenly stars was more refined and could be mastered only by the mind. Tycho should determine the position, distance, and significance of the new star and then determine the locations of the fixed stars; the courses of the sun, moon, and planets; and the influence of all these heavenly bodies on meteorological phenomena. Promising him fame if he

---

[61] I, 65–70.

should accomplish this, the muse then vanished, leaving her divine inspiration behind her. The few who had been granted this vision of the riches of Olympus could never again fix their sights on the perishable trivia of this world. Their desire was the desire of the gods – to use sublime reason to conquer the sublime stars, forcing the celestial ether itself to yield to their commanding spirit.

The optimism of this poem suggests that Tycho had resolved all of his doubts about his future. In this poem he defiantly announces that he has made his choice and will go his own way. The rich imagery of Renaissance Latin poetry expresses a new confidence in his calling, as one who would seek command over nature itself, rather than merely over castles and provinces. Many years later, when Johannes Kepler remarked that if nothing else, the new star of 1572 heralded the appearance of a great astronomer, he spoke more truly than he knew. For in provoking Tycho to publish, the new star forced him through the crisis of his young manhood and put him on the path he was to follow for the rest of his life.

The manuscript acquired its final form in the last weeks of April. Pratensis wrote his part of the foreword in the form of the agreed-upon letter to Tycho, dated 3 May 1573. Consisting basically of a review of various contents of the work, it praised Tycho's knowledge as derived directly from the inexhaustible sources of Nature. It also (of course) urged Tycho to allow his manuscripts to be published, saying that their friendship, Tycho's genius, and both Urania and Hermes demanded it. At the front of the book, Pratensis attached a Latin poem – a standard component in the humanist format – by Professor Johannes Franciscus, praising Tycho's noble descent and even more noble learning and urging him to publish his treatise so that his reputation might shine like the new star itself. At the back of the volume, Pratensis placed Vedel's contribution, which dealt in powerful stanzas with the astrological significance of the new star as a parallel to the new star of Bethlehem and thus a harbinger of the Second Coming of Christ.

After some negotiation over the title (Pratensis had not liked Tycho's title of *Lucubrationes*, or "Nocturnal studies"), the manuscript was taken to Lorentz Benedicht, the foremost humanist printer in Denmark. Sometime in the latter half of 1573, the *Mathematical contemplation of Tycho Brahe of Denmark on the new and never previously seen star just now first observed in the month of November in the year of Our Lord 1572* made its way into the public domain. It is safe to say that it was not a best seller. Tycho described the printer as having been very "niggardly" about printing copies, meaning by this, no doubt, that the latter had been unwilling to produce on speculation any considerable number of copies over and above those that Tycho himself had surely been obliged to purchase in order to get the book printed at all.[62] Tycho gave his copies to friends, patrons, and European men of learning. If he gave any to his relatives, none is among the surviving twenty-odd copies.

Tycho was by no means the only one with cause to complain about the dissemination of his ideas. Michael Mästlin published a little tract to which Tycho reacted very appreciatively, but of which no copies at all are known to have survived.[63] Then, as ever, the popular market was for sensational prose, not competent analysis. Given its small circulation, one could scarcely expect Tycho's little book to have had much contemporary impact, and a study of the question suggests that it probably did not.[64] As the earliest substantial documentation of Tycho's professional thought on a number of issues, however, the book is important for its view of the young Tycho's interests and attitudes.

[62] III, 96. The situation seems to have been entirely typical of the eternal conflict between the great expectations of authors and the hard economic experiences of publishers. Even though the later fame of its author undoubtedly fostered the preservation of many more copies than would otherwise have survived, the book is very rare.

[63] Actually, the tract was not published individually, but as an appendix to Nicodemus Frischlin's *Consideratio novae stellae* of 1573. Fortunately, it was reprinted by Tycho: III, 58–62.

[64] C. D. Hellman, "The Gradual Abandonment of the Aristotelian Universe," *Mélanges Alexander Koyre* (Paris: Hermann, 1964), pp. 283–93.

Chapter 3
Becoming a Professional

A LTHOUGH Tycho's decision to publish *De stella nova* seems to have been based on a decision to leave Denmark, the publication itself, ironically, was probably largely responsible for keeping him in Denmark. Certainly in the short term it was. The trip abroad that Tycho had conceived as the springboard to emigration was first delayed by ill health[1] – probably the consequence of an overly zealous regimen of daytime writing and winter-night observing – and then postponed for a year, as the numerous details of publication kept Tycho occupied beyond the normal spring departure times.

Starting with the letters to Pratensis concerning the publication of *De stella nova* in the spring of 1573, the occasional references to Tycho's "address" are to Knudstrup rather than Herrevad. Whether this means that Tycho was actually occupying the main residence, living in the vicinity, or simply using it as his address because he was now entitled to be addressed as "Tycho Brahe of Knudstrup" is not clear. It is probable, however, that the principal occupant of the manor was Tycho's mother. She was certainly living there with her youngest children in the spring of 1574 and even wrote to a friend that Tycho was living with her.[2] Presumably, she should have said that she was living with Tycho, if he were now legally lord of the manor. But it does not appear that Steen ever took up official residence at Knudstrup, either, so Beate may well have had some kind of life tenancy on the manor house. In any case she and the younger Brahes surely enjoyed a moral and pragmatic claim on the house, just as Tycho was surely enjoying the incomes from his share of the estate, whether or not the legal procedures had been completed. Where Kirsten fit into the scheme, however, is far from clear. She and the little girl to which she gave birth on 10 October 1573 might have lived in the manor house, but Kirsten certainly would not have "carried the keys" to it.

Considering how little is known of Tycho's social life, it is fortunate that one of the few surviving documents from Tycho's early years is a letter from Pratensis inviting him to a party. Nominally it was a Martinmas feast, held every year on 11 November to celebrate the harvest. But Pratensis was not one to forgo an oppor-

---

[1] III, 95. For a list of abbreviations of frequently used sources, see Appendix 1.
[2] G. L. Wad, ed., *Breve til og fra Herluf Trolle og Birgitte Gjöe*, vol. 2 (Copenhagen, 1893), pp. 309–10.

tunity to display his humanist learning and wit, so he embroidered the invitation to Tycho by satirizing the academic admonition he had received – to hold to Hippocrates and Galen in his teachings, even though he was a Paracelsan – when he had been elected professor of medicine two years earlier. Now, he confessed, he had indeed obtained "fair and rich flowers from foreign gardens by means of the secret and ingenious art of Proteus" and had planted them in place of "Galen's weeds."[3] In penance for this duplicity, he promised to give his friends eighty bottles of wine, as well as sugar, almonds, chestnuts, a goose, and a suckling pig, in short, a Martinmas feast. And to those who would object to this penance, he wished "cold feet by night, headache by day, impotence, and an unbearable hatred of girls." Coincidentally, the occasion marked the anniversary of Tycho's discovery of the new star. Whatever discussion it or the current, much-reduced appearance of the star received, Tycho must have been moved to reflect on how much his circumstances had changed – and yet how much they had remained the same – during the year.

Tycho followed the decay of the new star until its final disappearance sometime in March 1574 but otherwise made very few observations. Much the most interesting of those are some made of the lunar eclipse of 3 December 1573,[4] concerning which Tycho had published predictions at the end of the *De stella nova*. They reveal, first, that a twenty-minute adjustment he had made in the calculation from the *Prutenic Tables* had succeeded beyond his wildest expectations and to his evident elation: "I myself cannot sufficiently marvel over the fact that at this early age, only twenty-six, and without the aid of numerous and accurate observations of the motions of the sun and moon, I should have been able to obtain such precise results." Second, they show Tycho using his first assistant – his seventeen-year-old sister Sophie – and his first "professional"-looking instrument.

By the end of 1573, Tycho had been using astronomical instruments for ten years. Although there had been several of them, of various sizes and capabilities, they all had been constructed essentially of wood and strictly for business. The new instrument he now put into service was a manifestation of Tycho's status as an heir. As Figure 3.1 shows, it was a work of art.[5] A solid quarter-circle of

---

[3] The invitation survives only in the copy sent to Peder Sørensen: See E. Bastholm, ed. and H. Skov, trans., *Petrus Severinus og hans Idea medicinae philosophicae: en dansk paracelsist* (Odense, Odense Universitets Forlag 1979), pp. 42–4.

[4] X, 38–9. I, 131–2.

[5] Tycho's description and depiction are in V, 12–15. *Dreyer* (101) states that the instrument was made at Uraniborg, but the eclipse observation of 8 December 1573 (X, 38–39) contains an unambiguous reference to it.

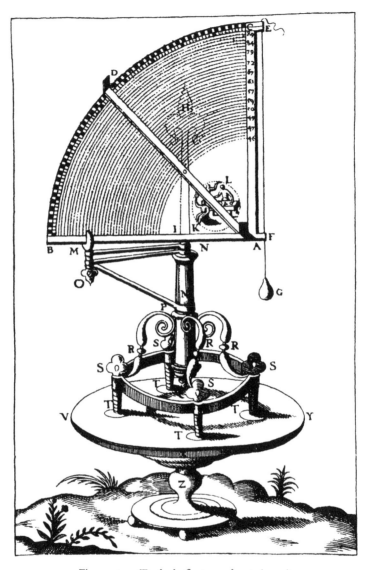

Figure 3.1. Tycho's first quadrant (1573).

brass, gilded with an amalgam of gold, it was decorated with an allegorical painting reflecting the mood expressed by Tycho in his *Elegy* at the time the instrument was constructed during the previous spring. The central figure of the painting was a tree, representing the dichotomy between the life Tycho wanted to lead and the life he was supposed to lead. The latter was illustrated to the right of the tree, by a table laden with coffers symbolizing pecuniary gain, scepters, and coats of arms representing political advancement, and the fine dress, goblets, dice, and cards associated with the good life to which Tycho had been born. Surrounding the table were symbols of the futility of this life-style. The figure standing over that table was a skeleton, and the branches above it were withered, and even the roots on that side of the tree were dead. Only on the other side of the tree, where a young man was shown seated on lush grass under leafy boughs contemplating a book and a celestial globe, did life flourish and existence have any meaning. Above the latter (left) side was inscribed the moral "By the spirit we live," and over the right was written "the rest belongs to death." In addition to depicting the frivolity and evanescence of all human activity outside the realm of ideas, Tycho's motto also summed up the essence of Christianity: "In Christ we live, the rest belongs to death."

Although the quadrant no doubt looked much prettier than Tycho's other instruments, it seems to have been scientifically disappointing right from the beginning. The basic shortcoming was its small size. A mere cubit (approximately 40 cm) in radius, it had been divided down only to intervals of five minutes of arc, which was much coarser graduation than the one minute he had achieved on his four-cubit sextant. Perhaps Tycho had assumed that an all-metal instrument would offer enough subtle virtues in construction to compensate for its inferior size: Perhaps he had thought that the novel system of additional, specially divided arcs (visible in the illustration) would prove more useful than they turned out to be.

Most likely, however, Tycho was simply making the best of a bad situation. He needed a quadrant. Although he probably ordered it too late to use it for observations of the new star, he surely wanted to get some experience in constructing and using that type of instrument, anyway. At the same time, he doubtless saw metal as the wave of the future and felt that he wanted to start testing it for his own purposes. It is not unlikely that one cubit was simply the largest anyone would make for him, but it is also possible that it was a pragmatic compromise for an instrument that Tycho regarded as experimental and that had to be transportable. If Tycho expected any great performance from it, he did not work very hard to get it, for his logs show very few observations using this quadrant.[6]

[6] Tycho's logs contain numerous references to a *Q. min.* which one might (with Dreyer, as in note 5) assume was Tycho's small(est) quadrant. However, the observations with the *Q. min.* always show altitude readings to individual minutes, as well as some azimuth readings that would simply have been unobtainable with his first quadrant. Observations with the *Q. min.* are, therefore, made with what Tycho labeled the *Quadrans mediocralis* in the *Mechanica* (V, 16–19) but never cited by that name in his log.

Sometime during the summer of 1574,[7] Tycho appears to have moved his family to Copenhagen. Kirsten was probably pregnant at the time. Before the end of the year she presented little Kirsten with a baby sister, baptized Magdalene, presumably after an aunt of Tycho's who had died tragically in 1571 after being gored by a bull. Unlike her namesake, however, this Magdalene would not have the noble surname of Brahe but would have to be called simply Magdalene Tygesdatter, because she was not of noble birth.

There were, of course, any number of things that could have inspired Tycho's move to Copenhagen. Because Kirsten could scarcely have been comfortable at Knudstrup, it is easy to imagine that Tycho just finally moved away from his mother's presence – if we could be certain that Kirsten ever lived at Knudstrup or even moved into Copenhagen with Tycho. Likewise, it seems possible that Tycho decided he wanted to have more, or a different kind of, intellectual company from what had been available to him at Herrevad. Most likely, however, the move was motivated by pragmatic considerations. When Tycho had been living a long day's journey from Copenhagen, he had had to ask Pratensis to look after not only the everyday details of the construction of his one-cubit quadrant and the gilded celestial sphere he had commissioned in one of the shops there[8] but even the printing of his book. Tycho still had a two-meter wooden globe on order from the famed Christoph Schissler of Augsburg and may have left orders for an armillary and an astronomical ring, too, when he had last been abroad.[9]

Now, probably sometime in 1574, Tycho designed a new sextant to be made completely of metal, mostly steel. If it was fabricated in Copenhagen (as opposed to Herrevad), he probably would have wanted to oversee its construction himself.

By mid-1574 Tycho was no longer just a noble dilettante in the Copenhagen intellectual circle. His publication on the new star seems to have established him as an authority in astronomical matters and as a serious professional as well. His name and credentials appeared prominently in a list of distinguished Danish nobles published as a dedicatory epistle to King Frederick,[10] by the prolific author Professor Erasmus Laetus.

Sometime during the summer, apparently, Tycho's expertise was recognized when it was arranged for him to deliver lectures at the

---

[7] As late as March and April, Tycho made observations at Knudstrup, Herrevad, and Copenhagen (X, 40–1).

[8] X, 10–11.

[9] These came to him in 1575 but may have been gifts from the Hainzels and Wolf: VII, 18.

[10] *Norlind*, 70–1.

university on some higher aspects of astronomy not ordinarily included in the curriculum. What passed for basic astronomy was still being taught by the same Professor Scavenius who had taught it a dozen years earlier in Tycho's day, and he was in no sense a specialist in the subject. Certain members of the faculty – notably Pratensis – also had an interest in Tycho's lecturing, as they knew that he would promote a general view of the world that they shared with him.

This view was a minority one, as the strictures on Pratensis's lectures suggest, and professor of theology Niels Hemmingsen, the most influential member of the faculty, had recently threatened its status further by publishing a tract opposing astrology.[11] Nonetheless the opportunity to present a Paracelsan view of the utility of astronomy was one that Tycho did not want to miss.

But the proposal also raised problems. One was that Tycho had never considered an academic career and had therefore never obtained a master's degree which entitled a person to be a master. Peder Sørensen had presented lectures in meteorology without being a member of the faculty, but he had at least had an M.A. degree.[12] A more serious problem was that Tycho was a noble.

One of the first rules of sixteenth-century society was that each rank was expected to stay in its proper place. The Danish nobility was a class of warriors, landlords, entrepreneurs, and high governmental administrators. In return for extensive privileges, they were expected to render their services to society in those areas only and to stay out of areas reserved for the other estates. Since the Reformation, the church and the university had been included in the sphere of the learned middle classes. There had been no noble bishops, parish clergy, teachers, or professors in Denmark during Tycho's lifetime and there were not to be any for years to come. It thus would not have been appropriate for Tycho as a nobleman to lecture at a university for this would have been seen as intruding into the privileges of another estate.

As one might expect, however, where there was a noble will there was a noble way. It may have been Dançey who conceived the means of circumventing the problem. If it was the lectures were probably held in the great reception hall of the French embassy. Whatever the case, a second solution was also needed and conceived. A petition was drawn up, signed by a number of noble students, and presented to the king. When he added his "request" to the invitation, the lectures could be construed as having been requested by and given

[11] See Dreyer's note to I, 170 on I, 313.
[12] Bastholm and Skov, *Petrus Severinus*, p. 3.

primarily for Tycho's peers, although they would be open to anyone who wished to come.[13]

Following the custom of the university, Tycho inaugurated his lectures on 23 September with a formal hour-long oration. Because this introduction carried a sustained argument and included extended quotations from classical poets, to say nothing of the fact that it was his first lecture and would have an audience that he expected would include most of the professors, Tycho wrote out his speech in full. And because the occasion was a memorable experience for him, and the theme of his lecture remained long afterwards one that he was planning to develop for publication, he took the trouble not only to preserve the manuscript but also to record at the end of it his impressions of the reaction to his talk, the intellectual exchange after it, and the general content of his subsequent lectures.[14] From a modern standpoint, the speech was about what one expects to get when a scientist waxes philosophical or historical.[15] But it nevertheless provides an invaluable glimpse of Tycho's general view of his chosen profession as he approached the age of thirty.

For the modern reader, the most striking feature of Tycho's *Oration* is how it reflects the religious ethic of medieval and early modern Europe. To a degree not found in any of his later writings, the *Oration* demonstrates Tycho's tendency to interpret every form of human activity as being within a framework of theological considerations. Thus, his outline of the history of astronomy begins with Seth and Moses; his argument for the utility of astronomy is based on its power to free the human mind from mundane concerns and direct it toward the heavens; his discussion of astrology is focused on rebutting the theological objections to it; and even his understanding of science itself is colored by his interpretation of relevant scriptural passages.

But even in the *Oration*, it is clear that scriptural influence stopped short of the domination that it exercised on most of his contempor-

---

[13] The factual skeleton from which John Christianson recreated this chain of events is at the beginning (I, 145) and end (I, 170–1) of Tycho's lecture. The French embassy is currently some distance from the university. I have been unable to ascertain where it was in Tycho's time.

[14] Tycho gave a copy with theologically motivated emendations to one of his students, Cort Axelson, who published it in 1610 (Copenhagen) under the title *De disciplinis mathematicis oratio*. It was reprinted in Hamburg in 1621 (*Dreyer*, 73–4). The text published in I, 144–73 is from a manuscript found in Vienna, which, contrary to *Norlind* (63), really cannot have been anything but Tycho's actual talk.

[15] Nicholas Jardine, *The Birth of History and Philosophy of Science* (Cambridge, England: Cambridge University Press, 1984), pp. 263–4, read some twenty-five such productions of Renaissance scientists and described Tycho's (and most others) as being "entirely devoid of originality in its themes and organization."

aries. Thus, Tycho's historical sketch quickly moved to Hipparchus, Ptolemy, and Copernicus as the principal architects of astronomy as the sixteenth century knew it. Tycho's assessment of Copernican theory as being contrary to sound physical principles was partly a product of theological bias – but only partly so, and did not blind him to the fact that Copernicus had demonstrated true genius and made outstanding contributions to the science of astronomy.

Indeed, astronomy was a science – an empirical rather than a theological enterprise – and offered such valuable practical services as keeping calendrical track of human affairs, forecasting the weather, and even predicting human events. The utility and efficacy of timekeeping were too obvious to require comment. Weather prognostication could not, however, be taken for granted in the same way. A century's worth of annual almanac forecasts had depressed the activity to such low esteem that the professor at Copenhagen officially charged with issuing an almanac had not done so for many years. Thus Tycho felt obliged to conduct a brief defense of astrological weather forecasts.

Starting from the known responsibility of the sun for the seasons of the year, and the generally accepted influence of the moon on the tides, Tycho proceeded to the association of certain stars with stormy weather, and the much more complicated effects of the various configurations of the planets. As in the rest of the *Oration*, he refrained from mentioning his own achievements and aspirations in any way, but because many in the audience were familiar with his book on the new star, they would have known that Tycho had already tried his own hand at this endeavor.

The greatest problem was horoscope astrology. Many people who believed that the stars could affect the elements of nature balked at the notion that they could influence people. In particular, theologians since Augustine had opposed astrology because of what they took to be its inimical implications for Christianity. Martin Luther had joined the side of orthodoxy, although with less militance than ridicule. The equally revered Philipp Melanchthon, however, had developed a sufficiently strong belief in the efficiency of macrocosmic influences to convince him that he should never visit Denmark because his horoscope ruled against it. Niels Hemmingsen had spent his own student days with Melanchthon in Wittenberg and had become the foremost spokesman of Danish Philippism, but he had recently chosen to attack his old master's position on astrology. Hemmingsen was in Tycho's audience. In his notes, Tycho wrote that when he began this section of his lecture, all eyes turned to the old theologian, and he acknowledged the attention by smiling and tipping his academic biretta. Warming to the challenge, Tycho now launched into the primary theme of his lecture, to reconcile his view of the universe with the variety of Philippist theology that prevailed, under Hemmingsen's influence, at the University of Copenhagen.

Tycho began by defending the Philippist position that one could scarcely imagine an omnipotent God having created the vast and complex wonders of the heavens unless they had some purpose. He then developed the Paracelsan position (although without mentioning Paracelsus's name) that because humans were made up of elements and absorbed them daily in their life processes, they must also be subject to the planets' influence. Tycho then described how planetary conjunctions produced storms and briefly mentioned his theory of the lunar octads, before arguing the Paracelsan view that the fixed stars were like sterile women who brought forth nothing in this world until they were aroused and impregnated by the seven wandering planets. He illustrated this by showing how the great conjunction of Saturn and Jupiter near the beginning of Leo in the year 1563 had brought a universal plague to Europe in the following years. He went on to assert that individual human beings were also subject to sidereal influences because the human soul was a part of heaven itself and the human body was a microcosm whose major organs were analogous to the seven planets. This argument Tycho couched in terms reminiscent of Galen, but he moved beyond the Galenic humors to describe how the great diversity of human nature and ability derived from celestial influences:

> Some investigate in solitude lofty matters which are far beyond the grasp of common people: Saturn, the highest star, had formed them felicitously. Some are more interested in judicial and political affairs, over which the splendor of Jupiter shines. Some breathe nothing but war, slaughter, tumult, and quarrels: these the fervor of Mars agitates. Others seek honors, dignity, and dominion over affairs because of the ambitious influence of the sun. There are those who spend their lives in loves, pleasures, the fine arts, and other pleasant pursuits: Venus, the enticing star, has enchanted these. Others dedicate themselves totally to practicing ingenious crafts or even commerce: these have been stimulated by Mercury. Some, influenced by the lunar nature, spend the course of their lives in popular affairs, pilgrimages, voyages, fishing, and the like. In this way, the great variety of temperaments can be seen to be like the influences from the seven wandering stars. Many people are affected by diverse combinations of these planets, and they pursue different types of activities at diverse times in their lives, at one time being occupied with this business and at another time with another, as they are subject at various times to the secret influxes of this or that planet.[16]

[16] I, 158. I have benefited from the use of an unpublished translation by Jeremiah Reedy.

Having established the positive aspects of his intellectual position, Tycho moved to refute the various criticisms of astrology. No one could deny that plagues and wars killed off large numbers of people who had different horoscopes, but any responsible astrologer would leave room in his predictions for the possibility of general calamities that had nothing to do with the specific fate of the individual. Nor did the fact that people could be born at the same instant but meet different ends discredit astrology, for the stars did not determine the basic circumstances of life but, rather, produced the variations that distinguished the fates of people who lived in the same basic circumstances. Twins, who shared both horoscope and circumstances, were actually born at slightly different times, and one was always weaker than the other. Most important was the fact that astral influences were influences, not determinants. That is, they could be altered by God or countered by man through the exercise of free will. Thus the ancient objection that prognostications were not even desirable, as they merely diluted the joy of happy events and added worry to the grief of sad events, was forestalled by the possibility of resisting the influences working to produce undesired situations.

It was this allowance for the force of individual human will that constituted the crux of Tycho's attempt to reconcile astrology to Philippist doctrine. He began by declaring that proper education, discipline, and other human factors could deflect the influence of the stars, and he then turned directly to a discussion of free will, using the earthy imagery common to sixteenth-century scholars:

> The free will of man is by no means subject to the stars. Through the will, guided by reason, man is able to do many things that are beyond the influence of the stars, if he wills to do so.... Astrologers do not bind the will of man to the stars but grant that there is something in man that has been raised above all the stars. Because of this, man, if he wishes to live as a true, supermundane person, can overcome any malevolent inclinations whatsoever from the stars. But if a person chooses to lead a brute's life, dominated by blind desires and fornicating with beasts, God must not be considered the author of this error. God so created man that he can overcome all malevolent influences of the stars if he wills to do so.[17]

As a defense of astrology on empirical and theological grounds, Tycho's stance was a good one. Indeed, Tycho noted that it overcame Hemmingsen's theological objections. And to the present day the principle that the stars impel rather than compel has remained the standard explanation of unfulfilled astrological predictions. But in

[17] I, 163.

making it easier to explain away erroneous predictions, Tycho was simultaneously making it harder to advance the discipline; for the capacity to recognize and label something as wrong or discordant is the source of progress in science, if not the very characteristic that distinguishes those ideas that constitute a science from those that do not.

Although Tycho closed his oration by stating that he had chosen to discuss astrology, not because he valued it above the other mathematical sciences, but because it was much the most maligned of the mathematical sciences, it is clear that he was fully committed to the potentialities of astrology. If in later years his zeal abated, it was because he believed that the restoration of astronomy was indispensable to the advancement of astrology, not because he had lost faith in the potential of astrology.

Immediately after the lecture, Albert Knoppert, professor of law, came up to Tycho and Dançey and said, "When I heard your attacks on the philosophers and physicians, and even the theologians, I was afraid that you would also launch into us jurists – so afraid that I broke into a sweat." Tycho remembered enough from his legal studies to guess that Knoppert was alluding to a well-known section of the Justinian code (Roman law) entitled "Against Mischief-Makers and Mathematicians (Astrologers)." So he replied (in the same collegial vein) that he had once studied law himself in Leipzig but that he had abandoned it when he discovered that jurists did not know the difference between mischief and mathematics.

Following the lecture, Dançey entertained Tycho and all the professors at a leisurely meal over which they discussed further ramifications of Tycho's ideas. During the conversation, Dançey expressed his residual feeling that despite Tycho's clever arguments, astrology was a hindrance to evangelical teaching. Tycho replied that astrology was not a threat if it was handled soberly and circumspectly, did not succumb to superstition, and avoided giving predictions a political significance. When Niels Hemmingsen heard this conversation, he assumed that it was directed at him, as it was, and so he joined in. He stated that he had no objections to the views that Tycho had expressed in the oration, so long as Tycho did not deny either that God works and acts with absolute, unrestricted freedom or that human beings have completely free will. Tycho replied that no astrologer except an atheist or an Epicurean would deny that God works and acts freely, in complete independence of all creatures. But at the same time, he believed that human beings could not only liberate themselves from the influences of the stars by means of their free will but that they could also raise themselves above the stars and, with the help of God, make themselves lords over them.[18] When

---

[18] I, 171–2.

Hemmingsen pronounced himself satisfied, Tycho had fulfilled the purpose of his oration: He had reconciled his views with the mainstream of Danish intellectual thought, as represented by the dominant figure at the only university in the realm.

On the day after his oration, Tycho began the substance of his lectures. The choice of subject matter must have been difficult, for few of the students could be expected to be ready for anything that would be intellectually challenging to Tycho. Renaissance astronomy was traditionally dichotomized into studies of the *primum* and *secundum mobile*. The *primum* dealt with the most fundamental phenomena, the nightly risings and settings of the celestial sphere as a whole, and the annual cycling of the portion that could be seen on any given night. Strictly speaking, it was a course in spherical trigonometry, and anyone who wanted to achieve a sound understanding of the subject had to master what was then the highest form of mathematics that had been developed. But because the mathematics was very difficult, the material was generally discussed in simple qualitative terms, just as the so-called doctrine of the sphere is, in today's introductory college courses. Planetary theory could have been taught on the same basis, for it was possible to extract planetary positions from astronomical tables without knowing trigonometry. But the line had to be drawn somewhere, and through the years the doctrine of the *secundum mobile* had become established as an advanced course for which students should finally learn trigonometry. In conjunction with the complexities of the astronomy involved, it was a formidable course, and the lack of student demand must have been at least partially responsible for the occasional complaint by sixteenth-century astronomers that the universities generally lacked faculty competent to give instruction in it.

In Copenhagen, this course was normally the domain of Tycho's friend, Dr. Johannes Franciscus. He began with Euclid's *Geometry* and then moved on to trigonometry and Ptolemaic planetary theory. Tycho's course seems to have been a substitute for Franciscus's during the autumn term of 1574. But there was one important difference: Tycho based his lectures on Copernicus, not Ptolemy.

Tycho appears to have lectured from nine to ten in the morning on Monday, Tuesday, Thursday, and Friday of each week, according to the normal lecture schedule of the University of Copenhagen, from late September until just before Christmas. There is no record of the details of his presentation, probably because Tycho was in sufficient command of the material to require very little in the way of notes. Tycho did record, in the previously mentioned epilogue, however, that he covered the theories of the sun and moon "according to the models and parameters of Copernicus," even to the point of supplying copies of the relevant portions of the *Prutenic Tables* to those who could not afford to buy them.[19] He had originally intended to lecture in a similar manner on the planets during the spring term, but his plans changed during the fall. As his concluding theme, therefore, he presented the general model of Copernicus's planetary theory and then showed how this theory could be "adapted to the stability of the earth."

This general approach was well established in the Philippist universities. It stemmed from Melanchthon himself, who had skipped rather lightly over the first book of Copernicus, which presented a heliostatic cosmology, and read the rest of the work with admiration. Melanchthon found in Copernicus a set of mathematical theories and tables that could be tremendously useful to astronomers making the traditional geostatic assumptions. He and others of his school simply ignored the cosmological conflict of Copernicus with Ptolemy and exploited the mathematical aspects of both. Erasmus Reinhold, Melanchthon's younger colleague at Wittenberg, had worked out the *Prutenic Tables* on the basis of this view.[20] Homelius had brought it to Leipzig, as other disciples of Melanchthon brought it to other Lutheran universities. Melanchthon's son-in-law, Caspar Peucer, had lectured on Copernican astronomy, referring to an immobile earth as early as 1559 in Wittenberg, and his lectures were published in three different editions between 1568 and 1573.

Virtually all the astronomers Tycho had known since his student days belonged to this school that drew on the theories and tables of both Copernicus and Ptolemy without any sense of a conflict between the two. The major exception would have been Georg Joachim Rheticus, who had left the Wittenberg faculty to seek out Copernicus and urge him to publish his manuscripts. Rheticus remained a fervent advocate of the heliostatic cosmology.

By the 1570s, however, there was a growing interest among younger men in the implications of the Copernican cosmology. The problem was that if the earth moved, there should be an annual parallax or shift in the apparent positions of the stars. If this annual parallax could not be detected, it would imply that the universe and the individual stars were so immense as to stagger the imagination. The same line of thought led to a reconsideration of the order and distances of the planets and even to speculation about the infinite size of the universe and a plurality of worlds. The appearance in 1572 of the new star also raised the possibility of a dynamic rather than a

---

[19] I, 172–3.
[20] For discussions of Reinhold's and Gemma's views, see O. Gingerich, "The Role of Erasmus Reinhold and the Prutenic Tables in the Dissemination of Copernican Theory," *Studia Copernicana* 6 (1973) 51, 55–6.

static universe. Even more than Copernicus, Paracelsus encouraged Tycho and others of his persuasion to think of the universe in physical as well as mathematical terms, as did the strong general interest in Platonic and Hermetic thought. A casual mixing of Copernican and Ptolemaic models, however, now began to seem absurd, as if to say that the world could have two forms at the same time. Some way would have to be found to integrate all the old and new ideas into a single coherent picture of the universe. And as far as Tycho was concerned, that picture would have to begin with a stable earth.

That the earth is in continual motion is one of the basic scientific cornerstones of the modern worldview. So often is it taught or alluded to in even the most general education that virtually everyone regards it as a given, even though few people can cite even one scientific argument for the earth's motion, and almost no nonastronomer has any idea how complicated it actually is.[21] So ingrained is the concept that it rarely occurs to people sitting quietly in their chairs and contemplating their surroundings that nothing is more obvious than the "fact" that they are *not* moving. For people who have not been indoctrinated to the contrary, this "fact" is transcendant. They scarcely know how to react to the proposition that they are actually rushing through space with a manifold (threefold, in Copernicus's theory) motion that dwarfs any speed they have ever knowingly experienced.

This conflict with common sense was not the only hurdle. All of science as it was then known rested on a distinction between terrestrial elements, whose natural motions were rectilinear, and the celestial ether, whose natural motion was circular. In these terms, circular motions for the earth (which was the aggregate of all the terrestrial elements) were as inconceivable as velocities exceeding the speed of light are for modern physics. If they were to be accepted, all of Aristotelian physics would have to be rejected, and this is eventually what happened. But such fundamental changes take time, even when there are good reasons for making them, and the truth is that there was no good reason in the sixteenth century for assuming that the earth was in motion. The only empirical test lay in seeking astronomical parallax. Fortunately for the development of science, this test was single edged. It could prove the motion of the earth but not disprove it, as the failure to find the earth's motion reflected in an annual displacement of the stars could be explained away (and

<hr>

[21] For a list of nine motions of the earth that can be described for an introductory astronomy course, see George O. Abell, *Exploration of the Universe*, 4th ed. (New York: Holt, Rinehart and Winston, 1982). p. 115.

had to be, for three hundred years)[22] by postulating stellar distances too remote to allow detection of the displacement. But the fact remained that the one astronomical check available provided negative evidence.

In additon to the scientific objections available, there were various scriptural passages that seemed to rule out a moving earth. They too, could be explained away. But at a time when Europe was being ripped apart by a Reformation rationalized on the departure of the established Church from literal interpretation of the Scriptures, such a move would not have been popular. These passages alone, therefore, constituted powerful counterarguments to which even Tycho usually alluded in any discussion of the earth's motion.

Last but by no means least of the problems associated with the motion of the earth was the fact that it was not clear that it was even being seriously proposed. Copernicus, it is true, opened his great work with an explicit argument for the motion of the earth and presented his planetary theories in heliocentric form: Modern readers would never even question his commitment to the physical reality of heliocentrism. But in the sixteenth century the notion that astronomical theory might actually represent the structure of the heavens was almost completely unprecedented. Aristotle had distinguished the mathematical sciences from physics on precisely the grounds that the former could not involve the search for causes that characterized physics. In the short run, astronomy was well rid of having to elaborate Aristotelian causes for the celestial motions. By Ptolemy's day, this freedom had come to be interpreted as a license to invent whatever mechanisms were necessary to generate reliable predictions. And in the form of the "equant"[23] invented by Ptolemy, this license had fostered remarkable results.

In the thirteenth and early fourteenth centuries, when Greek learning was being introduced to the Latin West, this "likely story" interpretation was gradually extended to all of science, as the price of accommodation with the medieval church.[24] It had purchased the freedom to consider almost any topic, but at the cost of ensuring that

---

[22] Parallax was finally demonstrated by three different astronomers, Bessel, Struve, and Henderson, in three different observatories, between 1837 and 1840.

[23] The *equant* was the traditional conception of uniform motion in a circle modified by the provision that the motion was uniform with respect to some point other than the center of the circle. (For comparison, see the glossary of technical terms in Appendix 2.) It is commonly believed that such combinations of circles, like Fourier analysis, can represent any kind of path. For examples to the contrary, see R. C. Riddell, "Parameter disposition in pre-Newtonian planetary theories," *Archive for History of Exact Sciences* 23.

[24] For an extended argument of this position, see chaps. 3 and 6 of Edward Grant's *Physical Science in the Middle Ages* (New York: Wiley 1971).

no conclusion would be taken seriously. Thus, by Copernicus's time, the motion of the earth had already been pondered by Renaissance scholars for two hundred years, as a result of the renowned Buridan's having shown that the question "If the earth rotated under us daily instead of the heavens moving above us, would we be able to tell the difference?" had to be answered "Probably not."[25] But it had been entertained as a purely philosophical problem, by people who were seeking not to determine whether the earth was physically rotating but, rather, to exonerate themselves from the embarrassment of their not being able to prove what everybody already knew anyway.[26]

As a result of attempts to eliminate Ptolemy's equant mechanism, Copernicus arrived at a system that suggested that the earth was moving around the sun.[27] We do not know what role Buridan's puzzle might have played in inducing Copernicus to interpret his result literally. The important thing, however, was that he did interpret it literally and assert that the earth was moving, instead of merely posing the annual revolution as another hypothetical motion that could not be disproved. This assertion eventuated in the scientific revolution of the seventeenth century. But in the sixteenth century, it was so contrary to the prevailing view of scientific thought that it would have been interpreted figuratively by some readers under the best of circumstances.

As it turned out, all readers were directed toward this interpre-

---

[25] For an English translation of Buridan's arguments, see Edward Grant, *Source Book in Medieval Science* (Cambridge, Mass: Harvard University Press, 1974), pp. 500–3. See also the discussion of Buridan and Oresme in Grant, *Source Book*, pp. 64–70; and J. D. North, "The Medieval Background to Copernicus," *Vistas* 17, 10–12.

[26] An instructive parallel can be seen in twentieth-century reaction to the problem of universal nocturnal expansion. Most important (and most likely to be overlooked) is the fact that most of the population will live and die without ever hearing about the issue. Second, and scarcely less flattering from an intellectual standpoint, is that for most of those who do hear about it, the news that some philosophers are exercising themselves over the possibility that the universe might suddenly change size is more likely to inspire caustic comment about philosophers than to stimulate serious thought about the issue at hand. Third, those who bother to pursue the matter will find (probably not to their disappointment) that for the philosophers involved, the inquiry is entirely hypothetical – that what is being considered is the question "If the universe were to expand (or contract) overnight by a uniform factor, would we be able to detect it?"

[27] For detailed discussion of manuscripts that seem to show Copernicus moving geometrically from the Ptolemaic equant to a geostatic epicyclic planetary mechanism, and an argument that he might thence have moved through a "Tychonic" system to the Copernican system – as Dreyer, *History of Planetary Theory from Thales to Kepler* [Cambridge, England: Cambridge University Press, 1906] p. 364, had speculated – see Noel Swerdlow, "The Derivation and First Draft of Copernicus' Planetary Theory: A Translation of the *Commentariolus* with Commentary," *Proceedings of the American Philosophical Society* 117 (1973): 471–8.

tation by a preface interpolated by a Lutheran clergyman named Andreas Osiander, who supervised *De revolutionibus* through the press (in Nuremburg, far from Copernicus's home in Poland). Animated by a conviction that theology was the only source of truth and frustrated in a previous attempt to get Copernicus to say so, Osiander arbitrarily inserted an unsigned preface conveying his position.[28] The result was that Copernicus's arguments for the physical reality of the earth's motion were preempted by the explanation that astronomers had long been accustomed to doing whatever had to be done to account for the appearances and that the reader should not be upset if he found a few improbable assumptions in the work, because no one ever took mathematical hypotheses seriously, anyway. Sixteenth-century readers had no way of knowing that this preface was not from Copernicus himself, although Rheticus crossed it out in his two copies of *De revolutionibus*.[29] All they knew was that whoever might have told them that some eccentric mathematician was proclaiming the motion of the earth had not read the fine print confirming the nature of the enterprise.

Unlike most of his contemporaries, Tycho both understood that Copernicus had presented his system as a physical reality and agreed that astronomy could aspire to physical reality. But because he could not believe that the earth was moving, he was bound to reject that feature of the Copernican system. He could not simply dismiss Copernicus, because, like most of his contemporaries, he regarded him as far and away the foremost mathematician of the century.[30]

Uncritical commentators have assumed that because the sixteenth century ignored or rejected heliocentrism, it also ignored or rejected Copernicus. But among professionals, at least, this was not true. From an inspirational standpoint, alone, Copernicus's impact was enormous. At a time when the concept of progress was just beginning to emerge, when Europeans were just beginning to apprehend

---

[28] For details see Edward Rosen, *3 Copernican Treatises* (New York: Columbia University Press, 1939, reprinted New York: Dover, 1959), pp. 22–26. For an argument that Osiander's primary goal was to promote acceptance of Copernicus by reducing conflict with theology, see Bruce Wrightsman, "Andreas Osiander's Contribution to the Copernican Achievement," *Westman*, 213–43.

[29] According to a quotation by F. R. Johnson, *Astronomical Thought in Renaissance England* (Baltimore: Johns Hopkins University Press, 1937), p. 162, from the appendix to the 1576 edition of Leonard Digges', *Prognostication Everlasting*, Thomas Digges was "in no way deceived" by Osiander's preface. Similarly, R. S. Westman, "Three Responses to the Copernican Theory: Johannes Praetorius, Tycho Brahe, and Michael Maestlin," *Westman*, 331, shows that "Maestlin possessed strong suspicions even before he learned in 1589 of the true authorship of *Ad Lectorem*." But only in 1609, at the beginning of Kepler's *Astronomia nova*, was the situation made public.

[30] Tycho printed poems (dated 2 October 1584) praising Copernicus. He published them in 1596 along with his correspondence with the *landgrave* (IV, 270–1).

that they were not condemned to remain, at best, a pale reflection of the golden ages of classical antiquity,[31] *De revolutionibus* provided virtually the sole example of intellectual parity with the ancients.

To no one was this role model more important than to Tycho, who was to be one of the few figures in the sixteenth century to aspire to surpass the achievements of antiquity. From a pragmatic standpoint, too, Copernicus was influential. During Tycho's youth, *De revolutionibus* was reprinted (1566); Reinhold's *Prutenic (Prussian) Tables* were printed and reprinted (1511, 1562); and ephemerides computed from Copernican theory were widely distributed in the astronomical marketplace. Tycho himself had been using Copernican ephemerides for over ten years. Obviously, it was one thing to deny the motion of the earth and quite another to refuse to have anything to do with astronomical predictions based on it.

No one has satisfactorily explained the interest in Copernicus's predictions. Modern analysis had shown that they were, indeed, slightly better than the contemporary Ptolemaic ones.[32] But modern scholarship has raised considerable doubt that anyone in the sixteenth century (except Tycho in his later years) did enough serious observation and comparison to know that they were superior. Still, there seems to have been a general feeling among astronomers that Copernicus's redetermination of all the constants involved, the first since the (much less complete) one done for the *Alfonsine Tables* nearly three hundred years earlier, had been long overdue. Thus, even some astronomers who were convinced Ptolemaics had issued reworkings of Ptolemy's theories utilizing the constants determined by Copernicus.[33] Tycho did not find this compromise appealing because he felt that Copernicus's models were as superior to Ptolemy's as his constants were, and probably more so, for he had decided as far back as Leipzig that Copernicus was correct in regarding the equant as an abomination. On the other hand, Tycho had also probably concluded that Copernicus's observations left plenty of room for improvement.

[31] The classic formulation of this theme is in J. B. Bury, *The Idea of Progress: An Inquiry into Its Origin and Growth* (London: Macmillan, 1920). According to an unpublished paper by Edward Grant ("Was there an Idea of Progress in the Middle Ages?"), attempts to document the idea among thinkers of the Middle Ages have been generally unconvincing.

[32] It was easy to find instances in which either Ptolemaic- or Copernican-based ephemerides, or both, were egregiously erroneous. What was difficult in that prestatistical era was to decide generally which set of predictions was better, even if one had a substantial number of observations and comparisons in hand. For reasons probably more ideological than empirical, Tycho was already committed to Copernicus (I, 23, 172, 185–90). For one example of computer-comparisons, see Gingerich, "The Role of Erasmus Reinhold."

[33] Works by Dasypodius and Peucer appeared in 1568, 1571, and 1573. See ibid, pp. 59–60. Tycho knew them at the time of his *Oration* (I, 173).

To reconcile these various considerations, Tycho extracted the basic model of Copernicus's planetary theory from its unacceptable cosmological system and attempted to sketch, for those long-suffering devotees who were still with him in December 1574, how its equantless mechanism could be used in a scheme that did not imply a moving earth.[34]

As a form of recognition in his mother country, the opportunity to give the lectures must have been very welcome to Tycho. The lectures also gave him the opportunity to examine Copernican astronomy systematically and to compare it in detail with the traditional corpus of Ptolemaic astronomy. Although Tycho had now been studying astronomy for over ten years, having to lecture four times a week and to explain his materials to students must have helped him clarify many matters in his own mind. In the evenings when the sky was clear, he offered to instruct the students in observational astronomy. In November 1574 one of his students published a small humanist treatise dedicated to Tycho. A little pamphlet containing a dialogue in Latin hexameters on the lunar eclipse of the previous year, it was printed by the same man who had printed Tycho's tract on the new star.[35]

Sometime during the fall, Tycho reactivated his plans for emigration and canceled the second half of his lectures. Perhaps it was a decision that had been half made when he moved out of Knudstrup; perhaps it developed from a discovery that he and Kirsten were not significantly more comfortable in Copenhagen than at Knudstrup. Perhaps it was the appearance of a wayward student nephew of the Hainzel brothers that reminded Tycho of other places and more congenial circles. Paul Hainzel and Hieronymus Wolf both wrote in March 1575, forwarding an armillary that Tycho had admired at the house of Oswald Schreckenfuchs[36] and thanking Tycho for his kindness toward Hainzel's nephew. By the time this package arrived, however, Tycho was probably on his way south.

As Tycho embarked on what was to be the last of his trips abroad before emigrating he traveled as a grand seigneur, accompanied by servants and a train of baggage. He was by now a seasoned traveler and during the boredom of the journey may have reflected on how

[34] I, 173. "Indicavi nihilominus generali quadam expositione, quomodo in reliquis Planetis haec ipsa etiam essent intelligenda, & qua ratione illorum apparentiae ad terrae stabilitatem, monentibus Copernici numeris, adaptaer possent, insinuavi."

[35] According to *Dreyer* (117–18), the author, Peder Jacobsen Flemløse, who was later to join Tycho on Hven as a coworker, "believed (the eclipse) to mean that the second coming of Christ was soon to take place."

[36] VII, 17–19. Oswald Schreckenfuchs (1511–79) was professor of mathematics at the University of Freiburg. He was not known as an instrument maker: see *Zinner*, 530.

different each of his successive trips had been . First he had gone to Leipzig, for broad exposure to the culture and curriculum available from court and lectern. He then had gone to Wittenberg for more of the same, but in pursuit of fewer interests. If it had been largely accidental that he had spent only a few months first at Wittenberg and then at Rostock, the fact remained that his knowledge of science had then already begun to reach the frontiers of sixteenth-century achievement. Short terms at Rostock, perhaps Wittenberg, and Basel on his third trip had driven home the fact that he had exhausted the resources of the university. Since then, circles of intellectuals occupied with doing science (or at least talking about doing science) had comprised the framework of his activity. Five years and one book after leaving Augsburg in 1570, Tycho was a professional, intending to take one last look around Europe before finally committing himself to a place to settle down to work.

Although Tycho certainly planned to include Augsburg in his itinerary, and probably even entertained some thought of settling there among his old friends, there were other places to be considered first. First among them in both interest and location was Cassel, where the ruling *landgrave*, Wilhelm of Hesse, conducted what Petrus Ramus had praised in print (and probably mentioned to Tycho in person) as a veritable Alexandrian academy for the advancement of astronomy. Some fifteen years older than Tycho, the *landgrave* of Hesse had been actively involved with astronomy for nearly twenty years when Tycho arrived there in 1575.

Over the years Wilhelm had commissioned a succession of artisans to build him practically every kind of astronomical instrument, planetary clockworks, and miscellaneous "gadgets" known to the sixteenth century.[37] Still ahead of him were employment of the mechanic who would make the best clocks of the century and the astronomer who, with Tycho and the *landgrave* himself, would participate in the most celebrated astronomical correspondence of the century. Of course, such projects were simply a matter of will for a man of his means. But Wilhelm was different in kind as well as in degree from the wealthy dabblers who supported the extensive mechanical craft industry of Germany, for Wilhelm had always done his own observing. From as early as the appearance of a comet in 1558 and observations of the sun in 1561, he had collected data that exceeded both in quantity and quality anything being done by any of

[37] The work of Wilhelm (1532–92) and his instrument makers is described by *Zinner*, 585–9, and elaborated by Bruce T. Moran, "Princes, Machines and the Valuation of Precision in the Sixteenth Century," *Sudhoffs Archive* 61 (1977), 209–28. The reference by Ramus was in his *Proëmium* and is quoted in *Gassendi*, 26.

his contemporaries. Thus, the only thing at all mysterious about Tycho's visit was why he had not made it five years earlier on his way home from Augsburg.

In fact, however, the intervening years had been lean ones for the *landgrave*, astronomically speaking. The one real deficiency in his program was that he had always done the observing himself and that, since his accession to power in 1569, he had had little time for serious astronomy. When Tycho arrived, however, Wilhelm postponed or delegated, affairs of state to make time. High on the list of interests for both must have been the elegant quadrant that Wilhelm had recently commissioned (Figure 3.2). Tycho, in turn, must have described his own smaller quadrant and the sextants he had made over the years, if, indeed, he did not have one of them with him.[38]

Wilhelm had observed the new star, too, and had reached the same contra-Aristotelian conclusion that Tycho had – that it was beyond the moon. But Wilhelm thought he had found a little bit of parallax for it (3′) and regarded its gradual extinction in 1573–4 as evidence that the star had moved rectilinearly away from the earth until it disappeared, so even that subject required some discussion. They also talked at some length about Wilhelm's long-term ambition to prepare an accurate star catalogue, and Tycho came away with the results he had obtained to that date. In return, he left not only the suggestion that Wilhelm ought to hire an assistant to help with his observing but perhaps also a hint as to where to go for one, for before the year was out, Wilhelm had hired one Hans Buch from Augsburg.

Observing by night and conversing by day, on topics ranging from such technical matters as astronomical refraction to such idiosyncracies as Wilhelm's having once insisted on finishing an observation of the new star while his servants extinguished a fire that had broken out elsewhere on the premises, the two astronomers visited for eight to ten days.[39] At that time, however, Wilhelm's infant daughter died, leaving Tycho obliged to withdraw his presence from the family in its time of grief. Circumstances were to prevent their ever meeting again. But the one encounter was destined to be bountiful enough by itself. Inspired to rise above the cares of office and renew his commitment to astronomy, Wilhelm began to look for a mathematician to execute his plans for a star catalogue. In the meantime, he arranged to commemorate Tycho's visit by having the

---

[38] In describing his steel sextant (V. 76–9), Tycho implied that he took it to Germany with him on his travels. It is argued in Chapter 5, however, that the instrument he carried on that trip must have been his light wooden sextant.
[39] *Gassendi*, 26–8.

Figure 3.2. Quadrant used at Cassel in the observatory of Wilhelm IV, *Landgrave* of Hesse.

court painter prepare a small picture of Tycho as background for the next official portrait.[40] If Tycho, on his part, did not actually need such external stimulus, he must nevertheless have found the intellectual comradeship extremely gratifying. And a few months later, he would learn that there had been extrinsic returns, as well, from his encounter with the *landgrave*.

From Cassel, Tycho went south to Frankfurt, where each spring and fall the booksellers of Europe gathered to vend their wares. Books on the new star were still coming out, and Tycho enlarged his collection on that subject until he had well over a dozen titles. He also bought many other books. Between the ones that have been identified because they still exist with his name and year of purchase on the binding, and those to which he referred at one time or another in his extensive correspondence, it is possible to form a good idea of the breadth of Tycho's literary interests. Except for Conrad Gesner's pioneering work on bibliography and a new edition of a Greek poet, Tycho's known purchases were all in astronomy, mathematics, astrology, and alchemy. He bought a few ancient authors, including Proclus, astrological works ascribed to Ptolemy, and four works of the Hermetic philosopher, Synesius of Cyrene. He also acquired a printed edition of a treatise by the thirteenth-century astrologer Guido Bonatti. The bulk of his purchases consisted of recent imprints by contemporary authors, including a few textbooks of astronomy, some polemical works, and a number of advanced treatises. German Philippist authors like Brucaeus, Peucer, and Christopher Encelius were well represented, as were Italian and French authors. Works by French scholars included Mizaldus on meteorology, as well as a new treatise by the Paracelsan physician, Joseph du Chesne, and five mathematical works by Oronce Fine. There were also works on mining and minerals relating to Tycho's interest in alchemy.[41]

From Frankfurt, Tycho continued southwest to the Swiss city of Basel. If the city had displayed any special charms in 1568, they had not been sufficient to keep him there very long. Now, however, its virtues impressed him greatly. Perhaps after talking with Peder Oxe, Charles Dançey, Steen Bille, and Hans Pratensis, all of whom had had extensive schooling in France, he felt a need to be near its intellectual pulse. At any rate, he made particular note of Basel's location near the border between France and Germany, far enough south also to provide easy access to scholars and ideas from Italy.[42]

---

[40] Norlind, 65–7.
[41] See *Norlind*'s survey of Tycho's professional library, 335–6.
[42] See Tycho's autobiographical remarks in V, 108–9, and VII, 25.

Besides the University of Basel, it had an intellectual tradition that had already commended it to Erasmus as the ideal place to live. In addition, Basel was blessed with a mild climate and graced by a style of life that Tycho found very agreeable. How long it took for these considerations to materialize into a decision Tycho does not say. Possibly, however, the issue was still unresolved when he left Basel, for instead of turning homeward by way of Augsburg, Tycho continued south into Italy.

Concerning his route and purpose, Tycho mentioned only that he got as far south as Venice. Most likely he crossed the Alps into the Po valley and meandered across northern Italy to Venice. Peder Sørensen had spent some time in Venice, and the glassmakers of Herrevad had been from there too, so Tycho probably had some contacts. He was invited to the learned gatherings and "academies" that flourished among the patricians of the Venetian republic and included scholars from nearby Padua. Perhaps he visited some of the splendid new villas of the Venetian architect, Palladio. Tycho found the atmosphere stimulating. He enjoyed the sophistication of the Venetian patricians, and as a nobleman, he admired the aristocratic, oligarchic government of the Venetian republic. But Tycho did not stay there for more than a few weeks. Before the fall snows would make the Brenner Pass uncertain, he slipped through to Innsbruck and, from there, a few days north to Augsburg.

Tycho's second stay at Augsburg seems to have been anticlimactic in several respects. Of course, his old friends were there, and they doubtless had experiences to share that had not been reported in their letters but few of their experiences had been astronomical. The Hainzels had done no observing with the great quadrant while Tycho was gone and, indeed, had apparently not repaired the damage it had suffered in a great windstorm the previous December. Wolf's interest had never gone much further than astrology, although he and Paul Hainzel had obtained books and some minor astronomical devices for Tycho. Even the great globe that Tycho had ordered from Schissler was disappointing, as Tycho found that it was not only imperfectly spherical but also marred by cracks. How long Tycho stayed in Augsburg – or in Basel or Venice, for that matter – is not known, but by the time he left he must have decided to settle in Basel.

It was already late October when Tycho started north across Germany. He had lingered to attend the coronation of Rudolph of Hapsburg as king of the Romans and heir apparent to the Holy Roman Empire. The festivities were being held at Regensburg, just down river from Augsburg, and Tycho probably expected to find Wilhelm there for the occasion. Unfortunately, the *landgrave* did not appear, but someone else did who turned out to be almost as stimulating and just as important a contact, a Bohemian physician named Thaddeus Hayek.

In their few days together, Tycho and Hagecius as he latinized his name, formed a lifelong bond, based on a community of knowledge across the entire spectrum of learned discourse.

Hayek had been interested in things astronomical since his undergraduate days. During the long years of medical studies that prepared him for an illustrious career as a physician, he had lectured at Prague on mathematics, published a series of astrological calendars, and even written a tract (in Czech) on the comet of 1556. Twenty years older than Tycho, Hayek had parlayed twenty-five minor astronomical publications and fifteen years of medical practice into an appointment as personal physician to the Hapsburg emperor and elevation to the nobility.[43] Given his medical interests, there can be little doubt that some of their conversation dealt with medical theory. Tycho's firsthand knowledge of Petrus Severinus's thoughts on Paracelsus probably even gave him some kind of intellectual parity in the exchange. Paracelsism would have led naturally into discussions of alchemy, in which Hayek's interests were even stronger than Tycho's. (In 1584 Hayek recorded watching John Dee and an accomplice transform mercury into gold.)

But astronomy was not an afterthought for Hayek. Even though he was not the professional astronomer Tycho was, he had nevertheless made observations (though not very accurate ones) of the new star and published a 176-page book to convey his findings. Moreover, Hayek was sufficiently interested in theoretical matters to have appreciated the value of an unpublished manuscript by Copernicus that had happened to cross his path. Known as the *Commentariolus* since its rediscovery in the nineteenth century, it was a description of Copernicus's new system (and mechanisms), circulated to his friends thirty years before the publication of *De revolutionibus*.

Although Tycho was, in the conventional sense, a non-Copernican, just as Hayek was, his reaction to the manuscript is indicative of his general appreciation of Copernicus's work. Not only was he extremely interested in it himself, but he expected any other astronomer to be, too, and, throughout his career, proudly distributed copies of the manuscript to astronomical colleagues as symbols of

---

[43] I am indebted to Jerzy Dobrzycki for details of Hayek's life and publications. For a bibliography of Hayek's works, see Jiri Bousaka, ed., *Tadeas Hajak z Hajku 1525–1600* (Prague: Univerzita Karlova, 1976). Hayek's birth year is variously cited as 1525. But in a letter written to Tycho on 7 September 1589, he stated clearly that he would be sixty-three on 1 October, VII, 194.

favor. Despite Tycho's interest, however, the forty-page document sank into oblivion and was brought to the attention of modern scholarship only by the discovery in 1878 of a copy originally presented by Tycho to one of his students.[44]

There were other things to talk about as well. Hayek, for example, had been a regular correspondent with Melanchthon. He was also sufficiently interested in astrology to have on hand and be able to give to Tycho a recent work by Offusius, which greatly impressed Tycho when he read it. Tycho was able to reciprocate with a copy of du Chesne's new book just purchased at Frankfurt. Some of their bibliographical discussions may have been inspired by the presence of Hugo Blotius, who had collaborated with Tycho in a quadrant-building project six years earlier as a student at Basel, and who was now curator of the emperor's library.[45] Other discussions were conducted in the company of an artist named Tobias Gemperlin, who seems to have been persuaded by Tycho at Augsburg (rightly, as it turned out) that it would be worth seeking his fortune in Denmark.[46]

After Rudolph II was crowned on November 1, Tycho started for home. He appears to have stopped in Nuremburg long enough to see Joachim Camerarius, the physician-son of the famed humanist of the same name. If, indeed, he visited only once this great center of the German clock and instrument trade, this was the occasion on which he established the contacts that would bring one of its artisans to Denmark in 1577 to engineer a fountain for Kronborg Castle and design the system of running water for Tycho's house on Hven.

From Nuremburg Tycho went to Saalfeld, where he met the younger Erasmus Reinhold, physician-son of the famed author of the *Prutenic Tables*. The younger Reinhold had likewise published on the new star, but had interpreted it as a heavenly miracle, rather than as a physical phenomenon of astronomical concern. But because he had at least attributed a clearly supralunary parallax to it (3'), he could in some sense be regarded by Tycho as an ally. Certainly the visit was cordial, for Tycho still remembered years later being shown the senior Reinhold's manuscripts, including some painsta-

---

[44] The first copy found in the modern era (in Vienna in 1878) was a copy of Tycho's copy, made by his student Longomontonus and presented to another student Johannes Eriksen. See Norlind's note in *Gassendi*, 230. The *Commentariolus* has been translated in Rosen, *3 Copernican Treatises* and Swerdlow, "Derivation and First Draft."

[45] VII, 328.

[46] Gemperlin seems to have been a native of Augsburg (VII, 46), and so Tycho may have known him from 1569–70. While he was in Denmark, he painted the portrait of Tycho that dominated the mural quadrant (see Chapter 5), as well as a portrait of King Frederick. Tycho reported (IV, 235) that Gemperlin died shortly after helping him make his last observation of the comet of 1577.

100 The Lord of Uraniborg

kingly computed tabulations of planetary equations for every 10' of anomaly.[47]

Last was the obligatory stop at Wittenberg, where Tycho probably still had friends from previous stays. On this occasion, he found professional interest as well, for in that city, too, the new star had been observed, by the well-known astronomers Wolfgang Schuler and Caspar Peucer. Schuler had at first found a parallax of 19', using an old wooden quadrant whose only recommendation was that it had been used by Reinhold. When Schuler learned that Wilhelm had found virtually no parallax, however, he hastily had a new and very large triquetum built, probably by the astronomer and instrument maker Johannes Praetorius. Around Wittenberg it was still regarded as a great achievement when Tycho was there in 1575. But although Tycho had to applaud the fact that Schuler and Praetorius had found little or no parallax with the triquetum, he could only marvel that respected astronomers might seriously regard such a mean instrument – to say nothing of Reinhold's quadrant – as sufficient to the needs of astronomical science.[48]

If he had not heard the news earlier in his travels, Tycho certainly would have learned in Wittenberg of the excitement at the university during the previous year. It had started with Elector (of Saxony) Augustus's discovery that crypto-Calvinism was rife among the theologians at Wittenberg. Whether out of fear of political consequences (according to the Peace of Augsburg of 1555, only Catholicism and Lutheranism were officially tolerated within the Holy Roman Empire) or the zeal of a convert (he had embraced Lutheranism as a result of his marriage to Anne, King Frederick's sister), the elector had reacted vigorously. All of the offending theologians were still in prison, even Melanchthon's son-in-law, Caspar Peucer.

Interrogations during the winter of 1574–5 got to the heart of the matter. In the spring of 1575 the elector conveyed through a returning Danish scholar, master Jørgen Dybvad, an accusatory letter to King Frederick. All the culprits had defended themselves by saying that they had obtained their ideas from Niels Hemmingsen, during the festivities at which King Frederick had married Queen Sophie in 1572. And indeed, shortly thereafter, Hemmingsen had published a work presenting a thoroughly Calvinist interpretation of the Eucharist (although with no mention of Calvin), while a similar work had appeared, anonymously and almost simultaneously in Saxony. It all was too much for mere coincidence, in Augustus's view.[49]

---

[47] III, 213.
[48] Tycho discussed the matter at some length in III, 143–57.
[49] For more discussion, see Robert Kolb, *Caspar Peucer's Library: Portrait of a Wittenberg*

The Eucharist was only one point of dispute between the Philippists and an emerging party of so-called Gnesio-Lutherans in Germany. The religious articles of the Peace of Augsburg did not apply to Denmark, and so there was no political need to distinguish so sharply between Lutheranism and Calvinism in that kingdom. In general, the main religious aim of the Danish government had been to achieve a solid Protestant consensus within the Danish realm, and that had been accomplished under the leadership of Niels Hemmingsen. Moreover, many princes in Germany continued to support the Philippist view. Among them were Queen Sophie's father, Duke Ulrich of Mecklenburg, who maintained close ties with Denmark, and Landgrave Wilhelm IV of Hesse-Kassel, whose sister had married an uncle of King Frederick. Saxony and other German states leaning toward Gnesio-Lutheranism were important allies of Denmark, however, and their criticism of Hemmingsen's crypto-Calvinism could cause considerable embarrassment to the Danish crown.

Although Tycho may have learned of the details only on his return to Denmark, Frederick had already decided he had to act. Early on the morning of 15 June 1575, all endowed professors of the University of Copenhagen, all pastors of Copenhagen churches, and the bishop of Roskilde had been summoned to the castle of Copenhagen to answer the elector's charges. Besides Niels Hemmingsen, this group included Anders Sørensen Vedel, Johannes Franciscus, and Johannes Pratensis. They had been examined by a commission of three of the most powerful noblemen in the royal government: Chancellor Niels Kaas, Lord Treasurer Peder Oxe, and his brother-in-law, *rigsraad* Jørgen Rosenkrantz. Hemmingsen spoke up and defended what he described as the unity of the Danish Church. German theologians were numerous, he had said, and they all leapt about like cooks trying to please the palates of their respective lords. If the Danes listened to them, there would be utter confusion in Denmark, instead of the unity of belief and religious practice that now prevailed. Rosenkrantz had not been satisfied and continued to ask harsh questions, but Peder Oxe, when the hearing concluded, had reassured Hemmingsen, "No misfortune will come upon you for my sake."[50]

Tycho probably was home by Christmas. As he began to think

Professor of the Mid-Sixteenth Century (St. Louis: Center for Reformation Research, 1976). The offending work was Hemmingsen's *Syntagma institutionum Christianarum* (Copenhagen, 1574), reprinted in Geneva in 1578, in Antwerp in 1581, and in Leiden in 1585.

50 John Christianson, "Tycho Brahe's German Treatise on the Comet of 1577: A Study in Science and Politics," *Isis* 70 (1979): 115.

seriously about liquidating his assets for a move to Basel, he went to
court to pay his respects to the king. The court was at Sorø Abbey
for the Christmas season, and Inger Oxe was probably there too, as
she had had the honor of being in charge of the queen's chamber
since the coronation. King Frederick's attention had been called to
Tycho by an emissary from the *landgrave* of Hesse, conveying
Wilhelm's personal endorsement of Tycho's potential and a strong
recommendation that he be encouraged. Frederick might even have
suspected that Tycho wanted to emigrate, for he received Tycho
with exceptional graciousness and asked him to state his requirements in terms of fiefs and honors. When even some specific
suggestions could not elicit a reply from the astonished Tycho,
Frederick told him to think it over and let either him or his chancellor
know his decision.

   Tycho went to Copenhagen to trade news of the year's events
with Pratensis and Dançey. In a separate discussion with Dançey,
which may have been motivated by the desire to inquire privately
into the apparently dangerous state of Pratensis's health, Tycho told
his old friend, who after decades as an ambassador was essentially
an émigré himself, of his plans to leave Denmark. In his agitation,
however, Tycho was sufficiently transparent to allow Pratensis to
divine his inclinations, too, for Pratensis wrote at the end of January
assuring Tycho that his interests were indeed respected in Denmark
and were reflected in the king's offer to him. At the same time, he
alluded to his own pleasant memories from Basel, and begged Tycho
not to leave without first answering some technical questions for
him.

   Tycho must have wished for one last chance to consult Peder Oxe,
but that old statesman had died in October. Tycho thus went to
Herrevad to discuss things with Steen Bille. The problem was that if
Tycho were going to obtain the means of buying an estate, so that he
could live like a gentleman in Basel, he would have to sell his share of
Knudstrup. How could he do that without involving his mother and
brother, who not only had cointerests in the estate but were actually
living there? (Steen had returned from abroad and in 1575 married
a niece of Jørgen Rosenkrantz.) Even if Tycho sold his share to
someone else in the family, how would it reflect on his relations if he
were to turn his back on the king's generous offer and slink out of the
country? Yet Tycho could not bring himself to accept one of the
properties offered by the king, because any of them would involve
not only the political duties he himself had long abhorred but also
social obligations that he and Kirsten simply could not meet.

   Not feeling that he could press the king for more preferential
treatment than he was already receiving, Tycho merely left the

matter hanging and continued his preparations for departure. Wittingly or unwittingly, he left the delicate business of bargaining with the king to be conducted by (or through) his Uncle Steen. Probably through old connections in the chancellery (although a prominent vassal like Steen certainly had direct access to the king), Steen let Frederick know that Tycho was considering emigration and that the reason he was doing so was that the standard form of royal favor thus far offered to him involved duties that would "interfere with his work." Steen may also have mentioned the island of Hven as an alternative that might forestall Tycho's inclination to leave the country, although the king may have actually remembered on his own (as he told Tycho he did) a conversation with Steen a year earlier in which it was mentioned that the island had some special appeal to Tycho.[51] In any case the king resolved the situation with truly royal flair.

As Tycho excitedly recounted things in a letter to Pratensis,[52] he was lying in bed two hours before sunup one morning, contemplating his position from every possible angle, when a messenger appeared with a summons from the king. The page was a kinsman of Tycho's and mentioned that he had been commanded to travel day and night until he could put the message into Tycho's hands, and so Tycho perceived that haste was in order for him, too. Although he was at Knudstrup, he made it to Frederick's hunting lodge outside Copenhagen that evening. There the king told him that he had heard of Tycho's plans to go to Germany and now realized what the problem was – concern that political and social responsibilities would hamper his research efforts. But when he had been at Helsingør recently, checking on the construction of Kronborg Castle Frederick's glance had happened to fall on the little island of Hven, on the southeast horizon (some fifteen kilometers away). This, he thought, was a perfect place for Tycho: isolated, unassociated with any administrative obligations, and (therefore) unbound to any noble in fief. If the royal exchequer were tapped for the expenses of founding and maintaining a proper establishment, was there anything that Tycho hoped to do abroad that he could not do here, where it would redound to the credit of his country, his king, and himself? Would Tycho at least promise to think it over and give him an answer before running off to Germany?

Even though Frederick had more than matched any expectations for support abroad that Tycho could rationally have hoped for, and met every argument against working in Denmark that Tycho could explicitly advance, Tycho still hesitated. Only after riding home to

---

[51] VII, 27.   [52] VII, 25-9.

Knudstrup the next day (12 February) and waiting another day did Tycho narrate the developments of Pratensis in Copenhagen. Even then he was so far from decided on the issue that he asked Pratensis to burn his letter (after showing it to Dançey) and then stand ready to advise him.

Pratensis and Dançey urged him to accept the offer. As Pratensis wrote, "Apollo desires it, Urania recommends it, Mercury commands it with his staff."[53] He also pointed out that Tycho's relatives and the other nobles who believed he should follow the same political careers they pursued would be amazed to learn that the king wanted to promote Tycho's studies.

What Pratensis did not touch on but what must have been as important to Tycho as was immunity from the petty politicking of aristocratic careerism was Frederick's implied acceptance of his relationship with Kirsten. The gossip picked up by one of his students twenty years later and subsequently reported by Gassendi was that "all of Tycho's relatives were very disturbed by the diminished esteem the family suffered because of Kirsten's low birth, so that there were hard feelings toward Tycho that were put to rest only when the king intervened."[54]

Frederick certainly did not intervene in any explicit sense, and it is extremely unlikely that Kirsten was ever mentioned in their conversation. But at the same time, the obvious and singular mark of royal favor bestowed on Tycho must have been perceived as at least balancing out whatever loss of face Tycho's *slegfred* wife – if she had even yet fulfilled the requirements for that status – was costing the family. Apparently Tycho envisioned this possibility fairly quickly, because by the eighteenth of the month, he had consulted his friends, responded to the king, and accepted a pension. In return for an initial cash grant of five hundred dalers per year, he pledged loyalty to his liege lord and service "according to his abilities and conditions."[55] They were the same words his ancestors had recited for generations, as they had entered the service of the Danish crown, but for Tycho Brahe of Knudstrup, they signified unprecedented intentions.

[53] VII, 30–1. It was written the day Pratensis would have received Tycho's letter, 15 February 1576.
[54] *Gassendi*, 23–4.
[55] See the royal grant dated 18 February 1576, on XIV, 4–5.

Chapter 4

The First Years on Hven: 1576–1579

As of Tycho's day, the island of Hven had played no role in Danish history for a long time. Several folktales associated with it were still in circulation, and the ruins of four forts could still be discerned at strategic points on the island, but nothing important had happened there since the Norwegian king, Eric the Priest Hater, had reportedly destroyed the forts in 1288.[1] Through the years, some forty families had together tilled the land, grazed a few animals on the less tractable areas, and shared their meager yields with the crown through a series of provincial governors living on the mainland.

In 1576, however, the scene changed radically. On May 23 Frederick II signed a document conferring "to our beloved Tyge Brahe ... our land of Hven, with all our and the crown's tenants and servants who live thereon, and with all the rent and duty which comes from it ... to have, enjoy, use, and hold; free and clear, without any rent, all the days of his life."[2]

In fact, by that date the new landlord was probably already a notorious figure on the island. By·then he must have been almost accustomed to the two-hour boat trip from Landskrona harbor to the landing on the north side of the island and familiar with the 150-foot bluffs that greeted the eye from any other perspective of the island. Even at the landing, it was a good climb up to the plateau and then a few minutes' walk to the village of Tuna, where the inhabitants of the island lived in thatched and half-timbered buildings. The three great fields of the village, plowed in many long strips, covered all of the island except for the commons around the central heights and far to the south. On a high point beyond the village, in the direction of the coast of Sjaelland, was the village church of St. Ibb. High on a hillock near the village was a great post windmill.

Except for the Lutheran pastor, the miller, and perhaps a blacksmith, all the villagers were peasants and rural laborers. The peasants were freeholders who owned their own land. They farmed their strips in common and formed a guild to set down their bylaws and carry out their daily activities, the way things were done in peasant villages throughout Denmark. Contacts with the crown were few and would aways have been conducted through the pastor or the

---

[1] For a more extended narration of the folklore and history of Hven, see *Dreyer*, 88–92. Abbreviations of frequently cited sources are in Appendix 1.

[2] XIV, 5. For an English rendition of the entire text, see *Dreyer*, 86–7.

constable. The constable was typically a prominent villager appointed by the governor to maintain the peace and collect the annual taxes in coin and kind. Social and economic contacts with the mainland were through islanders who sold their produce in towns along the sound, grazed their swine in the forests of Skaane during the summer, went to the high street markets, and fraternized and intermarried with peasant families in Skaane and even Sjaelland.

What initially recommended the island to Tycho was its isolation. But he appears not to have allowed that criterion to be decisive. Three months elapsed between Tycho's acceptance of the king's pension and his acceptance of Hven as his fief, and we can assume that Tycho used the time to make sure that the island would be suitable in every respect. The logistics of getting building supplies to his projects would be difficult but not insurmountable. The income from forty farms was a puny endowment, but it was the trade-off that had to be made to obtain the privacy that would not be available on larger estates. A more serious problem would be labor. Tycho was entitled to two workdays, from sunup to sundown, per week from each farm, which was not much labor, either skilled or unskilled, for the magnitude of the establishment he envisioned. The last hurdle was a site and a building plan. For this Tycho turned to his books on architecture.

The fount of classical knowledge on architecture was Vitruvius. The greatest technical production of Roman culture, his *Ten Books on Architecture*[3] was a work with which Tycho was thoroughly familiar. But Tycho was also strongly impressed by the more recent architecture he had seen on his travels in northern Italy and surely owned the magnificently illustrated editions of his contemporaries, Serlio and Palladio. The Vitruvian ideal, shared by all Renaissance builders, was that an architect must be what is now termed a "renaissance man," a "man of letters, a skillful draftsman, a mathematician, familiar with scientific inquiries, a diligent student of Philosophy, acquainted with music, not ignorant of medicine, learned in the responses of juriconsults, familiar with astronomy and astronomical calculations."[4]

The reason for this requirement was that a great house had many functions: Tycho's was to be a noble residence, observatory, alchemical laboratory, and administrative center for his fief. Each of the parts had to have its place in the architectural scheme, and all the parts had to be integrated into the whole with symmetry and

---

[3] On Vitruvius, see *DSB* XV, 514–18.
[4] Vitruvius, *On Architecture*, trans. F. Granger (Cambridge, Mass.: Loeb Classical Library, 1945–70), bk. I, chaps. 1 and 3.

harmony. The Palladian ideal was that a country house was really a small city, and a city was really a great house. In another sense, each was a reflection of the cosmos. According to Palladio, architecture, like all the other arts, imitated nature. By this, he meant that architecture imitated the geometrical sense of order that constituted the hidden framework of the universe. If an astronomer like Tycho had achieved a certain understanding of the universe, he could express that understanding through architecture. Vitruvius had emphasized the human dimension of this cosmic unity by describing it in terms of the microcosm. His standard of symmetry and proportion was the human body, the famous Vitruvian man that intrigued the artists and architects of the Renaissance:

> The foot is a sixth of the height of the body; the cubit a quarter, the breast also a quarter. The other limbs have their own proportionate measurements. And by using these, ancient painters and famous sculptors have attained great and unbounded distinction. In like fashion, the members of temples ought to have dimensions of their several parts answering suitably to the general sum of their whole magnitude. Now the navel is naturally the exact center of the body. For if a man lies on his back with hands and feet outspread, and the center of the circle is placed on his navel, his fingers and toes will be touched by the circumference. Also a square will be found described within the figure, in the same way as a round figure is produced.... Therefore if Nature has planned the human body so that the members correspond in their proportions to its complete configuration, the ancients seem to have had reason in determining that in the execution of their works they should observe an exact adjustment of the several members to the general pattern of the plan.[5]

Palladio's *Four Books of Architecture*, published in Venice in 1570, was the most thorough possible application of microcosmic proportion and symmetry to architecture. His system of proportions included the use of harmonic ratios derived from musical theory, and this in turn related his architecture directly to the wellsprings of the Pythagorean tradition. It had a powerful appeal to mathematicians like Tycho, trained in the Philippist tradition of humanist learning.

Palladio (1518–80) established systems of proportions regarding the length, breadth, and height of rooms, the size of one room to another, and of the central building to the porticos, wings, and even gardens, in a manner that integrated plan to elevation, room to room, interior to exterior, and building to site. There was a clear, hierarchical symmetry throughout, with central elements dominat-

[5] Vitruvius, Book III, chap. 1.

ing peripheral ones. Like the human body, Palladio's buildings were symmetrical to a central axis, with single elements along the axis – like the nose, mouth, or navel of the body – and lateral elements in pairs – like the eyes, ears, and arms of the body.

Some of Palladio's designs were centralized. Such buildings were symmetrical along four axes extending at right angles from a central point, and all four sides were the same. Fundamental to the microcosmic plan of these structures was the Vitruvian figure of human proportion within homocentric squares and circles. Here in its most simple and monumental form, the unity of microcosm and macrocosm, the central role of humanity in the universe, and the merging of God's spirit with the world's mathematical structure could be expressed in a building that was at once a work of art, an emblem, and a human dwelling.[6] In the case of Tycho's manor on Hven, it would also be a temple for the worship of Apollo, Mercury, and the muses.

Much of Tycho's time during the spring of 1576 must have been spent in drawing circles and squares in harmonious proportion, as he pondered the complex theories of architectural geometry in the plan of his house on Hven. His friends doubtless helped him, although the one who could probably have helped most, Tobias Gemperlin, had apparently returned to Augsburg, perhaps in order to settle his affairs before moving to Denmark.[7] Tycho may also have consulted Hans van Paaschen, the royal architect, and obtained designs from a sculptor known to scholars as the Alabaster Master of Copenhagen. This unknown master had designed several large sepulchral monuments for the Oxe family, including the tomb of Peder Oxe. He had either visited Italy or employed an Italian journeyman in his shop, for he sometimes decorated columns with festoons in the Venetian style of the early Renaissance and was also the first Danish artist to use Italian Mannerist elements like the broken pediment. In short, he introduced the High Renaissance style to Denmark. Features reminiscent of his work appear in the woodcuts of Tycho's manor on Hven.[8]

Slowly the plan took form. The basic structure of the house would be a square, bisected in each direction by corridors that subdivided it into four square areas of equal size. On the north and south sides of

---

[6] James S. Ackerman, *Palladio* (Baltimore: Penguin, 1966), pp. 68–73, 160–71. See also Rudolf Wittkower, *Architectural Principles in the Age of Humanism* (London: Alec Tiranti, 1949), pp. 1–28.

[7] VII, 47.

[8] See Francis Beckett and Charles Christensen, *Uraniborg og Stjaerneborg* (London: Oxford University Press, 1921). Their ideas have been elaborated by C. A. Jensen, *Danske adelige gravsten fra sengotikens og renaissancens tid: Studier over vaerksteder og kunstnere II* (Copenhagen: A. F. Høst, 1952), pp. 13–32.

this living area Tycho planned large rounded towers appended to the square (Figure 4.1) to form the working areas of the house, and at the ends of the other corridor he added smaller portal towers to formalize the two entrances.[9] The plan thus possessed what would today be called *axial symmetry*. In Tycho's day it was termed *correspondence*. *Symmetry* meant something else, and Tycho's house would have it, too.

Symmetry is a concept that has been sought and elaborated since the ancient Pythagoreans discovered consonant ratios in nature by studying the sounds emitted by vibrating strings. With the same tension and size, a string of a given length and another twice as long produce the harmony called an *octave*. Strings in the ratio of two to three produce the harmony of the fifth, and those of three to four, the fourth. Three strings in the ratio of 1:2:3 produce an octave and a fifth, and in the ratio of 1:2:4, they produce two octaves. The progression 1:2:3:4 contains all five of these basic Pythagorean consonances of the Greek musical system. Plato applied the concept of musical harmony to the cosmic order, and Vitruvius described its application to the architecture of the theater. More generally, Vitruvius stated, it consisted in a harmonious effect arising from proportion, which in turn consisted of "taking a fixed module, both for the parts of a building and for the whole,"[10] in the way that all the parts of the human body are related to the hand. Since the work of the great artist Leon Battista Alberti in the mid-fifteenth century, the Pythagorean system of musical harmony had been fundamental to the architectural theory of the Renaissance.

There is some disparity between Tycho's various descriptions of the symmetry of his manor, Uraniborg, and modern archaeological excavations of the site. It appears, however, that the size of Tycho's architectural unit, or fixed module, was fifteen Danish feet (each about 259 mm) and that the module itself was the portal tower fifteen feet wide and fifteen feet long on each side of the house.[11] The height of the facade was thirty feet (Figure 4.1) the peak of the roof forty-five feet, and the side of the central square sixty feet. The symmetry between exterior and interior was more complex and suggests that Tycho may have been familiar with some of the more esoteric developments of Renaissance architecture. In later years, as Tycho enclosed the manor grounds with a great earthen wall and constructed outbuildings at its four corners (Figure 4.2), all of the additions were planned to harmonize with the theme of the manor.

---

[9] The woodcut was commissioned by Tycho and printed in his *Mechanica* (V, 142).
[10] Vitruvius, Book III, chap. 1.
[11] Wilhelm Wanscher, *Arkitekturens Historie* (Copenhagen, 1931), pp. 244–6. Carl Henrik Jern, *Uraniborg: Herresäte och Himlaborg* (Lund, Sweden: Studentlitteratur, 1976), pp. 25–6, 37–9, 110. On the so-called Tychonic foot, see *Raeder*, 9.

ORTHOGRAPHIA PRÆCIPVÆ DOMVS

ARCIS VRANIBVRGI IN INSV-
SIA VULGO HVENNA,
RANDÆ GRATIA CIRCA AN-
EXÆDI-

LA PORTHMI DANICI VENV-
ASTRONOMIÆ INSTAV-
NVM 1580 A TYCHONE BRAHE
FICATÆ.

a

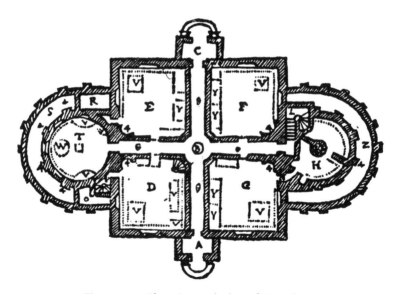

b

Figure 4.1. Elevation and plan of Uraniborg.

Figure 4.2. Grounds of Uraniborg after about 1590.

Details of the system may have escaped modern commentators, but the general intent is clear: The house Tycho designed for himself would be built to reflect in its proportions the order of the heavens and the earth, both the great cosmos and the cosmic dimensions of humanity. This relationship, which seems quite esoteric to the modern observer, would have been obvious to any painter, printer, architect, Philippist clergyman, or other learned person in Tycho's day, including many of the kings and great aristocrats who were patrons of the Renaissance style.[12] What would have been more striking to the latter group, however, would have been the very modest size of the building. For, whether from dictates of economy

[12] Wittkower, *Architectural Principles*, pp. 91–3.

or style, or a pragmatism that placed actual need above the desire for ostentation, the house/observatory was smaller than any one of Knudstrup's four wings and was insignificant in comparison with the many mighty châteaux being built by Tycho's peers.

As spring advanced and the plan developed, Tycho began to arrange for the numerous practical matters involved in the building. For the site of his manor house he chose the highest spot in the center of the island. After surveying the site and setting out stakes to mark the north–south and east–west axes of the house, he doubtless informed the elders of the village that he, rather than the royal governor, would soon be their overlord and that in case they had not already guessed it, he would be establishing his residence on the island. The good aspect of this news was that the peasants would now share their lord's exemption from the extraordinary taxes levied in the kingdom. The bad news was that they would instead be obligated to render their labor services to their lord, without pay.

In addition to providing two man-days of labor a week, they would have to supply tools and bring draft animals and wagons a certain number of days each year. The burden would fall on the landholding peasant households rather than the landless laborers and cottagers and usually involved a son, younger brother, or hired hand rather than the head of the household. The peasants would not have been pleased with these arrangements, but they were Danish law and facts of life in feudal society.[13] Kings and queens commanded nobles in their service, and noble lords ruled over peasants, who in turn commanded their own households of family and hired hands.

With the formal granting of Hven, the work started. The work levies came to the center of Hven at sunrise. With spades, wheelbarrows, and wagons they excavated a square foundation with sides running exactly north–south and east–west. They worked all day until the sun went down. On the northern and southern sides of the square, they excavated circular extensions, and at the center of the northern circle, they dug straight down until they struck water and then lined their excavation to make a well. Two hundred feet from the center of the square foundation, at each of the cardinal points, the workmen marked out the foundations for four other buildings. From each of these outer foundations, great earthworks were to be thrown up to form a square rampart around the whole site. Gradually, the plan of Tycho Brahe's celestial palace was revealed.[14]

---

[13] Fridlev Skrubbeltrang, *Det Danske Landbosamfund 1500–1800* (Odense: Den danske historiske forening, 1978), pp. 55–7, 72–3, 85–8.

[14] The construction can be inferred from Tycho's plan and various statements in XIV, 17–19. The remains of the estate are so slight that reconstructing it would be hopeless without Tycho's woodcuts: see Appendix 6, note 4. For the social circumstances of the peasants, see Skrubbeltrang, pp. 34–95.

With the excavation under way, Tycho began to hire the skilled labor that would be required. Ordinarily this would have meant going to Copenhagen, but the king was building a great castle at Helsingør (the Elsinore castle that was to seize Shakespeare's imagination), and many craftsmen were working there. Ordinarily the work force available would have been mostly Danish, but the disturbances in the Low Countries had generally hampered construction there and driven many Dutch and Flemish artisans to Denmark.

In the proper order and with the proper lead time, Tycho would need masons, numerous stone carvers, carpenters, tile workers, and even a hydraulic engineer. Most of them seem to have been Dutch or Flemish, but the design shows some elements of Italian influence, too. The latter may have come directly from Tycho's sojourn in Italy, but it is more likely to have been due to an Italian stonecarver employed by the Alabaster Master. In at least one case, Tycho seems actually to have imported his artisan, albeit indirectly. He convinced Frederick II, or his architect, that the person to build an elaborate bronze fountain for the central courtyard of Kronberg Castle was George Laubenwolf, whom he had met in his travels through Nuremburg. Once Laubenwolf was in Denmark, Tycho commissioned him to provide running water in his manor house through "pipes reaching in all directions . . . to the various rooms, both in the upper and lower story," including a supply for a splendid fountain in the central hall.[15]

Tycho traveled busily between Skaane, Sjaelland, and the isle of Hven during the spring and summer of 1576, crossing the sound in small boats and riding with his entourage across the country. He was in Skaane, early in the month of June, when a shattering message caught up with him. Pratensis had been lecturing at the University of Copenhagen on 1 June 1576, when he collapsed, coughing up blood. His life had ebbed out as he lay there in the arms of his students. His last words were reported to Tycho: "Jesus, son of the living God, have mercy upon me and receive my spirit."[16] Tycho was deeply grieved to learn of his best friend's death. Nobody had been more witty and warmhearted, and Pratensis's fertile mind had been a tremendous stimulant to Tycho's own intellectual development. Now he was gone, buried in the church of Our Lady in Copenhagen at the age of thirty-two. In his mourning, Tycho composed a Latin epitaph of sixty lines.[17] Sometime later, Peder Sørensen wrote

---

[15] See the allusions to Laubenwolf's departure from Nuremburg in Tycho's letter to Camerarius of 21 November 1576 (VII, 42) and Tycho's reference to his plumbing in V, 142.

[16] VII, 31–2.

[17] Preserved and published in IX, 176–7.

Tycho lamenting the death of the friend he had probably known (through their travels together) even better than Tycho did. Without mentioning his own poem, Tycho suggested that Pratensis deserved a gravestone and that if Peder would compose an epitaph for it, he would pay for its inscription.[18]

When Tycho began his excavations, Dançey had asked if he could have the honor of providing the cornerstone for Tycho's establishment. Pratensis had volunteered to compose a suitable inscription and plan the ceremony, but now, as Dançey wrote on 26 June, someone else would have to take over, and he had drafted a text for Tycho's consideration.[19] When he wrote again on 14 July, he had accepted Tycho's suggested changes and was having the inscription carved. He wanted only to recommend that Professor Johannes Franciscus be added to the guest list, as he taught astronomy at the university, and Dançey foresaw misfortune if he were not invited.[20]

Concerns for astrological implications did not end there. When the foundation was laid and progress on the basement walls allowed Tycho to think about a date for laying the cornerstone, Tycho consulted the planetary positions to ascertain when all the aspects would be propitious. And when the date of the ceremony had to be postponed a week, he undoubtedly made sure that the signs would not deteriorate in the meantime.

Finally, a day or so before the eighth of August, a party of distinguished high government officials, noble relatives of Tycho, and professors from the university began to gather on Hven.[21] The lord of the island arranged for them to enjoy the hospitality of the village. At sunrise on the eighth, all gathered to watch Dançey solemnly cement into place a stone inscribed to proclaim to the world the dedication of the building to the contemplation of philosophy, especially astronomy. If not by this time, within the next year, Tycho named his temple of the muses, Uraniborg.[22]

Although Tycho had no way of knowing that it would be over four years before his house was ready for occupancy, he must have learned very early that such projects always took a good deal of time. And because it is inconceivable that he would not have found it necessary to supervise the building and landscaping on a virtually daily basis, establishing temporary quarters on the island must have been an early priority for him. It is most unlikely that anything already existing there was suitable, but he may have commandeered the parsonage or the best farmhouse on the island. The oldest

---

[18] VII, 38–9.   [19] VII, 34.   [20] VII, 35.
[21] V, 143. See also *Norlind*, 73–4.
[22] See the name mentioned by Tycho's uncle Steen Bille in VII, 45.

manuscript map of Hven, drawn some ten years later, shows a large farm at the western end of the village, with three wings of buildings around a courtyard that was walled on the fourth side. Tycho might have lived in such a place, or he may even have built something for his personal use in the short term, converting it for an alternative purpose in later years (such as C, in Figure 4.2, which later became quarters for his servants). Until September he probably commuted from Helsingborg, for until then his family was living there.

At that time, however, an epidemic swept through Skaane and carried away Tycho's eldest daughter, Kirsten. After burying her in the church at Helsingborg, Tycho commissioned a bronze plaque (which is still in the wall of the church there) with a Latin epigraph for her gravestone, noting that she was his *filiola naturalis* and that she had lived only two years, eleven months, eleven days, and eleven hours.[23] Kirsten was pregnant again, and Tycho decided he should move her and little Magdalene to a remote area up the coast near Väsby, where they would be less vulnerable to whatever had struck Kirsten. Tycho himself probably went south to Landskrona and commuted across to Hven, unless he already had a place on Hven that would accommodate him but not the rest of the family.[24] By the end of the year, he clearly did, and as the pace of construction slackened with the onset of cold weather, he began to have time for a bit of observing. On 14 December 1576 he noted that he was making his first observation of the sun on Hven and doing it on his (thirtieth) birthday.[25]

From brief observing sessions on 24 and 25 December, it seems that Tycho spent the holidays on Hven, rather than wherever the Brahe clan may have gathered in the vicinity. Such observations were to remain the rule, although it is possible that in later years someone else was making them. If Kirsten was on the island with him, she was back in Väsby by 2 January 1577, when she gave birth to a son.

Convention dictated that this firstborn[26] son be named Otte: The

---

[23] IX, 174–5.

[24] Tycho's letters to Severinus (3 September) and Camerarius (21 November) were sent respectively from Knudstrup and "Our Island of Hven."

[25] X, 42.

[26] The modern genealogy of the Brahe family (*DAA* V, 97–115) credits Tycho with having sired a stillborn son, named Otte, before his first daughter was born. No Tycho scholars, however, know the authority for that assertion, and at least one has speculated that it must be a mistake. Even if it were not, Tycho could have named his "second" son Otte. If it were not a mistake, it would be another piece of evidence for Tycho's association with Kirsten as early as 1571. It is interesting that the private collection of Tychoniana accumulated by the Counts Wachtmeister, who have owned Knudstrup since 1771, includes a neat, handwritten genealogy of Tycho that features a firstborn son named Otte.

three of Tycho's four brothers who had children were all to honor
their father in this way. Tycho, however, was apparently unable to
do so. Even though the baby lived for only six days, it was as Claus –
after Beate Bille's father – that he was buried, when, for the second
time in four months, Tycho had the melancholy task of making
funeral arrangements for a child. Baby Claus was interred at Väsby
church, under a gravestone that likewise proclaimed him to be the
"natural son" of a father, Tycho Brahe, who had never seen him.[27]

During the rest of the winter, Tycho made systematic observations for the first time in his life. He did not spend a lot of time at
them, but he did get out for a few minutes, either at noon for a
meridian attitude of the sun or in the evening for a couple of
positionings of a planet, on ten to fifteen occasions a month.[28]
There was correspondence to catch up on, too. Tycho had already
answered a letter from Severinus, who was tired of the ceaseless
peregrinations of the court and the many daily burdens placed on
him as the royal physician, and expressed a longing for freedom to
pursue his scholarly activities in peace.

Tycho could certainly sympathize with Severinus's complaints
about the interference of court life with scholarly activity, but he
warned that everyone had cares and troubles in this life, and that
imprudent action could make things worse. The court accepted one
and all with flattery and benevolence, but it showed displeasure at
letting people go. Tycho advised patience and intelligent behavior,
pointing out that time always brought unexpected changes and that
it would not do to run upon the rock (Scylla) in attempting to avoid
the whirlpool (Charybdis). Here Tycho was reflecting the familiar
view of the alchemists, that the path to wisdom is always difficult
and full of hindrances. He urged Severinus not to abandon his studies
and spend the rest of his life in work that would achieve him no
renown. Throughout this letter is a tone of intimacy and love of
knowledge reminiscent of Tycho's correspondence with Pratensis.
Severinus, like Pratensis, posed challenging questions to Tycho,
growing out of his own restless quest for understanding.

In an earlier letter, Severinus had asked Tycho to explain Copernicus's theory of a triple terrestrial motion. Tycho had duly explained
the theories of the diurnal rotation of the earth on its axis, the annual
revolution of the earth around the sun, and the third motion that
Copernicus had deemed necessary to maintain the axis of the earth

Unfortunately, the current generation of Wachtmeisters, who kindly gave me access to the
collection there and to the memorabilia at Kaagerød church, has no idea of the provenance
of the document.
[27] XV, 3.   [28] X, 44–50.

in the same orientation with respect to the sun and stars, thereby causing the earth's seasons. In so doing, Tycho had expressed boundless admiration for the skill and ingenuity of Copernicus as an artificer, saying that because the Copernican theories were so complex, he hoped someday to build a mechanism to clarify them by showing them in motion all at once.[29]

During the winter, Tycho wrote to all of his correspondents, telling them of the recent turn in his fortunes and, of course, answering previous inquiries. Johannes Major had written from Augsburg about a huge portable globe that Christopher Schissler had made for Marcus Fugger, an Augsburg millionaire with whom Tycho was apparently acquainted. The globe could be disassembled into eight parts. Instead of sending the old, cracked wooden globe to Denmark, Schissler had proposed to Major that he make a globe of this new type for Tycho, even larger than Fugger's. Having already been provided with one useless globe, however, Tycho was not about to contract for another, particularly for a thousand guilders. He therefore told Major that if he would just have the present globe transferred to Tycho's old rooming-house landlord, Master Laurentius the Dane, Tycho would arrange to have it shipped back to Denmark from there.[30]

Hayek had started a correspondence from Prague that was to extend over twenty years. He described some of his scholarly activities and enclosed a copy of a polemical work he had just published on the new star of 1572. Henrich Brucaeus had written from Rostock, asking Tycho to advance the career of one Master Lucas from Hamburg by finding him an academic position in Denmark. Brucaeus also took to task his good friend and former student, Tycho, for being too critical of academic astronomers in an earlier letter. He pointed out that Peurbach, Regiomontanus, Stoeffler, Reinhold, and Rheticus all had been professors and that even Copernicus had lectured in Rome and assisted a professor in Bologna. Much of the rest of the letter dealt with alchemical problems of interest to both correspondents.[31]

Tycho's international ties and high favor at court almost immediately combined to draw him into political activity. One of the people to whom he had sent a copy of his book on the new star was the famed Scottish humanist, George Buchanan, teacher of Montaigne and James VI of Scotland. The phrasing of the thank-you note he sent to Tycho suggests that they may have met at some time, probably when Buchanan was at the Danish court in 1571. Now

---

[29] VII, 39–40. Severinus had made a similar complaint to a German friend: *DBL* XXIII, 308.
[30] VII, 46, 56.   [31] VII, 43–4.

Buchanan wrote to Dancey asking him to appeal to Tycho to aid the case of a Scots sea captain with King Frederick II.[32] Soon there were other requests. In the consistory of the University of Copenhagen on 18 May 1577, Professor Niels Hemmingsen nominated Tycho Brahe to the position of rector magnificus. Although the office was normally held by a professor, elected annually by his colleagues in rotation among the four faculties of the university, Tycho was elected. The rector was the actual administrative and ceremonial leader of the academic community, ultimately responsible to the chancellor of the university, who was the royal chancellor, Niels Kaas. Tycho's lectures on astronomy two years earlier had established his identity with the university, but they did not constitute a reason to make him rector. What actually motivated the call was the fact that Niels Hemmingsen was in doctrinal trouble again, and he hoped that the newest favorite of King Frederick, the young lord of Hven, would be able to help him.

Ever since Jørgen Dybvad's return from Saxony in 1575, Hemmingsen had been in disfavor.[33] Most of the Danish theologians had stood by him, and Anders Sørensen Vedel had translated one of his vernacular works into Latin to demonstrate abroad that it was harmless to Lutheran orthodoxy. But the pressure from Saxony was relentless. Frederick yielded somewhat: first by giving the elector's spokesman, Dybvad, an extraordinary professorship of theology at the university and then, in April 1576, by forcing Hemmingsen to sign a formal retraction of all his offensive views and forbidding all further debate on the subject of the Eucharist. In the ensuing year he had pushed the matter no further.

Although the enthusiastic elector twice requested that Frederick have Danish theologians "debate"(and, of course, adopt) a special confessional statement just drawn up by a conference of Gnesio-Lutheran theologians, the king refused to disrupt the "consensus" that Hemmingsen's retraction had wrought in Denmark. His inaction was not well received in Saxony, and the Gnesio-Lutherans of Germany continued their attacks on Hemmingsen. With Dybvad entrenched in the University of Copenhagen and the possibility that Elector Augustus would return to Denmark for the baptism of the prince who had just been born, Hemmingsen felt himself under intense pressure in the spring of 1577.

Thus by nominating Tycho Brahe as rector of the university,

---

[32] VII, 21–2, 40–1. The captain apparently owed some money to Dançey.
[33] The following discussion is from a much more detailed one in John Christianson, "Tycho Brahe's German Treatise on the Comet of 1577: A Study in Science and politics," *Isis* 70 (1979).

Hemmingsen hoped to acquire a powerful advocate at court, one who, with the particular authority of a nobleman, could bridle the pernicious energy of Dybvad within the university. Although Tycho was doubtless enormously gratified by the unanimous election and would probably have been willing to try to help if he had thought Hemmingsen was in any real danger,[34] the situation represented all the worst aspects of the life-style that Tycho had so long, so earnestly, and at such great personal cost sought to avoid.

Whether or not the risk of offending Councillor Jørgen Rosen-krantz, one of Hemmingsen's critics, also figured into the decision, it was fortunate that Tycho did not put himself in the position of having to cross words with a man who would be so influential in the future of Hven. On 21 May, Tycho wrote a short letter to the university faculty thanking them graciously for an honor that he must regretfully decline because of the press of affairs associated with "making something habitable out of what has until now been wilderness."[35]

But not all such situations could be avoided. On 12 April 1577, under the care of Inger Oxe and three midwives, Queen Sophie gave birth to a son, the prospective and long-awaited heir to the throne. Early in May the nobles of the realm received an invitation to come to Copenhagen on 1 June to attend the baptism of the prince the next day. Tycho's mother, Uncle Steen, and brother Steen were on a more select list of nobles who received invitations to be godparents. But whichever invitation one got, it was more of a command than an option, and so Tycho's spring program of observations came to an end on 25 May.

After going to Knudstrup to get his court attire, Tycho set off for Copenhagen with a full complement of servants. Kirsten probably stayed wherever home was at the time, for there is no reason to believe that she could ever have showed her face at court. Tycho's mother and Steen Bille's wife were already in Copenhagen as part of the small group of noblewomen in charge of decorating the castle. Steen Bille and Steen Brahe were also there with all of their squires and their best horses, ready to escort the young prince into the city. Perhaps because of the current theological tensions with Denmark, the elector and electress of Saxony did not attend in person, though other princely relatives did, as well as the cream of the aristocracy of Denmark, Norway, and Schleswig-Holstein.

---

[34] In fact, after an unauthorized edition of his *Syntagma* was republished at Geneva, Hemmingsen was removed from his professorship in mid-1579 (*DBL* X, 60). He lived an honored and comfortable life in retirement until he died at the age of eighty-seven in 1600.

[35] VII, 45–6. For the university's nomination, see XIV, 6.

The baptism was celebrated in high style on Trinity Sunday in Our Lady's Church with solemn Lutheran ceremony, followed by festivities that lasted for more than two weeks in Copenhagen and eventually at Frederiksborg and even in Roskilde.[36] Tycho probably did not have to stay for the entire series of events, because he now had his first assignment from the king: the proud parents naturally wanted to know what the stars had to say about the outlook for their heir and kingdom.

One can imagine Tycho's being reasonably happy about the opportunity to reciprocate the patronage so lavishly bestowed on him. Whatever the spirit with which he entered into the venture, it is clear that in carrying out the task Tycho summoned every resource at his command.[37] He began by computing the positions of each planet at the time of the prince's birth, according to both the Prutenic (Copernican) and Alphonsine (Ptolemaic) tables. To press home the fact that the king was getting more than just another horoscope, he then used the observations made during the winter (probably in expectation of the task at hand) to emend the places of the sun and the three "middle" planets. Accordingly, only for Saturn (which had been above the sun), Mercury, and the moon (which moves too swiftly to be interpolated with any confidence) did he have to use the tabular (Prutenic) values.

From these data Tycho generated a *figura natalis* (Figure 4.3) of the peculiar round (rather than square) form for which he had already shown a preference in his astrological analysis of the new star. Further computation produced the many directions of the various planets, and the ascendant (the point of the ecliptic on the east horizon at the time of birth).[38] After these twenty-seven pages of more or less objective work, Tycho had to do the remaining forty-four pages of "judgments" on his own. There were principles for predicting such things as a serious illness in the twelfth year or the unlikelihood that the subject would survive his fifty-sixth and fifty-seventh years, and Tycho naturally provided the astrological reasoning behind such dire results. But there was enough ambiguity in both the system and its application to leave plenty of outlet for the practitioner's creativity and Tycho was able to predict that the prince would be well formed, charitable, righteous, nimble of body for the manly exercises of hunting and warfare, and quick of mind for wide

---

[36] See the contemporary records of events in MS. Rigsarkivet, Copenhagen: Sjaellandske tegnelser, nr. XIII: 1575–7, 267b, and in KB 1576–9, 167–8, 183–4.

[37] The horoscope is published in I, 183–208.

[38] The natal figure is reproduced on I, 90. For a more detailed discussion of Tycho's procedure and conclusions see *Dreyer*, 146–8.

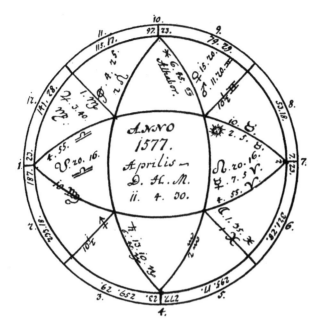

Figure 4.3. Tycho's horoscope for Prince Christian.

cultural and intellectual interests. On the negative side, the prince would be somewhat too fond of sensual pleasure,[39] would be prone to danger in matters of religion, and would have to overcome adversity in order to achieve honor and riches. (In short, he would be what every father wanted – a son he could regard as a chip off the old block.)

If it were accurate, this sixty-year horoscope would be a tremendously important document for a prince to possess, because it would allow him to anticipate all of the great crises, personal and political, during his lifetime. And Tycho pointed out that he had done his best to make it as accurate as possibe, even though an error of as little as four minutes in the royal clock used to establish the time of birth would vitiate the whole enterprise. But at the beginning of the text, in phrases that merged the view he had expressed in his oration of 1574 with the comments of Niels Hemmingsen after the oration, he preempted any such complaints with a warning that also held out the prospect of both divine and human intervention to avoid the adverse influences of the stars:

---

[39] If Christian's eighteen illegitimate children bore out this prediction, they did not square so well with Tycho's prediction that Christian would have few children.

It must be remembered that everything does not have to be as described here, because the Lord God rules and wields power over his creatures according to his own pleasure, and human beings have a free will as a gift of God, according to which they can break or turn away the signs of nature, if they choose to live otherwise in the true light of reason.[40]

The final stages of the task were completed in the last week of June. Tycho pulled all his thoughts together, wrote out the results of his deliberations in High German (still the only language with which Queen Sophie was really comfortable) and dated the document 1 July 1577. On the next day, when the king was scheduled to inspect construction at Kronborg Castle, Tycho probably sailed over to Helsingør and presented his handiwork in person.

During the second building season, Uraniborg began to take form. Every day except Sunday, the peasants toiled at sculpturing an estate out of the bare plateau. The foundation hole from which the brick walls of the house were now rising had been only the beginning of the digging. The next project had been the excavation and flooding of the sixty fish ponds of which Tycho later was to boast. And it was not merely digging. The dirt was being hauled to form a great square wall, nearly five meters thick at its base, which would have a perimeter of three hundred meters and would reach who-knew-what height. The area inside had been leveled, and some of the landscaping was doubtless being done, for Tycho would have known how long it takes for trees and shrubs to achieve their effects. If there was clay on the island, the peasants were digging and hauling it, too, to make brick for the house and outbuildings.

Whether it was available on the island or the brick was being made on the mainland, the firing required a considerable quantity of wood. Because there was essentially no wood on Hven, woodcutting forays were made to the mainland – not just for firewood but for construction lumber as well. One could not cut wood just anywhere. As an important natural resource it was all owned by someone who guarded his cutting rights jealously. When Tycho had taken Kirsten to Väsby to escape the plague, he had noticed that there was plenty of wood there in the great forests surrounding the royal manor of Kullagaard and that it would be relatively easy to get it down from the heights, onto boats, and over to Hven. In some way Tycho managed to present his requirements to the crown, and on 28 August

---

[40] MS. Royal Library, Copenhagen: GKS 1821, 4°, "Tycho Brahe: Christian IV's horoscop," sign. 74. The Latin portion of the horoscope is in I, 183–208, with a similar rejoinder on 208, cf. 183–4.

1577, the area was granted to him in fee for the duration of the royal pleasure.[41]

In the midst of his activities, Tycho found time for some astronomical ventures. In the spring of 1577 he obtained his first serious clock, that is, one that displayed not only hours and minutes, but also seconds. It may have been ordered from the Copenhagen clockmaker, Steffen Brenner, who is known to have worked for both the Danish crown and the Bille family and indeed had already produced a globe for Tycho in 1573. But whoever made it, it was clearly inadequate, for in the next four years Tycho acquired three more clocks.

A parallel effort to produce his first armillary brought similar grief to Tycho. Somewhere, very likely from Brenner, Tycho commissioned the construction and engraving of copper rings three to four feet in diameter, made so they could nest in one another in a certain sequence. He then sent the rings to Steen Bille, with the request that one of the smiths at Herrevad make an iron mounting for them according to Tycho's written instructions. Steen conveyed the work, but that was as far as it got. Whether the smith could not understand Tycho's instructions, as Steen seemed to think, or whether either Tycho's instructions or the smith's skills were inadequate, Tycho did not get his armillary at that time. The rings were probably made into a usable instrument eventually, but not before November 1581, at the earliest.[42]

A second quadrant, probably commissioned in Copenhagen, was more successful. Tycho observed lunar eclipses with it in April and September and continued to use it for several years before other, larger instruments rendered it obsolete.[43] There was not a great deal of time for other observing, at least until after the evening of 13 November 1577, when Tycho experienced the second of the two great astronomical events of his lifetime.

It was dusk on a Wednesday afternoon. Tycho was out by one of his ponds catching fish for dinner when he noticed a bright star in the western sky. Because it was too bright to be Saturn, the only planet in the evening sky at the time, Tycho watched it as the dusk turned to darkness. As the star grew a long ruddy tail, it became clear that the comet Tycho had wanted to see ever since the appearance of the new star, five years earlier, had arrived.[44]

Not surprisingly, Tycho's first thoughts seem to have focused on

---

[41] XIV, 6.
[42] VII, 45. On the completion of the Ptolemaic armillary, see Chapter 5.
[43] For more on this *Quadrans mediocralis*, see Chapter 5.   [44] IV, 5–6.

the physical characteristics of the spectacular apparition. Only after noting that its head was a bluish white resembling the color of Saturn, 7' or 8' in diameter, and its tail slightly reddish, like a flame penetrating smoke, did Tycho break off from his drawing to record distances from a couple of prominent stars so that he could determine its position.[45] Presumably the question of determining whether the comet displayed any parallax occurred to Tycho fairly early.

Unfortunately, however, the task was nowhere near so easy for the comet as it had been for the new star. First, the comet was located near the sun and therefore was visible only for an hour or so right after sunset. In such a short period of time, the diurnal rotation provided a very small baseline for any perspective on the comet. Even more problematical was that comets were known to be moving bodies. However anomalous the visitation of the new star had otherwise been, its resemblance to the "fixed" stars conveyed an a priori suggestion that any movement it displayed would be a manifestation of parallax. For the comet, any identification of a parallactic motion depended on a preassessment of its intrinsic motion.

With the necessity of waiting for a longer evening's visibility, suffering through cloudy weather, and arriving at an estimate of the comet's own motion, it was ten days before Tycho made his first attempt at parallax. Observations made three hours and five minutes apart showed that the comet had moved twelve minutes. During the preceding ten days, however, the comet had moved an average of about three degrees daily, which worked out to seven and a half minutes per hour, or about twenty-three minutes of proper motion during the given interval. It looked, therefore, as if there were about eleven minutes of motion missing, which could be accounted for only by parallax.[46]

It would be interesting to know how far Tycho pursued these eleven minutes. At first sight, of course, they seem sufficiently small in comparison with the sixty-odd minutes produced by the moon to promise an exciting supralunary distance. But because they were obtained from a relatively small period of rotation, and that one near the horizon, where the cosine (of the altitude) function that describes the parallax changes very slowly, they extrapolate to a horizontal parallax that would be much larger. Later, Tycho was to demonstrate in his formal write-up that the moon would show almost the same parallactic motion between the two altitudes in question.[47] If Tycho worked it out at the time, therefore, he would have obtained the least satisfying answer possible: too close to the moon to tell whether it was above or below.

[45] XIII, 289.   [46] XIII, 291–2.   [47] IV, 95–101.

In fact, however, Tycho soon disposed of the entire eleven minutes, by reconsidering the comet's daily motion and finding that it was closer to 2 degrees than 3. This meant that the twelve-minute observed change of position was sufficiently close to the pro-rata proper motion ($\frac{3}{24}$ of 2 degrees) to leave virtually nothing to attribute to parallax.[48] Rechecks at the end of the month, when the daily motion had decreased even further, provided a completely satisfactory account of the comet's positions. Similar checks at the end of December provided final confirmation that the comet showed essentially no parallax.[49]

Already by Christmas the comet's initially glorious tail had become exceedingly tenuous. With the next waxing of the moon, the comet itself was drowned out.[50] Tycho saw it one last time on 26 January when he took Gemperlin out to show him where the comet had been when it had last been visible two weeks previously.[51] By this time, he had essentially formulated his conclusions regarding the comet and was beginning to write them up.

The result, which dwarfed all the others in significance, was that the comet was indisputably above the moon. How far above, Tycho was not prepared to specify. But he was certain that the horizontal parallax could not have exceeded fifteen minutes, so that the comet had to be at least 230 earth radii (e.r.) away.[52] Because the lower bound of the moon's sphere (and hence the upper bound of terrestrial bodies) was 52 e.r., there was no doubt that the comet was a celestial body, contrary to the teaching of Aristotle.

This finding and a number of lesser ones combined to lead Tycho to a conclusion that was almost as important as the comet's distance. First, the comet had, throughout its brief existence, moved in the direction in which the planets normally move, that is, in the direction opposite that followed nightly by the rotation of the vault of stars. This movement had for the first week carried the comet out in front of the sun very swiftly, but thereafter the comet's elongation had crept slowly (for two weeks) up to 59°55' and then had begun to diminish again.[53] During all this time the comet had progressively faded, suggesting that it was moving away from the earth. These data virtually cried out for an orbit circling the sun, and that is how Tycho accounted for them, in some diagrams (Figure 4.4) entered at the back of the log containing his observations of the comet. It was a move destined to have significant consequences.

There can be little doubt that Tycho was greatly excited by the

---

[48] See several remarks to this effect attached to Tycho's original observations in XIII, 292–3.
[49] XIII, 294–5, 300.   [50] XIII, 298–303.   [51] IV, 235.   [52] IV, 387.
[53] See IV, 177–9 for an idealized daily tabulation of the comet's elongations.

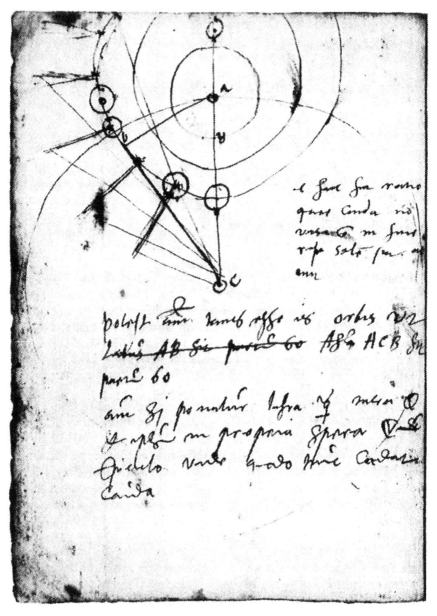

Figure 4.4. Tycho's working hypothesis of the (retrograde) orbit of the comet of 1577 around the sun (a), and also around the orbits of Mercury and Venus.

comet's corroboration of the anti-Aristotelian implications of the new star. For if the putative supralunary status would not have the novelty value that his claim for the new star had had, Tycho probably expected it to have, ipso facto, a greater persuasive impact – all the more so as he had implicitly predicted in his tract on the new star that comets would prove to be supralunary.[54]

Almost immediately, therefore, Tycho made plans for a book to publicize his findings. From his experiences with the new star, however, he knew how close minded and muddle headed most of the rest of the commentators on the comet would be and how difficult it would be to prevail against them among a public who tended to share these intellectual handicaps. So he seems from the beginning to have conceived a book considerably more ambitious than his ineffective work on the new star had been,[55] one that would stand out from its competitors in rigor, in detail, and in sheer weight, if nothing else. While he was still observing the comet, therefore, he began observing the sun (at winter solstice) to determine his own value for the obliquity of the ecliptic, for computations involving the comet.[56] And before the comet disappeared, he began a notebook of star observations so that he could determine his own coordinates for the various reference stars, instead of accepting the received catalogue values.[57]

Before Tycho could begin the exhaustive analysis of his results, which was eventually to be published as *De mundi aetherei recentioribus phaenomenis*, he had to tend to the modest private obligation associated with his lavish patronage – a report on the comet to the king. The day after its appearance he had written to Peder Sørensen, asking him to arrange affairs with their patron so that Tycho could observe without being disturbed.[58] Ironically, Frederick had seen the comet two days before his highly paid expert did, for someone at Sorø Abbey, where he was holding court at the time, had noticed it.[59]

The abbot of Sorø was Ivar Bertelsen, a former professor from Tycho's days at the University of Copenhagen, who had published a pamphlet using the solar eclipse of 1560 to predict the apocalypse, in phrases of poetic fury. After a checkered career that had included three years in rags as a prisoner in the very abbey over which he now presided, Master Ivar was so great a royal favorite that he was

---

[54] I, 23.
[55] C. Doris Hellman, "Was Tycho Brahe As Influential As He Thought?" *British Journal for the History of Science* 1 (1963): 295–324.
[56] X, 52–3, 55, 59.   [57] X, 69.   [58] VII, 47.
[59] Christianson, "Tycho Brahe's German Treatise," p. 119.

married to a noblewoman (who would one day be one of Tycho's close acquaintances).[60]

There can be little doubt that not only the king but probably everyone else at that St. Martin's Day feast (11 November) heard the abbot's views on the comet. Among the king's retinue, besides Sørensen, was that restless, irrepressible extraordinary professor, Jørgen Dybvad. Within five weeks he published a pamphlet on the comet,[61] through the same Copenhagen printer who had done Tycho's pamphlet on the new star and most of Niels Hemmingsen's theological tracts.

Dybvad's was not an insignificant publication. Dybvad was a good scholar, whose writings on astronomy, meteorology, and mathematics reveal a tendency toward the new, anti-Aristotelian patterns of thought. He had been sufficiently able and energetic in his youth to have won the patronage of Peder Oxe and also the royal travel stipend previously enjoyed by Petrus Severinus and Johannes Pratensis. He had been the first Dane to publish a commentary on Copernicus[62] and had been willing to accept the new star as evidence of celestial mutability. But he was not an observer, and there is no evidence that he was even interested in the kinds of problems that intrigued Tycho. Rather, what concerned Dybvad were the astrological/theological/political implications of "this gruesome comet."

As Dybvad argued at some length, from the experience of nearly fifty previous comets over the course of two thousand years, such apparitions were always followed by great changes in weather and politics. According to the present comet, he thought, Denmark would have a cold, snow-laden winter, followed by a hot, dry summer, with tempests, crop failure, and "gruesome treachery in affairs of religion."[63] Other realms would experience similar trials from pestilence, war, or lesser effects of the comet, but they were not really Denmark's concern. In fact, it is not clear why Dybvad should have been worried about the comet's specific effects even on Denmark, for his general thesis was that this "terrible great comet" was only one of many signs that "the day of the Lord ... is at hand." Against this prospect, the only effective action was "that we seriously turn to the LORD." That, indeed, was what Dybvad did, conclud-

[60] *DBL* II, 548–9. On Ivar's wife, see Chapters 11 and 12.

[61] Jørgen Christoffersen Dybvad, *En nyttig Vnderuissning om den COMET, som dette Aar 1577. in Nouembrj først haffuer ladet sig* (Copenhagen: Laurentz Benedicht, 1578).

[62] Kristian Peder Moesgaard, "How Copernicanism Took Root in Denmark and Norway," in *Dobrzycki*, pp. 117–19.

[63] Christianson, "Tycho Brahe's German Treatise," p. 120.

ing his treatise with a prayer for mercy and divine protection over Denmark and the Danish royal family.

By the time the comet disappeared, Dybvad's pamphlet was out, and Tycho undoubtedly had a copy. Unlike Dybvad's previous fulminations, this one could not be ignored. Dybvad was from a different social stratum than Tycho was, but his ambition was unbounded and his political instincts were strong. In the past few years he had dedicated many writings to influential courtiers and learned noblemen, had married the daughter of the mayor of Vejle, and had ingratiated himself at court with his reports from Saxony. These exertions did not go unrewarded. Early in 1578 a rather inactive professor in Copenhagen was exiled to the chapter of Lund so that Dybvad could be appointed to his chair.[64]

Tycho did not have the stake in this threatening precedent that the outraged incumbent professors did, but he could scarcely view with equanimity the fact that Dybvad was granted a monopoly on the publication of almanacs as a perquisite of his chair. Here, clearly, was a man who had won the ear of the king. This fact alone, considering the advantages of birth, wealth, social status, and kinship within the ruling oligarchy enjoyed by Tycho, made him a rival. Tycho could not acknowledge Dybvad specifically, but neither could he allow him to challenge his authority.

Nor was it merely personal considerations that were at stake. During the previous few months the opponents of the Gnesio-Lutheran activists had banded together to mount a counterattack. Denmark stood with a foot in each Protestant camp while the two camps drifted steadily apart. Soon it would be necessary to jump in one direction or the other, which would entail choices in foreign policy as well as in religion. Which way the king would jump, how the Church would respond, and how the university would be affected all might be influenced by the very report that offered Tycho the opportunity to put Dybvad in his place.

Sometime in the late winter of 1577–8, therefore, Tycho wrote his own document for the king. Only in the present century was this manuscript discovered, and only very recently was it correctly identified as Tycho's confidential analysis for Frederick.[65]

Tycho's report to the king generally paralleled his pamphlet on the

---

[64] Dybvad eventually went too far. After thirty years of inharmonious relations at the university, he issued a general criticism of religious procedures in Denmark that caused his removal in 1607 in favor of one of Tycho's former students, Cort Axelsen.

[65] Christianson, "Tycho Brahe's German Treatise," pp. 127–8. The original version in Tycho's sixteenth-century German is published in IV, 381–96. Citations in the following discussion are to Christianson's English translation.

new star. The first half dealt with technical and descriptive aspects of the comet, all of which Tycho said he would treat in considerably more detail in a formal Latin publication on the subject. If most of it was superfluous from the king's standpoint, Tycho was not averse to showing where his own priorities lay and probably hoped that the aura of expertise established by the objective material would lend authority to his astrological interpretations.

Tycho began these interpretations by confronting those astrologers like Dybvad who compared the significance of comets with the significance of regular, predictable celestial events, because comets, as "new and supernatural" creations of God, actually worked in opposition to the "natural courses of the heavens." Thus, neither the recent comet nor any other celestial sign could presage the apocalypse. And those "pseudoprophets who have thought so [like Bertelsen and Dybvad], have mounted too high in their arrogance, and have not walked in divine wisdom [and] will be punished." On the other hand, this did not mean that there was nothing to fear from the comet. Historical experience "too long to recite here" has taught that comets bring climatic catastrophe of various kinds "from which usually follows great scarcity . . . many fiery illnesses and pestilence and also poisonings of the air by which many people lose their lives quickly." They also signify "great disunity among reigning potentates, from which follows violent warfare and bloodshed and sometimes the demise of certain mighty chieftains and secular rulers." And regrettably, this comet bodes worse than usual, for both its position in the sky and its physical characteristics augur "an exceptionally great mortality among mankind."

Tycho's expectations from the comet were as fearsome in their way as Dybvad's were, but his response to the expectations was totally different. Unlike Dybvad, whose instincts as a member of the politically impotent middle class were to cower and pray in the face of great majesty, Tycho reacted as a man accustomed to regarding practically any problem as soluble in human terms. And whereas many of his contemporaries looked for magical solutions to political or cosmic problems, Tycho's response was that of a politically mature aristocrat tempered by the cool rationality of Melanchthon. What he would emphasize, therefore, was what he had claimed both in his *Oration* at the university and in his horoscope of Prince Christian – that a rational exercise of free will could moderate or control the predictable effects of the comet and other cosmic events. Rather than taking refuge in anguished prayers for general deliverance from divine wrath, Tycho wished to treat those specific forebodings of the comet that might be mitigated by appropriate policies.

In regard to this view, there was little point in paralleling Dyb-vad's doomsday roll call of places and plagues. If pestilence were going to strike Lithuania, so be it. But if the comet would bring warfare and bloodshed to the Muscovites and Tartars in 1579–80, and even cause the downfall of tyrannical Ivan the Terrible by 1583, it would pay to be prepared to benefit from whatever happened there. Closer to home, if the comet would have special "significance over the Spanish lands and their reigning lords" and "others of the Spanish stem" (Rudolph II, of Hapsburg, the Holy Roman em-peror), it would behoove one to beware Spanish treachery in the Netherlands (where at least three of Tycho's four brothers had fought with the Dutch rebels) and to anticipate "great disunity to arise in Germany" (where Frederick had some of his best allies and was himself the duke of Holstein).

These predictions, however much they may have fallen short of the precision Tycho hoped for in his own studies, nevertheless translated fairly directly into political action. Obviously, the political action implied in them reflected both Tycho's own biases and those he knew to be held by the king.

Among Tycho's biases, the most pronounced and most interesting is the contrast with the apocalyptic violence of Dybvad, specifically, and Renaissance thought, generally. Consistent with his aversion to the implicit struggles of court life, but not so consistent with some other aspects of his behavior, Tycho displayed a Philippist longing for peace and harmony: The revolt in Russia, if it should come, would be "well-deserved punishment [for] inhumane tyranny." Those "who were associated with [warfare] will be assailed, and those who are always on the prowl will cause great injury to others, but must also expect to receive like measure upon themselves." "Those who deal with [politics] will also be much stifled, and their honor, dignity, and goods will suffer great diminution in worth. . . . Monks, priests, and others of the Popish religion . . . might expect truly to be repaid . . . for the ruthlessness, murder, and pain they have inflicted." And although the comet as well as the new star and a great conjunction of the superior planets augur "great alterations . . . both of spiritual and secular regimes," Tycho was prepared to believe that these "may even bode more for the better of Christen-dom than for the worse," and that "it might [even] be presumed that the eternal Sabbath of all Creation is at hand."

Whether Tycho might have been referring to the Second Coming by this strange locution is difficult to ascertain: He doubtless elabo-rated on it and other conceptions in confidential consultation. But neither are the specific tenets of his political views particularly important. What is important is the aggregate view they offer of Tycho, not as the isolated, ivory-tower scientist that he wanted to be, but as a figure subject to the various currents of his intellectual milieu.

Tycho completed his German treatise on the comet during the spring of 1578. The Danish court came to Kronborg Castle from Copenhagen on 5 April, and Tycho was there at least part of the following week, and so he may have presented his report to Frederick and discussed it with him or his ministers at that time. It did not take a trained astronomer to perceive that Tycho's treatise was technically far superior to Dybvad's. Presumably the clarity and thoroughness of the astrological and political analysis commended themselves to the king, and perhaps they even influenced Danish foreign policy. But it was too late to affect Dybvad, at least at the time.

The king had just learned that the professors at the University of Copenhagen had still not dismissed Andreas Coagius nor allowed Dybvad to assume the chair of mathematics. This led Frederick to send an official admonition to the professors on his way out of Copenhagen on 5 April. But even though the king accused them of being more concerned with the welfare of their colleagues than with the quality of instruction, he tactfully avoided the issues of personal and religious enmity toward Dybvad and resentment of royal intrusion into academic affairs. Instead, he assured them that the crown had provided a good income for their evicted colleague as the canon of Lund Cathedral and accordingly commanded that Coagius retire immediately to Lund so that Dybvad could assume the professorship of mathematics. After all, consideration had to be given to "what reputation it would give to the school if foreigners came here and found this chair not properly occupied, besides what neglect the young people might suffer."[66] This would be especially embarrassing if the "foreigners" happened to come from the court of Saxony.

Whatever conversation Tycho had at Kronborg, it was not all on his prognostication, and it was by no means in vain. Over the winter, some of the peasants on Hven had actually abandoned their farms in order to escape the onerous and unceasing work details. On 10 April 1578, the king, obviously at Tycho's request, issued an open letter forbidding the peasants to leave the island.[67] As part of the case for his labor requirements on Hven, Tycho probably mentioned that it was very time-consuming to send his laborers back and forth to

[66] H. F. Rørdam, *Kjøbenhavns Universitets Historie fra 1537–1621* (Copenhagen, 1868–79), vol. 2, pp 182–3, 563–7.
[67] XIV, 7.

Kullen to cut wood. Five weeks later Tycho went to the king's palace at Fredericksborg to make the point again. On 18 May his complaints bore fruit: Frederick supplemented his fief of Kullagaard with rights to the rents and labor from eleven farms in the area. Nor was Frederick's largesse yet exhausted. He promised Tycho the incomes from the Chapel of the Magi in Roskilde Cathedral following the death of the incumbent. As it stood, it was really not much more than a renewal of the promise Tycho had received ten years earlier. But Tycho was now in line for one of the most lucrative canonries in the realm, and in the meantime, the king granted him the fief of Nordfjord, quit and free of all dues to the crown.[68]

Nordfjord was a large, mountainous district on the west coast of Norway, far from the island of Hven. Tycho would have to engage a bailiff to administer it, and he soon found a man named Christopher Peopler, who was apparently a Dutch refugee. The dues of the fief were collected in dried fish, hides and skins, butter, cheese, livestock, tar, and some coin. If it were well administered and the market were good for most of these products, it could give the lord as much as one thousand dalers per annum. Having negotiated a significant increase in his annual income, Tycho turned with new vigor to the tasks of construction on Hven during the summer of 1578.

During one of his visits to Helsingør, Tycho became acquainted with a young master mason named Hans van Steenwinkel. Born in Antwerp, Steenwinkel had grown up in the Frisian city of Emden, where his father was the master builder of the great town hall.[69] During the winter of 1577–81, he had come to work at Kronborg. Like Palladio, a stonecarver by trade, Steenwinkel began his career as a skilled craftsman with a wide range of interests but without the formal education of a humanist.

Tycho brought Steenwinkel to Hven, instructed him in astronomy and geometry, and put him to work building Uraniborg. Their talents were complementary. Tycho expained the complex symmetrical plan of the building and grounds. Steenwinkel was a clever draftsman, skilled at sketching designs in the Renaissance style for portals, windows, spires, domes, and other architectural elements. He quickly learned the more theoretical aspects of architecture as well. Before long, Tycho began to refer to him as "my architect."[70]

Steenwinkel was not Tycho's only intellectual companion on Hven. Toward the end of 1577, Tycho arranged to add a university

[68] XIV, 8–9.
[69] See the biographical sketch of Steenwinkel in *DBL* XXII, 484–8.
[70] X, 95, 122, 153, 156. XIII, 309.

graduate to his staff, with the specific assignment of assisting him in his work in astronomy, alchemy, and other learned disciplines. The individual he chose was Peter Jacobsen Flemløse, who had studied in Copenhagen and Leipzig and attended Tycho's lectures on astronomy in 1574. While Tycho had been abroad in 1575, Flemløse had published a Danish translation of Simon Musaeus's medical treatise on melancholy.[71] He was now twenty-three years old, clever, quick-witted, and versatile.

By January 1578, Tycho had taught Flemløse how to use both the cross staff and the sextant and had put him in charge of compiling the new catalogue of reference stars for the comet. Flemløse also liked to sketch. He decorated the cover of the star catalogue with a sketch of himself, symbolically clad in ancient Roman dress, standing under the open skies of Hven and observing the stars with Tycho's iron sextant. From then on, whimsical drawings decorated Tycho's observational journals: the moon with a frowning face during an eclipse, the constellation Pisces as a plump carp from Tycho's ponds, and Cygnus as a flying swan.[72]

Under the supervision of Tycho, Steenwinkel, and Tycho's bailiff on Hven, the work on Uraniborg progressed. A great stone entablature with architrave, frieze, and cornice now ran around the central square of the building, just below the second-story windows. Its height was such that it divided the facade into perfect squares on each side of the portal towers. In the center of each square was a window of many small panes set in leaden frames. The glass undoubtedly came from Herrevad. The sills were simple, but above each window was a complex, bracketed pediment formed of corresponding volutes. This design seems to have been adapted from Serlio's illustrations by Steenwinkel, the Alabaster Master, or some other skilled craftsman. Second-story windows of the same design were in line with those of the first story and rested on the entablature.

Each side of the building's central square was surmounted by a commanding, tripartite gable in the Venetian style.[73] At the corners were obelisks, resting on lupine gargoyles thirty Tychonic feet above the ground. Sweeping arcs with radii equaling the width of the portal tower rose to an elevated central section that supported a powerful, raised, and bracketed pediment. The pediment was broken in the new, Baroque manner of the elderly Michelangelo and from it rose the base of a cornice and finial. This was a dynamic and original architecture, breaking through the restrained conventions of the Renaissance.

[71] What little is known about Flemløse is summarized in *DBL* VII, 105–6.
[72] See X, 59, 67, 69–73.
[73] Jern, *Uraniborg*, pp. 6, 54–5, 75–7.

On the ridges of the roof, a balcony with balustrades was under construction, leading out to open cupolas in each of the four directions and surrounding the base of a great central dome. From the round towers on the northern and southern sides of the building, high pointed roofs would be erected to cover the second-story observatories. Little by little it approached the form depicted by Tycho's engraver in Figure 4.1a. Smaller buildings of the same design were built at the northern and southern corners of the walls. One of these buildings may have served as the temporary seat of the lord and his household. Portal houses in a Tuscan style were constructed at the eastern and western corners, and plans were being made to construct a grange south of the compound.

Although the summer months left little time for astronomy, Tycho must have enjoyed the intellectual, artistic, and administrative challenges of building. He threw himself into this labor with energy and creativity, linking the construction of Uraniborg and Kronborg Castle through the exchange of craftsmen, materials, and ideas. In this way, as well as through his horoscopes and prognostications, he linked his own fortune to that of King Frederick II. Like all of his forebears for untold generations and all of his kinsmen in his own day, he found his vocation in service to the Danish crown in the highest of offices. At the same time, he was building the first research institute in postclassical Europe.[74]

After supervising the construction activities through the summer, Tycho turned to the task of preparing his definitive analysis of the comet. As the initial step in what would be a completely unprecedented means of conducting astronomical discourse, Tycho gathered all of the comet observations and meteorological descriptions of observing conditions and listed them seriatim as Chapter I.[75] Because all of these observations were distances from fixed stars, Tycho's next task was to use those stars to determine the successive positions of the comet. This required (Chapter II) a redetermination, from his own observations, of the positions of the twelve reference stars involved.[76] Having done that, Tycho compared his positions with those given by Ptolemaic and Copernican theory, to show why he had not wanted to use catalogue values.[77]

[74] For a different view of Tycho's goal, see Owen Hannaway, "Laboratory Design and the Aim of Science: Andre as Libavius Versus Tycho Brahe," *Isis* 77 (1986): 584–610.

[75] IV, 11–20. The observations do not generally agree with the raw measurements found in XIII, 288–303, because of the corrections Tycho had to make for the optical parallaxes of his cross staff and sextant.

[76] IV, 21–32.

[77] The term *theory* is used advisedly, because both the Ptolemaic and Copernican positions rested on observations made in the second century B.C. and had to be precessed theoretically (and nonlinearly, in Copernicus's catalogue) up to Tycho's day. The results averaged a 16′ difference from Tycho's latitudes and half again that much from his longitudes: IV, 32.

Finally, when many years elapsed before these results were finally published, Tycho appended a note explaining that he had in the meantime reexamined his positions many times and now provided his mature results to show how good his original determinations had been.[78] After determining the longitudes and latitudes of his reference stars, Tycho could convert the comet's distances into daily longitudes and latitudes. This he did for all twenty-four days on which the conditions of observation warranted taking his results seriously.

Following a routine exercise in spherical trigonometry, converting the twenty-four pairs of ecliptic coordinates to equatorial coordinates (Chapter IV),[79] Tycho plotted the comet's trajectory across the sky (Chapter V). Using two positions of the comet, he determined the inclination of its path to the ecliptic (29°13′) and the longitude at which the comet would have crossed the ecliptic traveling in that path (260°52′). He then followed with no less than six similar computations using other pairs of data and resolved the scatter by generalizing the seven results to an inclination of 29°15′ and a nodal point of 260°55′.[80] After repeating the whole process to find elements with respect to the equator,[81] he concluded the chapter with a great ephemeris showing nightly positions of the comet (interpolated for nights that were cloudy at Hven) for the entirety of its duration.[82]

When Tycho eventually got his manuscript ready for printing, the five chapters of data processing thus far described would occupy ninety pages. Little in them is of any interest at all, per se. The observations were not particularly accurate (at least by Tycho's standards) because he did not yet have any of his mature instruments. Nor were his observations particularly numerous. Over the three months during which the comet was visible, Tycho had managed to find only some thirty nights during which the moon was not too bright or the clouds too thick to allow good sightings. And although all of these occasions had permitted at least two or three measure-

---

[78] IV, 33–7. Most of the differences were one or two minutes, but there were five differences in longitude of 6′.

[79] IV, 62–9.

[80] IV, 70–3. These coordinates are *geocentric* and very close to those obtained by several of Tycho's contemporaries (see note 61, Chapter 8). Unfortunately, Tycho conceived of the comet as *heliocentric*. Maestlin made the same mistake, as did Copernicus, in his treatment of the inferior planets. It points up the importance of Kepler's recognition of the problem.

[81] IV, 73–8. Tycho's generalization of seven trials was an inclination of 33°45′ and an ascending node of 299°50′.

[82] IV, 78–81. Tycho knew the comet had been visible before he saw it, not only from Dybvad's pamphlet, but also from another one written by an old acquaintance from his student days in Copenhagen, Jens Nilsson, who was now bishop of Oslo and Hamar in Norway. *Dreyer* (158) mentions that it was seen on 1 November in Peru and 2 November in London.

ments, none had inspired more than eight or ten. So, as was to be the case with his later observations of the planets, it was the persistence with which Tycho attended the comet, rather than the sheer number of repetitions, that made his efforts remarkable.

At least as important as his collection of data, however, was Tycho's willingness to display it to present more data to the reader than were strictly and minimally required to justify the mathematical conclusion. Nobody until Gauss, at the beginning of the nineteenth century, was able to use rigorously more than one, two, or three data at a time to determine, respectively, a point, a line, or a circle on the celestial sphere. But nobody before Tycho even catered to the ordinary rules of evidence, whereby several determinations of a result would be deemed to provide more credibility than one or (rarely) two did. The problem with such redundancy, no doubt, was that it required concomitantly modern attitudes toward empirical data.

The willingness to acknowledge error and the capacity to analyze its origins were the crux of Tycho's concern for and success in the construction of instruments, so it is not surprising to see these attributes manifested in his handling of data. Yet, even in this context, Tycho demonstrated a candor in Chapter III that can be described in historical terms only as unprecedented.

On the first night of his observations, Tycho found that the distances from his stars give the comet a position 18′ away from the one derived by measuring its distance from the moon. Although he could easily have excised the offending observation, Tycho, in what must be the earliest such display in the history of science, permitted the conflict to stand and tried to explain it away by taking refuge in refraction.[83] Only in 1587, when the printing was being completed, did he discover the source of his problem (the theoretical position he was using for the moon) and insert an explanatory "Annotation by the author derived from later observations of the moon."[84]

When he had completed his data reduction preliminaries, Tycho was ready to close in on his main objective – to prove that the comet "ran its course high above the sphere of the moon in the Aether itself," contrary to what the followers of Aristotle had maintained for so long with nothing but subtle arguments.[85] He would not, he announced, try to use the method of Regiomontanus to find parallax, because it assumed that the object in question has no proper motion, that it can be observed on the meridian, and that the times of observation can be determined perfectly. Because none of these three conditions obtained, he would have to use other means to establish

[83] IV, 41–4.   [84] IV, 42–3.   [85] IV, 83.

supralunarity. First, the analysis of its path: The constant $29\frac{1}{4}°$ inclination to the ecliptic argued to Tycho that the path of the comet was a great circle, like those of the sun and moon. And if its motion was not as regular as those of the sun, moon, and planets, neither was it totally irregular. There were no fits and starts, just a gradual loss of velocity. Moreover, even the initial velocity was far less than the moon's, which corroborated the other indications that it was a celestial body and also suggested that it was considerably farther away than the moon.[86] Finally, the comet ran from beginning to end a total course of exactly 90° in longitude and ranged in declination almost from one tropic to the other, associations that would be most peculiar in a terrestrial object.[87]

But for Tycho, the basic evidence had to be numbers. He drew them from the observations of 23 November, which showed the comet progressing twelve minutes during a three-hour period when its proper motion should have been fifteen minutes. To point out just how small the resulting three minutes of parallax was, Tycho performed the elaborate computations necessary to show that if the comet were at the highest point of the sublunary sphere, it would have shown a parallax over four times as large.[88] In fact, when everything was considered precisely, the comet was probably 300 e.r. away, nearly six times the minimum distance of the moon. And this was an extreme case; consideration of several other sets of observations showed no parallax at all.

This seems to have been as far as Tycho got during the winter of 1578–9. No doubt the project was affected somewhat by the birth of a second prince in December 1578 and the consequent necessity of working up a second horoscope.[89] But at the same time, Tycho was probably not in any hurry. Having already sampled the literature on the comet, Tycho knew he would have to do something special to make his views stand above the bewildering profusion of commentary that would be issued in nearly two hundred documents from the ubiquitous presses of Europe.[90]

What Tycho already envisioned as a means to that end was using some of the critical techniques of the Italian humanists. To do so, he had to know what his competitors were saying, and he had already begun to collect, through his friends abroad, every reasonable publication he

---

[86] IV, 84–93.
[87] IV, 93. Tycho was referring here to the longitudes and latitudes given in his table on IV, 78–81.
[88] IV, 94–104.
[89] The horoscope is printed in I, 209–50.
[90] For a census of writings on the comet, see C. Doris Hellman, *The Comet of 1577: Its place in the History of Astronomy* (New York: Columbia University Press, 1944).

could find on the comet. In December 1578 Brucaeus wrote saying that he had received a catalogue of books for sale at the Frankfurt book fair and that there were innumerable publications on the comet of 1577, including one by Master Michael Mästlin of Tübingen. In Brucaeus's opinion, however, the task of reading, and perhaps refuting, all of them seemed overwhelming.[91]

In March, 1579, Paul Hainzel sent Tycho a collection of works on the comet and some unpublished observations by such astronomers as Gemma, Roeslin, Hayek, and the *landgrave*, which had circulated through Augsburg. He could not send any from Tycho's *Quadrans maximus* because it was no longer in working order.[92] Hieronymous Wolf sent a letter with the same messenger and included excerpts from a letter he had rcently received from Mästlin. When Tycho replied to Wolf, he mentioned that he found Mästlin's report on the comet to be the most interesting.[93]

To be sure, Mästlin's observational methods were medieval compared with the modern methods that Tycho had developed, for Mästlin had observed the comet by holding up a piece of string to line up reference stars and then had found the positions of the reference stars by looking them up in Copernicus. But it was Mästlin's analysis of his results, rather than his methods of observation, that interested Tycho. If Mästlin had the means to acquire a good large metal instrument, Tycho wrote to Wolf, he could do much to determine and correct the ephemerides of the celestial bodies. Mästlin's views, as far as Tycho could tell from Hainzel's letter, were similar to the conclusions that Tycho himself had reached on the basis of many observations. Tycho asked Wolf to send him copies of ephemerides or anything else published by Mästlin, and then with the grace and charm that he could muster when it was needed, Tycho added,

> If you are well acquainted with him, send him my greetings as from a stranger, and bid him write to me when he has occasion, and send me, if he will, his observations, if he has any more exact; I for my part will do him the same service in reply, and I shall exert myself to promote him in all ways possible.

After asking Wolf to send him any new publications he could find on the comet, Tycho explained that his own treatise on the comet had been delayed by various tasks he had performed for the king and for friends, and especially by his work of building on and organizing the island of Hven. He hoped to work on his treatise again during the coming winter, God willing, and he also intended to correct errors

---

[91] VII, 48–9.   [92] VII, 49–50.   [93] VII, 51–3.

concerning other astronomical matters. For this work, he wrote to Wolf,

I have suitable instruments, partly now ready to be erected, partly under construction, which I believe to be inferior to none, either ancient or modern, in size, craftsmanship, accordance with their purpose, great cost and labor incurred in their construction, and in their extraordinary accuracy. I have planned a building that is to be as well suited as possible for these instruments and for the observation of the stars in comfort. For this reason, I have withdrawn to this isle in order to devote myself to philosophical and astronomical studies without disturbance.

Tycho concluded his letter to Wolf by asking one last favor. His good friend Vedel had recently lost his young wife of less than two years and was planning a trip abroad for a change of scene. He would also seek out lost and rare historical books and manuscripts for his own research and would doubtless be commissioned to buy books for Tycho, too. Tycho now asked Wolf, who was familiar with the libraries of the Fuggers and other Augsburg patricians, to receive Vedel well (and, of course, introduce him around) when he arrived there.

The social year began early in 1579. Already at the end of February the Danish aristocracy gathered at Koldinghus Castle for a double wedding hosted by the king. One of the grooms, Niels Parsberg, was a brother of Tycho's erstwhile duelling opponent, and the other was a son of the mighty viceroy of Schleswig-Holstein, Heinrich Rantzov. Rantzov was widely known as a close friend of King Frederick and a fabulously wealthy bibliophile, astrologer, and alchemist and within a few years was in fairly frequent contact with Tycho. Whether they met on this occasion, however, is doubtful. Although Tycho's mother and at least two of his brothers and their wives attended, Tycho seems to have exercised his right not to. At any rate, the sun's meridian altitude was recorded at Hven on the day in question, and it is unlikely that it was done in Tycho's absence.

On the other hand, Tycho almost surely went to one wedding that year, that of his youngest sister, Sophie. She married Otte Thott of Eriksholm, whose family had long been closely allied to the Brahes and who had two sisters married to Sophie's second cousins, Ove and Henrik Brahe. During the fifteenth century the Thotts had been so powerful in Skaane that they seemed to rival the kings of Denmark and Sweden, like the dukes of Burgundy between France and Germany. Even though the numbers and influence of the family had waned since then, the tradition remained. When the wedding was over, twenty-year-old Sophie moved into Eriksholm Castle some ten kilometers southeast of Knudstrup, to live in great style with her thirty-six-year-old husband.

Whether Kirsten attended even this wedding is unlikely. But she was, by now, surely into her second year of residence on Hven and, as Tycho's good common-law wife, was doubtless keeping the keys to whatever temporary residence they were occupying. The daughter to which she gave birth on 4 August 1578 was named Sophie after Tycho's just-married sister and his paternal grandmother. A third (surviving) daughter, born during the summer of 1579, was baptized Elisabeth, after Tycho's deceased older sister (and his maternal grandmother). The rest of Tycho's family was thriving too. Brother Steen had been elected a councillor of the realm in 1578, and Axel, Jørgen, and Knud all were in service at court. At least, all had been gentlemen-in-waiting to King Frederick until the previous holiday season, when Knud had left the country suddenly.

Barely back in Denmark after several years in the Dutch wars, Knud had met, wooed, but not wed a young noblewoman, Sophie Mormand, residing at his cousin's house in Copenhagen. When she learned she was pregnant, they both were in trouble, for their conduct had been a serious breach of the courtier's code, and the punishments prescribed by that code were draconian.[94] Although Knud was nine years younger than Tycho, they had become well acquainted when Tycho lived at Herrevad Abbey while Knud had been a pupil and page there. When Knud decided that a hasty departure from the court was in order before Frederick got wind of Sophie's condition, he turned to Tycho for help. There is clear evidence that Tycho provided it – if not the horses and servants for the escape, at least advice about places abroad that Knud had never visited and recommendations to friends in Augsburg, Italy, and elsewhere.[95] Sophie was not so fortunate. She retired in disgrace to her father's manor to await the birth of the noble bastard and later married a commoner.[96]

If the king ever suspected Tycho's complicity, he did not challenge him on it. Tycho continued his own work, observing, among other things, lunar and solar eclipses in September and February.[97] In the fall he responded negatively to the king's query as to whether another new star had appeared,[98] and during the spring he produced a horoscope for Frederick's second princeling. By May Tycho had a small volume "bound in pale green velvet with gilt edges, containing

---

[94] *DBL* III, 571. See also Frederick Rosenkrantz's case in Chapter 13.
[95] See, for example, Tycho's request of Vedel (VII, 56) to see whether Knud had touched base with Laurentius the Dane in Augsburg.
[96] Sophie's misfortunes are related in *DAA* XXI, 307.
[97] X, 61–3, 75–6.   [98] XIV, 9.

about three hundred pages all written in [his] own hand" ready to
deliver to the king.[99] The following month Tycho finally received
the canonry in Roskilde Cathedral that had so long been promised to
him. The noble incumbent of the Chapel of the Magi, Henrik Holck,
had died that spring, and on June 10 Tycho was enfeoffed with his
former holding.[100]

The reason that that particular canonry had been held for Tycho
was probably that the king wished to keep a nobleman in charge
of it, for its chapel was one of the most prestigious ones in the
cathedral. It had been the chapel of a fifteenth-century royal order of
knighthood and was the site on which the freestanding tomb of King
Christian III, Frderick's father, was still under construction at the
very time when it was given in fee to Tycho Brahe.

The prebend included a residence in Roskilde, the mill of Karlebo,
over fifty farms grouped mainly around the churches of Store
Heddinge and Udby on Sjaelland, the right to appoint (and pay) the
clergymen of both those churches, and the manor of Gundsø. In
return for all the incomes and privileges of this extensive prebend,
Tycho Brahe assumed the obligation of keeping all the properties in
good repair, maintaining two choristers among the pupils of Roskilde Cathedral school to sing psalms daily in the chapel with the
vicars, and providing food, ale, and clothing for two students at the
University of Copenhagen.[101] The annual incomes of the prebend
were paid mainly in barley, rye, and oats, which had been rising
steadily in price in recent years. In assuming this canonry, Tycho
Brahe became a member of the Lutheran cathedral chapter of Roskilde, but this carried no obligation to enter holy orders nor even to
reside in the city, and of course Tycho had no intention of doing
either. To Tycho, the canonry of the Chapel of the Magi in Roskilde
Cathedral was little more than a splendid addition to his sources of
income.

When Tycho had received the fief of Nordfjord, a year earlier, it
had been clearly stated that he would hold this fief until he received
the Chapel of the Magi in Roskilde. Consequently, a letter was now
sent to the governor of Bergen, instructing him to take over Nordfjord from Tycho Brahe's bailiff.[102] Tycho protested, however, that
he needed to retain Nordfjord, even though he had just received the
Chapel of the Magi, citing, no doubt, the tremendous cost of his
instruments and construction on Hven. The king therefore acceded

[99] I, 209–50.   [100] XIV, 10.
[101] J. O. Arhnung, ed., *Roskilde Kapitals Jordebog 1568* (Copenhagen: Munkegard, 1965), pp.
      12–16, 202–4.
[102] XIV, 9.

to his request and restored Nordfjord to him for the duration of the royal pleasure.[103] Tycho now held Hven, Nordfjord, Kullagaard, and the Chapel of the Magi as fiefs of the Danish crown, and he also received an annual pension of five hundred dalers.

This was far more than the income of any other man of learning in Europe, and even in aristocratic terms, it was a substantial income. A fief like Herrevad Abbey, to be sure, was larger than all of Tycho's fiefs combined, but Steen Bille never held it without expensive obligations to the crown. Tycho was never to achieve the great wealth of his father, but he did set a new European standard for the financial support of scientific research. As of 1579, however, he was still using that support primarily to build the establishment in which he would conduct his research.

[103] XIV, 11.

Chapter 5
Urania's Castle

A T the beginning of July 1579, Tycho wrote letters to Vedel and Dançey announcing that his house was now far enough along to be worth seeing.[1] It was far from complete. Tycho did not move into the house for another eighteen months and was not to pronounce it "finished" until a year after that, in the late fall of 1581.[2] Moreover, he would be adding outbuildings right up to the end of the decade. But with the exterior complete except for ornamentation, the rough framing done inside, and the grounds generally laid out, Uraniborg had assumed enough form to permit Tycho to convey to his friends a good idea of how it would eventually look.

What first greeted the visitor to Uraniborg was Tycho's only concession to the medieval tradition of noble residences as fortified bastions of defense: the wall. Five and a half meters high and nearly five meters thick at its base, this stone-veneered earthen edifice completely enclosed the seventy-eight-meter-square area[3] that constituted the heart of Tycho's island estate. The square was oriented astronomically, with its principal avenue running from the main gate at the east corner (see Figure 4.2) due west to the other portal at D.[4] Through its other diagonal was a north-south path servicing the servants' quarters at the north extremity of the compound (C) and what would become Tycho's printing establishment at the south corner (B). Inside the great walls was another square, constructed of wooden paling, which separated the outer area of orchards and ornamental trees from the inner area containing geometrically shaped beds of herbs and flowers. In addition, the paling lined the four paths and outlined the central circle containing the house itself.

Like the grounds, the house was also oriented astronomically. The long axis ran north to south, so that the visitor entering the grounds at either gate got a full-length view of the house, which must have looked very much as imagined by the nineteenth-century Danish painter, Henrik Hansen (Figure 5.1). Most of the house's exterior was red brick, with carved sandstone trim and ornamentation. But

---

[1] VII, 55–6. A list of abbrevations of frequently used sources is provided in Appendix 1.
[2] Letters to Hayek (4 November 1580: VII, 59) and Schultz (12 October 1581: VII, 62).
[3] Tycho gives (V, 140) 300 feet. Analysis of various statements and artifacts from Tycho suggests that the unit was equal to 0.259 meters: see *Raeder*, 9.
[4] Tycho's diagram and discussion are from V, 138–41, which is a reprint of his *Astronomiae instauratae mechanica* of 1598.

Figure 5.1. Painting of Uraniborg by Henrik Hanson (1862, as imagined from the sixteenth-century woodcut); hanging in Fredericksborg Castle.

wood was also prominent, in the roofs and parapets of both the primary observatories (see O and R of Figure 4.1a) and the two ancillary turrets (N and S) constructed later. The pyramid-style wooden observatory covers combined with the rest of the various chimneys, spires, arches, domes, and figurines to provide a roofline that was more exotic than elegant, but they were functional as well because the triangular sections could be removed individually to give access to any part of the sky. The octagonally shaped railing, on the other hand, blended in nicely with the two half-octagonal entrance ways and the four small domes and single large pavilion capping the main part of the house. In the east and west sections of the pavilion were clock dials, and atop it all, nineteen meters off the ground, was a cupola containing the clock chimes and a gilt weathervane in the shape of Pegasus.

In the early years, the steps of the entry to Uraniborg converged in a corridor that conducted people immediately to the center of the building (see A → θ, in Figure 4.1),[5] where, if they were visitors of any note, they were treated to the spectacle of "a fountain with a water-carrying figure rotating around and throwing water into the air in all directions." In later years, however, the wall that formed the left side of the passageway was ripped out, so that those who climbed the steps of the entry up to the first floor (above the level of the basement windows) came immediately into the room (D) around which all the institutional activity at Hven revolved.

In the description accompanying the woodcuts of Uraniborg published in 1598, Tycho called the room "the winter dining room or the heating installation." And because every room had a fireplace (Tycho mentions a total of fourteen), such a description can only mean that this room was the one that was actually kept reasonably well heated in the winter and that was therefore the gathering place for meals, for general reading and study, family, and for Tycho and the ten-odd students and assistants who, after the early years, were always in residence at Hven. From this room, one could go across the other corridor to room E, where Tycho's mural quadrant was (after 1581) fastened at the south end of the west wall, or into the circular library (T) where the great globe (W) and most of Tycho's books reposed, or (by the staircase off the south wall of the room) to the south observatory above the library or the alchemical laboratory below the library. In fact, the lab work could be even closer than that, for, in subsequent romodeling, Tycho installed five furnaces and some ancillary equipment right in the dining room so that extended distillation processes could be conducted without having to carry a lantern down the stairs at regular intervals to check their progress.

Visitors would have been taken down those steps to see Tycho's sixteen-furnace (nine different kinds) laboratory, because alchemy surely had at least as much appeal to laymen as astronomy did. At the other end of the basement there were various cellars and pantries associated with the kitchen and the deep well that furnished the water for the house. Guests who saw them were probably taken up the staircase at the north end of the house, through the kitchen and then up to the second floor.

Like the first floor, the second was divided so that the central square was devoted to accommodations and the two towers were used for work. The towers actually were nothing more than platforms surrounded by a low wall and covered by removable roof boards.

Tycho's guests were probably conducted to the observatories, shown whatever instruments constituted Tycho's latest pride and joy, and led out to whatever portion of the galleries and small observatories happened to be complete when they were visiting. They might have been shown the red and blue chambers comprising the east side of the floor, where Tycho's family probably lived.[6] If it were summer – as it was when most of Tycho's visitors came – they probably would have eaten at least one meal in the fifteen-meter-long summer dining room, which constituted the west half of the second floor. Here, in what Tycho called the green room, because of the pictures of numerous plants painted on the ceiling, Tycho's household and guests could enjoy, over Uraniborg's west wall, the view of the many ships that passed daily through the busy toll gate at Helsingør.

From somewhere on the second floor, another flight of stairs spiraled upward, most likely through the center of the house (see Figure 5.2). Perhaps the axis around which it spiraled was open, right up to the dome capping the house. Such an architecturally interesting plan would have allowed the dome to function as a skylight and ventilator and also revealed the third clock face in the ceiling of the dome, which registered the time and wind direction shown by Tycho's clock and weathervane outside.[7] If the octagonal pavilion at the top of the stairs lacked a true floor, however, it at least had a gallery around the wall, for Tycho mentioned having a "free view in all directions" from up there. A door to the outside provided similar access to a gallery "on the top of the house itself" (see the arches). Squeezed between the second floor and the dome, and presumably reached by the same staircase, was a garret Tycho called "the upper story" or "the very top of the house, where some round windows [X] are visible, [containing] eight bedrooms for the collaborators." Built into the walls, somehow, were some cords connected to small bells in the rooms, by means of which Tycho could summon particular students. Tycho apparently enjoyed astonishing his guests

---

[6] In materials sent to the *landgrave* of Hesse in 1586, Tycho called those rooms the king's and queen's chambers (VI, 348a). In a house the modest size of Uraniborg, this appellation must have been purely figurative; for, since the rooms designated as guest rooms on the first floor show "desks for the collaborators," it seems doubtful that the family lived in those rooms. The upstairs suite actually consisted of three rooms, as Tycho mentioned an octagonal "yellow room" over the front entry.

[7] Two hundred years later, these items were regarded as ingenious features of Thomas Jefferson's Monticello.

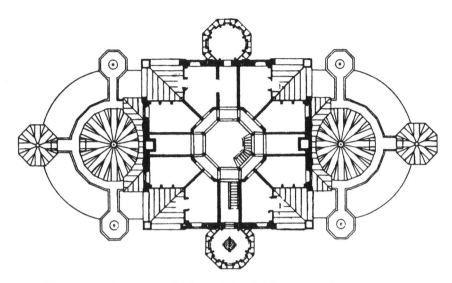

Figure 5.2. Reconstructed plan of the third floor student garrets at
Uraniborg.

by pulling a cord surreptitiously, then calling out softly the student's name, and making him appear, as if by magic.[8]

In later years, of course, there would be much more to show:
elegant decorations inside; aviaries, herb gardens, and gazebos outside; and a multitude of curious facilities – instrument works, underground observatory, printing press, paper mill – on the extended
estate outside the wall. Neither the conception nor any of the individual features were original.

Manorial desmesnes had been developed around the conception of
self-sufficiency since late Roman times. And if expectations of life
had risen considerably during the early Renaissance period, magnates
such as Peder Oxe had responded to the challenge by cultivating
imported fruit trees, exotic herbs, freshwater fish, and various small
game birds or animals. Moreover, during a century when the cash
incomes of the aristocracy had risen considerably, many estates were
being created virtually from nothing, in a short period of time, just
as Uraniborg was. But although most of them were much larger
than Uraniborg, none of them attained the degree of either technical or artistic development that Uraniborg did. Even though the
technical aspects had been partially anticipated at Herrevad, and the
artistic aspects had been variously anticipated in Holland and northern Italy, the combination was, and long remained, unique. Indeed,
discerning visitors would have recognized a third component in

[8] *Gassendi*, 180.

Uraniborg, especially those whose backgrounds included some university education. For, in the crowded conditions of life and the gentle hum of scholarly activity, they would have noticed an atmosphere that resembled more a university professor's boarding house than a noble diplomat's manor house, let alone a knight's fortified castle.

As important as the building of Uraniborg was to Tycho's personal and professional life, it was dwarfed in significance by the fact that he was at last settling down somewhere. Now he could begin to make definite plans for carrying out his long-envisioned restoration of astronomy. Fundamental to those plans was the acquisition of suitable instruments. Those that Tycho had so far collected fell so far short of his ideals that there was scarcely any point in making observations with them: The only one he ever used to any extent after his move to Hven was the cross staff, and the only good thing he ever had to say about it was that it was light and easy to handle.[9] As Tycho's plans for Uraniborg began to take shape, therefore, he turned to the crucial task of providing instruments for his work.

With his enfeoffment on Hven, Tycho was able for the first time to contemplate the construction of instruments that would neither have to be moved nor have to be paid for out of his patrimony. Unfortunately, however, there still remained one serious constraint, the ability of the artisans to convert Tycho's designs into working reality. His instruments usually had been produced in commercial shops. But those in which he had even a modicum of pride were already considerably beyond the size and sophistication of instruments that were routinely built[10] and so would have to have been special orders.

The arrangements for this construction must have been extremely unsatisfactory. For the average artisan, the degree of perfection that Tycho demanded must have been almost impossible to comprehend, let alone attain. Even for those few who were capable of satisfying Tycho, it must still have required constant monitoring by him and occasional reworking by the harassed smiths, to achieve Tycho's ideals. The possibilities for misunderstanding in the pecuniary terms of such a project would have been legion.

Tycho seems to have realized that instruments of the kind he envisioned required a different standard of accuracy from that ordinarily achieved by the woodworkers, armorers, smiths, and engrav-

---

[9] III, 185; IV, 464; V, 96; X, 156. John D. Roche, "The Radius Astronomicus in England," *Annals of Science* 38 (1981): 1–32, has argued persuasively that the problems of the cross staff were dealt with more or less successfully by Thomas Digges but that Tycho's "dis"-recommendation destroyed the reputation of a perfectly reasonable (and relatively inexpensive) instrument.

[10] See *Zinner*, passim.

ers on whom he had to depend for his work; and that even an instrument maker of the caliber of Schissler, at Augsburg, dealt primarily in such trivia as sun dials and only rarely in serious observational instruments.[11] As early as his Herrevad days, therefore, Tycho probably began to dream of having his own instrument shop, where he could convert skilled craftsmen into specialists, by hiring them to devote their time exclusively to his projects and to work where he could conveniently supervise every phase of their activity.

During the late 1570s, Tycho achieved this ambition. Concerning the establishment of his "workshop for the artisans," he was almost completely silent, limiting his remarks about it to giving its location, stating that it was equipped with "mills driven partly by horses and partly by water power . . . (although this procedure was generally useless)"[12] and affirming that it was very expensive. He never mentioned the names or even the principal competences of his artisans but said only that some of his instruments took five or six people and three years to make and that by 1590 he had let them all go except for one or two who made repairs.[13]

Tycho did, however, say a great deal about his instruments, for if he learned anything from the controversy over the observation and interpretation of the new star, it was that instrumental accuracy had not only to be achieved but had also to be seen (by his readers) to have been achieved. In his later works on the new star and the comet he would provide thorough discussions of the specific instruments on which his conclusions rested.[14]

Because Tycho's ambition was nothing less than to reestablish the empirical foundations of the science of astronomy, he decided fairly early that a published account confirming the general superiority of all his instruments would be very valuable.[15] The press of his numerous activities delayed the project. But unlike several other works for which he also conceived plans in the 1580s, it was eventually completed, in 1598. In this *Mechanica*, or *Instruments for the Restoration of Astronomy*, Tycho printed large, detailed pictures and extensive descriptions of the construction and use of each of his twenty-two most important instruments. In conveying a feel for the quality of Tycho's instruments, these discussions leave almost nothing to be desired. And because they display Tycho's characteristic ingenuity

---

[11] See the catalogue of Schissler's work in *Zinner*, 504–20.
[12] V, 19, 151. On Tycho's map of Hven, an "Officina artificum, Astronomica Instrumenta, & alia fabricantium" is shown about fifty meters outside the northwest wall of the grounds. V, 151 (see Figure 6.1).
[13] VII, 273.   [14] II, 330–5; IV, 369–76.
[15] By August 1585, he had formulated plans for such a work and had two diagrams in print to send Hayek (VII, 96).

and attention to detail, they invariably are described at length in accounts of Tycho's life and work.[16]

Tycho said very little about when his instruments became available to him. The great authority on the history of astronomical instruments, J. A. Repsold, concluded that the two dozen instruments described by Tycho were developed and used more or less simultaneously, so that Tycho must have been making instruments simply (or at least primarily) to keep his artisans occupied.[17] However, the chronology that at once refutes Repsold's impression and provides the historical element missing in Tycho's accounts is available in his observation logs.

When an instrument was ready for service, it had to be evaluated according to the quality of the observations made with it. Moreover, once the accuracy of an instrument was established, it was important to be able to identify subsequently all observations made with it. If Tycho was going to have a hope of keeping track of things, conscientious labeling was imperative. From this labeling system and his occasional historical remarks, it is possible to reconstruct with considerable confidence the development of Tycho's instrument-building program on Hven.[18] What emerges most dramatically from this reconstruction is the sheer amount of time that elapsed between the nominal beginning of Tycho's work (with the granting of Hven) and the real beginning associated with the completion of his later instruments.

Although Tycho probably envisioned the establishment of his own instrument shop in his initial plan for Hven, he knew it would be a long time before that shop turned out any instruments. And although he had attempted since 1573 to obtain good instruments, the one-cubit quadrant was almost useless, and the heavy steel sextant (Figure 5.3a) was sufficiently uninspiring that Tycho frequently reverted to his cross staff for distance measurements during the following years.[19] Accordingly, Tycho made one last call on his

[16] See *Dreyer* (325–32) and *Norlind* (270–82). Actually there is no better exposition than the one by Tycho himself readily accessible in English translation as *Tycho Brahe's Description of His Instruments and Scientific Work:* see Raeder.

[17] "Man hat den Eindruck, dass Instrumente gebaut werden, nur um Arbeit zu schaffen." J. A. Repsold, *Zur Geschichte der Astronomische Messwerkzeuge* (Leipzig, 1906), vol. 1, p. 29.

[18] This task was carried out in great detail in Victor E. Thoren, "New Light on Tycho's Instruments," *Journal for the History of Astronomy* 4 (1973): 25–45.

[19] Tycho's descriptions of the instrument are in IV, 369–71, and V, 76–9. In the latter he seems to imply that he already had it for his travels in Germany in 1575. A reference in VII, 11, to an "instrumentum ferreum" may even indicate that it was already under construction in the spring of 1573. On the other hand, Tycho's description of it to the *landgrave* (VI, 10, 253) is based on the assumption that the *landgrave* did not see it when they observed together. We shall see that the resolution of the discrepancy is probably in the new sights and divisions Tycho put on it after his travels.

SEXTANS CHALYBEVS PRO DISTAN-
TIIS PER VNICVM OBSERVATOREM
DIMETIENDIS.

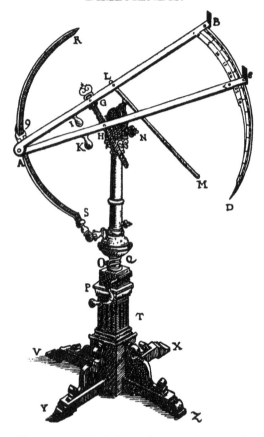

Figure 5.3a. Tycho's steel sextant (ca. 1574).

commercial sources. This time the quadrant that emerged (Figure 5.3b) was a huge success, even though it was probably made by the same smith who had made his one-cubit model[20] and was only half again as large as that piece's forty-centimeter radius. The features that made it a landmark in the development of Tycho's instruments and allowed him to use it long after he had much larger instruments to work with, were its advanced sights and divisions.

Until Tycho's day, the best astronomical aiming was done through pinhole sights attached to the ends of an alidade. In theory, this

---

[20] Compare the two sets of depictions and descriptions in V, 16–19 and 12–15.

QVADRANS MEDIOCRIS ORICHAL-
CICVS AZIMVTHALIS.

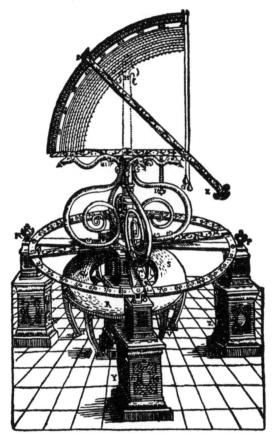

Figure 5.3b. Tycho's "*Q. min,*" finished in 1580.

method was excellent. In practice, however, Tycho found it extremely difficult to locate and sight on stars through pinholes, if they were really pinholes. For, if the holes were large enough to see through easily, then the sighting would not be precisely central in the holes, and the position could be off by a considerable fraction of a degree – at least, a fraction that was considerable to Tycho, for whom it was a matter of some surprise that previous astronomers had not dealt with the problem.[21]

This was the issue: that nobody before Tycho had both cared

---

[21] V, 46, 155.

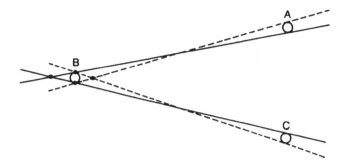

Figure 5.4. Tycho's analysis of the parallax inherent in "post" sights.

about accuracy and taken enough other steps to achieve it, to make
such a minor matter worthy of concern. Tycho's first attempt at
remedying the situation had been an adjustable slit sight, consisting
of two posts, one at the front and one at the back of the alidade. As
the observer aligned the pointer on a star, the width of the slit
diminished, until perfect alignment was achieved at the vanishing
point of the star. Unfortunately, this extremely sophisticated device
had a severe and inherent shortcoming: Although the posts could be
positioned to provide one accurate alignment, any attempt to use the
pivot post for a second sighting necessarily introduced an eccentricity into the reading (Figure 5.4).

Tycho had recognized and made corrections for this parallax. But
in the construction of his second quadrant, he inaugurated a new
apparatus that solved his sighting problems once and for all. As
Tycho's depiction of it shows (Figure 5.5),[22] it was another slit sight.
And the slits were still adjustable, as the slits AB, BC, CD, DA were
spring-loaded in such a way that they could be widened or narrowed
simultaneously by manipulating a single screw on the inner side of a
plate). But the basic idea of the device was new: parallel sightings.
The alidade was pointed correctly when the image could be seen
both along $BC \rightarrow FG$ and $AD \rightarrow HE$. On the quadrant, where the
alidade lay flat on a vertical plate, these two sightings gave the
object's altitude. When the azimuth was desired too, the quadrant
had to be rotated until the image could be seen along $CD \rightarrow GH$ and
$AB \rightarrow FE$. Finally, for observations of the sun, a hole was provided
in the front plate so that the sun's image could be passed through and
lined up to fall exactly on a circle drawn on the inner side of $ABCD$.
(If Galileo had known that working astronomers of the day did not

----

[22] Tycho first published a description of his sights in 1588 (IV, 373–4). The same discussion is
found in the *Mechanica* (V, 154–5).

Figure 5.5. Detail of Tycho's sighting mechanism.

observe the sun directly even with the unaided eye, he might have avoided damaging his eyes when he turned his telescope on the sun.)

Just as the problems of sighting were largely unrecognized before the demands placed on instruments by Tycho, so too were those of instrument division. Dividing a quadrant into 270 reasonably equal intervals of twenty minutes each (to estimate the nearest ten minutes, which is generally regarded as the best accuracy achieved before Tycho's time)[23] would not have been easy. But neither should it have been particularly challenging on any arc more than about a quarter of a meter in radius, for which the divisions would have been six or seven to the centimeter. Tycho, however, wanted to read individual minutes. Using straightforward divisions, the best he had been able

---

[23] Asger Aaboe and Derek Price, "Qualitative Measurement in Antiquity," *Mélanges Alexandre Koyre* (Paris: Hermann, 1964), p. 2, suggested that it would have required an instrument of nearly two meters to allow graduation to five minutes of arc. Hermann Vogt, "Versuch einer Wiederherstellung von Hipparchs Fixsternverzeichnis," *Astronomische Nachrichten*, no. 5354 (1925): cols. 39–43, contended, on the grounds of frequency of occurrence of various fractions, that Ptolemy's instruments were divided down to twenty minutes, so that he could estimate the nearest ten minutes. R. R. Newton, *The Crime of Claudius Ptolemy* (Baltimore: Johns Hopkins University Press, 1977), p. 247, argued that the evidence really implies graduation to only whole degrees.

to obtain on his first quadrant was divisions at five-minute intervals. In an effort to obtain finer readings, Tycho had, on the first quadrant, tried out a notion advanced by Pedro Nuñez in 1542.[24] The idea was to inscribe a series of concentric arcs inside the graduated arc but to divide those inner arcs successively into 89, 88, ... 46 parts, so that the alidade would always be lined up on a graduation point of one of the forty-five arcs (compare Figures 3.1 and 5.3b). Proportional calculations (e.g., $33/73 = x/90$) would then give a precise reading.

There were, however, several problems that limited the method's effectiveness. Even assuming that the goldsmith, or whoever was doing Tycho's work for him, managed to divide quadrants into fifty-nine or eighty-three equal parts, it must have been very difficult for the observer to look through forty-five arcs and decide which one had a division mark nearest the alidade, especially when taking account of the inevitable imperfections in the edge of the alidade.[25]

For his second quadrant, therefore, Tycho already modified the scheme. And sometime during the construction or first years of use of this "small" quadrant, he devised another means of improving his primary graduations. The idea had originated in 1564, when Tycho had learned from Schultz that the divisions on his cross staff could be made much finer by using diagonal lines to produce an artificial lengthening of the interval that needed graduating (Figure 5.6).

Although the method is now known to date back to Levi ben Gerson, it seems not to have had a wide circulation in Tycho's day, most likely because its restriction to rectilinear scales limited its astronomical application.[26] Tycho realized that such geometrical fastidiousness was an astronomical irrelevance, that the mathematical precision lost by applying transversal divisions to curved scales was more than compensated by the gain in technical utility.[27] But even in Tycho the mental block stood for over a decade, so that it was only on the small quadrant, and possibly belatedly, in mid-1580,[28] that

[24] Petri Nonii Solaciensis, *De crepusculis* (Lisbon, 1542). Tycho's reference to it is in V, 12.
[25] J.-B. J. Delambre offered an extended critique of the shortcomings of the scheme in his *Histoire de l'astronomie au moyen age* (Paris, 1819), pp. 402–6.
[26] For an outline of the history of linear transversal divisions, see *Zinner*, 224. Tycho's story of his introduction to them is given in V, 108. It is now known that Levi also used them on arcs: See Bernard R. Goldstein, "Levi ben Gerson on Instrumental Errors and the Transversal Scale," *Journal for the History of Astronomy* 8 (1977): 102–12.
[27] Tycho first described and depicted his transversals in the *De mundi* of 1588 (IV, 372). In the *Mechanica* (V, 153–4), he provided a rigorous defense of them by showing that the error involved was a maximum of slightly over three seconds.
[28] On 2 May 1580, Tycho noted (X, 80) that the meridian altitude of the sun was "52 18 juxta superiorem instrumenti divisionem, sed respectu inferioris divisionis fuit 52 18½." Later notations to the "supremam circumferentiam" (X, 84, 99) and "novam quadrantis di-

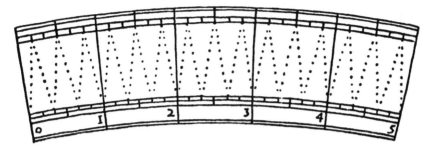

Figure 5.6. Tycho's representation of his transversal sights.

he tried the new method. So successful was it, however, that he substituted transversals for the Nonian divisions on his steel sextant and dispensed with all further experimentation.

In addition to his old and new sextants and quadrants, Tycho brought other items to Hven that if they were not instruments in the strictest sense, were nevertheless important additions to his observatory. The ones for which Tycho had the highest hopes were his clocks. Already on 2 April 1577 he used a clock to time a lunar eclipse, and by August 1579 he had a second one available. For the comet of 1580 he had at least three, and by August 1581, when Tycho made his first attempt at precision timing, he had four clocks from which to choose.[29]

Some of the later clocks may have been made in Tycho's shop, but there are good reasons for doubting it. One is that Tycho said little about them. In contrast with his minute descriptions of his instruments, he merely mentioned hurriedly that he had four clocks and that the largest wheel on the largest clock was solid brass, three-quarters of a meter in diameter, and geared with twelve hundred teeth.[30] Another reason is that there was a thriving commercial trade in intricate clocks for church steeples and town hall towers. This meant that expertise already existed and did not have to be established (and subsidized) by Tycho. Of course, there would not have been a great demand for clocks that read seconds and remained scrupulously accurate, so at least some of Tycho's clocks must have

visionem" (X, 80, 85) also suggest that at this time transversals had just been ruled or reruled.

[29] X, 46, 77, 110. See the reference to the newness of the large and small clocks in October 1580, in XIII, 312.

[30] Tycho discussed his clocks only indirectly, in connection with his mural quadrant (V, 29–30), reserving more detailed description for an occasion that never presented itself.

been custom-made according to Tycho's specifications, but it is not very likely that Tycho had much to say about the way in which the specifications were to be met.

As Tycho's persistence with his clocks suggests, he was after something more than a convenient way of ensuring that events proceeded on schedule at Uraniborg. What he wanted was nothing less than a precision instrument with which to record the exact times at which stars crossed the meridian. If he could do it accurately and thereby determine the differences in the right ascensions of several stars from the differences in their transit times, the labor in compiling his star catalogue would be reduced enormously. Meridian clockings could be recorded and subtracted serially with much more facility than interstellar (angular) distances could be measured. (The meridian altitude would provide the second observation required in either case.)

With so much at stake, Tycho worked hard on the project. Gradually he adjusted the clocks through the breaking-in period during which errors sometimes accumulated into hours over an interval of a few days. It soon became part of Tycho's routine to reset the clocks by the noon sun and record the error every day. Once the clocks were adjusted, the daily drift was generally only a few minutes, but even the frequent gain or loss of twenty minutes looked as if it should be amenable to compensation (by distributing the error proportionally over the twenty-four-hour period) in the way that Tycho was accustomed to doing with his other instruments.

Tycho tried using two clocks to improve his timings. He made a determined run in the fall of 1581 and another one in the spring of 1583. Both times the results were unsatisfactory. Tycho never articulated his dissatisfaction, but his logs show what he was up against. Among the group of stars that he timed during his first trial, the interval between the stars representing "the mouth of Pegasus" and "the right shoulder of Aquarius" is recorded more than twenty times. The differences in transit time range from 19^m41^s to $21^m 30^s$,[31] with the average observation showing an absolute deviation of 20^s from the average of all the timings (20^m30^s).

During Tycho's second trial, some twenty odd timings of the interval between Sirius and Procyon (among many others) show an average deviation of $11\frac{1}{2}^s$.[32] Tycho was unable to conceptualize the results in these terms. But an error of 4^s of time was equivalent to an error of $1'$ of arc, and Tycho did not have to scrutinize the scatter of 61^s in the intervals between Sirius and Procyon to realize

---

[31] X, 110–15.   [32] X, 265–8.

that he was dealing with observations that were nowhere near the quality he was already getting from his instruments.[33]

Tycho, with great reluctance, thus had to abandon his clocks. Despite the expenditure of a great deal of money, and the dedication of a great deal of time, he was unable to develop them to the status of being instruments and therefore had to content himself with using them as handy auxiliaries. What added frustration to his disappointment was that he could not even determine specifically why they were not working properly. He knew the "changes in the air and the wind" that necessitated constant adjustment from one season to the next could also be considerable over a given night, but even keeping his clocks in heated rooms did not solve the problem. All he could conclude was that no matter how well clocks were made, they were inherently unstable, either from undetectable irregularities in a gear tooth or a cog wheel or from inconstancies of drive due to the lengthening and stretching of the cord during the fall of the driving weight.[34] In an effort to salvage something from the method of timing meridian transits, Tycho even tried the ancient clepsydra, or water clock, modified to use mercury and maintained at a constant pool depth to ensure a regular flow. It gave better results than his clocks did, but Tycho does not say how much better. And because the handling and precise weighing of small quantities of mercury (or, in an analogous hour glass device, lead oxide powder) could not have been less difficult than measuring angles, Tycho chose to stay with more traditional methods.

The most significant item Tycho acquired while he was setting up his shop, in respect to both the pains of production and the value of the final result was his great one-and-a-half-meter globe. Begun in 1570 at Augsburg by Schissler, it was not seen (let alone overseen) by Tycho until his return there in 1575. By that time, the rigors of storage had caused enough warpage in the wood to produce not only distortions of the (hollow) form, but even openings, here and there,

---

[33] In 1583, even before the completion of his best instruments, Tycho was getting consistent agreement within one minute of arc on meridian altitudes taken with three or four different instruments (X, 232–5).

[34] This was Tycho's mature judgment, expressed several times in the later 1580s (IV, 34). Initially he was prepared to accept the *landgrave*'s statement (VI, 51) that his clock did not depart by a minute over twenty-four hours and to content himself with warning his readers (IV, 194) that those lacking the *landgrave*'s resources would be ill advised to trust their clocks. But after Tycho learned that he and the *landgrave* differed by eighteen minutes in their timings of the comet of 1580, and 6′ in the longitudes of all their stars, he took a much harder stand against clocks in statements composed between 1589 and 1591 (II, 156–8, 282: III, 17, 221). As late as 1590 Tycho was still making determined attempts to get his clocks working accurately. See XII, 3, 35, 48, 61. See also Tycho's early attempts to experiment with changes in the driving weights of the clocks, in XIII, 309, 311.

between the individual pieces of wood.[35] Even when it was new, however, it is unlikely to have had anything like the perfection to which Tycho aspired. Tycho described Schissler as the clever and ingenious craftsman he had sought in vain in other places. But could Schissler ever have imagined that Tycho, after filling the cracks and shaving the bulges, would begin gluing on, and sanding down, hundreds of sheets of parchment, to produce perfect sphericity? Or that he would then monitor its shape through two years of seasonal changes before proceeding to the finished work? He would probably not have been surprised to learn that Tycho in 1579 had it covered with brass sheets "with such great care and accuracy that one might believe the globe to be of solid brass" (Figure 5.7). He would surely have been astounded to learn that Tycho went on to have (by December 1580) the zodiac and equator etched onto the brass and divided into individual minutes of arc (by means of transversals, four minutes per millimeter), so that it could be used to transform trigonometrical coordinates in a day when such computations were much more tedious than they are with modern formulas.[36]

A decade later, when Tycho was doing his star cataloguing, he entered on the globe each completed position, precessed to its coordinates in 1600, By 1595 he claimed to have one thousand stars located on it. Long before that time, the globe was the conversation piece of Uraniborg, and it is clear from Tycho's references to it that he derived as much general satisfaction from it as he did from any project he ever undertook. If the thousand or more dalers he spent on it represented a truly incredible expense,[37] it was, as he begged forgiveness for boasting, a unique production and "a huge and splendid piece of work." Not only was it the envy of all of his visitors, but it actually worked, allowing him "to determine mechanically, with very little trouble and without difficult calculations, all the details concerning the doctrine of the [sphere]." Nothing could be more appropriate than that Tycho should have coined, in defense of the twenty-five-year development of this superb auxiliary, the phrase that deserves to be regarded as his general credo: "If it has been done well enough, it has been done quickly enough."

---

[35] Tycho appropriately discussed his great globe in the *Mechanica* (V, 102–5) with (but after) the rest of his instruments. Some of the filling and sanding must have been done in Augsburg, because Tycho did not get the globe back to Uraniborg until the fall of 1579, at the earliest (VII, 56), whereas the globe was etched by December 1580 (X, 88).

[36] X, 85, 88–9.

[37] VII, 361 *Gassendi* (127) reports five thousand dalers as a figure given by Tycho in private conversation. That it might be the correct figure is suggested by the fact that Rudolf II paid six thousand thaler between 1587 and 1589 to an Augsburg clockmaker, Georg Roll, for a gilded brass globe less than 20 cm in diameter but driven by clockworks: *Zinner*, 492–3 and plate 60.

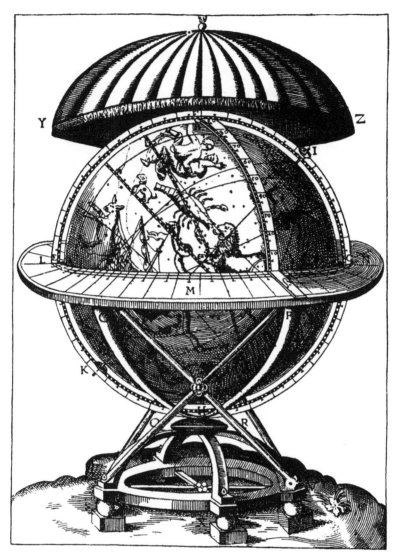

Figure 5.7. Tycho's Great Globe.

In addition to its intrinsic merits, Tycho's celestial globe had the
distinction of being the first product of extensive work in Tycho's
own shop. It seems plausible to assume, then, that because Tycho
did not send for the globe until August 1579, it was only then that he
began to anticipate being able to work on it when it arrived. By late
1580, when the globe had been covered with brass plates and etched
with coordinate markings, Tycho's shop must have been in full

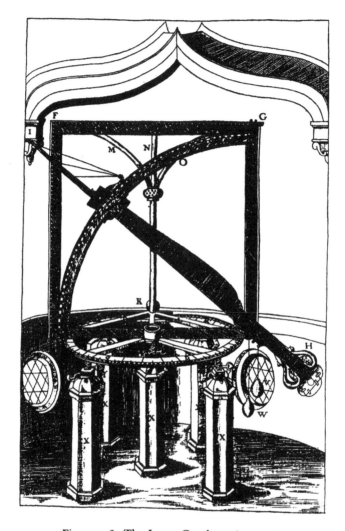

Figure 5.8. The Large Quadrant (ca. 1581).

production, for during the next two and a half years it and Tycho's artisans produced eight of the large instruments that have come to be so intimately associated with Tycho's name.

The first instrument, inaugurated on 21 March 1581, did not augur well for Tycho's shop. As Figure 5.8 shows, it was a large quadrant, denominated variously in his logs as the *Q. maius* or *Q. max.*,[38] to distinguish it from the one-and-a-half-cubit quadrant he had been

---

[38] V, 92–3; X, 92, 132, 181, 185. The simultaneous reading on p. 189 is an editorial mistake.

using (which thenceforth had to be called the small quadrant, or *Q. min.*).[39] An imposing five cubits – nearly two meters – in radius, the instrument's steel frame was sufficiently large to allow divisions to sixths of a minute on the brass strips that lined the arc and the circumscribed square. Unfortunately, however, the pillars Tycho designed to support it were so unsatisfactory that it "could hardly be rotated without setting the whole instrument in motion." The result was that Tycho made so little use of it that even his decision to describe it in his *Mechanica* was obviously an afterthought. But he did not simply write it off, and when he began to analyze its shortcomings and to redesign it for useful service, Tycho would soon conceive of and start building a new kind of observatory. In the meantime, there were tests to run and then observations to make with still another quadrant, which emerged from his shop in June 1582.[40]

If there is any one instrument that is particularly associated with Tycho, it is the mural quadrant. From a historical standpoint this fact seems somewhat anomalous, as the element of Tycho's originality in that instrument was considerably less than those in, say, the sextant or the equatorial armillary. To explain, we need look no further than the oft-reproduced depiction in Figure 5.9.[41] Like the great globe, the mural quadrant was not only a very practical part of Tycho's equipment but a stunning "parlor piece" as well. Whether it was already envisaged in the initial specifications for Uraniborg, its basic component, a section of wall oriented and constructed along an astronomically precise north-south line, was essentially built into Tycho's house on general principles. Even if it was not preplanned, therefore, the only alteration required would have been removing a piece of wall for the sighting hole. All that then remained was to forge a solid brass quadrant two meters in radius, thirteen millimeters broad, and five millimeters thick;[42] divide it by transversals

---

[39] Tycho called it the "medium-sized" quadrant in his description (V, 16–19), but there is no such denomination in his log. Presumably the one he described (V, 12–15) as the "small" quadrant in his *Mechanica* was so far obsolete by 1580 that he never again used it.

[40] X, 134. In view of the consistent use of *mur* in reference to the mural quadrant, there is little excuse for believing with Dreyer (101) that Tycho's observations in April 1581 *per magnum instrumentum* (which are identical in style to those made with the Q. *maius*) "were probably also made with [the mural] quadrant."

[41] The mural quadrant is described and depicted in V, 28–31.

[42] Dreyer (101) follows Tycho in giving five inches and two inches for the breadth and thickness. But these are Tycho's archaic units. They can be approximately converted by reference to Tycho's depiction and his statement (V, 153) that he usually made his transversals one-forty-eighth of the instrument's radius. According to the picture, the inside and outside borders on the arc each were roughly equal to half the size of the transversal strip itself. This would indicate that the five-inch width of the whole arc would have been one-twenty-fourth of the width of its five-cubit radius, so that Tycho's cubit was equal to twenty-four of his inches. In this argument lies the first indication that Tycho was using the traditional relationship of sixteen inches = one foot = two-thirds cubit.

Figure 5.9. Tycho's famed Mural Quadrant.

down to sixths of a minute; mount the arc on the wall in perfect
position;[43] and fit sights to it.

The rear sights, which show up rather badly in the perspective of
the figure, were identical to the rear slit-plates already displayed in
Tycho's diagram of his sights. But instead of being attached to an

---

[43] In 1583 Tycho was still adjusting the quadrant on the wall: X, 135.

alidade, they (both the original plate and a second one added in 1584) were merely clamped onto the arc. With no physical connection between the front and rear sights, a flat front sight-plate similar to those on the azimuth quadrants would have shown a different apparent size with every shift in position of the rear sight, thereby frustrating parallel sighting. Tycho solved this problem by using a round front sight – a cylinder cemented lengthwise and horizontally into the opening in the wall.

Concurrently with his experimentation on the quadrants, Tycho was also exploring the possibilities of another traditional family of instruments, the armillaries. Because of the convenience they offered in coping with the ecliptic coordinates involved in planetary observations, armillaries (or their cousins, see Figure 5.10a and b) had been among the earliest instruments invented. Because Tycho had from the beginning contemplated extensive planetary observations, he had already in 1577 begun work on an instrument that would give him his planetary positions directly. A somewhat abbreviated model of less than sixty centimeters outside radius, and four rings instead of the usual five or six, it must have looked very much like Figure 5.11.

Yet despite its relatively small size, its massive steel meridian, strong solid iron base, and iron braces, and even though the three inner suspended rings were made of wood for lightness, Tycho found that the instrument was subject to flexures. These flexures, being variable according to the manifold possible combinations of position of the rings, were not even subject to tabulation and compensation. Because a larger instrument would have had even greater distortions, and a small one would give him readings so coarse as to be worthless, Tycho had reluctantly concluded that the armillary, at least in its zodiacal form, represented a blind alley.[44] By the time of the lunar eclipse of November 1581, therefore, Tycho had reset the rings in an equatorial form, to explore another option.[45]

One of the significant accomplishments of Tycho's early years on Hven was the development of his sextant into its mature form. When he had invented it in 1569, he had thought of it as an alternative to the light, handy cross staff. With the increase in weight attending its successive development from wooden half-sextant to steel sextant, the instrument had reached the point of requiring a stand. Subsequent experience apparently convinced Tycho that making delicate

---

[44] X, 52–5.
[45] X, 100. Tycho depicted the remounted set, too, as "another equatorial armillary" (V, 60–3). He said nothing about the duplication except that the rings are "the same size." But he gave a "working" description of the rings under the second ("another") discussion and, for the zodiacal armillary (V, 52–5) resorted to the primarily historical critique of its effectiveness, already related.

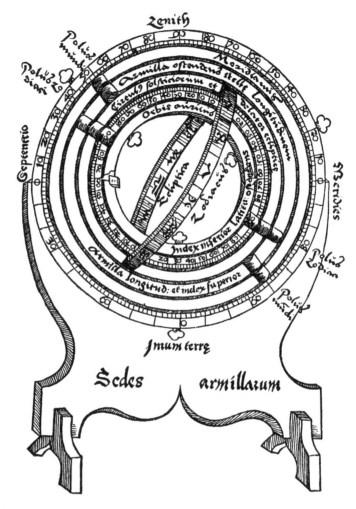

Figure 5.10a. Armillary from Regiontanus's "Scripta" (1544), after Repsold.

sighting adjustments on a pointer one and a half meters long would be easier if the observer were situated at the end that had to be moved. In the final metamorphosis of his sextant, therefore, he reversed the direction of the sighting. He also had to change the mode of sighting, to utilize his newly conceived parallel sights. Because his previous sextants had already had a pivot post analogous to the cylinder of the mural quadrant, all that was required was to substitute his new slit-sights for the old rear posts. Indeed, the similarity of the slit/cylinder arrangement of the early sextants to the

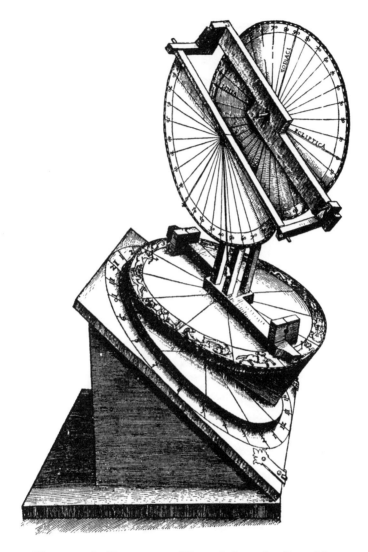

Figure 5.10b. Torquetum of Peter Apian, after Repsold.

final form of Tycho's sights suggests that the cylinder sight might well have been developed in connection with the sextant.

The chronology of Tycho's logs is consistent with this supposition. Already on 16 October 1581, eight months before the beginning of references to his mural quadrant, he made his first observations with "the new sextant not admitting parallax."[46] It was not the instrument of Figure 5.12, but a virtual twin, having two movable

[46] X, 125, 127.

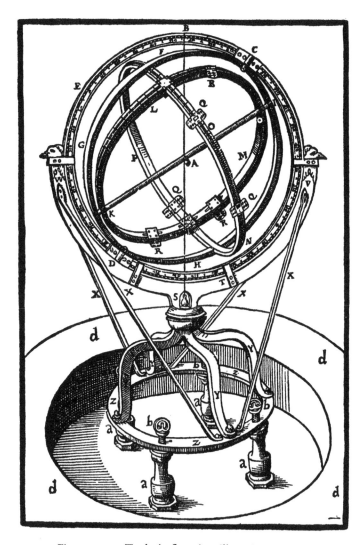

Figure 5.11. Tycho's first Armillary (1576–1580?).

alidades, which Tycho named in the *Mechanica* his bifurcated sextant.[47] Apparently it was not a happy experiment and Tycho abandoned the instrument almost at once in favor of the single alidade model shown.

The most noteworthy feature of these sextants was the ingenious and practical globe (or ball and socket) mounting, which allowed

[47] The sextans bifurcatus is described briefly in V, 96. Citations of the sextantem biformen (X, 103, 129) indicate that the dual-alidade model was the first one out of the shop.

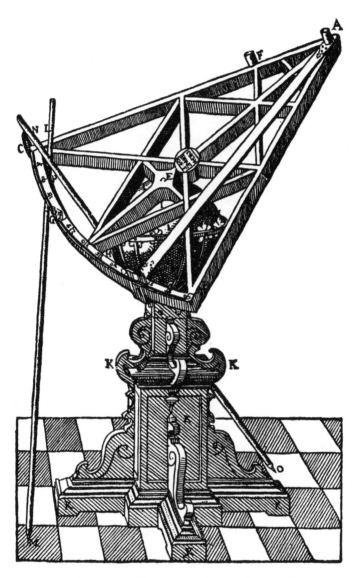

Figure 5.12. The mature Sextant (ca. 1582).

easy universal positioning of the instrument. One contingency that Tycho does not seem to have foreseen, however, was the possibility that the two observers might get in each other's way when they had to measure small interstellar distances. To cope with this problem, which was particularly pressing because of his need to determine accurately the positions of the stars in Cassiopeia for his discussion of

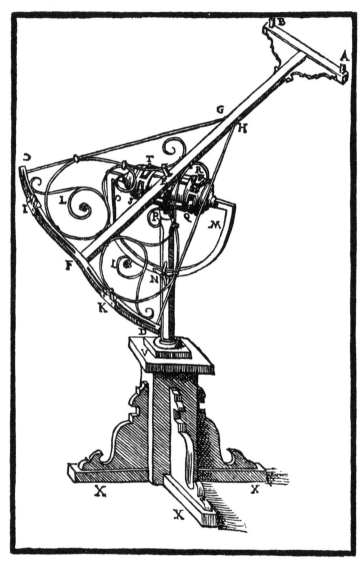

Figure 5.13. Bipartite Arc (1583).

the new star, he invented the bipartite arc shown in Figure 5.13.[48]
While constructing it, he seems to have realized that the principle
of the new instrument – splitting the axis of the angle into two
parallel "zero points" – could be adapted for use on the sextant itself.

[48] V, 68–71.

Accordingly, he added an extra cylinder (F) and a detachable sight plate (G) to the sextant, placed so as to determine a line parallel to the central axis, AE. The right-hand half could then be used as a half-sextant, with one observer sighting along the alidade while the other one sighted not at AE but at its optical equivalent, FG. This modification was first used at the beginning of 1583, almost simultaneously with the appearance of the first observations with the bipartite arc.[49]

The year 1583 also saw the completion of the largest instrument Tycho ever made himself – the large triquetrum, or rulers, shown in Figure 5.14.[50] Based on an idea dating back to Ptolemy, it consisted of a three-and-a-quarter-meter main rule and two hinged half-rulers. In addition, it included a considerable auxiliary framework to (1) keep the rulers coplanar, (2) counterbalance the extended main beam, and (3) simplify the task of adjusting the alidade for the person doing the sighting. The weight of the framework dictated, in turn, that the main beam be made of metal; in fact, Tycho made all three rulers of solid brass, probably to forestall problems from warpage. The result was an instrument so cumbersome that azimuth measurements were almost hopeless, and altitude measurements were either so meaningless or so infelicitous that Tycho bothered to record only a handful of them. Naturally, he was not entirely candid in his public assessment of the great rulers, avowing that they were essentially as accurate as his large quadrants were. But because he was kind enough to suggest that azimuths were best taken by moving the main rule to a selected azimuth and then waiting for the star to cross it, his audience would have understood what the situation was.[51]

The completion of the large rulers marked approximately the halfway point in Tycho's lifetime production of instruments. His progress at that stage is shown in Table 5.1. Even the least of his

[49] The chronology is reconstructed from the following evidence: The sums at the bottom of X, 275 establish the *arcum astronomicum* as Tycho's log notation for the bipartite arc. References to this instrument first appear at the beginning of 1583 (X, 239, 246). References to an *arcum parallaticum*, which must have been the sextant's extra sighting apparatus, appear at the same time (X, 275). In his discussion of the arc, Tycho stated that it was built specifically to perform a task that "cannot very easily be done with . . . a sextant" and then mentioned, almost parenthetically, that "this drawback has been removed, it is true, by another expedient."

[50] V, 48–51. Observations are first found in X, 243, 236.

[51] In a brief statement in his *Progymnasmata* of 1602 (II, 153), however, Tycho expressed himself rather bluntly on the shortcomings of the rulers. One and a half centuries later, Sawai Jai Singh tried vainly to introduce precision metal instruments into India. V. N. Sharma's analysis of his failure "The Astronomical Efforts of Sawai Jai Singh" in G. Swarup, Bag, Shukla, ed., *History of Oriental Astronomy* (Cambridge, England: Cambridge University Press, 1987), pp. 233–40, is that the instruments were too heavy.

Figure 5.14. Large Rulers (ca. 1583).

productions to that date would have been a source of pride for any other astronomer of his era, except the *landgrave*. But to Tycho, even the best several of them did not look like the basis for a renovation of astronomy. He had no usable armillary and no means of checking the results of the one large quadrant he could use. In addition, he needed good, portable instruments for the task of mapping Denmark that had been suggested by Frederick, and Tycho still had a few new ideas to experiment with as well. For the next several years, therefore, the design, production, and testing of instruments continued to be a major part of his activity. In fact, the period between 1584 and 1588 was the culmination of his instrument-making career. The point of departure was his invention of the equatorial armillary.

From its inception in antiquity, the raison d'être of the astronomical ring and its more complex cousins had been to cope with the complications of zodiacal phenomena. Ptolemy had claimed that he used the armillary in making his star catalogue. And because Tycho planned to emulate him in using the sun's and moon's positions in the zodiac as references,[52] he started along the same path. But he

---

[52] Ptolemy, *Almagest*, trans. and annot. G. J. Toomer (New York: 1984), pp. 339–40. For a good argument, however, that Ptolemy did not even make the (bulk of the) observations

Table 5.1. *Genealogical Chart of Tycho's Instruments*

Year				
1569	Half-sextant			
1572	First sextant			
1573			First quadrant	
1576	Steel sextant			
1577			Q. *min.*	
1580	Globe			Zodiacal armillary
1581		Bifurcated sextant	Q. *maj.*	
1582		Triangular sextant	Mural quadrant & Q. *max.*	
1583	Large rulers	Bipartite arc	Portable quadrant	
1584		Astronomical sextant		Equatorial armillaries (2)
1585	Small rulers			Large armillary
1586			Revolving quadrant	
1588	Azimuth semicircle	*Semicirculus*	Steel quadrant	
1589			Icelandic quadrant	
1591	New canal			Portable ring armillary

seems to have decided so quickly that the zodiacal form of the armillary would not work that he did not bother to record any observations made with it until after he had converted it to equatorial form. The idea for the conversion may have come from a sketch in a book published by Gemma Frisius[53] in 1534, which Tycho had read. But it was one thing to see a pamphlet describing every possible combination of rings and discussing a few of the tricks that could be performed "with the aid of such small toy instruments"[54] and quite another to envisage the practical possibilities. It thus seems more likely that Tycho evolved the idea for the one form from the failure of the other. In any case, Tycho's initial results with the first equatorial armillary in history were sufficiently promising to induce him to start warping new rings – only three this time – one-third again as large. By the time that second armillary was ready in the summer of 1584,[55] he had planned a third, which was to become the single most important instrument on Hven.

In December 1584, Tycho laid the foundation for the axis of the large armillary depicted in Figure 5.15 and by the following summer solstice had it ready for use.[56] In simplicity it rivaled the mural quadrant. The only complete ring was a declination circle of one-and-a-third-meter radius. To minimize flexures and optimize handling, it was made of wood (except for the brass alidades and the brass graduation strips) and pivoted on a hollow steel axis. The equatorial armilla was reduced to a half-circle; and all that remained of the meridian circle was the lower pivot point, resting on a half-buried stone pillar one and a half meters long, and the upper wishbone support, likewise so massive that its stone pillars flanked the entrance to its crypt. The sighting on the declination circle (of all three armillaries) was entirely analogous to the plate–cylinder arrangement on the mature forms of the sextant. The readout of right ascensions followed the adaptation invented for the mural quadrant: individual rear sight-plates on the equatorial semicircle and a cylindrical front sight consisting, in this case, of the axis of the declination circle. With the completion of the large armillary, Tycho fulfilled the ambition he had expressed in his youth. Between the one-and-a-

---

for his star catalogue, see Newton, *The Crime of Claudius Ptolemy*, pp. 211–56. For a counterargument, see James Evans, "On the Origin of the Ptolemaic Star Catalogue," *Journal for the History of Astronomy* 18 (1987): 155–72, 233–78.

[53] *Tractatus de annulo astronomica* (Louvain, 1534), as cited in *Dreyer*, 316.
[54] Tycho's general dismissal of rings (after he bought one, modified it, and made another) and his specific reference to Gemma occur in a brief discussion in V, 98.
[55] X, 292, 363. The two small equatorial armillaries are discussed and depicted in V, 56–63.
[56] X, 302, 351. Tycho's description is in V, 64–7.

Figure 5.15. The Large Armillary (ca. 1585).

third-meter armillary and the two-meter mural quadrant, graduated to quarters and sixths of a minute, respectively, he possessed the best equipment relative to the state of his discipline that any scientist has ever enjoyed. In addition, he had a quantity of backup instruments (any one of which would have been, in the absence of the others, the

Figure 5.16. The Revolving Quadrant (1586).

technical marvel of the day) sufficient for the most rigorous demands
of his activities, both observational and instructional.

Several times in the spring of 1586 Tycho used seven or eight
instruments to register the position of the noonday sun.[57] Yet this
very refuge in numbers is a measure of his insecurity. What he
needed was a means of checking the performance of his two major

[57] XI, 3-4.

instruments, and only the corroboration of other instruments of the same quality would satisfy him. One, the one-and-a-half-meter wood quadrant shown in Figure 5.16, became available at the turn of 1585–6.[58] Probably an experimental analogue to the mural quadrant, it was kept rigid by lightweight struts, instead of anchors to a wall, so that it could be turned to any azimuth desired. The second, remarkably similar to the revolving quadrant in both function and design, entered service two years later.[59] Somewhat larger (one and three-quarters meters) and made of steel instead of wood, it was actually the old $Q. max$ (Figure 5.8) recycled into useful form after having been given a completly stable base (Figure 5.17). The installation of the two large quadrants completed the essential furnishings at Hven. With the security they provided, Tycho no longer needed to use a multitude of lesser instruments to check his daily meridian altitudes. With agreement from his four large instruments, he could be absolutely confident that he had the perfectly accurate solar observations without which his dream of a redintegrated astronomy would be a chimera.

Simultaneously with these important productions Tycho's shop turned out a variety of less useful ones. Some, such as the portable quadrant[60] and portable sextant[61] of 1583 and 1584, were simply built with less than maximal aspirations, for specific limited purposes. To this group belong also a new Icelandic quadrant (1589),[62] a new wood canal (camera obscura, for solar observations, 1591),[63] and a portable single-ring armillary (1591).[64] Tycho's semicirculus,[65] made in about 1588 and operated on the globe-shaped sextant stands as essentially a triple-sextant. It allowed the measurement of angular distances between any two covisible stars, but turned out to have very limited utility. Only such devices as a small replica of Copernicus's rulers and possibly the azimuth semicircle of 1588 (Figure 5.17b),[66] can be categorized as capricious, and by the time he made

---

[58] X, 427; V, 32–5.   [59] XI, 1, 244; V, 36–9.

[60] X, 232; V, 20–3.   [61] X, 289–95; V, 24–7.

[62] XI, 319, 360. Presumably a quadrant made for, or according to a design suggested by, Odd Einarson, one of Tycho's short-term collaborators who was also the bishop of Skalholt in Iceland.

[63] XII, 108–18.

[64] XII, 110. The *armilla portatalis* is described in V, 98–100. Its radius was sixty centimeters, and Tycho mentioned that it was useful even at Stjerneborg for low-altitude observations, where the walls of the crypts interfered with observations by the larger armillaries.

[65] Described but not depicted in V, 96–7. Its radius was one and a half meters.

[66] V, 40–3; XI, 251. In fact, the azimuth semicircle is shown mounted on the pillars and azimuth circle made for the old $Q. max$. See Tycho's allusion to the mounting's shortcomings in V, 43. Moreover, the azimuth semicircle and the semicirculus might have been the same instrument, with two different pivot points and different graduated strips on the two sides of the frame. Tycho listed the size of each as six cubits.

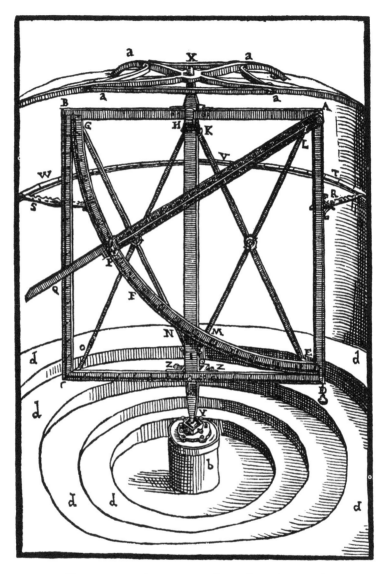

Figure 5.17a. The Great Steel Quadrant (ca. 1588).

them, Tycho was phasing out his construction program. When his two most special colleagues asked him to make instruments for them, he sent Brucaeus his old small quadrant and then had to tell Hayek (in 1590) that he no longer had the work force required to make instruments.[67]

[67]   VI, 101; VII, 273.

Figure 5.17b. The Azimuth Semicircle (ca. 1588).

On the whole, therefore, the instruments that so bewildered Repsold in their number and variety can be rationalized as individual elements in several series of experiments (schematized in Table 5.1), designed to produce a gradual evolution in astronomical accuracy. In such a program, many of the first- and second-generation instruments in the quadrant, sextant, and armillary families would necessarily be obsolete by the time their successors were refined. Some other first-generation instruments would look so hopeless as to convince Tycho to abandon further attempts to improve them. In still other instances, Tycho pursued an idea to the point of developing overspecialized instruments. But the production of less than perfectly successful instruments, or even failures, scarcely justifies the conclusion that the experiments should not have been tried. If Tycho indulged his fascination with machinery in some respects, particularly in his later instruments, what better way was there to do it? If the marginal value of some of the instruments was small, so was the marginal cost.

The man who seems to have been Tycho's chief technician, Hans Crol, was also one of his most trusted observers.[68] It would have been hard to let him go even if Tycho could have been certain that none of his instruments would ever need further modification or repair. And Tycho never did. Crol died on Hven in November 1591,[69] which happens to have been the last year in which Tycho registered any new instruments. But as long as Crol was there and had access to the facilities and experience accumulated during a decade of work, even the most esoteric instruments should have been built on Hven if they were ever to be built at all. Scarcely less important as a reason for Tycho to build them was his unique capacity for communicating the results of his experiments to others, in an era preceding the advent of scientific periodicals. If anyone ever retried any of the instruments Tycho rejected in the *Mechanica*, it is unlikely to have been from ignorance, and it was certainly to no avail.

By 1583, at the latest, it was obvious that Tycho needed more space for his work. The large rulers completed that year occupied one entire observing platform of the two originally designed into the house, and the various other instruments in use or under construction promised to fill up even the ancillary "pillar" observatories being built adjacent to the two main ones. Tycho was still planning new instruments, large ones, for which it would be desirable to have not only 360° access to the sky but also enough isolation to guarantee that the observers who manned them would work independently of each other. Adding further to the house was not really feasible aesthetically, and constructing a separate building anywhere inside the wall would have destroyed the symmetry of the grounds. The wall itself was a possibility and so were the buildings at its four corners.

Tycho's observing experience now, however, pointed to the necessity of having a ground-level observatory. He already had one instrument on his hands – the Q. *max.* – that was useless because of the instability of its mounting, and he hoped to build other, still larger instruments. Only on the ground itself could he secure a

---

[68] See XI, 292, and XII, 30, 39, 44, 46.
[69] Tycho himself recorded Crol's death in his *Meteorological Diary* (IX, 106–7).

foundation that would not be the weak link in his quest for fraction-of-a-minute accuracy. Associated with the problem of stability for his large instruments was the physical difficulty of observing with them. To observe at all altitudes between the horizon and the zenith with an instrument of a one-and-a-half-meter radius, an observer had to be prepared to cope with a variation of one and a half meters in the height of the rear sight of the instrument. A ground-level observatory offered the possibility of digging out a hemisphere around the instrument stand so that the position of the observer's eye would automatically be accommodated to changes in the instrument's elevation (see Figures 5.15, 5.16, and 5.17). Moreover, the excavation of the pit provided some shielding from the winter winds for both the observers and the instruments whose readings were sufficiently sensitive to be affected by gusts. Accordingly, Tycho selected for his new observatory a spot about thirty meters from the south corner of the wall, where a slight hillock offered an enhanced view of the horizon and from where the wall and the house would block only a relatively small and low portion of the least significant (north) aspect of the heavens.

Excavation for the observatory seems to have begun in the spring of 1584.[70] Steps (A in Figure 5.18b) were cut out and a hole was dug for the "warming room" (B) which would be the focus of activity in the observatory. By modern standards it would be a small room, only some three meters on each side but set deep enough that its walls would extrude above the ground only enough to accommodate four low windows. During the summer (probably), the digging was continued to produce the passage to and outline of crypt C and the steps back up to and the shallow pits that would become crypts D and E. The hemispherical pits were then terraced in concentric brick circles (Figures 5.15, 5.16, and 5.17). The floor and walls of the warming room and the walls of the three crypts were more straight-forward and may well have been finished that summer, too. But because the zodiacal armillary for crypt E was the only instrument ready for installation,[71] it is clear that the new observatory was not used during its first winter.

In December the foundation stone for the great equatorial armillary was implanted in crypt C.[72] By the summer of 1585, the ar-

---

[70] The details concerning Stjerneborg were given by Tycho in V, 147–9, and VI, 250–86. The diagram is in V, 146.

[71] Presumably Tycho reconverted his first (smaller) equatorial armillary into its zodiacal mounting (Figure 5.11) so that he would have one available in that form. He confessed (V, 55) that "we did not use this kind of armillary instrument very often, and particularly not when the greatest precision was required."

[72] X, 302.

ORTHOGRAPHIA STELLÆBVRGI
EXTRA ARCEM VRANIÆ SITI.

a

ICHNOGRAPHIA STELLÆBVRGI.

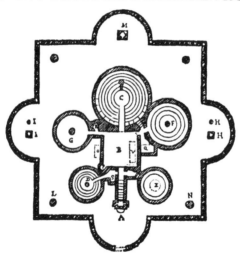

b

Figure 5.18. Elevation (a) and plan (b) of Tycho's underground observatory, Stjerneborg; constructed 1584–6.

millary was in operation, and by the beginning of 1586, the wooden revolving quadrant was being oriented in crypt D.[73] Because all three instruments by this time must have been provided with cover from the elements and because crypts F and G would have such low profiles as to make little difference whether or not they were completed, the facility would have had essentially the silhouette displayed in Figure 5.18a by its second winter. By this time it doubtless also had its name: Stellaeburg (Stjerneborg in Danish), or star town.

Refinements were added during the following two or three years. Whether as original design or as a modification, the roofs of the crypt turrets were

> made of small, smooth beams ingeniously joined together and connected, below the horizontal top of the wall and outside the azimuth circle, by a strong, round wooden ring. Hidden inside this ring (were) wheels, placed opposite each other in four places. With the aid of these wheels the roof (could) be turned around, with little effort, as (might) be desired. In this way, the two oblong windows, which (were) placed in the roof opposite each other ... (could) be turned toward any star that (was) to be observed.[74]

Crypts F and G were added in 1587 or 1588[75] to provide a solid mounting for the great steel quadrant and for one of the sextants that had previously been used more or less portably, on the various "globe" mountings (Figure 5.12) situated strategically on the platforms up at the house and at each corner of Stjerneborg. The sextant was probably mounted permanently in anticipation of the great star catalogue project that moved into high gear in about 1587 and continued for most of five winters. Extra stability for the instrument was no doubt one motivation for the move, but the discomfort of midwinter open-air observation was surely a factor, too.

The shelter provided by the partial interment of crypts C, D, and E may by this time have been perceived to be insufficient protection against the winter windchill, for crypts F and G were buried up to their (rotatable) windows. Alternatively, Tycho may have been willing to forgo all low-elevation observation from those two crypts in order to minimize the obstacles to observation from the other three crypts. But there can be little doubt that many winter nights were

---

[73] X, 351, 427.   [74] V, 35.

[75] *Dreyer* (104) assumed that because crypts D and E were higher than the entrance tunnel – whereas crypts F and G were level with it – crypts D and E must have been built later. The chronology of the instruments and the order of Tycho's lettering of the crypts both argue the opposite. The orientation of the quadrant is noted in XI, 1, 244.

thoroughly unpleasant.[76] Even the roof of the warming room was covered with dirt and sod so that it looked like a little hill (except for windows to admit light and emit smoke), and yet the little niches with beds for Tycho (O) and his assistants (Q) were probably used as much for warming up as for resting between observations. And even these expedients may have failed to solve the problem, for eventually Tycho was to start his peasants digging a tunnel under the wall and all the way back to the house.

Some of the less important amenities of Stjerneborg must have been completed after 1590. But over the years, Tycho was able to convert the cramped little facility into a most unlikely showpiece. Above the entrance, three stone lions called attention to Tycho's Latin motto proclaiming, "Neither wealth nor power, but only knowledge, alone, endures."[77] Other figures set off the Brahe coat of arms. On the back of the tunnel housing was a dedication tablet similar to the one set in Uraniborg, explaining that this facility had been constructed at enormous expense for the advancement of astronomy and asking that posterity preserve it for that purpose and for the glory of God and the honor of Denmark.[78] Even the one feature of Stjerneborg's exterior that did not lend itself readily to decoration, the sod roof of the warming room, was dignified as a miniature Parnassus, the mount of the Muses, and capped with a small brass statue of Mercury turned by (probably) a windup spring.[79]

Although the crypts themselves were all business, the warming room was blessed with most of the comforts of home. Over the underground entrance to it was another slab expressing in Latin verse the surprise of Urania at having access to her stars even from a cave in the bowels of the earth.[80] Inside, the ceiling was painted with circles representing Tycho's system of the world, and the spaces between the instruments mounted on the walls were filled with pictures of eight astronomers. Under each of the first six pictures – of Timocharis, Hipparchus, Ptolemy, Albattani, Alfonso, and Copernicus – was a label listing the credentials of the honoree. The lines under Tycho's picture left the judgment of his work to posterity, but the picture suggests what Tycho thought the basis of that judgment would be, by showing him pointing up at a depiction of his world system and asking if this were not the way the world is constructed. The lines under the last picture provided Tycho's confident answer to the question. Referring to an unborn astronomer, a descendant

---

[76] The French astronomer Picard visited Hven a century later in September, October, and November (1671) and found the weather completely untenable. See *Dreyer*, 105, and *Norlind*, 84.

[77] VI, 272.   [78] VI, 273.   [79] VI, 287.   [80] VI, 273.

of Tycho, they expressed the hope that "Tychonides" would be worthy of his great ancestor.[81]

For all its elegance, however, Stjerneborg was probably used only by those who absolutely had to be there. Although it was more comfortable than the laboratory or the parapet observatories, it was considerably less spacious. Those not actually collecting data, therefore, must have worked in the library, except during the cold months, when they may have preferred to join the less advanced members of the group in the well-heated winter dining room.

In the library, accordingly, Tycho kept his collection of over two hundred professional books,[82] as well as his cabinets of spheres, globes, drafting instruments, and other equipment. The tone of the decor was set by his spectacular one-and-a-half-meter brass globe, with its delicate silk dust cover. With such a focal piece, Tycho could scarcely fail to add paintings commemorating (in fact, supposedly depicting) various heroes of astronomical progress or Renaissance enlightenment. And the remaining space – Tycho could not tolerate wasted space – between cabinets, windows, and pictures was filled with framed poems and mottos exhorting and inspiring Tycho and his students to greater heights.[83]

Stjerneborg was only one of several construction projects undertaken during the 1580s. Sometime after the wooden fence that defined the shape was raised, Tycho decided that the semicircles breaking up each eighteen-meter side constituted a nice touch and that the earthen walls surrounding the grounds of Uraniborg could benefit from the same feature. No doubt the resulting burden for the peasants on Hven was one reason that the underground passage to Stjerneborg was never finished.

As Tycho began to anticipate completing his great tome on the comet, another project presented itself: getting the book printed. Although printing was an enterprise well into its second century, it was still basically a cottage industry. With a relatively small supply of type, a printer could set a few pages, run proof sheets from them, correct the errors, print the required number of copies, break down the galleys, and proceed to the next few pages. It was very efficient in every respect, except the proofreading. Unless the material was simple enough to be checked by the printer, a proofreader had to be on hand for an hour or two every few days to keep the process rolling.

Tycho's material definitely was not simple. Thus, to have the

[81] VI, 274–6.
[82] See *Norlind's Bilaga* I (333–66) for a list of the books that Tycho is known or presumed to have owned.
[83] VI, 268–71.

printing done at Copenhagen would have required Tycho to live there as long as the book was being printed. The alternative of having the printer hold each set of galleys while the proofs were carried out to Hven and back would have been even more expensive. In addition, Tycho would surely have had to buy any printer extra type to set the *numbers* for his highly quantitative material, anyway. He probably realized during his first years on Hven that the only reasonable solution to these problems was to establish his own press.

By the middle of 1584, his preparations had begun, and by the fall of that year he had a press set up in the small building that had been constructed into the south corner of the wall. On 27 November 1584,[84] Tycho noted in his observation log that he had on that day inaugurated his press by printing a poem honoring his friend Erik Lange. And because it was almost a year before he had anything more serious to print on his press, the productions that followed – epitaphs for the friends of his youth, Pratensis and Johannes Franciscus, and poems to ex-*rigsraad* Jacob Ulfeld, Chancellor Niels Kaas, and Governor (of Holstein) Heinrich Rantzov[85] – were equally peripheral.

Tycho was candid enough to mention in one of these epitaphs that the printing was inspired by the need to give his printer something to do, but such humanist enterprises were part of the scholarly ethic of Tycho's day. The poem to Kaas was 288 lines of Latin. Such productions served to advertise the merits not only of the dedicatee – Ulfeld, Kaas, and Rantzov all were renowned as patrons who were learned in their own right – but also of the dedicator. Tycho had already paid for the printing of such productions as an ode to his still-born twin brother and would continue to issue others in the future.

After this initial period of underemployment, Tycho's press was in almost continual use. At the end of 1585, a meteorological calendar compiled by one of his students and augmented by Tycho's observations of the comet of 1585, was printed.[86] It was followed almost immediately by the printing of fifteen hundred copies of his book on the comet and then by the preparation of three other books before Tycho packed up his press and took it with him when he left Hven in 1597. So thoroughly occupied was the press, in fact, that Tycho almost immediately had difficulty obtaining paper for it. After having his printing held up several times during 1586 and 1587

[84] X, 302.
[85] IX, 176–90. The poem to Lange was found after volume IX went to press and is printed in XV, 3–5. Ulfeld had been expelled from the Rigsraad for negotiating an unsatisfactory treaty with the Russians.
[86] "Diarium astrologicum et meteorologicum anni a nato Christo 1586." IV, 512–13.

while he searched frantically for paper,[87] Tycho decided that the only thing to do was to build his own paper mill. That, however, would be an enterprise of the 1590s.[88]

Just as significant as the new facilities were the new amenities added to Uraniborg during the 1580s. If the work on the family rooms became sufficiently completed to allow occupancy only at the end of 1580, it is hard to believe that the fitting of trim, laying of tiles, painting, and decorating of the rest of the house did not continue for another couple of years, particularly in the student garrets. With the construction of the pillar observatories and the gallery around them, and the expansion of the winter dining room, there could not have been much time during the first half of the decade when some part of the house was not being worked on. The result was that even the primary decoration of Tycho's study – the mural behind his mural quadrant – was not painted until 1586,[89] even though the instrument itself had been in constant use since 1581. This mural (Figure 5.9) is both a remarkable source of information about Uraniborg and an apparently representative example of Tycho's concern for beauty and detail, which he displayed in furnishing and decorating his home.

As the banner across the top of the mural proclaims, the portrait depicts Tycho during his fortieth year. Tycho stated it was generally regarded as a good likeness and mentioned proudly that it was done by Tobias Gemperlin of Augsburg, whom he himself had encouraged many years earlier to seek a career in Denmark. The landscape at the top of the picture was done by the king's painter in residence, Hans Knieper of Antwerp.[90] The much more interesting scene below the arch was not part of the mural at all, but a contribution by the engraver of Tycho's copperplate, carefully designed to fill in the available space without impinging on the mural itself. The latter was painted by the royal architect, Hans van Steenwinckel, who had worked with Tycho from the middle stages of Uraniborg through the completion of Stjerneborg and whose interest in actually making observations himself was, as we have seen, documented by several references in Tycho's log.

Perhaps the most noticeable feature of Steenwinckel's contribution is the "wall" created behind Tycho's back by the clever use

---

[87] VII, 106, 118–19, 385.   [88] See Chapter 10.

[89] The portrait is dated 1587. But as it also purports to depict Tycho in his fortieth year and Tycho mentioned in his book on the comet (IV, 235), which was printed in 1587, that the painter "died not long ago from the plague in Copenhagen," 1586 seems a more reasonable date. In 1581 Knieper painted a great tapestry (that is still extant) for Kronborg Castle. In the background of the tapestry's portrait of Frederick are two noblemen, one of whom is Tycho. See Harold Mortensen, "Portrætter af Tycho Brahe," *Cassiopeia*, 1946, p. 55.

[90] Tycho's description of the mural is in V, 29–31.

of perspective. The vertical "corner" line and slanting "bookcases" almost suffice in themselves. The miniatures of King Frederick and Queen Sophie hanging (or painted) at appropriately different levels complete the illusion. The niche between them was real and probably not sloped or shaped to conform to the perspective, but the "frame" around it may well have been painted in to enhance the overall effect. The globe was real, too, but small and spherical and hence no problem. Tycho said that he designed it as an automaton, to reproduce the motions of the sun and moon and even represent the phases of the moon. Almost as interesting as an artistic device is the humorous contrast between the lazy repose of Tycho's dog and the industry of Tycho's assistants in the scenes above him. The scenes themselves portray a north-south section through the middle of the "working" (south) wing of Uraniborg but add little to the knowledge imparted by Tycho's other diagrams. In conjunction with those other diagrams, however, Tycho offers enough information to provide a good idea of the surroundings in which he and his associates were working by the late 1580s.

By all indications, the chemical laboratory was, and remained, basement. The fact that Tycho installed some small furnaces in the winter dining room must have reduced the actual use of the basement facility to those occasions on which particularly extensive or particularly noxious experiments were contemplated. Similarly, the upstairs observatories were, and remained, basically roof. The removable triangular sections of wood comprising the conical covers would have provided little protection against the elements and no place at all for even the storage of equipment, let alone the introduction of amenities.

When Stjerneborg was completed, therefore, most of the observational work would probably have been shifted to it even if it had not housed the largest and newest instruments (except for the mural quadrant). Steenwinckel's scene confirms that this shift had not been completed as of 1586, for it shows (top left) the unmistakable form of the steel quadrant, apparently in use or ready for use before crypt F of Stjerneborg was ready (1588) to receive it.

That the creation of an entirely new standard for astronomical facilities and instrumentation was very expensive, was something that Tycho mentioned repeatedly, both privately and publicly. Much of the expense was borne by others: The most easily reckoned form of aid was the various grants awarded to Tycho by the king. All together they provided an annual income of about 2,400 dalers – about 1 percent of the crown's total revenues – with a purchasing power about eight times that of the highest paid professors at the university (the Theologians) and perhaps thirty times the salary of the lowest paid (mathematicians, typically).[91]

The people who really contributed this money, given the power structure of the day, were Tycho's noble peers. For although the king dispensed the monies from the state, they were basically destined to reward one noble contribution or another and thus were bound to be spent privately on one noble pursuit or another. Tycho was sufficiently conscious of this fact to be very proud of the status conferred on him with these award. What he was much less conscious of was the contribution to his efforts by the peasants of Hven.

Even though they may not have done any more than their obligation under the law and even though many other peasants in Denmark were suffering under the same demands, if Tycho had chosen to settle somewhere else, the lot of the farmers on Hven would have been much easier.

The costs of which Tycho was most aware were those that he paid. They included even the money spent from his grants, as there was no precedent obligating him to spend the funds in any particular way, and because, like most recipients of public largesse, he undoubtedly regarded them as no more than just fruits of his own virtues and labors, anyway. But it is extremely unlikely that Tycho kept separate books on his enterprises. Proceeds from his personal estate surely went into the construction of Uraniborg as fast as they became available.

Tycho also alluded frequently to the price he paid in professional time, but it is doubtful that even he realized just how great that price was. It was not one that can be reckoned simply as the total amount of time he devoted to establishing the shop, designing the instruments and observatories, and supervising every step of their production and construction. Each of these expenditures had to be multiplied by the inefficiency factor inherent in any protracted task, for there was a period of at least ten years during which the construction of his facilities and the production of his instruments could never have been completely out of his thoughts. But the design and fabrication of Tycho's instruments was only part of his responsibility for them.

---

[91]   Eric Warburg, "Was the university of Christian III a twopenny university, or was the salary demanded by Leonhart Fuchs unreasonable? The professors' pay in the sixteenth century," *Centaurus* 15 (1970), 72–106. After the reorganization of the University of Copenhagen in 1571 the two professors of Theology received 300 dalers per year. Scholars with the M. A. who needed to travel abroad for advanced degrees were deemed to be able to live on 30 dalers per year. Preparatory school teachers typically received 20 dalers and board. The royal architect's remuneration was one hundred dalers plus quarters and various quantities of foodstuffs (*Gade*, 63).

Putting them into service on stable, perfectly oriented mountings in circumstances conducive to good observing was just as important and probably just as time-consuming. Moreover, after each completed instrument had been given a careful initial orientation, it had to be rechecked periodically to make sure that it continued to provide accurate results as long as Tycho was still using it.

Tycho's logs are littered with annotations on everything from minor adjustments to major alterations, each tantamount to an obituary for all observations with the instrument in question, and all testimony to Tycho's uncompromising determination to settle for nothing less than the very best performance obtainable. Time and again an instrument disappears from the log entries only to return months later with the explanation for the disappearance "post rectificatum," to bear out Tycho's claim that "we have remade most of (our) instruments more than once ... (and) have indeed had instruments completely rebuilt."[92]

Even more time-consuming than the fact that the design, production, and performance of his instruments bred a series of intermittent diversions for his creative energies through virtually his entire career, was the time lost through sheer waiting. As long as Tycho had visions of access to better instruments, there was little point in making observations (except to test instruments) and even less in trying to work seriously with them. Thus, the continuous improvement of his instruments forced Tycho to postpone almost every aspect of his work. Even his publications on the new star and the comet had to wait for the most exact possible locations of the background stars, which could come only after 1585, when the great armillary and the revolving quadrant became available to check the results of his mural quadrant.

That Tycho was entirely justified in both the pessimism with which he regarded his early observations and the esteem in which he held his later ones, has been shown by modern analyses of his observations. From the instruments available to him before his shop began to produce them in 1581, he obtained observations averaging errors of about 4′.[93] Although this was better than anyone before him had done, it was far short of the 1′ accuracy he demanded. By the time of the comet of 1585, however, Tycho had achieved his goal.[94] Because the instruments and procedures he used on the comet

[92] V, 19. Tycho told Kurtz that he got a late start in his work because of the founding of Uraniborg (VII, 260).

[93] This result was based on a statistical determination of the orbit of the comet done by F. Woldstedt, *De Gradu praecisionis positionum cometae* 1577 (Copenhagen, 1844).

[94] C. A. F. Peters, "Bestimmung der Bahn des Cometen von 1585 nach den ... Originalbeobachtungen Tycho's," *Astronomische Nachrichten* 39 (1849): 109–276.

were essentially those he used for all his subsequent planetary and stellar positions, this figure must be representative of the accuracy of Tycho's later observations. It is certainly much more reasonable than the 4' so frequently cited in secondary reports of Tycho's work.[95]

These typical results could be refined considerably with multiple observation. For example, the distances between the stars of Cassiopeia (which were important as reference stars for the nova) were determined with the bipartite arc to an accuracy of ±41",[96] and the fundamental stars for Tycho's catalogue were positioned within ±25", from observations with the mural quadrant and the great armillary.[97] In the case of meridian altitudes, in which observations could be made without moving the instruments appreciably from one object to the next, Tycho could obtain such accuracy even in fairly routine daily work. Hundreds of meridian observations of the sun show average errors of about 47" for the years 1582–5, 38" for 1586, and 21" for 1587–90.[98] In the context of the developmental history of Tycho's quadrants, from which these observations are primarily derived, this progress requires no elaboration. It stands both as a typical piece of evidence of the purpose to and efficacy with which Tycho employed his artisans and as eloquent testimony to the fact that if Tycho himself personally paid a high price for astronomical accuracy, he at least got what he paid for.

[95] Dreyer (357) rightly objected to the citation of the four minute figure on the grounds that it stemmed from observations that included "the cross-staff, which Tycho always mentions as an untrustworthy instrument," without realizing that the instrument used for most of the rest of the observations was the *steel* sextant (IV, 369–71), with the old slit-sights and pretransversal divisions. In a recent study of the individual accuracy of many of Tycho's instruments, Walter Wesley, "The Accuracy of Tycho Brahe's Instruments," *Journal for the History of Astronomy* 9 (1978): 42–53, showed that one minute is a good estimate of the average observational error of the best several of them. In another study, V. Bialas, "Bayerischen Akademie der Wissenschaften, Abhandlungen," *Math-Nat. Klasse* 148 (1971): 106–18, examined 107 observations of Jupiter and found errors whose average absolute value was 2'16".

[96] F. W. A. Argelander, "Über den Neuen Stern vom Jahre 1572," *Astronomische Nachrichten* 62 (1864): 273–8.

[97] *Dreyer*, 351–2.

[98] G. L. Tupman, "A Comparison of Tycho Brahe's Meridian Observations of the Sun with Leverrier's Solar Tables," *The Observatory* 23 (1900): 132–5, 165–71.

Chapter 6
The Flowering of Uraniborg

TOWARD the end of the summer of 1581 as the finishing touches were being applied to Uraniborg, Kirsten gave birth to a son. In a family that to that point consisted solely of daughters, it must have been a great occasion, no matter how elevated the rights and status of Danish women may have been relative to those in other contemporary cultures. After having disposed of his father's name (Otte) and used his maternal grandfather's name (Claus), Tycho now baptized his third son Tyge, after his paternal grandfather. When another son was born in 1583, he was named Jørgen, presumably after Tycho's uncle and stepfather but perhaps after Kirsten's father as well.

Tycho's sons were only the vanguard of a considerable increase in the Brahe household. From at least as early as the planning of the eight-room garret on the top floor of Uraniborg, Tycho had envisioned having a significant number of assistants to work with him. As space, instruments, and time became available to Tycho in the early 1580s, therefore, he began to select collaborators until he accumulated a group of eight to twelve[1] members of varying degrees of permanence and competence. Thanks to the curiosity of an anonymous inmate who compiled a fragmentary list of his fellows[2] toward the end of the 1580s, we can obtain a glimpse of at least the upper half of the spectrum of Tycho's assistants.

For the most part, Tycho's students were Danes, drawn from the highways and byways of the realm as far away as Norway and Iceland. There also were several from Holland and Germany and at least one three-month visitor from England. With the exception of two or three who were acquaintances from Tycho's own student days and an occasional "older" traveling scholar, the students were in their early twenties. They were veterans of the university and were, in theory at least, striving for excellence in some form of higher learning, just as Tycho had after his first trip abroad. Some quickly discovered that a full-time scholarly life – at least under the restrictions of an island habitat, as opposed to the license of a university town – was not what they had thought it would be. Already by the time the aforementioned list was compiled, eight of the thirty-two listed – including the two immediately following Flemløse – could

---

[1] *Gassendi*, 45. For a list of abbreviations of commonly used sources, see Appendix 1.
[2] The complete list is in XIV, 44–5. See also *Dreyer*, 381–4.

literally not be remembered by name. On the other hand, one who was recruited after them, Gellius Sascerides, was associated with Tycho for almost as long as the ten years that Flemløse was.

The son of a Dutch-born professor of Hebrew at the University of Copenhagen, Gellius had been born and raised in Copenhagen and had attended the Universities of Copenhagen and Wittenberg before coming to Hven.[3] He seems to have arrived before Christmas of 1581, because the "Andreas of Viborg" who came after Gellius is credited in Tycho's logs with having made an observation on 24 December 1581. If the latter is the same Andreas who helped Tycho[4] in the fall of 1584, he was at Uraniborg for most of three years. But by the time the list was drawn up, Andreas was back in Viborg as a parish priest.

Sometime in 1582, probably, Jacob Hegelund joined the crew. He was at Uraniborg long enough to be remembered as a good writer, musician, and a lively intellect, but he was probably not there for more than a year or so. The younger brother of an old school companion of Tycho's at Leipzig and Rostock, he did not rise as far in the establishment as his bishop brother, Peder,[5] did. But he did obtain a teaching position at Denmark's elite Sorø Academy and thus met Tycho's sons when they enrolled there in 1590 and 1593.[6] The headmaster at Sorø during those years was another veteran of Hven, Johannes Stephanius, who was with Tycho for a couple of months in the early spring of 1582 and then again from 5 December 1583 to 15 October 1584. Either Stephanius's status was different from that of the other people on the list, or he was simply forgotten when the list was compiled. We know of his presence on Hven only because Stephanius had at the age of twenty-two already acquired the habits and inclinations that would advance him to the position of professor at the university (1597) and royal Danish historiographer, and thus he recorded, in addition to his own arrivals and departures, many valuable data concerning Tycho's life that have not been preserved elsewhere.[7] Gassendi, too, managed to record a few anecdotes in the late 1640s when he was interviewing people for the first biography of Tycho. Through him we are told that Tycho's establishment boasted for a while a dwarf named Jeppe, whom Tycho believed had some kind of prescience[8] and that Tycho was superstitious about

[3] *DBL* XX, 580–1.   [4] *Gassendi*, 292.   [5] *DBL* IX, 528–9.

[6] *Norlind*, 96. According to Tycho's meteorological diary, little Tycho left to start Sorø on 16 July 1590 (IX, 88), and Jørgen went on 28 May 1593 (IX, 119).

[7] See *DBL* XXII, 578; and J. R. Christianson, "Tycho Brahe's Facts of Life," *Fund og Forskning* 27 (1970): 21–8.

[8] *Gassendi* (181) reported that Tycho is supposed to have asked Jeppe for advice on numerous

both rabbits and old ladies and would return home if either crossed his path when he set out on errands.

Another source of information on a much broader variety of subjects is the meteorological diary that Tycho began on 1 October 1582. Long interested in the problem of predicting at least the gross departures from the normal weather patterns, Tycho apparently decided after fifteen years that he should work more inductively from the phenomena. For virtually all of his remaining five thousand days on Hven, Tycho had various of his associates record in Danish (or, occasionally, German) each day's fair, cloudy, windy, or rainy conditions, along with any unusual meteorological appearances.[9]

By October 1584,[10] this undertaking was expanded to include the recording of other news. Arrivals and departures of island personnel and visitors were the most common notations, but a fierce storm (30 January 1592), a house fire (6 May 1587), a body washed up on the island (20 May 1596), a shipwreck (three times), an incarceration (13–16 October 1590), and deaths (too numerous to cite) all were deemed worthy of mention. Over the years they accumulated to some seven hundred items, usually entered in Latin, which provide interesting tidbits of daily life on Hven.

Through the early years, a frequent keeper of the log was a Dane from the island of Mors by the name of Elias Olsen. His hand first appears in the entries for April 1583, and his place on the list suggests that he cannot have come long before that time. Olsen remained on Hven until his death in 1590.[11] He seems to have established himself quickly as Tycho's most capable assistant, perhaps because he was more interested in astronomy per se than Flemløse and Gellius were.

In the spring of 1584 when Tycho decided to send someone to Poland to confirm the latitude of Copernicus's observatory, it was Olsen who went. The theoretical background of the expedition was

matters. On one such occasion, when the peasants on Hven had been insubordinate, Jeppe counseled that they be assembled and given all the beer they could drink, instead of being punished. The historian Ole Worm, who related the tale, suggested that Tycho would have done better to have heeded Jeppe's advice (*Norlind*, 88). *Norlind* (89) also contains a picture (painted in 1691) of a woman who was supposed to have served Tycho in her youth and lived to the age of 124. She apparently claimed to have received access to Tycho's medical knowledge through Sophie Brahe and made a reputation (and a living) as a healer.

[9] IX, 5–146.

[10] For reasons that are not clear, *Dreyer* (122), and *Norlind* (87) following him, cite April 1585 as the beginning of the notations.

[11] IX, 82. Because of a diary reference to an(other) Elias Olsen in 1596, *Dreyer* (123) believed that the reference to his "obiit" intended to convey his "abiit" (departure). However, the exact recording of the hour (11–1/2 *noct.*) and the unlikelihood that anyone would leave at midnight for a two- or three-hour boat trip to the mainland, argue strongly for "obiit."

the discovery by Tycho, during the previous winter, of the effects of astronomical refraction. This discovery led Tycho to conclude that the obliquity of the ecliptic was appreciably greater than the 23°28′ found by Copernicus[12] and by practically everyone else in Renaissance Europe and that the latitude of Hven was also greater than that shown by his determinations from the (refracted) meridian altitudes of the sun. Both of these constants were bound up with the solar theory, on which Tycho was just then in the process of making almost-final adjustments. Tycho's version of the theory involved constants that differed rather substantially from those used by Copernicus. And although Tycho was not inclined to worry about proclaiming results that disagreed with those sanctified by either time or authority, he was naturally curious as to how Copernicus might have achieved his results.

At the beginning of 1584, a rare opportunity to look into the matter presented itself to Tycho: An ambassador from Prussia and his entourage were leaving the Danish court and returning to Danzig in Danish ships. Because Copernicus's obervatory at Frauenburg was only a short distance from Danzig, all that was required for a check of the latitude that Copernicus used to reduce his observations of the sun was passage on, and the use of, one of the royal ships for a few weeks. The arrangements were made, and Olsen was selected to make the trip. Because he happened to be tending Tycho's meteorological diary at the time, he took it with him and recorded the weather on his trip. Whatever this may indicate about his and Tycho's understanding of the geographical extent of weather patterns, it preserved for posterity an account of Olsen's journey and results (and perhaps initiated the idea of using the diary to record nonmeteorological events).[13]

Sailing with the wind, the party took ten days to reach Danzig, whereupon Olsen's ship left the others and took him east to the Frisches Haff, arriving at Frauenburg three days later. Some three weeks of observations there confirmed (when the computations were eventually done) Tycho's hunch that Copernicus, through his ignorance of the effects of refraction, had indeed underestimated his latitude by more than two minutes. At Frauenburg, Olsen was asked whether he might also take the time to determine the latitude of the great port at the other end of the Frisches Haff, Königsberg. After three weeks in that task, he returned to Frauenburg for five days, went to Danzig briefly, and then finished the journey with two weeks' sailing back to Denmark.

[12] Although Tycho mentions 23°28′, Copernicus actually settled on 23°28$\frac{2}{5}$′ (Book III; chaps. 2, 6, 10).

[13] X, 345–8.

When Olsen returned to Hven, he had with him more than just the measurements for which he had been sent. The canons at Frauenburg had been so impressed either by the technical level of the expedition itself or the attention it focused on their illustrious predecessor that they had sent to Tycho a wooden triquetrum (rulers) reputed to have been made and used by Copernicus himself. It was a truly remarkable souvenir, and Tycho was genuinely awed by it. So inspired was he that he placed it on permanent display in the North Observatory and sat down to compose an impromptu Latin poem to commemorate the occasion.[14] In time, however, Tycho's reverence was overcome by objectivity. As he lived with the instrument daily, he found himself wondering more and more how one could make decent observations with rulers only eight feet long, made out of pine, and scaled by pen and ink. In the end, although he had already constructed a larger version out of brass and found it wanting, he could not help duplicating Copernicus's model in his shop, just to see how much better he could do with the same design. However superior his result may have been, it was not enough to remedy the shortcomings inherent in the basic design of such instruments. And because Tycho was able to circulate his findings so extensively in his *Mechanica*, the replica of Copernicus's rulers may well have been the last such instrument built.

As word of the enterprise on Uraniborg spread, Tycho began to see a flood of would-be assistants. How many he interviewed is not recorded, but he accepted at least fourteen between late 1582 and late 1584. If he had any criteria for selection, they do not seem to have served him well, because most of the fourteen were on Hven for very short tenures. One group of six came together and left together after only a month or so. A parish priest from Jutland and an older cleric from Iceland seem likewise to have stayed for only about a month. The latter, Odd Einarsson, appears to have persuaded Tycho to build some special kind of quadrant, for Tycho mentioned having made an "Icelandic quadrant" of unspecified characteristics, and Odd returned to Hven for a few days in 1589 – this time as bishop of Iceland – perhaps to pick up his quadrant.[15] One assistant, said to have been highly gifted, was killed in Copenhagen by a university student. The younger brother of Gellius Sascerides left after six months. Only

---

[14] V, 44–7. Part of the poem has been translated by Edward Rosen, in Jerzy Dobrzycki, ed., *Nicholas Copernicus on the Revolutions* (Baltimore: Johns Hopkins University Press, 1978), p. 412. Five years later Tycho expressed his attitude toward Copernicus's instruments (and theoretical prowess) with the words "by means of these puny cudgels he surmounted the lofty Olympus": VI, 253, 265–7.

[15] The quadrant was tested on 23 and 24 April 1589 (XI, 319, 360). Odd arrived on Hven on the twelfth and seems to have left on the sixteenth: IX, 71–2.

Rudolph of Groningen (Holland) stayed for more than the three years[16] that Tycho regarded as the period necessary to recover the time and energy required to train an assistant, and Rudolph spent some of his time in Copenhagen, where Tycho arranged his admission to the university.

Given the rapid turnover in Tycho's staff, it is clear that either finding or keeping good collaborators was at least as inefficient as obtaining good instruments was. And given the frequently blustery, not to say overbearing, tone of Tycho's correspondence and publications, it would be easy to conclude that Tycho's personality was at least part of the problem – if there were not ample evidence that such behavior was more or less the norm for the Danish nobility.

Most of the personnel problems that Tycho experienced, however, were due to the essential incommensurability of the university education of the day and the research Tycho was doing on Hven. The credentials that the students brought to Hven were of very little value to Tycho. And because the training Tycho could provide had little value to the students beyond the role it could play in the research at Uraniborg, a student was worth training only as long as he seemed to have the motivation and aptitude to contribute to Tycho's (or his own) research. As Tycho expressed his recruiting problems to Rothmann in 1589:

> If Victorinus Schönfeld [professor of mathematics at Marburg] wants to send his son here to stay with me, that is all right with me. But as soon as his son or I feel that it is appropriate for him to leave, either of us must be free to bring it about. Whether or not he has obtained the M.A. degree, is immaterial to me. I would prefer that he really be a master of arts, rather than just have the degree. But that is no easy matter, so it will suffice if he is a serious student.[17]

For the most part, therefore, all Tycho could do was to accept students and see how they developed. By late 1584, when the parish priest on Hven, Jacob Lollicke, expressed an interest in joining the professional staff, Tycho came up with a new wrinkle: a three-year contract that would not only test the applicant's commitment but also make sure that Tycho got an appropriate return on the investment of time necessary to produce a skilled assistant. Although Tycho seems to have retained this indenture scheme, it does not appear to have worked out particularly well in Lollicke's case. For when he left Hven, after only a year and a half, to accept another

---

[16] Rudolph is listed as an observer of the comet of 1585: XIII, 345. His departure is registered in the diary for 8 June 1588.

[17] VI, 198.

parish call, it was at a parish so close to Knudstrup that Tycho could probably have arranged to have the offer withdrawn if he had valued Lollicke's services.

As nearly as can be determined, the most important technical assistant Tycho had during his long tenure on Hven was not one of his students at all but, rather, one of the artisans from his shop, Hans Crol. Crol was a goldsmith, lured from Germany (or some land to the east of it) sometime before 1585, when his name first appears beside an observation in Tycho's log.[18] By 1590, Crol had done enough observing to establish a reputation for having very keen eyesight and, accordingly, to be singled out by Tycho as the one to make a series of important measurements of the angular diameters of the various planets.[19]

All that Tycho said about Crol's term on Hven is that when he died and was buried there on 4 December 1591,[20] he had "had charge of my instruments for many years, and had even built some of them himself."[21] If this description suggests that Crol probably did not arrive early enough to participate during the ·crucial period of the early 1580s – when the largest number of Tycho's most important instruments were conceived and fabricated – it also shows that he provided other services that removed him from the status of a mere technician. Indeed, Tycho's previous biographers accorded Crol the status of not only a student but also the particular student who cared enough about the tradition of the program to record the list of its members.[22] The problem with this assumption, which is based on comparisons of handwriting, is that Crol is known to have been with Tycho before – and probably well before – the fall of 1585, whereas the "I" that occurs as number twenty-nine on the list of students must have arrived at Hven after the summer of 1587.[23]

In the spring of 1586, Tycho found another promising assistant, nineteen-year-old Christian Johansson of Ribe, who arrived the day that the Reverend Jacob Lollicke left. "Ripensis" was to work with Tycho for four years before leaving to start a career in the church that was to take him through a professorship at the University of

[18] X, 373. Carl L. Jansson's speculation, "Några spridda notiser angående Tycho Brahes lärjungar," *Cassiopeia* 8 (1946): 112–17, that he was the son of University of Copenhagen professor Hans Thomesen Gullsmed seems to founder on the fact that Aurifaber (Crol) was buried at Hven. If he were a "local," it seems likely that he would have been buried in a family plot.
[19] XI, 292.
[20] IX, 106, 107. A son of Crol's had died on the island a year and a half earlier: IX, 85.
[21] VI, 299, 371.   [22] See *Dreyer*, 381–2; and *Norlind*, 94, 98.
[23] Number twenty-seven on the list, John Hammond, is said to have been on Hven a quarter of a year (XIV, 45) and to have left there on 2 November 1587 (IX, 58). Sebastian, number twenty-six on the list, is mentioned in the log for 10 March 1587: IX, 52.

Copenhagen to the rank of bishop. Long after leaving Hven, he would send back to Tycho observations of the comet of 1593 and the solar eclipse of 1598.[24] Christian was followed by the usual collection of short-termers, including a German by the name of Sebastian, who is credited with having translated a manuscript into German for Tycho's press,[25] and thirty-six-year-old John Hammond of Kent (England), who later was to become personal physician to James I and his son.[26]

Because Tycho was on Hven for some nine years after the list of his first thirty-two students was compiled, one must assume that another thirty aspirants rotated through Uraniborg during that period. No doubt the second batch resembled the first, in including a large percentage of short-termers. The second list also included a relative of a former fellow student of Tycho's – the oldest son of Jens Nilsson, the bishop of Oslo[27] – and a long-term student who contracted to marry one of Tycho's daughters. At least two assistants served terms of three or more years without attaining any special competence, and two others so distinguished themselves in the seventeenth century that they must be mentioned in more detail.

When Christian Sørensen arrived at Hven in the summer of 1590,[28] Tycho had been without the services of Flemløse and Gellius for two to three years. Within a short time, Sørensen was probably giving Tycho more astronomical help than the two departed assistants had together. Longomontanus, as he later styled himself in Latin, because he came from the little Danish village of Longberg, came from a family of such limited means that he had had to support his studies by intermittent terms of manual labor. When he was finally recommended to Tycho by a professor at the University of Copenhagen, he was already twenty-eight.[29] Sørensen remained in Tycho's service for most of the following decade and eventually became the only one of Tycho's disciples to attain a professorship in astronomy. During his career at the University of Copenhagen (1605–47), he published and republished the ponderous but very well circulated *Astronomia danica* (1622, 1643, and 1663), which was universally regarded as Tycho's final testament.

---

[24] *DBL* XIX, 478–80; XIII, 388, 118.
[25] The book was surely "Flemløse's" book of folk signs about the weather, published in both Danish and German on Tycho's press in 1591.
[26] *Norlind*, 98.   [27] *Norlind*, 100.
[28] A notebook from Hven cited by *Norlind* (102) sets Longomontanus's arrival in the summer of 1590. The traditional 1589 date stems from Tycho's letter of recommendation in 1597 (VII, 384), which refers to eight years of service. It is not impossible that Tycho either misremembered or counted the portions of both 1590 and 1597 as "years."
[29] *DBL* XIV, 445–8.

In many respects even more prominent in the seventeenth century was a Dutch student named Willem Janszoon Blaeu. Blaeu was about twenty-five when he wintered on Hven in 1595–6, and he seems already to have been committed to a technical career as a producer of maps and globes. Whether as the result of his own genius or of something he saw in Tycho's print shop, Blaeu made an independent mark as the inventor of the so-called Dutch Press. Such commercial interests were by no means beneath the dignity of royal mathematicians or even of professors of mathematics. But they cou..1 also indicate background or interests that were more technical than academic, and that seems to have been Blaeu's situation. Although he was sufficiently fortunate in his family circumstances to have had access to a university education, he spent the years immediately before his tenure on Hven as a general apprentice in a family mercantile office.[30]

Whatever the deficiencies of his formal education may have been, Blaeu made the most of his time with Tycho. Displaying the entrepreneurial instincts that propelled his firm to great prominence in the seventeenth century, Blaeu sought out and completed a project of his own – a study of the path of the comet of 1580 from observations by Tycho, Wittich, and Flemløse many years earlier.[31] It was not the exhaustive analysis that Tycho had made of the comet of 1577, but it was a nice student's exercise, relevant to the work that Blaeu would be doing for the rest of his life. And the experience was obviously very stimulating for him. Fifty years later, his celestial globes were still advertising the stellar positions of Tycho's catalogue, and his *Atlas* included a map of Hven, describing the activities there and telling anecdotes about Tycho not found in other sources.[32]

There can be little doubt that the common denominator of schooling and service for all the people who set foot on Hven was astronomy, the raison d'être of Uraniborg. So rare were exceptions to this that the name of Sebastian on the list of students is accompanied by a notation that he did not study mathematics. What this probably meant was that he had done nothing in the sciences of arithmetic, astronomy, geometry, and music beyond the introductory concepts required for a B.A. – and that he had no intention of doing anything. In fact, there is no good reason for believing that many of Tycho's students were accomplished mathematicians or astronomers. That, presumably, was one of the goals that motivated apprenticeship on Hven. Certainly, command of the spherical trig-

---

[30] *DSB* II, 185.   [31] XIII, 325–31.
[32] Blaeu also did enough observing to discover (and depict on his globes in 1600) the variable star 34 Cygni: III, 407. On his maps and globes, see *Zinner*, 249–52.

onometry that constitutes the basis of all astronomical computation was so far from routine that there was instruction in it at Uraniborg: A notebook containing exercises in it can be found in Copenhagen.[33] But if computational aid could be expected only from advanced students, observational help was another matter.

Prior to the granting of Hven, Tycho made his observations either by himself or (occasionally) with the aid of a servant. By the time Uraniborg was completed, however, circumstances reduced considerably Tycho's role in gathering his observations. First, many of his more sophisticated instruments required two people just to sight them. Nor could it have taken Tycho long to discover that having someone other than the sighter come with a lantern to read positions off the instrument would facilitate the next observation (by preserving, as we would now say, the dark-adaptation of the observer's eyes). In due time, in fact, Tycho was probably using a recording crew of three: one to hold the lantern and read out sightings, another to sit by (or carry) the log and make entries, and a third to stand by the clock and call out time as readings were recorded.[34] And because such a crew could have worked much faster than any instrument could be sighted, it could easily have serviced two or three sighting crews at once. When combined with the facts that good observing conditions seem not to have been something that could be taken for granted[35] and that Tycho had plenty of things to do with his time, it was probably inevitable that the observations would come to be relegated to the assistants.

How long Tycho even continued to operate an instrument himself is open to question. In later years there are several notations in the log stating that he has performed or verified a particular alignment himself,[36] which suggests, conversely, that he was not doing so with very many of the others. In fact, even the appearance of his handwriting in the log, which at least documents his presence at the observing sessions, becomes less and less frequent after 1585. The implications of these facts for Tycho's contribution to his observing program are ambiguous. In all likelihood, Tycho was somewhere in the area during most sessions, providing occasional supervision and just generally exercising a constraining influence. Even that would have been no mean achievement, for through most of the twenty-year existence of Uraniborg there were about 85 observing sessions a year.

---

[33] *Norlind*, 101.
[34] This is the scene depicted on Tycho's mural quadrant (Figure 5.9).
[35] See X, 231, and Tycho's meteorological diary (IX, 5–146), passim, for numerous references to days as gray or cloudy.
[36] X, 127, 243, 363; XI, 58, 103, 163, 279, 287; XII, 18, 50, 72, 121, 282, 301, 345.

Tycho must, in general, have been working on his own writings: From the beginning of 1587 to the end of 1591, for example, he composed almost a thousand pages of (eventually) published Latin prose. Whether he could have worked effectively in the warming room of Stjerneborg amidst the hustle and bustle of the observing is doubtful. Perhaps going out there from his library to check up on his assistants periodically was the primary purpose for which he conceived the never-completed tunnel. On at least one occasion, he regretted having failed to supervise closely: After a late-night notation by one of his assistants that observations had been terminated by the onset of clouds, Tycho added in an angry hand "and laziness."[37] But however often his staff got away with being lazy, if Tycho had had to do all the work himself, some aspect of his renovation of astronomy would have made notably less progress.

Although it was probably not "researched" with the same self-consciousness that astronomy was, the status of theology as the foundation of European culture and the core of Renaissance education must have made it almost as important as astronomy in the spectrum of intellectual concern at Uraniborg.

The foundation of Tycho's worldview was a sort of holism derived from a concept Garstein[38] termed *Platonic immanence.* According to this philosophy, the two spheres of matter and spirit overlapped or interpenetrated to such a degree that comprehension of either one was impossible without a thorough understanding of the other. In this view, Nature and the Bible were equally authoritative theologically. With this underpinning, Tycho and his circle – notably the prominent theologian, Cort Axelsen – rationalized the search for God within nature. For Tycho, this justificatory aspect seems to have constituted most of his interest in theology: There is little other evidence that he was otherwise religiously inclined. In fact, he expressed in print his view that theological issues were the most divisive ones known, that the resulting sectarian disputes were almost more numerous and worse than wars with weapons, and that in fighting them churchmen routinely committed most of the sins they so zealously instructed their flocks to avoid.[39] Whatever the personal beliefs of Tycho's students, however, they had an additional incentive for taking theology seriously. Organized religion was the basis of nearly every intellectual job in Renaissance society.

As we have seen, life was by no means all study and drudgery for Tycho's assistants at Hven. Those who established their competence

---

[37] XII, 275.

[38] See the Ph.D. dissertation (in Norwegian) by O. Garstein "Cort Aslakssøn," Oslo, 1953.

[39] III, 313. In III, 289, Tycho suggested that the litigants in religious quarrels were guilty of sophistry, hypocrisy, and even lying.

were given opportunities for intellectually challenging work not available anywhere outside Uraniborg. Yet there is a limit to the time that can be spent in the short-term pursuit of self-satisfaction. Sooner or later it becomes necessary to look to the future. For most of Tycho's students, who were not only free agents, but obviously a carefully selected, strongly motivated group, some reasonable anticipation of long-term return for their labors was an important consideration. Future employment as an astronomer was not a reasonable expectation, although one of Tycho's students, Longomontanus, attained that goal. But general employment in the "white collar" establishment of the day was a reasonable expectation and one that was routinely achieved by "graduates" of Uraniborg. The usual point of entry to the system after university was a post as parish priest somewhere in the provinces. With the support of Tycho's special status and connections, however, one could expect a more certain and probably more favorable entry into the system and also aspire to rise further in it.

Two of Tycho's students from the 1590s exemplify this advantage. One was Cort Axelsen, who, after two and a half years of service on Hven, was subsidized for seven years of study in foreign universities by employment as traveling tutor to one of Tycho's brother's sons. After his return, he became the first Norwegian to secure an appointment as a professor (of theology) at the University of Copenhagen. And the basis of his career there was the utilization of the new Tychonic cosmological ideas to form a "scientific" theology, one based on the book of nature, rather than exclusively on biblical exegesis.[40] A few years later, another three-year veteran of Uraniborg, Johannes Isaacsen Pontanus, accompanied another of Tycho's nephews on five years of similar travels. The son of King Frederick's (Dutch) factor in Amsterdam, Pontanus became a professor in Holland,[41] but he also was one of Tycho's two students to serve a term in Vedel's old post as royal Danish historiographer. The other one, Johannes Stephanius, became a professor at Copenhagen.

The basis on which Tycho recruited (or, at least, retained) Flemløse was the royal promise of a canonry at Roskilde;[42] he later placed him in the post of physician to the governor of Norway. Finally, on Tycho's recommendation, Gellius received a royal stipend to support his studies.[43] But even those who won no direct and conspicuous rewards for service seem to have benefited from participating at Hven. No fewer than four of Tycho's students who entered the

[40] *DBL* I, 545–7. In 1610 Axelson edited Tycho's oration on astronomy from his Copenhagen lectures of 1574 and dedicated the publication to his former patron, Steen Brahe.
[41] *DBL* XVIII, 448–50.   [42] XIV, 11.   [43] XIV, 27.

Church rose to the rank of bishop. At the very least, Tycho can be said to have chosen his long-term students very perceptively.

Concerning the financial aspects of study on Hven, there is no documentation whatsoever. Institutional tuition was not completely unprecedented in Europe, and private tutorial fees were routine at all universities. Yet it seems unlikely that either Tycho or King Frederick ever even considered casting Tycho in the economic (and, by implication, social) role of a professor. We may presume, therefore, that only actual living expenses could have remained as an issue. A significant fraction of those had already been underwritten by Tycho (or Frederick) in the capital expenditure for the construction of Uraniborg. Free bed was surely assumed by anyone who visited Uraniborg, as an act of noblesse oblige if nothing else. So, too, no doubt, was board, at least for a day or two.

For longer terms, however, some kind of understanding would have been necessary. Again, in order to avoid putting Tycho in the position of haggling over bills like an innkeeper, the understanding was probably that Tycho's royal stipend would cover the boarding costs of his students, at least those from Denmark. The advanced ones, those who could provide real professional help, must even have received some money for incidental expenses. For although many of them had personal resources for such requirements, Longomontanus, for one, almost surely did not. In fact, Tycho's conception of support clearly extended beyond the walls of Uraniborg, for when the *landgrave*'s mathematician, Christoph Rothmann, visited Hven in 1590 Tycho provided his room and board for a month and could only express his amazement that the *landgrave* had not done the same for Flemløse when Tycho had sent him to Cassel for a few days' visit.[44]

The dual role as dispenser of both knowledge and patronage at Uraniborg certainly gave Tycho a decided advantage in his relationships with his students, and there is nothing in Tycho's character to suggest that he would have yielded it. In addition, he was a noble, someone to be referred to – and deferred to – as "the *junker*." It was a situation that left very little room for nonsense, and there is no reason to believe that there was any. The best information available is from Tycho's dealings with other people, both above and below the status of his assistants, and from the domineering behavior of the Danish nobility as a class, all of which suggests that Tycho ran a tight ship.

How many students found the atmosphere oppressive we have no way of knowing. Only one, a German named Frobenius who came to Hven in about 1591 with an M.A. and a recommendation from

_____
[44] VIII, 295.

Caspar Peucer, actually registered his unwillingness to work under Tycho. Johannes Kepler's celebrated problems with Tycho were due at least as much to the fact that Kepler had a family (with special requirements) and did not fit into the "college" mold, as to professional considerations. On the other hand, there is considerable evidence that Tycho's autocracy was essentially benevolent, or at least guided by enlightened self-interest. Those who were at Uraniborg to do serious work and did not take Tycho's dominance personally got along with him very well.

No fewer than a dozen assistants are known to have stayed with Tycho for two and a half years or longer. Blaeu (who was not one of them) and Pontanus published fond recollections of Uraniborg and Tycho a generation after the demise of both institution and founder.[45] Even Frobenius arranged to publish a Tychonic "Ode to Ptolemy" sixty years after its composition by Tycho.[46] Perhaps the most vivid counterexample of the frequent occasions on which Tycho appeared as a curmudgeon or tyrant is found in a letter to Hayek, written in late 1591. At issue was Peter Jachinow, who had appeared in Prague soliciting work as a mechanician (specializing in the building of odometers) and using Tycho's name as a reference. On being questioned by Hayek, Tycho affirmed at considerable length that Jachinow had, indeed, rigged odometers to the coaches of many nobles (including German princes and the Danish king) and that

> when he was last here he adapted an automaton of this type to my own carriage in which I regularly travel about this island with my friends. This machine indicates the whole miles and also their different subdivisions and their divisions into sixty parts by central pointers, and makes it clear by striking distinct sounds with two bells. If you will recommend this man, in such a way that ... he may be able to hire out his labor ... and, by this means acquire a small amount of money, you will thereby be doing me a favor; for I love this man for his evident honesty and trustworthiness, and I am sorry for him because in his advanced age he has to travel up and down to support himself and his family. Don't let his efforts down – I do not doubt you will do it for my sake – so that he may not complain whenever he returns here that my little recommendation, such as it is, turned out to be fruitless for him.[47]

---

[45] Norlind, 99–100.

[46] Mogens Brønsted, "Et ukendt digt af Tycho Brahe," *Fund og Forskning* 9 (1962): 71–8. The "unknown poem" [*ukendt digt*] was actually published by Tycho in his *Epistolarum astronomicarum* of 1596 and by Dreyer in VI, 269–70.

[47] VII, 320. The translation is by Professor Derek Price, as published in the *Horological Journal*, May 1955.

Jachinow was only one of some 275 visitors to Hven who was entered into Tycho's log over the years. Most of them probably visited only once, but some of them came often enough to earn consideration as contributing members of Tycho's circle. Perhaps the most marginal of such contributors was Tycho's (and Frederick's) architect, Steenwinckel, who obviously enjoyed helping with observations when the opportunity arose but probably had no other interactions with Tycho's group. Another was an obscure "Paul, the pharmacist," whose appearances on Hven are documented several times from 1590 to 1597.[48] Much the most frequent visitor to Hven was Tycho's youngest sister Sophie who, after the death of her husband in 1588, appeared at Uraniborg four or five times annually for visits that ranged from a few days to a few weeks. Sophie's great interest was in horoscope astrology, which she took seriously enough to collect notations comparing the observed and predicted fates of various friends.[49] She was also sufficiently conversant in medical chemistry to emulate her brother in dispensing free prescriptions to the poor. Presumably she did some experimenting on this subject someplace, and she probably did at least some of it at Hven – especially as by 1590 she had become engaged to Erik Lange, who came to Hven regularly to consult Tycho on alchemical problems.

Lange was what Tycho described as "a kinsman of mine" (a fourth cousin, strictly speaking). Their acquaintance dated back at least to the arrangements for the 1584 marriage of Lange's sister to Tycho's brother Knud. Lange was sufficiently engrossed in alchemy that he was squandering his fortune on his studies.[50] He therefore did not remain in Tycho's circle very long but had to flee the country in 1591 to escape his creditors. Only after a decade of exile and several years after Tycho had left Denmark was Lange able to meet up with and marry Sophie.

The most frequent, and apparently the most congenial, visitor Tycho entertained over the years was his erstwhile tutor, Anders Vedel. Relative to his station in life, Vedel had prospered as much as Tycho had. In additon to having established himself as a theologian of some repute, he was pursuing the historical studies that earned him a place in the pantheon of Denmark's literary figures. Under the patronage of powerful politicians and with the support of a canonry at Ribe Cathedral (since 1573), Vedel had completed an epoch-making Danish translation of Saxo Grammaticus's chronicles of medieval Denmark (1575) and embarked, as royal historiographer,

---

[48] IX, 118, 120, 141, 143, 146.   [49] *Dreyer*, 201.
[50] VIII, 47; *DBL* XIII, 549–50. In a letter dated August 1583 (VII, 75) Tycho asked Dançey to greet Lange for him.

on an ambitious attempt to write a complete political history of the realm. His appointment as vice-bishop of Ribe (1585) even permitted him to build a landmark residence there (which is still called Little Uraniborg) and to set up a printing press for the publication of his projected works.[51]

Vedel's activity is of more than passing interest because it seems to have had some influence on Tycho's circle. Tycho's gesture of publishing a Latin poem in 1575 exhorting the women of Denmark to contribute rags for the paper to print Vedel's translation of Saxo need not have represented anything more than support for a friend.[52] And the assignment of Flemløse and Olsen to mapping activities associated with Vedel's history might likewise have been merely the pursuit of a project requested of Tycho by the king. But the fact that two of Tycho's students would occupy themselves sufficiently with historical writing – in an era when history was not yet recognized as an academic discipline – to win terms as Denmark's official historiographer suggests strongly that Vedel's interests at least came under discussion at Uraniborg. After 1590 we may be almost certain that they did, because another historian, Professor (at Copenhagen) Niels Krag, began to spend a lot of time with Tycho, both on and off the island.

Although supervising the instruction and research of his students and assistants imposed another set of concerns, it was surely one from which Tycho realized a considerable net advantage. On the one hand, it is extremely unlikely that Tycho provided any routine instruction himself. Remedial work in astronomy, and even the lessons in spherical trigonometry, must have been handled by assistants or senior students. Aside from the responsibility of organizing such activities, therefore, Tycho's teaching must have been done almost exclusively at higher levels, in much the same way that postgraduate scientific research teams operate in the modern university. The most visible aspect of this was the astronomical work. But Tycho had other concerns, too, in such areas as alchemy and astrology, where he had not only strong intellectual interests of his own but also an undeniable political interest in catering to the enthusiasms of King Frederick and thereby justifying the great sums

---

[51] *DBL* XXV, 183–92. Vedel planned twenty-two volumes, which would not only be written in Danish but also focused on the people rather than the state. Given the magnitude of his enterprise, Vedel would never have finished it. His task also was complicated by his decision to edit and print the immortal first edition of Danish folk songs, which he completed in 1591.

[52] *Dreyer*, 79. Vedel did not hesitate to ask Tycho to try to get historical questions answered for him. VII, 79.

of public money that were continually being added to Tycho's endowment during the late 1570s.

These other subjects were much less thoroughly developed than astronomy was and therefore more appropriate to general intellectual discourse, more amenable to the immediate pursuit of sudden interests, and more suitable as research assignments when Tycho needed help on something. It would be strange, indeed, if the contribution of Tycho's students to these auxiliary ventures was not considerably greater (proportionately, at least) than their share in his astronomical program. Unfortunately, documentation on the subject is even poorer than it is for the astronomical program. So, with a few exceptions, the results of these marginal activities can only be attributed to Tycho.

Among the earliest records of business at Uraniborg are some papers labeled "Geographical observations made on the island of Hven." Collected, apparently, before the end of 1579,[53] they consist basically of triangulations between various landmarks on Hven and the surrounding mainland. Angular measurements to outside points allowed Tycho to establish the location of Hven relative to prominent nearby cities, and triangulations (and pacings, of course) on the island itself provided the data necessary for mapping Tycho's new domain. Already in 1586, Tycho was asked for a map of the island by Heinrich Rantzov, the governor of Holstein, who intended to include it in the collection of plates he was sending for reproduction in the North German section of Braun's six-volume *Civitatis orbis terrarum*. Tycho begged off, pleading the unavailability of his engraver and more pressing demands for his own time, but possibly having mixed feelings about the prospect of advertising the existence of what was, after all, supposed to be a scientific retreat.[54] Rantzov persevered, anyway, however, so that when the volume appeared in 1588, Tycho saw his little empire described in glowing terms and depicted in the full-spread (quarto) plate shown in Figure 6.1,[55] in one of the most prominent publications of the day.

[53] The observations (V, 294–300) were bound as a group into Tycho's log of astronomical observations, following the observations for the year 1579 (V, 338), suggesting that the last of them was made before the beginning of 1580.
[54] VII, 102. A crude, hand-drawn map of unknown vintage (but postdating 1584, when Stjerneborg was built) is reproduced by *Dreyer* on V, 293. Something like it may have been sent to Rantzov, for the map published by Braun, *Civitatis orbis terrarum*, vol. 4, 1588, map 27, shows a marked similarity, in its poorly rendered outline at least, to the drawing.
[55] In fact, Hven is displayed twice. Plate 26 (ibid.), featuring Frederick's Kronborg Castle but covering the whole north end of the Øresund, shows Hven prominently and even names Uraniborg (see inside front cover). The house, grounds, and instruments for the main map (Figure 6.1) were made (recut in reduced size for the insets) from woodcuts that Tycho had already had printed on his press and was circulating to his friends (VII, 96, 104). They were

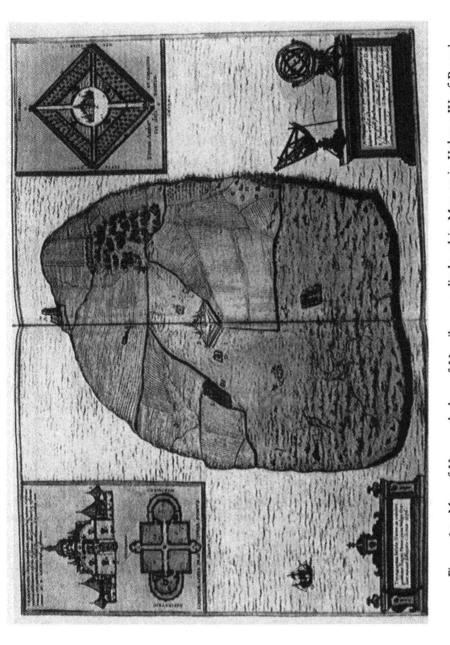

Figure 6.1. Map of Hven and plans of Uraniborg as displayed in Map 27 in Volume IV of Braun's *Civitatis Orbis Terrarum* (Atlas of European Cities) (1588).

In response perhaps to some needs perceived by Vedel for his history of Denmark, Tycho talked to King Frederick in 1585 about providing an improved map of the entire kingdom and obtained permission to use all the old maps and charts of the realm preserved in the state archives.[56] Nothing seems to have been done on the project, however, until 1589, when Tycho sent Elias Olsen with Vedel on a tour of Denmark to determine the latitudes of various places, apparently intending to use them for both a geographical table of European cities and an improved map of Denmark.[57]

In the same year and again in 1590 and 1592, Flemløse made surveys in Norway.[58] But the work was never extended southward to Tycho's satisfaction. The few eclipse observations he could solicit from various correspondents and the more-readily obtained reports of latitudes were eventually compiled into a table of longitudes and latitudes, but the result was not something that Tycho was willing to include in his catchall *Progymnasmata*, so the table remained in manuscript until Longomontanus published it.[59]

The mapping got only as far as an improved chart of Hven, which Tycho had printed by 1592 and eventually published in the *Mechanica*.[60] Although these results were not enough to establish any major reputation in geographical studies, they were far from negligible. Tycho's map was the first one of any part of the north that was based on actual measurement and, to one authority[61] at least, "makes a wholly modern impression," sufficient with his use of the method of triangulation to earn him recognition "as one of the pioneers in the technique of cartographic measurement."

Next to astronomy, alchemy was Tycho's greatest intellectual passion. Already during his student days at Rostock he was introduced to the conception of Paracelsus that every aspect of the natural

---

eventually published in the *Mechanica* (V, 60, 72, 138, 142). The illustration of what was essentially a country estate among plates that generally depicted cities made Hven very conspicuous, and Tycho seems to have been pleased (and perhaps surprised) with the results (VII, 386).

[56] XIV, 30, 31.   [57] V, 301–4, 342; VII, 219.   [58] V, 342.

[59] V, 309–13. Dreyer noted (V, 343) that the place names are written in Dutch rather than Danish, without suggesting that the list might therefore have been compiled by Blaeu, who was at Hven over the winter of 1595–6. Tycho's reference to the list as being essentially unfinished in 1597 is in V, 116. Although Kepler's "Catalogue of Principal Places of Europe" in the *Rudolphine Tables* (33–6) must have been drawn from Tycho's list, it is sufficiently altered and extended to constitute a new work: V. Bialas, "Data Processing in the Rudolphine Tables," *Vistas* 18, 749–69, vol. 18, pp. 749–67.

[60] VI, 295; V, 150.

[61] See H. Richter, "Wilhem Janszoon Blaeu with Tycho Brahe," *Imago mundi* 3 (1939): 53–60. Richter showed, incidentally, that all of Tycho's interesting geographical work was done before Blaeu came to Hven.

world was interconnected in subtle ways by agents whose operations were ultimately chemical in character. By the time of his stay in Augsburg (1569) Tycho was devoting considerable attention to these ideas, and in the succeeding years at Herrevad Abbey, up to the appearance of the new star at the end of 1572, they actually crowded out his astronomical interests.[62]

While Tycho was planning Uraniborg, he expressed an equal commitment to both the higher and lower realms of philosophy and consequently built sixteen alchemical furnaces in his basement.[63] That he must have followed through on this commitment to some degree at least is borne out by his alteration of the winter dining room in Uraniborg to include a small laboratory with five furnaces to reduce the time spent checking on the progress of some of the experiments in the basement[64] and by his claim late in life to "have been occupied by [chemistry] as much as by celestial studies since [his] twenty-third year."

But in contrast with the reams of records he accumulated in his astronomical research, Tycho left not a single document describing any alchemical experiment or observation. All he said about them was that the problems involved were somewhat analogous to those in the heavens, so that he called this science terrestrial astronomy, and that it had yielded up to him "a great many findings with regard to metals and minerals as well as precious stones and plants and other similar substances."[65] It seems safe to assume, however, that he followed the dictates of Paracelsus in believing that the object of chemistry was to make medicines, not gold, and accordingly focused his energies on what would now be judged to have been the more respectable, less spectacular, aspects of alchemy.

At least this is the picture conveyed by the intellectual company Tycho kept and the results he obtained. When he heard in 1583 that Erik Lange had found a way to multiply flour threefold alchemically, he enthusiastically characterized it as a much more important discovery than any analogous development with gold would be.[66]

Both Pratensis and Ripensis (with Dançey, his best friends in Copenhagen) were professors of medicine. Brucaeus and Hayek (over the years Tycho's most faithful correspondents) were respec-

[62] V, 108.
[63] "... spero me commoditatem non exiguam adepturum philosophandi tum in superiori et coelesti Astronomia, tum etiam in hac inferiori Spagyrica rerum praeparatione" (VII, 42). The furnaces are described in V, 145.
[64] V, 118, 142. In 1588 he volunteered a similar assertion to Rothmann: VI, 144–6. For evidence that Tycho was thinking seriously about alchemical (medical) matters at other times, see VII, 94–5, 238.
[65] V, 142.   [66] VII, 75–6.

tively professor of medicine at Rostock and personal physician to three Hapsburg emperors.[67] Brucaeus took Tycho's medical competence seriously enough to believe that Tycho might have a special competence in dealing with epilepsy and to argue doctrinal points concerning the efficacy of Tycho's chemical medicines relative to the traditional Galenic (vegetable) medicines.[68] Peder Sørensen, physician to two Danish kings, seems also to have regarded Tycho essentially as an adept.[69]

At Uraniborg, however, it seems as if the actual medical research must have been done by Flemløse and Gellius. Flemløse had already published a translation of a medical text when he came to Hven in 1577, and when he left there ten years later it was to become physician to the governor of Norway (Axel Gyldenstierne, Tycho's mother's cousin). He certainly did plenty of observing for Tycho and continued to help him by conducting surveys of Norway for his mapping project. In the fall of 1587, he and Gellius each supervised a crew for the detailed recording of a lunar eclipse.[70] Yet when Tycho wrote for Flemløse a letter of introduction to the *landgrave*, he described him as being skilled in "astronomy and pyronomics."[71] Adding all of this evidence to the fact that when Flemløse died suddenly in 1599 he was just about to obtain a medical degree from the University of Basel, it is not difficult to believe Longomontanus's statement that Flemløse came to Hven because of the relevance of astronomy to medicine.[72]

Gellius was at Uraniborg for over six years without making any notable astronomical contribution (beyond earning his spurs as an observer). When he left Tycho's circle in 1588, he studied at the University of Basel for an M.D. degree received in 1593. After an ill-fated return to Hven, he eventually served as professor of medicine at the University of Copenhagen from 1603 until his death in 1612. Thus, while it is true that mathematicians of the sixteenth century routinely took M.D.s to qualify for more lucrative professorships, Flemløse and Gellius seem to have had a true interest in things medical.

Whoever was doing the work, the results are not impressive from

[67] See R. J. W. Evans, *Rudolf II and His World* (Oxford, England: Oxford University Press, 1973), p. 152, for more biographical details about Hayek. Even Tycho's first mentor in astronomy, Schultz (Leipzig, 1563), was a Paracelsan of sufficiently strong persuasion to publish an edition of his *Vom Ursprung der Pestilenz* in 1573: ibid., 136.
[68] VII, 33, 101.
[69] However, the friendship of their youth became an adversarial relationship by the 1590s, when Sørensen registered strong objections to Tycho's habit of dispensing free medical preparations: Gassendi, 307–8.
[70] XI, 164.   [71] VI, 104.   [72] Dreyer, 118.

a modern point of view: two or three medications for plague and one for venereal disease. But Tycho might well have been as happy with them as he professed himself to be: One of his preparations was an elixir, good for anybody against anything, and one could scarcely wish for anything better than that. From the hindsight of history we can be sure that the medicines did not work as well as was hoped. But the medications Tycho dispensed from Hven were nonetheless highly regarded: Enough deaths were recorded in Tycho's diary to suggest that some of them may have been very ill patients who came to Hven looking for miracle cures. And although neither Tycho nor his confidants probably revealed the formulas of their cures, by the middle of the seventeenth century, recipes purporting to have come from Hven had circulated throughout Denmark and been published in the official Danish *Pharmacopea*.[73]

An essential aspect of the Paracelsan view of a universe run by unknown, and probably unknowable, forces was an "alchemical" link between the cyclings of the heavens and the vicissitudes of earthly life. Tycho's commitment to this concept is documented by two woodcuts he had carved in 1584, to print with the poems he distributed honoring his friends.[74] Each of them features the reclining figures of a man and a youth. In one, used on the title page of his *Astronomical letters* of 1598, the man is shown with the snake of Aesculapius coiled around his arm and some medicinal herbs in his hand, looking at the ground to represent the legend: *Despiciendo suspicio* (By looking down I see upward). In the other, used as the colophon for the *Astronomical letters*, the man is leaning on a celestial globe and looking at the heavens, to depict the reciprocal idea: *Suspiciendo despicio* (By looking up I see downward). In such a view there could not fail to be some kind of relationship between the positions of the planets and the fates of humanity.

The most obvious relationship was through the intermediation of the weather. Ptolemy had already argued that the sun, through its influence on climate, exercised a profound influence on human affairs.[75] Tycho was willing to follow him in concluding that the

---

[73] These were revealed only to his two most powerful benefactors, Heinrich Rantzov and Rudolph II, and to the two other people closest to him, sister Sophie and Holger Rosenkrantz. See the "formulas" in IX, 160–9; and Karen Figala, "Tycho Brahes Elexier," *Annals of Science* 28 (1972): 139–76. It seems to have been sister Sophie who imparted the "cures" to King Christian after Tycho's death.

[74] See Tycho's description and explanation in VI, 144–6.

[75] See Book I of Ptolemy's *Tetrabiblos* in the Loeb Classical Library, Harvard University. It is available in an English translation by F. E. Robbins, Harvard University Press, Cambridge, Mass, 1940.

moon caused the major departures from a strictly sun-controlled climate, whereas the planets refined the effects.[76] Although the failures of literally thousands of almanacs published during the first century of printing had left such predictions in low repute, Tycho was convinced that with a properly empirical approach he could realize some of the enormous potential of the project.[77] Already before the appearance of the new star he had begun not only to generate annual almanacs but also to keep track of their performance and attempt to improve it by such measures as adjusting the motions of the sun and moon.

From October 1582, Tycho kept daily records of the weather on Hven, eventually accumulating fifteen years of empirical data, which included such details as the sighting of the first lark or stork in the spring.[78] During some of this period, he may have produced more almanacs, for he is known to have been providing some kind of annual prognostication to King Frederick. The first book published on Hven, in fact, was the astrological calendar for 1586, which Tycho published under the name of "Elias Olsen of Denmark, apprentice in the practices of astronomy to the nobleman Tycho Brahe" and legitimized further with an appendix containing his observations of the comet of late 1585.[79]

Tycho did not need to be told that there were problems with the enterprise. He himself discovered a fundamental difficulty around 1580 while comparing observations made from various sites in Europe of the comet of 1577: Stations that were quite close together,

---

[76] I, 37–43; and John Christianson, "Tycho Brahe's Cosmology from the *Astrologia* of 1591," *Isis* 59 (1968): 316. Tycho even believed that the moon caused the tides, "like a magnet" (I, 155). Such belief (expressed by Kepler) in the moon's "dominion over the waters" and other occult causes shocked Galileo (Dialogues, Book 4).

[77] J. G. Hellmann, *Versuch einer Geschichte der Wettervorhersage im XVI Jahrhundert* (Berlin, 1924). Mary Ellen Bowden, "The English Revolution in Astrology: The English Reformers, 1558–1686" (Ph.D. diss., Yale University, 1974) showed that the same aspirations existed in England.

[78] *Norlind* (90) interpreted these notations as indicating that there was a bird watcher on Hven. But similar references to frogs and swallows (11 and 15 April 1589) and even hoes (27 February 1588 and 7 March 1590) suggest that for almanac predictions, someone was making empirical observations intended to connect the habits of animals with the return of spring.

[79] "Diarium astrologicum et metheorologicum anni a Christo 1586.... Per Eliam Olai Cimbrum, Nobili viro Tychoni Brahe in Astronomicis exercitiis inservientem. Ad Loci Longitudinem 37 Gr. Latitudinem 56 Gr. Excusem in Officina Uraniburgica." The book was described briefly by *Dreyer* (125). The account of the comet was largely astrological. But extant manuscripts in Tycho's hand accord with it so well that Dreyer felt (reluctantly) obliged to publish it as part of Tycho's collected works: IV, 399–414, 512. In fact, *Norlind* (354) saw enough similarity between the 1586 volume and Tycho's unpublished calendar for 1573 (I, 73–130) to assume (with Hayek, VII, 102) that it was essentially Tycho's work.

from an astronomical standpoint, could nevertheless be subject to very different meteorological conditions on a given night.[80] Tycho, therefore, seems to have been among the first people to note publicly just how limited in extent local weather patterns could be. But although this discovery impressed him sufficiently to be mentioned in both *De mundi* and his later correspondence with Rothmann, it does not seem to have destroyed his faith in the basic idea, at least not privately.

Although Hayek (who himself was a strong believer in astral influences) expressed the opinion that Tycho was wise to publish his calendar under Olsen's name, and Brucaeus had chided him for even sponsoring it (it reminded him, Brucaeus said, of Cato's query as to whether astrologers could keep from smirking at one another when they met),[81] Tycho followed it in 1591 with a general list of 399 aphorisms for predicting changes in the weather.[82]

This almanac, too, was published under the name of a disciple; Flemløse, this time. And according to the preface, it was written at Hven only because the old king (who died in 1588) had specifically asked Tycho for something of the sort, and it was printed there only because Tycho's presses happened to be temporarily idle.[83] But for all its defensiveness, this preface displays a firm commitment to the ideal of an astrological – or perhaps one should say, astronomical – meteorology. And although the commitment is ostensibly expressed by Flemløse, Longomontanus was later to testify that the preface was actually written by Tycho.[84] Thus, it seems safe to say that in regard

---

[80] IV, 113. From the distance of a planet – even the very modest distances assumed in the sixteenth century – places such as Hven and Prague would be indistinguishable and could therefore not be supposed to undergo different influences from the planet. The passage setting out this difficulty was probably written before 1580 and certainly before the end of 1584, when Tycho finally abandoned the obliquity and latitude used in the computations involved. In 1588, he referred to the same problem in correspondence with Rothmann: VI, 142.

[81] Evans, *Rudolf II*, p. 152; VII, 100. Brucaeus had objected in advance to Tycho's permitting such "trite and vulgar things" (91) to be printed on Hven and in a second letter expressed his fear that Tycho's sponsorship would "confirm the vanity of the art" (92). He promised to try to keep an open mind toward it but, as we have seen, was unable to restrain himself when he saw the results. Even Hayek, who was much more receptive to astrology, expressed his opinion that Tycho had been well advised to leave his name off the calendar, even though the section on the comet elevated the tract above the general run of such things: VII, 104.

[82] The work was described briefly by *Dreyer* on 118–19.

[83] This book in Danish was probably the source of the notion of the "Tycho Brahe Days" that are still figures of speech in Scandinavia. But *Dreyer* (154) found 32 such unlucky days in the popular folklore, without succeeding in attributing any of them to Tycho's authority.

[84] Longomontanus revealed this in the preface to the second edition of his *Astronomia danica*. Despite Tycho's shyness, he allowed the book to be dedicated to himself and circulated it in

to weather judgments, Tycho retained to the end both his faith in the ideal and his conviction that the ideal was sufficiently proximate to justify active research on it.

In contrast with his abiding interest in the rest of what would now be termed the *occult* sciences, Tycho experienced during his lifetime a steady loss of confidence in horoscope astrology. His initial naïveté was lost early, hastened no doubt by his peers' ridicule of his prediction in 1566 of the demise of a Turkish sultan who turned out to have been dead for some weeks before Tycho cast his horoscope;[85] but it was probably caused mainly by numerous less spectacular and unarticulated misjudgments. His fundamental belief that the stars had an influence on humans that could be divined if only one could find the correct procedure and knew precisely the positions of the planets, was still there at the time of his oration at the University of Copenhagen in 1574, albeit now heavily modified by Tycho's emphasis on human will as a possible counteragent.

Through these and the succeeding years Tycho continued to struggle with prognostications, and when he was installed on Hven he was still unshaken in his conviction that he was engaged in a semiviable enterprise.[86] In 1577, 1579, and 1583, he was called upon to provide horoscopes for the sons born to King Frederick. The spectacular comet of 1577 required some type of interpretation, too, and Tycho duly submitted an extended account of its portents.[87] Something of the same nature may have followed the brief appearances of comets in 1580 and 1582, if only in the annual prognostications Tycho was sending to Frederick,[88] and the more visible comet of 1585 was written up for publication in Olsen's calendar for 1586.

During this period, however, Tycho became progressively disillusioned so that in the end he was casting horoscopes purely out of duty as the king's astronomer and disliking even that. Already in the horoscope for the crown prince (1577) he seemed to be nervous about having his predictions taken seriously. Amidst all the scholarly motions (including even using planetary positions derived from observation, when he had data for them) undertaken to generate

both Danish and German. The German version of the preface was reprinted by John Christianson, "Tycho Brahe's German Treatise on the Comet of 1527," *Isis* (1959): 312–18, with a discussion of its circumstances and significance.

[85] I, 135; X, 13.

[86] Wolf, at Augsburg, alluded to Tycho's having cast his horoscope, probably on Tycho's travels in 1575: VII, 47–8. In his letter, Wolf implicitly raised a completely extraneous problem for horoscopes, by wondering whether Tycho would actually have foretold death for him if that was what the stars had revealed.

[87] IV, 389–96.

[88] Tycho's observations, covering two months and six days, respectively, are in XIII, 305–35. For a letter from the king reprimanding Tycho for being late with the year's astrological calendar, see XIV, 37.

lifelong expectations for baby Christian, Tycho mentioned time and again that the prognostications were neither precise nor immutable – that God could certainly alter the outcome and that in any case people's fates depended more on their own actions than on the configurations of the stars.[89] For the succeeding princelings, Tycho expressed his reservations even more forcefully. To recount the contents of these lengthy prognostications would be both tedious and futile,[90] but no more tedious and futile than it seems to have been for Tycho to make them in the first place.

How strongly he resented being responsible for things over which he did not believe he had any real control appears from a letter written in 1587 to a German nobleman, who was married to one of Tycho's cousins and had consulted Tycho on behalf of his liege, the duke of Mecklenburg. The problem was that the Duke had obtained two prognostications for 1588 that conflicted so radically that one forecast a year governed by two beneficent planets, the other by two malevolent ones. Tycho was able to account for the opposition by pointing out that one judgment was based on Ptolemaic and the other on Copernican tables, and he was willing to state further that because neither set of tables was accurate, neither set of predictions could be relied upon. In fact, Tycho went on to say, he did not willingly involve himself in astrological matters. He sent a prognostication to King Frederick every year, but only because the king expressly demanded one, for he really did not like to be associated with such doubtful predictions.

So far this could be a matter of pure pragmatism, a statement that astrology would not be worth doing until its astronomical basis was reformed. But Tycho went on to say that even if all astrologers used the same tables, very few pairs among a hundred prognostications would agree with each other, as astrologers used many different bases and procedures to form their judgments. That was why Tycho himself never placed any trust in them and wanted to restrict himself to astronomy, in which one could attain real truth.[91]

What is all the more remarkable about this frank document is that its ultimate addressee, the duke of Mecklenburg, was King Frederick's father-in-law (and brother-in-law), and Tycho even went so far as to tell him that if he wished to see what he had predicted for 1588 he would have to ask the king, because Tycho himself had not bothered to keep a copy of his prognostication. It can scarcely be doubted, therefore, that Tycho had already expressed at least some

[89] I, 185, 196, 205. On this last page, Tycho warned the king that if the time of birth were off by as much as four minutes, the whole judgment would be different.
[90] They were published in I, 179–280, and summarized at some length by both *Dreyer* (144–5) and *Norlind* (64–5).
[91] VII, 116–19.

of these opinions to his royal patron. Fortunately, Frederick's death early in 1588 relieved Tycho of the odium of prognostications until late in his life, when his new patron at Prague, Rudolph II, requested an astrological forecast for 1600 concerning the possibility of his being affected by an epidemic of plague.[92]

Yet Tycho could never quite subdue his belief that in a created universe the constituent parts would have to harmonize sufficiently to permit some kind of astrological prediction. Thus, after Frederick's death Tycho could not restrain himself from examining the king's horoscope for some clue from the stars that would have predicted his demise. Nor could he suppress the opportunity to quote to Caspar Peucer at about the same time a passage from the horoscope Tycho had cast for him some twenty years earlier during his student days that seemed subsequently to have been fulfilled.[93]

If Tycho explicitly eschewed the whole question of judgments when writing his extended critiques of the literature on the comet (*De mundi*), he could not resist some astrological second-guessing about the new star in the *Progymnasmata*.[94] So profound was the gulf between the intellectual allure of astrology and its operational value that Tycho often managed to denounce and defend it in virtually the same breath. Always, however, the denunciations concerned either the primitive state of the art or the incompetent practitioners of the art, rather than the foundations of the art itself.[95] Like most of his contemporaries, he found these just too compelling to give up.

Although one could wish for a more complete picture of the Uraniborg research effort in astrology and particularly alchemy, the outline of the available material accords well with the image that emerges much more sharply from the astronomical work on Hven. This picture is very different from the one generally associated with the occult sciences. Most significant was Tycho's great concern for empirical reliability. This is illustrated in the meteorology by his early comparisons of prediction and occurrence, the extensive data in his meteorological diary, and his willingness to submit his weather aphorisms to public trial. It can also be seen in Tycho's propensity for dispensing free medicine from Hven,[96] which may be more

---

[92] VIII, 240.   [93] I, 132–5; VII, 137.
[94] IV, 237. The last few pages (III, 309–19) of the 816-page *Progymnasmata* contain some astrological discussion.
[95] In III, 224, and IV, 361, Tycho criticized Leovitius and Dasypodius for being competent astronomers who could have accomplished something worthwhile but instead brought shame on themselves by issuing foolish prognostications. See also III, 309. In *De nova stella* (I, 43) Tycho stated that any fault in his predictions should be attributed to him rather than to the art, echoing Ptolemy's argument from antiquity. For some of Tycho's positive thoughts in later years, see II, 150; VII, 138; VI, 29.
[96] *Gassendi*, 240, 307.

reasonably interpreted as a means of testing the efficacy of his concoctions than as an indication of general humanitarian concern. Such an outlook – that no production could be any better than its empirical consequences – would also account for the manifest reluctance with which Tycho issued astrological judgments in his later life.

Almost as prominent in his research is Tycho's willingness to alter existing theory in order to achieve better correlations with the phenomena. There is no evidence that he found this necessary in his alchemical studies, except insofar as his leanings toward Paracelsus's new doctrines may have been influenced by empirical considerations. But for his astrological work he decided very early that something had to be done, and already by 1572 he had developed new methods (and written expositions of them) using horizon (instead of ecliptic) coordinates to define astrological houses, and the true (rather than the mean) motion of the sun to draw up the divisions of the zodiac.[97] Subsequently he appears to have decided to proceed almost purely inductively, with the meteorological daybook, while the horoscope astrology became so hopeless that he abandoned altogether his research on it.

In all of this work, we see Tycho struggling with phenomena that are immeasurably more complex than those of astronomy and that were too complicated to be resolved with the tools available in the sixteenth century.[98] But this does not alter the fact that Tycho's approach to them was methodologically sound. In only one respect – his adherence to the ethic of secrecy[99] – can he be faulted. And if that defect militated against the kind of written records that he kept for his astronomical researches – which he likewise zealously protected – it could not prevent him from keeping mental books on the success of his work. Over his lifetime this scientific attitude gradually forced him to modify, and in some respects even abandon, a set of propositions that he and most of his contemporaries found extremely attractive. For if Tycho was not above trying to fool others on occasion, he at least did not fool himself. The result was a concomitant shift during his years on Hven toward the area in which he was achieving results, the astronomical studies that were to immortalize him.

[97] I, 38–9. The originally unpublished charts survived to be printed by Dreyer on I, 75–130.

[98] More sophisticated techniques developed since then, however, have led some climatologists to deduce from them (and from other records) the existence of a period of gradual cooling from 1550 to 1700 that is sometimes termed the "little ice age." I am grateful to Dr. Mary Ellen Bowden for a reference to this issue in *The Climate of Europe: Past, Present and Future*, ed. Hermann Flohn and Roberto Fantechi, D. Reidel, Boston, 1984.

[99] Secrecy was the hallmark of the whole Hermetic tradition. Tycho's expression of his commitment to it makes it appear to have been a defense mechanism as much as anything else (V, 117–18), but the absence of records speaks for itself.

Chapter 7
First Renovations: The Solar Theory

A s the flood of activity associated with the initial planning of
Uraniborg began to ebb, Tycho was able to turn some of his
attention to astronomical matters. For the most part, that attention
was absorbed first by the necessity of exploiting the appearance of
the comet to the greatest possible degree and then by the need to put
into production instruments worthy of his long-term professional
aspirations. But already in 1578 Tycho was able to do enough observing both to obtain some insight into the problems that might
arise in doing more extended work on the island and to start
accumulating some useful observations. The latter seem to have consisted primarily of meridian altitudes of the sun, for Tycho continued to take about a hundred of them annually. A couple of hundred
distances recorded between various stars used as reference points for
the nova and the comet seem merely to have convinced Tycho that
such work was premature, given the instruments available to him for
after March 1578, he made very few observations of either the stars
or the planets for the next three years.

As Tycho began to settle into Uraniborg in 1581 and get access
to the instruments from his shop, he gradually shifted into being
a full-time professional astronomer. The most prominent manifestation of this transition was his expansion into serious nighttime
observation. Starting in the fall, Tycho established the annual pattern
he was to follow as long as he was on Hven: an average of eighty-five
sessions a year, most of them in the early evening hours, with half of
them during the cold months of December through March and
another third in the other dark months of September, October, and
November.

Through most of the 1580s, while Tycho's shop was still turning
out instruments, much of this work was experimental: More often
than not, Tycho was as interested in analyzing the performance of his
instruments or checking on the feasibility of an observing procedure
(such as timing meridian transits) as he was in recording positions of
the stars and planets. Almost none of these observations was carried
beyond the immediate task of registering the sighting. It was not a
matter of Tycho's being either ignorant of or disinterested in theoretical considerations. On the contrary, virtually all of his planetary
observations were made on nights when one planet or another was in
some theoretically significant position. It seems, rather, to have been

a consequence of Tycho's long-term desire to build up a collection of observations and the short-term demands on his time.

Tycho's most immediate concerns were his write-ups on the nova and the comet. Now that he was finally able to obtain precise coordinates for the reference stars against which he had observed them, it was imperative that he get his findings and conclusions on these crucial phenomena before the public. The result of these complementary interests was a decade of curiously dichotomized activity. While Tycho worked energetically and fruitfully during the 1580s, on both theory and observation, most of the theory was associated with observations made in the previous decade, and most of the observation was directed toward subjects with which Tycho would not be able to occupy himself theoretically until the next decade. Only for the sun did the two operations coincide significantly.

From at least the time of Hipparchus, it had been recognized that the motion of the sun could be represented satisfactorily by an eccentric circle – that it could be regarded as perfectly uniform and circular if only it were not assumed to be centered precisely on the earth. Given this model (Figure 7.1), the elements of the solar theory were simply the magnitude (eccentricity) and direction (longitude of apogee: arc VA) of the displacement of the center of the sun's orbit (C) from the earth (E).

Before Tycho, these parameters had always been determined by geometrical analysis of the unequal lengths of the astronomical seasons, that is, by developing rigorously the notion that if the sun takes longer to move through one season (SU) than others, it must be because that section of its orbit is longer and hence more remote than the others are. The data for such a determination could be obtained, in theory, from four observations: an initial one at the vernal equinox (V), subsequent ones at the summer solstice (S) and the autumnal equinox (U) for the length of two seasons, and a final one at the vernal equinox, again, to give the length of the year. Astronomers had always used other observations to check the values obtained, but these observations had always been for timings of the lengths of seasons.

Although Tycho seems never even to have considered the possibility that the sun's orbit might be anything but the eccentric circle used by his predecessors, he nevertheless made numerous observations of the sun in all areas of its orbit. In fact, if there is any object for which the sheer quantity of Tycho's observations conforms to the popular image of his work, it is the sun: From 1578 until well into the 1590s, he recorded meridian altitudes for more than a hundred noons a year, usually with several different instruments. This provided many more data than Tycho could use; not for more than two hundred

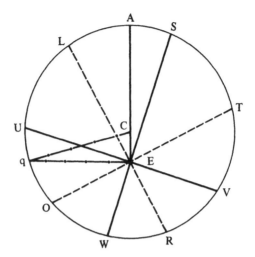

Figure 7.1. The eccentric model for the solar theory, displaying the
solstices (S, W), equinoxes (U, V), 45° points (T, L, O, R), apogee (A),
eccentricity (CE), and greatest correction (CqE).

years did astronomers learn how to utilize excess data to real advantage.[1]
  Therefore, Tycho had to use the traditional geometrical method to
present his results formally to fellow professionals; but through most
of the research phase of his work he proceeded much more empirically than his predecessors had. He began – probably in 1580 – by
making comparisons between the longitudes derived from his meridian altitudes over the years, and those predicted by the Alfonsine
and Copernican tables. These comparisons showed that the maximum correction (*CqE* in Figure 7.1) provided by Ptolemaic theory
(2°23′) was too large but that the one adopted by Copernicus (1°51′)
was too small. Already by the fall of 1580 Tycho had decided to try a
correction of about 2°, and by the end of 1581 he had computed and
sent to Hayek trial tables of his own using that value and a similarly

[1] On some occasions, however, Tycho came close enough to using the arithmetic mean as a
   method of combining observations to earn nomination by R. L. Plaskett, "Studies in the
   History of Probability and Statistics: The Principle of the Arithmetic Mean," *Biometrika* 45
   (1958): 130–5, as the inaugurator of the "technique of repeating and combining observations made on the same quantity to eliminate systematic errors." O. Gingerich, "Kepler's
   Treatment of Redundant Observations," in F. Krafft et al., ed. (Weil der Stadt: *Internationales Kepler-Symposium*, 1973), p. 314, showed that Kepler utilized extra data very well,
   by a "method" he described as "largely persistence." Volker Bialas, "Data Processing in
   the Rudolphine Tables," *Vistas* 18, 763–5, went a step further by claiming that Kepler
   "approached the method of least squares" in his use of Tycho's data.

estimated value for the longitude of apogee ($97\frac{1}{2}°$).[2] During 1582 and 1583 Tycho checked his theory against observations for over a hundred places of the sun. These "second approximation" comparisons, preserved in Tycho's logs,[3] display a pattern of error from which Tycho certainly could have obtained a very good revision of his elements.

Around the end of 1583, however, Tycho decided to see what kind of results he could get from a geometrical positioning of the orbit. This required knowing precisely the amount of time the sun took to traverse two sections of its orbit. For Hipparchus (and Ptolemy) one of these sections had been an entire season (VS), which entailed fixing the time of a solstice. But because solstices were notoriously difficult to establish precisely and because the geometry could be done just as well with a partial-season section, Copernicus had followed several Islamic astronomers in working with a half-season (i.e., half the interval from U to W on the tropical circle). Such consideration for the pragmatics of observation had a strong appeal for Tycho.

At some stage, however, Tycho began wondering why Copernicus had observed a 45° point around the winter solstice, where the effects of parallax and refraction were relatively large, instead of one of the higher-altitude points around the summer solstice where refraction, particularly, would be much less problematical. It may have been this train of thought that triggered the investigations of refraction he began on the winter solstice of 1583. At any rate, Tycho used two summer segments for his determination[4] and, on the basis of his findings, established what was essentially his final solar theory. Using his new equation ($2°3\frac{1}{2}'$) and apogee ($95\frac{3°}{4}$) he constructed new tables early in 1584 and generated ephemerides to check his theory against observation.[5] For four years, he checked and rechecked the intervals needed for the geometrical determination he

---

[2] In a letter written to Hayek on 4 November 1580 (VII, 60), Tycho stated that he had found the eccentricity to be $2^P5'$ (where CA equals 60^P) and the longitude of apogee to be 95°. By 12 October 1581, when he sent ephemerides to Hayek, Tycho had computed tables for a theory. The elements of those tables emerged from a letter of 23 September, 1582 (VII, 73) in which he said that he had used an eccentricity 9' greater than Copernicus's ($1^P56'$) and an apse of $97\frac{1}{2}°$. For $e = 2^P5'$, the maximum correction is just a bit less than 2°. For a glossary of technical terms see Appendix 2.

[3] X, 135–7, 237–9. For a list of abbreviations of frequently used sources, see Appendix 1.

[4] Hipparchus (and Ptolemy after him) had timed the arcs VS and VU (*Almagest*: III, 4). Copernicus used arcs UO and UV (*De Revolutionibus*: III, 16). Tycho used VT and LU (II, 24, 19). For a description of Tycho's orbit geometry, see J.-B. J. Delambre, *Histoire de l'astronomie moderne* (Paris, 1821), pp. 152–4.

[5] II, 25: X, 298.

would use to temper and justify his empirical results.[6] Only in the middle of 1588 did he make his final adjustments ($2°3\frac{3}{4}'$, $95\frac{1}{2}°$) and compute his final tables.[7]

Among Tycho's original reasons for giving the sun priority in his observational consideration was its utility for establishing his fundamental constants: the latitude of Hven and the obliquity of the ecliptic. Because they could be determined from solstitial meridian altitudes alone (Figure 7.2), Tycho began to make preliminary estimates of them during his first years on Hven, long before he was willing to venture any more complicated work with the sun. The same was true for the stars: Because merely averaging the meridian altitudes of a circumpolar star at the upper and lower culmination would provide his latitude, Tycho undertook such determinations (usually with the pole star) long before he tried any more serious placing of the stars.

Already on his birthday in 1576, in the first observation made after his move to Hven, Tycho began a series of what came to be annual year-end attempts to determine his latitude. Those of 1577 and 1578 seemed to converge on 55°53', and so Tycho used that value in writing up his remarks on the comet over the winter of 1577–8. For the obliquity of the ecliptic required in the same task, he compared observations of the winter solstice of 1577 and the summer solstice of 1578 to arrive at a value of 23°27'. Given the inherent limitations of the small quadrant, the frequent adjustments in orientation that must have been required in new (and not yet permanent) circumstances, and the gradual introduction of observations from new quadrants after 1581, Tycho was bound to get considerable scatter over the years: His logs show values ranging from $51\frac{1}{2}'$ to 54' (plus 55°, of course).[8]

Initially Tycho seems not to have realized that the latitudes he derived from observations of the sun were consistently lower[9] than those he got from the stars. But sometime after the summer solstice of 1583, when an extensive series of tests showed a latitude below $55°50\frac{1}{2}'$, Tycho recognized the tension.[10] He may well have

---

[6] Tycho's logs show only one attempt to "observe" an equinox in the traditional way, the vernal equinox of 1585 (X, 345). Apparently it did not work well, even though Tycho used both of his completed equatorial armillaries for the trial. All other equinoxes, as well as the times of 45° points, seem to have been interpolated from observations made a few days on each side of the event.

[7] II, 19–24. Tycho sent a five-year ephemeris computed from his old solar tables to the *landgrave* on 16 August 1588 (VI, 127). On 21 February 1589 he sent ephemerides for 1589–90 computed from his final elements (VI, 165).

[8] X, 42, 53, 59, 89–92, 108, 135, 183.

[9] X, 42 (55°52') and 55 (55°51$\frac{1}{2}$'). Observations implying 55°50$\frac{1}{2}$' remain unreduced in his logs: X, 92–3.

[10] X, 233–4.

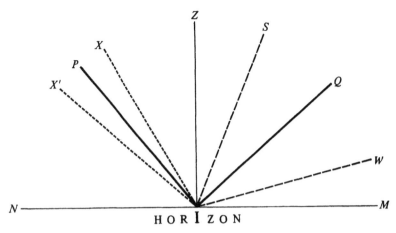

Figure 7.2. The meridian plane of the observer, determined by the celestial pole (P) and the observer's zenith (Z). Angles SIM and WIM are the meridian altitudes of the sun at summer and winter solstice, and QI represents the celestial equator. The obliquity of the ecliptic is SIQ (=WIQ): It is obtained by halving the *difference* between SIM and WIM. Halving the *sum* of SIM and WIM gives the altitude of the equator (QIM), which in turn gives the altitude of the pole (PIN), which is 90° from Q. The altitude of the pole can also be obtained as the average of the meridian altitudes (XIN and X'IN) of any star which is visible all year round. PIN is equal to the terrestrial latitude of the observer.

guessed what the problem was even before the autumnal equinox, because he made enough observations through the rest of the summer to allow him – when he eventually conducted his geometrical positioning of the solar orbit – to use "sections" chosen precisely because they were less subject to the effects of refraction than were Copernicus's winter sections. At any rate, in October (well before he usually made such trials), Tycho began observations of the pole star to verify the latitude given by it. When the sun reached the winter solstice in December, he was prepared not only to find a discrepancy but to explain it as well.[11] The latitude implied by the meridian altitude of the winter sun was, indeed, low, and Tycho expressed his hunch about the reason for it in a note written on the day of the determination (his birthday, again) – an artificially high altitude for the sun due to atmospheric refraction. During the following three weeks he verified it. Polaris indicated a latitude as high, perhaps, as even $55°54\frac{1}{2}'$, whereas the sun yielded at most $55°50\frac{1}{2}'$.[12] Some elaborate alternative computations involving other stars, done "for the sake of experiment,"

showed both that the higher value provided more consistent calculations and that the observations on which it was based seemed
themselves to be free from refraction.[13]

By early 1584, then, Tycho had achieved the first significant steps
(beyond the construction of his instruments) toward the establishment of precision astronomy. If neither his solar theory nor atmospheric refraction could be labeled discoveries in the strictest sense,
they were none the less important. Tycho's recognition of the
serious effects of refraction had at least three major consequences.

First, it established the latitude of his observatory. Once the low
values provided by the sun were removed from consideration,
Tycho quickly settled on the figure $(55°54\frac{1}{2}')$ that would essentially
remain the official latitude of Hven[14] and that compares very well
with the modern finding of $55°54'26''$. Second, it introduced a rather
radical change in the value that Tycho and everyone else in Renaissance Europe had been using for the obliquity of the ecliptic. Because
the latter was determined by halving the difference between the
solstitial meridian altitudes of the sun, it necessarily contained half
of the error in either observation. Thus, as long as Tycho neglected
the effects of refraction on the winter sun, as all his predecessors
had done, he consistently obtained values just $\frac{1}{2}'$ to $1'$ lower than
the $23°28'$ given by Regiomontanus, Werner, and Copernicus.[15] The
elimination of the winter solstice measurement (achieved by subtracting the colatitude of the pole star from the altitude of the
summer sun) increased his obliquity dramatically, to $23°31\frac{1}{2}'$.[16]

In view of the pride Tycho took in modifying so significantly this
fundamental and oft-determined constant, it is a pity that his result
did not turn out to be better than it was. Unfortunately, however,
refraction was only one of two corrections Tycho applied to his
observations of the sun. He was also correcting for solar parallax.

[13] X, 338–44.
[14] When his larger instruments became available after 1585, Tycho tended to obtain values 10''
to 15'' higher. Observationally (if not pragmatically) these were better figures, as Tycho
was assuming that refraction was null at such latitudes, when in fact it was raising his
results by about 40''. From 1585 (X, 427) until at least January 1587 (XI, 212, 222), he used
$55°54'40''$. A series of checks in 1588 and early 1589 (XI, 247–50, 303–5, 310–14, 325, 381)
produced results equaling or exceeding $55°54'45''$, so that Tycho was using that value at the
end of 1589. By March 1591, however, he was back (for good) to $55°54\frac{1}{2}'$ (XII, 107,
187, 270, 313, 374).
[15] X, 55, 235; IV, 69; II, 18. Tycho's lower results stemmed from the corrections he was
making for solar parallax. The "correction" for the winter solstice altitude raised it more
than the corresponding one raised the summer solstice altitude, thereby diminishing the
distance between them, which represented twice the obliquity.
[16] X, 292, 297. In computing tables for his solar theory in 1584 (for 1584 to 1588), Tycho used
$23°31'$ (II, 26), but he said later (VI, 87) that he always found it to be $23°31\frac{1}{2}'$, and his
logs bear him out.

Ever since antiquity, parallax had been understood, in theory, to affect observations of the sun. Relative to the accuracy of observation, however, it was so small that nobody had bothered to consider it (except in solar eclipse computations, which were sufficiently sensitive to several minor inaccuracies in the calculation to mask the problems caused by the excessive 3' value found in antiquity). For Tycho, of course, it was a matter of some pride to be establishing a standard of accuracy that finally justified, and even demanded, taking into consideration quantities as small as a couple of minutes of arc.[17]

Unfortunately, it seems never to have occurred to Tycho that the 3' value of his predecessors was based on observations (e.g., a ratio of 3 to 8 between the diameter of the moon and the cross section of the earth's shadow cone at lunar eclipse distance) that amounted virtually to guesswork compared with his own. Nor does he seem ever to have attempted to check it in any serious way. After injecting it into all of his calculations for many years, he appears simply to have regarded it as confirmed by the consistency of his various computations, particularly his eclipse reckonings. What he did not realize was the extent to which the other constants in his computations were themselves distorted to reflect the error in his parallaxes. By using the ancient values instead of neglecting the modern ones (based on 8.8″) that he would have failed to find in any legitimate (but admittedly very difficult) trial, he introduced into his value for the obliquity of the ecliptic an error of about $1\frac{1}{2}'$. In the solar theory itself he wrought even greater mischief.

In addition to modifying his fundamental constants to compensate for the effects of refraction, Tycho had to investigate the new phenomenon itself. Refraction per se was well known and must have been so for as long as people had been looking at oars, spears, or even stones in water. Its effects on light passing from air to water had already been quantified by Ptolemy and were certainly known to Tycho, even if not under Ptolemy's name.[18] Even atmospheric refraction had been noted in antiquity and used to explain an anomalous eclipse in which the sun and the partially eclipsed moon could be seen simultaneously on opposite horizons.[19] The great sources of European optical knowledge, Alhazen and Witelo, had described experiments for tabulating atmospheric refraction, and Bernard Walther had experienced its effects independently. Tycho was at least

---

[17] II, 18, 31, 89; VI, 69; VII, 279–80.
[18] See the classical investigation of A. Lejeune, *L'Optique de Claude Ptolemée* (Louvain: Bibliotheque de l'Université, 1956).
[19] See Cleomedes, "On the Circular Motion of the Heavenly Bodies," trans. T. L. Heath, in *Greek Astronomy* (Oxford, England: Oxford University Press, 1932).

generally familiar with the works relating these findings and had even
had the issue called to his attention more or less directly in 1575,
when the *landgrave* mentioned having noticed that the shadow of the
setting sun gave readings on his sundial that could not be reconciled
with the times on his mechanical clock.[20]

But no one had treated the issue thoroughly enough to leave a
lasting impression on Tycho.[21] Walther had dispensed with the problem, he thought, by simply avoiding horizon observations. Alhazen
and Witelo had reported experiments designed to quantify the
effects.[22] But they did not impress Tycho even on second reading,
for he was skeptical that the instruments described by them could
have been sufficiently precise to detect anything but the grossest
horizon effects. It therefore remained for Tycho to conduct the first
serious researches on the subject.

The method Tycho used was that described by Alhazen, of observing changes in the sun's declination as it rose and set. It was by
no means foolproof, but it was the simplest one available (given
Tycho's unwillingness to assume that all light coming through the
earth's atmosphere on a given path – whether from the sun, moon,
planets, or stars – would be refracted in the same way). It depended
on the fact that although the sun's declination swings from $23\frac{1}{2}°$N
to $23\frac{1}{2}°$S during the half year between solstices, it changes only very
slowly during any half day's movement from horizon to meridian.
In particular, it could be regarded as fixed when the sun was at the
extremes of its annual curve, the solstices. Thus, any observed
change in the declination of the rising or setting solstitial sun could
safely be regarded as purely apparent and attributed to differing
amounts of refraction at different altitudes.

Tycho began his work on refraction at the summer solstice in
1584. Through the evening of 16 June he recorded declinations as the
sun dropped to the horizon from an altitude of 25°. It must have been
very gratifying to watch the declinations increase from 23°26′ to 23°
$52\frac{1}{2}′$. A few hours later, as the sun rose, he watched the declinations

[20] Given the numerous allusions by Tycho to the work of Bernard Walther, it is clear that he
read him very carefully at an early stage of his career. This would have given him references
to Alhazen and Witelo, even if Tycho had not read the famed Reisner edition of their works
(1572) thoroughly enough to pick out the discussions of atmospheric refraction. The
landgrave's experience is related in IV, 41.

[21] It is interesting to see, however, that Tycho annotated an observation of 30 December 1581
(X, 103) with a reference to refraction and the writings of Alhazen and Witelo.

[22] Although Tycho was already citing Alhazen and Witelo as the great authority figures in
optics in his oration of 1574, he did not accept without reservation either their empirical
tests or their theoretical claims. See II, 76 for his doubts about their empirical tests, and VI,
92, for his rebuttal of their explanations of refractions.

reverse themselves.[23] This was as far as he tried to go in 1584. There was no point in attempting anything rigorous when he was expecting to have a better armillary nearly twice as large within a few months.

By the next summer the great armillary was indeed ready, and Tycho devoted several days in June, July, and September to determinations of refractions.[24] Rarely one to content himself with a minimum of evidence, he repeated the procedure at the next three summer solstices. And because his conception of the cause of refraction suggested that its effects might vary through the year, he checked his results at other times, even though his computations then had to allow for an inherent hourly change in the declination, independent of refraction.[25]

It was a difficult task, this construction of the first table of atmospheric refractions. Optical theory, even though it was one of the most thoroughly treated subjects in medieval science, could offer little real help. In fact, Tycho tended to let his measurements guide his theoretical conception of the phenomenon rather than vice versa. And because the major result of his experience with refraction was variability, he not unjustly concluded that refractions actually varied by as much as 1' for a given altitude, owing to changes in what he variously described as the serenity, purity, density, and temperature of the air.[26]

Tycho might have been bothered by the variations owing to differences in temperature and pressure near the horizon. But the most serious hindrance to his work was that he could not simply go out and "observe" refractions per se in controlled experiments; rather, they were bound up theoretically with his assumptions regarding the parallax of the sun, his determination of the obliquity of the ecliptic, and possibly his work on the solar theory itself.

To determine the amount of refraction that the sun's rays were incurring at a given altitude, Tycho had to compare the apparent (observed) place of the sun with its true place. This meant that Tycho's determinations would depend as much on his estimate of where the sun would appear if there were no refraction as on his actual observation of the sun's position. One way or another, the former had to come from theory. And one way or another, any theory that Tycho used had to depend on observed, refracted, positions of the sun. The situation was circular, but no more so than other scientific problems are. In fact, Tycho probably thought it was

---

[23] X, 293–4. [24] X, 351–61.
[25] II, 80. For examples see XI, 2, 5, and especially 133.
[26] II, 64, 89: VI, 125, 166: XI, 173, 313: XII, 135.

somewhat less so, for experience seemed to show that refraction was negligible above 45°.[27] This meant that he could obtain a true declination by direct observation of a (midsummer) meridian altitude and then compare it with declinations registered at the horizon to determine refractions at various low altitudes. It was not an unreasonable assumption, and it is not clear how Tycho could have proceeded if he had not made it. But the fact is that refraction still amounts to nearly 1′ at 45°.

More significantly, Tycho was also "correcting" his observations for the sun's supposed parallax. The minimum effect of these two misconceptions (at summer solstice) was to increase Tycho's "true" altitudes (and hence his "true" declinations) by about 2′7″: 1′30″ from the excessive solar parallax and 37″ from his failure to subtract for the refraction that still actually occurs at that altitude. This is almost precisely the amount by which Tycho's 23°31½′ determination for the obliquity of the ecliptic (obtained from just this kind of observation, by subtracting his colatitude) exceeded the true value (23°29′28″, obtained from modern theory). At lower altitudes the error introduced into his "true" places was correspondingly greater. Moreover, at other places in the sun's orbit, certain systematic features of the solar theory may also have influenced his computations; for Tycho's research on refraction was carried out during the very time (1584–8) that he was testing the almost-final elements for his solar theory. Insofar as the solar theory suffered from the effects of Tycho's ubiquitous "corrections" for solar parallax and other shortcomings, the refractions tended to take up the slack, because together they had to agree with Tycho's observed positions. Just what Tycho had to contend with was shown dramatically in a computer study conducted by Yasukatsu Maeyama.

According to Maeyama,[28] Tycho faced three sets of constraints in this work: limitations in observational accuracy imposed by his instruments, limitations on his data reduction process owing to imperfections in the constants used to prepare his raw observations for use, and limitations on the representational accuracy of his completed theory due to inherent shortcomings in the traditional geometrical model available to him.

<hr>

[27] In Tycho's published table of refractions, the refractions vanished only above 45° (II, 64); but Tycho frequently alludes to refractions being inconsiderable above 30° (XI, 15, 346; VI, 39).

[28] Yasukatsu Maeyama used a group of observations from 1590 that includes two very-low-altitude observations that Tycho himself would never have used for any purpose except testing refractions. The large errors of those two observations cancel, making a heuristic assessment of their effect on the group's average a matter of personal judgment ("The Historical Development of Solar Theories in the Late Sixteenth and Seventeenth Centuries," in Arthur Beer, ed. *Vistas in Astronomy* [London, 1974], vol. 16, pp. 35–60).

The first of these three limitations was much the least serious. For a random selection of eight days each in 1584 and 1590, involving readings from an average of more than three different instruments each day, Maeyama compared Tycho's raw observations with computer-generated meridian altitudes provided by modern theory. What he found is expressed in formal terms as a systematic error averaging 16″, with a standard deviation of ±38″. A less formal look at the same errors, grouping each day's sightings into one "most likely" meridian altitude, as Tycho himself would have had to do for his orbit work, shows that Tycho's positions averaged about 25″ low in 1584 (when he was using primarily the mural quadrant, the portable quadrant, and a sextant), and about half that in 1590 (when he was using his three large quadrants).

Raw observations, however good they might be, are only the first step. Unless they are treated to eliminate peculiarities (parallax and refraction) arising from the fact that any observation is made at a particular place and particular time, they will require a very complicated theory good only for that particular place and time. As a general rule, Tycho corrected for refraction simply by choosing observations made above 30°, where the refraction, as far as he was concerned, was negligible. But he always made formal corrections for parallax. Unfortunately, this correction was ridiculously exaggerated; not so much because Tycho thought it was a couple of minutes instead of a few seconds, as because the minor error in altitude so introduced could be expanded to major errors in longitude through the trigonometrical conversion of coordinates.[29] Moreover, the conversion trigonometry itself involved the constants for the latitude of the observer and the obliquity of the ecliptic, and was therefore affected by any errors in them. The cumulative result of these problems, as Maeyama has showed, was that by the time Tycho's observations were reduced to longitudes for comparison with theory, they were in error by 15′ for altitudes of 24° and progressively more at altitudes below that. In refusing to work with low-altitude observations, therefore, Tycho was wiser than he could have known.

The crux of the problem was Tycho's solar parallax "correction." Because of it alone, Tycho introduced errors into his "observed" longitudes that were systematically greater than 5′: See curve 1 of Figure 7.3.[30] At high altitudes, where the curve swings upward on the right, the errors could be rectified by exaggerating the obliquity.

[29] A parallax correction on observation S (see Figure A.4. in Appendix 4), which raises its altitude by a small amount h, will change its position to S′ and alter its longitude by the considerable amount. See also curve 1 of Figure 7.3. For Tycho's explicit statements about the sensitivity of this situation, see II, 16 and V, 39.

[30] Taken from Maeyama's curve 2 of Figure 4 in "The Historical Development," p. 49.

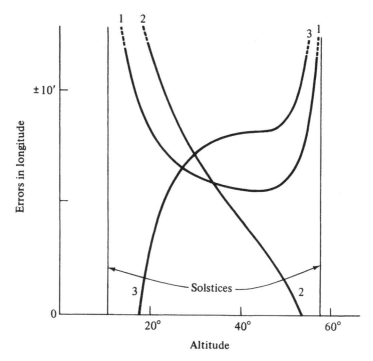

Figure 7.3. After Maeyama, showing the errors introduced by Tycho's solar parallax (curve 1), and the compensation for them provided by his exaggerated values for the obliquity (curve 2) and refraction (curve 3).

(In fact, as we have seen, the parallax correction was what produced the excess 2' of obliquity.) The resultant of the excess parallax and the excess obliquity was Maeyama's curve 2 (in Figure 7.3), which shows that the errors decreased sharply above the equinoctial altitude of 34°. At the low (left) end of curve 1, the errors could be canceled by using excessive corrections for refraction – which Tycho did, without knowing what he was doing, merely by combining his corrections so as to satisfy his observations. The result of the excess parallax and the excess refraction was Maeyama's curve 3, which shows the errors of curve 1 decreased sharply at low altitudes.

The result of introducing simultaneously the compensations for both high altitudes (by means of excess obliquity) and low altitudes (by means of excess refraction) is curve 4 of Figure 7.4.[31] Once Tycho adopted a solar parallax of 3', therefore, he was condemned – despite the heroic adjustments in his obliquity and refractions and the exquisite attention he paid to both the construction and use of his

[31] This is Maeyama's curve 6 (ibid.), which includes a correction for the 16″ systematic error in Tycho's meridian altitudes.

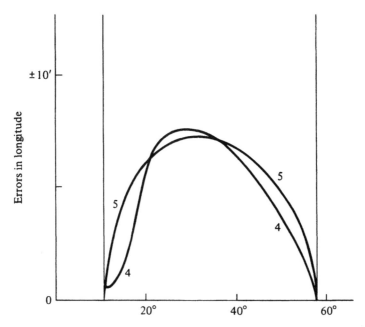

Figure 7.4. After Maeyama, showing the resultant errors (curve 4) introduced into Tycho's data by his imperfect values for parallax, obliquity, and refraction; and (curve 5) the inherent errors of uniform eccentric motion.

instruments – to obtaining solar longitudes that could be in error by nearly 8′.[32] These maximum errors, moreover, occurred at altitudes of around 30°, which was precisely the region (near the equinoxes and the mean distances) where most of the observations crucial to determining the orbit of the sun had to be made.

As Tycho saw his assignment, it was to produce a solar theory that would conform to the pattern of error displayed in curve 4. Given the peculiar nature of that pattern, one would not expect the task of constructing a theory that reproduced it to be easy, but it turned out to be almost automatic.

The reason for this rather curious fact lay in the orientation of the sun's orbit. With an accurate establishment of the line of apsides, the errors of almost any reasonable solar theory would take the shape of curve 5 in Figure 7.4,[33] being null at the apses and peaking in the

[32] Kepler would have been shocked to learn this! Curtis Wilson, "The Error in Kepler's Acronychal Data for Mars," *Centaurus* 13 (1969): 263–8 showed how it affected Kepler's work on Mars.

[33] See Maeyama's curve 1 of his Figure 4 and also Derek Price, "Contra-Copernicus: A Critical Re-estimation of the mathematical planetary theory of Ptolemy, Copernicus, and Kepler," in Marshall Clagett, ed., *Critical Problems in the History of Science* (Madison: Uni-

region halfway between the apses. For an orbit of the modest eccentricity of the sun's, the amplitude of the error would depend almost entirely on the constant chosen for the eccentricity. Tycho evaluated this constant primarily by observing it at its peak value. And because the sun's orbit was aligned so that its mean distances nearly coincided with the equinoxes, where the maximum uncompensated (by obliquity and refraction, as in curve 4) errors introduced by his solar parallax "correction" occurred, Tycho was destined to see a virtually perfect fit between theory and observation when he had actually produced a theory with errors that exceeded 7'. Tycho arrived at nearly the optimum eccentricity – only about 1' above the best fit he could have made. Only as his seventeenth-century successors gradually reduced their estimate of the sun's parallax could they converge on the modern value of the sun's eccentricity.[34]

Needless to say, the parallax correction had equally drastic repercussions in Tycho's table of refractions. In every single entry the specter of Tycho's 3' parallax appears. Dreyer could only express his disappointment that Tycho had "spoiled" his table in this way.[35] Delambre felt obliged to point out that even if the parallax component were ignored, there were many errors as great as one or even two minutes.[36] But if this first table of atmospheric refractions ever constructed was not a work of art from an aesthetic standpoint, it nevertheless got the job done. For Tycho, it was a table of solar refractions and would never have been used without compensating corrections from his table of solar parallaxes and his theory of solar motion. Just how neatly everything dovetailed is shown by Maeyama's computer study, which reveals that the difference between Tycho's observations and the predictions of this theory virtually is Tycho's table of refractions.[37]

In effect, therefore, the shortcomings of Tycho's table were, in fact, "merely" aesthetic. Insofar as there were significant objective errors – the larger departures to which Delambre alluded – they occurred at low latitudes which Tycho scrupulously avoided even

versity of Wisconsin Press, (1959), pp. 197–218. According to a personal communication from Dr. Maeyama, the inherent error arising from the failure to use a Keplerian orbit for the sun is only 48″. The rest of the error in Maeyama's curve 1 is due to improper eccentricity.

[34] Maeyama, "The Historical Development," pp. 51–60.
[35] Considering the problems Tycho faced in this enterprise, Dreyer (335) was rather uncharitable. In another allusion, he referred (123) to the one mistake compensating for the other.
[36] See Delambre, *Histoire de l'astronomie moderne*, pp. 156–8, for a comparison of Tycho's entire table with "modern" values.
[37] Maeyama, "The Historical Development," p. 48, curves 1 and 2.

after he established his tables of refraction, for he knew that his results were equivocal there.

In contrast with the solar theory itself, in which he believed he had achieved representation literally within a minute of arc, Tycho doubted that such a feat was possible for refractions, because of an inherent variability in the physical circumstances that produced them. Given the possibility that refractions varied from one locality to another and Tycho's conviction that they varied with the time of year and the vagaries of the weather,[38] he saw no alternative to printing with his table the disclaimer that its values might be in error by as much as a minute in a given situation.

Because his spurious parallax corrections for the sun had no analogue for the stars, Tycho found that starlight suffered less refraction than sunlight did. Sometime in 1590, therefore, he constructed a table of stellar refractions.[39] Apparently he rationalized the differences by assuming that refraction varied somehow because of the differing intensities of sunlight and starlight.[40] Such a theory would have led him to expect unique effects on rays coming from the planets, and as late as the turn of 1594–5, he was averaging the figures from his two tables to obtain refraction corrections for the planets.[41] Shortly thereafter, however, he decided that such special treatment was unnecessary, that the table of stellar refractions was sufficiently accurate by itself, for all planets except the moon. The latter, presumably because of its intense light, but actually because of its close computational links to the solar theory and Tycho's solar parallax corrections, did indeed require special treatment, which it duly received sometime after 1595.[42]

---

[38] II, 64, 89; VI, 125–66; XI, 173, 313; XII, 135. Tycho was willing to believe that the air of Egypt was purer and less susceptible to refractions, as some people claimed, but thought the proposition should be tested (II, 154–5). When he moved to Prague, one of the first things he did was to check the refractions (XIII, 193–4). Initially he thought he had found an instance in which the refraction was 2' greater than the corresponding one in Denmark. But he soon changed his mind, probably after deciding that there was some kind of error in the orientation of his quadrant.

[39] At the end of 1589, after some rather extensive testing of stellar refractions, Tycho constructed a partial table (XI, 380) that does not agree with the one he printed in the *Progymnasmata* (II, 287). Virtually all tests thereafter, however, were for the planets (XII, 153–5, 224–6).

[40] See XI, 145, where Tycho assumed that the moon must be encountering greater refraction than Aldebaran does at the same altitude and vowed to try to determine whether the crucial variable was the intensity of the source or the distance the light traveled.

[41] XII, 347, 463. Generally however, Tycho seems to have been using simply the stellar refractions: see X, 279, 468, 473.

[42] The table is on II, 136. Its values differ only slightly from those for the sun. As late as 1595 (XII, 388) Tycho was simply using solar refractions for the moon.

Chapter 8
The Tychonic System of the World

A s Uraniborg took shape, word of the remarkable new establishment began to circulate among the scholars who perambulated
across northern Europe. By the mid-1580s, Tycho was receiving not
only numerous applications from aspiring disciples, but also a steady
stream of overnight visitors from the summer circuit of touring
intellectuals and curiosity seekers. Among all of those who ever
appeared on Hven in either category, the one who probably made
the greatest impression on Tycho turned up in July 1580. The visitor
was a mathematician named Paul Wittich. He and Tycho seem to
have crossed paths already at Wittenberg during one of Tycho's
stops at the university there.[1] But that first meeting had obviously
been very brief, for when Wittich appeared at Uraniborg, he had
with him an introduction from Hayek. Of course, even without such
a formality, Wittich would have been more than welcome, for
Tycho at that time still had only Flemløse for observational help and
intellectual company.

Like Flemløse and all the rest of Tycho's later recruits, Wittich
came without much experience in the use of astronomical instruments: Tycho was later to report that his smiths were very amused
by Wittich's first attempts to observe. The complete theoretician,
Wittich not only knew none of the constellations but unrepentantly
argued that such knowledge was no more necessary for an astronomer than a knowledge of herbs was for a physician. Christopher
Rothmann later noted that Wittich had poor eyes and should have
stuck to geometry.[2] However, as a theoretician, Wittich soon displayed talents that made Tycho eager to recruit him as a long-term
associate. Whether during the recruitment or after he thought he had
obtained some kind of agreement from Wittich to stay and work for
him, Tycho revealed his whole enterprise to Wittich. He discussed

---

[1] Tycho's references to Wittich as a "youth" and his belief that he had recruited him as an
assistant led earlier writers to assume that Wittich was considerably younger than Tycho.
But Gingerich showed that Wittich was probably the same age as Tycho: See Owen
Gingerich and Robert Westman, "The Wittich Connection: Conflict and Priority in Late
Sixteenth-Century Cosmology," *Transactions of the American Philosophical Society* 78 (1988),
pt. 7, p. 11.

[2] IV, 454. Bruce Moran, "Christoph Rothmann, the Copernican Theory, and Institutional
and Technical Influences on the Criticism of Aristotelian Cosmology," *Sixteenth Century
Journal* 13 (1982) 85–108.

with him his technical work on the solar theory and his rectifications of the coordinates of the stars around the new star of 1572 and the comet of 1577 and probably showed him the observing facilities being built into Uraniborg, the ingeniously designed sights and divisions on his newest instruments, and even the novel features of the instruments that were under construction in his shop.[3]

What Wittich brought to Hven that earned him such high regard from Tycho was, first, a mathematical transformation that would, by the end of the century come to be known by the Greek term *prosthaphaeresis*, or addito-subtraction. The basic formula behind it originated (as far as Europe was concerned, anyway) in the work of Johannes Werner at the beginning of the century. In composing a treatise on spherical trigonometry, Werner had discovered that what has since come to be known as the law of cosines could be expressed in a form equivalent to the formula $2 \sin A \sin B = \cos(A - B) - \cos(A + B)$.[4] What this equation meant was that any multiplication of one sine by another could be transformed into an operation involving some combination of three additions or subtractions.

Werner's manuscript lay in obscurity in Nuremberg for twenty years, until Rheticus happened to come there in 1542 to arrange for the printing of Copernicus's *De revolutionibus*. Because Rheticus had just published a separate edition of Copernicus's trigonometric tables, he was told about Werner's manuscript and allowed to take a copy with him when he left to become a professor at Leipzig. By 1557, after having settled at Cracow, Rheticus had prepared the manuscript for printing.[5] However, it never was printed, and so the manuscript continued to be inaccessible except to those few people who came into contact with Rheticus at Cracow.

Wittich may have been one of those people. At any rate, among the three formulas that he might have discovered (the other two are for $\sin A \cos B$, and $\cos A \cos B$), Wittich reportedly had the one presented in Werner's manuscript – and reportedly had no proof for it. What Wittich does seem unambiguously to have contributed was the recognition that this formula could be used "artificially" as a substitute for the multiplication of any two numbers (since every number can be regarded as the sine of some angle, if one takes due account of decimal points). At Hven Wittich got from Tycho what was probably the first enthusiastic reception of his method. For although most mathematicians would live and die without doing

---

[3] VII, 63. The abbreviations used for frequently cited sources are listed in Appendix 1.
[4] On Werner, see *DSB* XIV, 272–7. On the details and fate of Wittich's ideas, see A. von Braunmühl, *Vorlesungen über Geschichte der Trigonometrie* (Leipzig, 1900), pp. 193–203.
[5] On Rheticus's life and work, see *DSB* XI, 395–8. See also *Zinner*, 358.

enough trigonometric calculations, or multiplication of large numbers more generally, to benefit greatly from the identities, Tycho
saw immediately that the process could have enormous practical
value for the operations on Hven. Not only was he going to be doing
many more calculations than anyone had ever dreamed of doing, but
he was going to be computing to a standard of accuracy (minute-
of-arc) that would require seven-digit calculations rather than the
four-digit ones that had sufficed until his time. Accordingly, he enthusiastically harnessed Wittich to the task of helping him compile a
manual of trigonometry that would direct his assistants through the
complexities of solving every possible case of plane and spherical
triangle.[6] And when it soon emerged that Wittich's formula could
not eliminate every multiplication and division Tycho needed for
his computations, he undoubtedly drove Wittich relentlessly – and
probably himself, too – toward achieving that end, But even though
Wittich was on Hven for about four months, the scheme was
incomplete when he left.

Wittich had some other results of interest, too, in an area of
mathematics in which Tycho himself had already done some dab-
bling: the juggling of planetary orbit mechanisms. Like the problem
of *prosthaphaereses*, it was one on which others had labored before
them. From virtually the first years of *De revolutionibus*'s existence,
various mathematicians had recognized with varying degrees of
clarity that it should be possible to use the fruits of Copernicus's
mathematical labors without subscribing to his physical ideas. But it
was not until just before Tycho lectured at the University of Copenhagen that Peucer and Dasypodius actually published Ptolemaic
tables that had been reestablished with new numbers from Copernicus's technical researches.[7]

Already a generation earlier, both Reinhold and Gemma had made
statements suggesting that they understood fully the possibility of
utilizing Copernican models for a cosmology that did not involve a
moving earth.[8] Tycho had been educated to the same concerns, in

---

[6] The only surviving manuscript is one copied by one of Tycho's students in 1591. It consists
of twenty sheets and is bound behind a copy of Rheticus's trigonometric tables. It is reproduced in I, 281–93. On the history of the manual, see Victor E. Thoren, "Prosthaphaeresis
Revisited," *Historia Mathematica* 15 (1988): 32–9.

[7] Caspar Peucer seems to have done this "translation" sometime well before 1568, when
Conrad Dasypodius published the results without knowing who their author was. See O.
Gingerich, "The Role of Erasmus Reinhold and the Prutenic Tables in the Dissemination of
Copernican Theory," *Studia Copernicana* 6 (1973): 59–60. Praetorius fully subscribed to the
same goals: See Robert S. Westman, "Three Responses to the Copernican Theory: Johannes
Praetorius, Tycho Brahe, and Michael Mästlin," *Westman*, 293ff.

[8] Gingerich, "The Role of Erasmus Reinhold, pp. 51, 58–9. Actually Reinhold may well
have conceived of the scheme before he died in 1553: See also the analysis of his manuscript

the tradition of Wittenberg, where first Reinhold and then Peucer had served as professors. When Tycho undertook in his lectures at Copenhagen the task that Pratensis quoted back to him as "casting aside the Ptolemaic assumptions and converting those of Copernicus to the stability of the earth,"[9] he was pursuing an old Wittenberg theme that he – and Wittich, too, for that matter – may well have heard proposed in a lecture by Peucer.

Not surprisingly, what Tycho had accomplished by the time of his lectures was probably rather limited. If Tycho had managed to conceive of his project by regarding the Copernican system as essentially correct and just looked for a way to eliminate its unsatisfactory feature – the motion of the earth – he might have grasped, in one inspired leap, what has since come to be called the "Tychonic system." But there is ample evidence, and Tycho's own statement, that he did not arrive at his system by simply inverting Copernicus's. What Tycho appears to have done was to regard the Ptolemaic system as basically correct and then look for ways to eliminate, piecemeal, its objectionable equants.

The technical way to do this for the superior planets was to substitute the Copernican mechanism for representing orbital irregularities ($\overline{T}C$ and EP in Figure 8.1) for the equivalent Ptolemaic equant (TCQ in Figure 8.2). All that had to be done was to change the center of motion from Q to C, to "move" the earth, T, out to one and a half times its former distance from C, and to decide whether to put the little epicyclet at M or at P. Ideally, therefore, in his lectures of 1574, "expounding the motions of the planets according to the models and parameters of Copernicus, but reducing everything to the stability of the earth," so as to avoid "both the mathematical absurdity of Ptolemy and the physical absurdity of Copernicus,"[10] Tycho was working with a generalized planetary mechanism that resembled Figure 8.3. Whether he had achieved every feature of the model before seeing it in a manuscript that Wittich brought to Hven can only be a matter of speculation. It is the diagram he was using for that purpose in 1585.[11]

Whatever Tycho did with the more complicated inferior planets in 1574, he soon simplified that portion of his task by adopting the so-called Cappellan variation of the Ptolemaic system in which

commentary on *De revolutionibus* by Christine (Jones) Schofield, "The Geoheliocentric Mathematical Hypothesis in Sixteenth-Century Planetary Theory," *British Journal for the History of Science* 2 (1965): 292–3. Tycho had access to Reinhold's papers in 1575 but obviously failed to find anything that would give him a shortcut to the system (III, 213).

[9] VII, 24.
[10] I, 172–3. For a glossary of technical terms, see Appendix 2.
[11] X, 284.

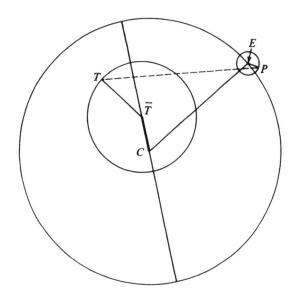

Figure 8.1. Copernican Theory for the superior planets.

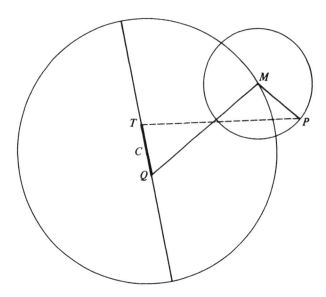

Figure 8.2. Ptolemaic Theory for the superior planets.

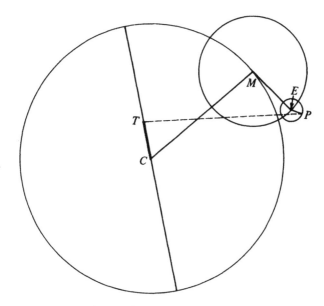

Figure 8.3. Tycho's hybrid theory for the superior planets.

Mercury and Venus revolved around the sun instead of circling between it and the earth. All that is known about either the cause or the date of his conversion is that he picked up on his travels in 1575 a book that depicted that arrangement[12] and that by the time he drew the diagrams illustrating his conceptualization of the path of the comet of 1577, he displayed Venus in what is clearly a heliocentric orbit.[13] Although Tycho did not include a drawing in his report to the king, he did mention that the great comet had probably moved in the sphere of Venus, that some people not implausibly imagined that Mercury and Venus circled the sun, and that in such a scheme the earth would not have to be moving.[14]

By 1577, therefore, Tycho seems to have taken the fundamental step toward what eventually was to be the Tychonic system. All that remains debatable is whether he had yet envisioned the possibility that the superior planets also might circle the sun. The answer to this question seems clearly negative: Historical statements by Tycho consistently point to a date in 1583 or 1584 for the origin of his system. As of 1578, then, Tycho's cosmos consisted of a stationary

---

[12] Westman, "Three Responses to the Copernican Theory," pp. 322–4.
[13] See Figure 4.4 of Chapter 4 and also John Christianson, "Tycho Brahe's German Treatise on the Comet of 1577," *Isis* (1979) 124–5.
[14] IV, 388.

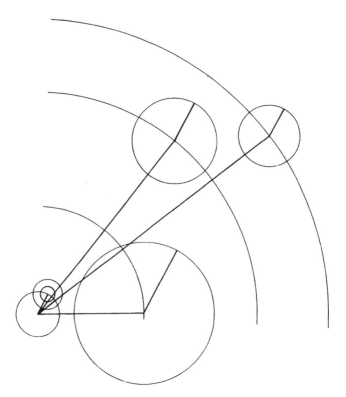

Figure 8.4. Hypothetical reconstruction of Tycho's amalgam of the
Capellan and Ptolemaic systems.

earth circled by the sun, with Mercury and Venus orbiting around
the sun (instead of between it and the earth), and Ptolemaic orbits for
the superior planets. Although the outward displacement of the
inferior planets would have allowed Tycho to move the sun closer to
the earth (without tangling Venus's machinery with the moon's), it
is clear from Tycho's mind-set on solar parallax (see Chapter 7) that
he would not have considered doing that. Because half of the orbital
mechanism of the inferior planets was now above the sun, Tycho
would have had to displace the orbits of the superior planets outward
by a corresponding amount and, of course, enlarge the Ptolemaic
epicycles by the same ratio. Any schematic he drew would have
looked very much like Figure 8.4.

At the time Wittich came to Hven, Tycho had probably not
thought about planetary theory or cosmology since his deliberations
on the comet. The same may have been true of Wittich, whose
contemplations on the subject had occurred, oddly enough, at almost

exactly the same time. But although Tycho's had been almost incidental thoughts, provoked only by his consideration of the comet,
Wittich's had been self-conscious, explicit explorations of the various
models and combinations thereof available for planetary theorizing.

In a series of twenty-six drawings only recently found, recognized, and published by Gingerich,[15] Wittich had permuted the
Copernican planetary mechanisms – both the double-epicycle (Figure 8.5) and double-hypocycle (Figure 8.6) forms of the *Commentariolus* and the epicycle–eccentric combination of *De revolutionibus* –
through various orientations to produce an abundance of options
that must have left Tycho astonished by both the power of the
methods and the ingenuity of Wittich. As a conclusion of his efforts,
he obtained a "Theory of the Three Superior [Planets] adapted to the
Immobility of the Earth."[16] Shown in Figure 8.7, the theory used
a Ptolemaic epicycle (*MP*) for the synodic phenomena and then
accounted for the anomalistic realities of the planet's orbit by making
the center of the orbit (*C*) ride on a double hypocycle (*TE* and *EC*)
around the stationary earth.

What probably did not seem as exciting at the time but proved
most significant in the long run was a final diagram that was strictly
not part of the series at all, in which Wittich had abandoned any
consideration of the anomalistic theme of the rest of the diagrams
and simply sketched a synodic schematic (Figure 8.8) of the entire
planetary system.[17] The similarity to Tycho's conception is striking,
but so is the difference: Wittich arbitrarily used epicycles for the
superior planets that were not only mutually equal in absolute size
but also equal to the size of the sun's orbit around the earth.

If Tycho was as blunt on this occasion as he showed himself to be
on most others involving scientific issues, he must have lost no time
in telling Wittich that his diagram was wrong. For although Wittich
certainly had the right to establish an arbitrary size for any epicycle,
he then had a corresponding obligation to make each deferent "fit"
its epicycle astronomically. In particular, as Saturn's (Copernican)
orbit is roughly ten times the size of the earth's, its deferent has to be
about ten times the size of its epicycle in order to account for the
phenomenon of Saturn's retrograde arc. Tycho would have sym-
pathized with Wittich's instinct for placing (and, therefore, sizing)
the deferents so that each successive "orb" was as close to its inside

[15] O. Gingerich, "Copernicus and Tycho," *Scientific American* 229 (1973). The diagrams are dated 22 January and 13 February 1578. At the time this article was written, Gingerich thought the diagrams were in Tycho's hand. For the correction of this assumption, see Gingerich and Westman, *The Wittich Connection*, pp. 5–9 and 23–6.
[16] Gingerich, "Copernicus and Tycho," p. 90.   [17] Ibid., 100.

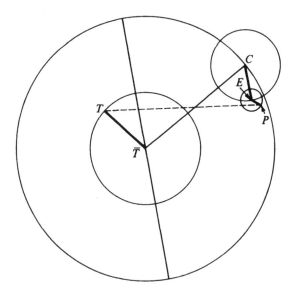

Figure 8.5. Double epicycle form of the Copernican Theory for the superior planets.

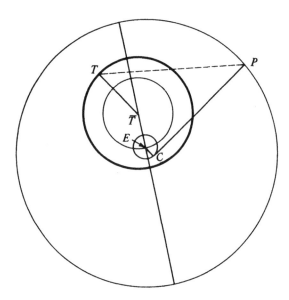

Figure 8.6. Double hypocycle of the Copernican Theory for the superior planets.

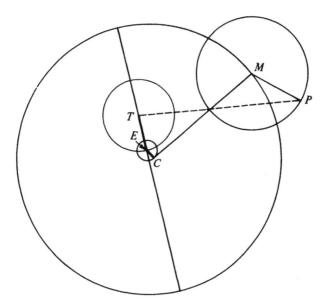

Figure 8.7. Wittich's "Theory of the Three Superiors adapted to the Immobility of the Earth," or Copernican hypocycle with Ptolemaic epicycle. Courtesy of Owen Gingerich and the Vatican Library.

neighbor as it could be without interfering with the operation of either assemblage.

Everyone since Ptolemy had followed this tradition,[18] and every astronomer of Tycho's era was thoroughly familiar with the picture that derived from it. The sphere of Mercury began just outside the sphere of the moon, about 64 earth radii away, according to the best astronomical information. When astronomers placed an appropriately eccentric deferent so that an appropriately sized epicycle carried Mercury just down to 64 e.r. at perigee of the epicycle and perigee of the eccentric, the same machinery carried Mercury at apogee of the epicycle and apogee of the eccentric to 167 e.r.[19] Similar computations for Venus extended its sphere from 167 to 1,160 e.r., and those for the sun took its sphere out to 1,260 e.r. For Mars, the epicycle

---

[18] The Islamic tradition attributing the original idea to Ptolemy was not verified until Bernard R. Goldstein, "The Arabic Version of Ptolemy's Planetary Hypotheses," *Transactions of the American Philosophical Society* 57 (1967), recognized the discussion in an Arabic text of Ptolemy's *Planetary Hypotheses*.

[19] The numbers used here are Willy Hartner's, "Mediaeval Views on Cosmic Dimensions," *Mélanges Alexandre Koyre* (Paris: Hermann, 1964) from the Islamic tradition. For the sixteenth-century use of similar numbers, see Westman, "Three Responses," p. 302; and Tycho (II, 417).

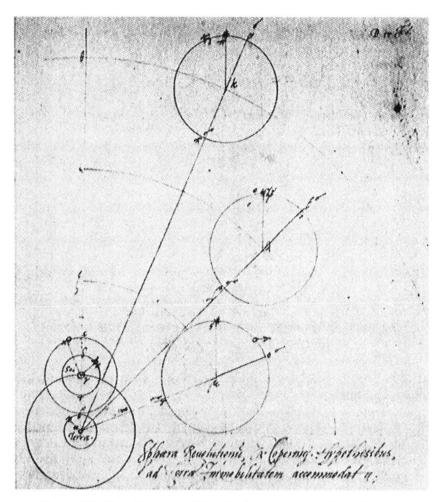

Figure 8.8. The Copernican Hypothesis accommodated to the immobility of the Earth, according to Wittich.

(corresponding to the earth's orbit) and deferent had to be in the ratio of approximately 100 to 152. For a simple circular deferent, this would have entailed a mean distance for Mars of 1,260 + 2,423 earth radii and a maximum distance of 1,260 + 2(2,423) e.r.[20] A more complicated adjustment for the eccentricity of Mars's orbit pushed its maximum distance out to 9,200 e.r. Jupiter's orbit occupied the space out to 14,400, and Saturn's extended out to about 20,000.

[20] In Figure A.4.2 of Appendix 4, the minimum distance, 1,260, must equal $R - r$, and r must be to R as 100 is to 152. So, $R - R (100/152) = 1,260 = 52/152 R$; thus $R = 3,683, r = 2,423$.

If these numbers seem arbitrary, they were only partly arbitrary. As Tycho probably reminded Wittich, one could specify only one item at a time: either the size of the epicycle, or the size of the deferent, but not both. If Wittich had not simply lost track of this basic astronomical fact, he would have defended himself by saying that he was simply schematizing the system and had not drawn the eccentrics to scale. They may even have discussed the implications of "opening up" the heavens to allow the space between the spheres that would be demanded by the proper scale. By using the Capellan arrangement for the inferior planets, both Tycho and Wittich had, in fact, already introduced some wasted space in the heavens, although Wittich had forgotten or refused to show it in his diagram. But whatever the degree of intent behind the shortcomings of Wittich's diagram, making all the epicycles the same size was a thought-provoking idea.

By inverting this isolated part of Copernicus's logic, Wittich unwittingly took the penultimate step toward the system that bears Tycho's name. All that remained to be done was to convert the epicycle–deferent mechanism of the three outer planets to their equivalent eccentrics. This step would add circles the sizes of the (appropriate) respective deferents, but centered on the sun, to the circles for Mercury and Venus already drawn around the sun. But it was a step that neither Wittich nor any of the other people who may have contemplated his diagrams during his travels was ever able to take. Even for Tycho it took at least a couple of years, and reconsideration in another context, to arrive at the Tychonic system of the world.

On 18 October 1580, the theoretical contemplations of Tycho and Wittich were disturbed by the appearance of a comet.[21] Halfway through the observations of the comet, Wittich told Tycho that he had to go home for a few weeks to collect an inheritance from a rich uncle.[22] Never one to pass up an opportunity to send mail with a traveler, Tycho composed a quick note to Hayek apologizing for his delay in responding to two earlier letters from Hayek and bringing him up to date on developments at Uraniborg. Then Tycho broached the matter that had been lying between them ever since he had received Hayek's book on the comet of 1577 – that Hayek had determined the comet to be sublunary. Whether because he had not yet analyzed Hayek's argument in detail or whether he was just being diplomatic, Tycho attributed the differences between their findings to shortcomings in Hayek's instruments and contented himself with asserting that he could "prove by many observations and computa-

---

[21] XIII, 305–33.   [22] VII, 63.

tions that neither [the last large comet nor the small one currently visible] existed in the elementary region but [that they] traced out their motions far above the moon in the ethereal abode."[23]

It would be interesting to know whether Tycho was able to exercise similar restraint in his dealings with Wittich. The appearance of the comet must have strained the relationship in two respects, first by subjecting Wittich to the unfamiliar and uncongenial labors of handling observational instruments and computing data and then by raising an issue that emphasized the differences in their philosophical orientation. Wittich's interests were clearly much more abstract than Tycho's; indeed, it is not clear that he should be regarded as an astronomer at all. With the exception of Wittich's planetary model juggling, his sole known foray into astronomical work before coming to Hven was some consultation with Hayek and Schultz that enabled the two of them independently to obtain, using a well-publicized method of Regiomontanus,[24] a large parallax for the comet of 1577.

Tycho had probably not yet developed his great contempt for this method. But his unshakable conviction that there had been no sensible parallax for the comet of 1577, the null determination that he reached for the comet of 1580 about a week before Wittich left Hven,[25] and Tycho's confident allusion to having changed Wittich's mind on the subject of comets[26] all add up to the certainty that Wittich discussed the issue with Tycho at some length, whether or not he wanted to. At the same time, it seems almost certain that the restless Wittich was destined to move on no matter how Tycho behaved. The only contemporary statement about his peripatetic life-style is a later letter by Andraeus Dudith stating that Wittich left the *landgrave*'s circle after a short stay in 1584 because he did not like the *landgrave*.[27]

After Wittich's departure, Tycho and Flemløse continued to observe the comet in the early morning sky until it disappeared into the sun in mid-December. Having mentioned to Hayek that he hoped to find leisure during the winter to finish his discussion of the comet of

[23] VII, 59.
[24] IV, 448–56. See also C. D. Hellman, *The Comet of 1577: Its Place in the History of Astronomy* (New York: 1944), p 204.
[25] XIII, 319.   [26] IV, 455–6.
[27] Gingerich and Westman, *The Wittich Connection*, p. 17. Dudith was a Catholic bishop who left the church in 1567 to marry a Polish noblewoman. In 1589 he reported that (the deceased) Paul Wittich had obtained a copy of Copernicus's "Commentariolus" from his uncle, Balthasar Sartorius, who presumably got it from Rheticus. This was apparently the pipeline through which Tycho got his copy from Hayek in 1575 (see Chapter 3). Jerzy Dobrzycki and Lech Szczucki "On the Transmissions of Copernicus's Commentariolus in the Sixteenth Century," *Journal for the History of Astronomy* 20 (1989): 25–7.

1577, Tycho probably spent at least some time on it. This may well have been the period during which he analyzed Hayek's and Schultz's misuse of Regiomontanus's method and added to his own Chapter VI the demonstration that the method could show a null parallax if used properly.[28] When he wrote to Schultz the following fall, at any rate, Tycho was able to catalogue his old friend's transgressions very specifically if also very diplomatically. But because that letter contains almost the same reference to the need for finding leisure to finish his manuscript that Tycho had made to Hayek a year earlier, it is clear that Tycho had not progressed very far in the meantime.[29]

Sometime during the next three or four winters, Tycho composed the next chapter, describing the comet itself.[30] In the German tract Tycho had sent to the king in 1578, this description had been very brief, merely mentioning that the comet's tail followed the general rule of being directed away from the sun and converting the apparent sizes of the head and tail to true sizes.[31] What was eventually published as Chapter VII of *De mundi* is a twenty-five-page day-by-day instantiation of Tycho's contention that contrary to the findings of Apian, Gemma, Fracastoro, and Cardan that comets' tails always oppose the sun, the tail of this comet was directed away from Venus.[32] Similarly, the discussion of the sizes of the comet's head and tail which was going to be an appendix to Chapter VII, involves numbers that are noticeably different[33] from those in Tycho's report to the king.

At this stage the book was finished, except for the task of reviewing critically all of the other publications on the comet (adding material that was destined to constitute Chapter VIII). Tycho started the printing of the book and even printed a title-quarto for it, whose table of contents describes in some detail the eight chapters.[34] If the book had been completed as planned, it would have been an eminently forgettable one, however much one might still wish to insist on the necessity of a monograph that would drive home the contradiction of Aristotelian cosmology embodied in the comet.

---

[28] IV, 123–34.   [29] VII, 61.

[30] IV, 107–34. The computations were made with the constants of Tycho's new solar theory, which became available in 1580 (VII, 60).

[31] IV, 386, 389.

[32] IV, 135–54 (original pagination, 158–84). The orientation from Venus arose from Tycho's faulty inclination: see J.-B. J. Delambre, *Histoire de l'astronomie moderne* (Paris, 1821), p. 222. Tycho later came to believe that the phenomenon was some kind of illusion or special effect of perspective (IV, 175; VI, 171).

[33] IV, 171–2, 497.

[34] IV, 491, 497. Tycho says in IV, 174, that his table of distances was derived after Chapter VII, on the tail of the comet, was written.

Sometime during the printing, however, Tycho added the chapter on his world system that has subsequently commanded virtually all the attention given to the book. Tycho would certainly have agreed with the unanimous assessment of this chapter as "the most important [one] in the whole book."[35] As Lagrange later said about Newton's work, it is given to only one person to discover the system of the world. Tycho thought he had done it. The question naturally arises, then, why was he so late in making plans to publish it? The answer appears to begin with Tycho's investigation of the parallax of Mars in the winter of 1582-3.

Exactly when Tycho first encountered the proposition that the earth was moving, he does not say. Lecture syllabi from Wittenberg suggest that he was probably introduced to the technical achievements of Copernicus without being told of the great mathematician's cosmological speculations.[36] But it is hard to believe that his innocence could have survived the period he spent with Schultz when he was seventeen. By the early 1580s, therefore, Tycho had presumably been aware of the issue for close to twenty years and clearly understood that Copernicus truly believed ("his" preface notwithstanding) that the earth was moving. Whether Tycho himself was ever able to take the proposition seriously, however, is doubtful. His few references to the motion of the earth all characterize it as physically absurd. And through a career in which he developed instruments whose accuracy rose higher and higher above any standard previously achieved, he never documented any attempt to detect the stellar parallax that alone could verify the annual motion of the earth.

There was, however, another kind of parallax that looked as if it should be more accessible: planetary parallax. Given the actual (modern) values of these parallaxes – about 25″ maximum for Mars – Tycho was not going to find any real evidence on the issue. But because he had adopted Ptolemy's ancient 3′ parallax for the sun, he thought he would be dealing with detectable quantities. At the very same time that this unworthy stepchild was working mischief with Tycho's solar theory and table of refractions, therefore, it would also do its best to cause him to misinterpret evidence on what was the much more significant scientific question of the system of the world.

A value of 3′ for the sun implied an even larger figure for the inferior planets, if they were truly inferior in the Ptolemaic sense. But as Tycho had already decided that they moved in Capellan orbits

---

[35] *Dreyer*, 167.

[36] See O. Gingerich, "From Copernicus to Kepler: Heliocentrism As Model and As Reality," *Proceedings of the American Philosophical Society* 117 (1973):516–20, for evidence "that Copernicus was well known and esteemed as a mathematician and astronomer" but that his heliocentric theory was generally ignored.

(which were indistinguishable from Copernican orbits), there was no cosmological distinction to be inferred from their parallaxes. For Mars, however, the situations were quite different. According to the Ptolemaic conception, what happened when Mars went through opposition (to the sun) was that the planet moved (counterclockwise) through perigee of its epicycle (as in Figure 8.8), producing both retrograde motion and Mars's nearest approach to the stationary earth. At that time, according to the Ptolemaic cosmology, Mars's distance was just a bit greater than the sun's maximum distance, for the scheme was based on the assumption that Mars's entire "orb" was situated above (i.e., outside) the sun's.

According to Copernicus's interpretation of Ptolemy's epicycles, the retrograde motion and closest approach of Mars were caused by the fact that the earth, in its orbital movement, was overtaking its slower-moving outer neighbor. Without reference to something outside the system, it was impossible to distinguish between these two views on the basis of motion alone. However, if one assumed, after Copernicus, that Mars's epicycle was actually the orbit of the earth, then the astronomically determined ratio of Mars's orbit (to its epicycle) implied that Mars, around opposition, was considerably closer to the earth than the sun was.

On the basis of a 3′ solar parallax, Tycho could expect Mars to show 4 $\frac{1}{2}$′. It was so obvious a test situation that Tycho probably did not give much thought to the fact that even on the Ptolemaic scheme Mars should show nearly 3′ of parallax. During the winter of 1582–3, at the first opposition to occur after the large instruments began to issue from his shop, Tycho looked for parallax. When he failed to find an amount that should have been very easily detectable, he concluded that the Copernican hypothesis was untenable. Given Tycho's opinion about the motion of the earth, one would expect this apparently nullifying result to have been very gratifying. But by the time Tycho reported his result in a letter to Brucaeus in the spring of 1584,[37] he may well have developed reasons for having a curious ambivalence about the result.

Because the Tychonic system is nothing but the rather trivial-looking inversion of the Copernican system depicted in Figure 8.9, it is rather startling to see the numerous attestations by Tycho that he discovered or invented it sometime around 1583.[38] Although it is tempting to take refuge in the fact that most of Tycho's professional contemporaries lived and died without seeing their way to the inversion, to account for its decade-long gestation period in Tycho

[37] VII, 80. Tycho's trials for parallax are on X, 174–8, 243–9, 283–8.
[38] IV, 155–6; V, 115; VI, 179; VII, 199; VIII, 205.

NOVA MVNDANI SYSTEMATIS HYPOTYPOSIS AB
AUTHORE NUPER ADINUENTA, QUA TUM VETUS ILLA PTOLEMAICA REDUNDANTIA & INCONCINNITAS, TUM ETIAM RECENS COPERNIANA IN MOTU TERRÆ PHYSICA ABSURDITAS, EXCLU- DUNTUR, OMNIAQUE APPAREN- TIIS CŒLESTIBUS APTISSIME CORRESPONDENT.

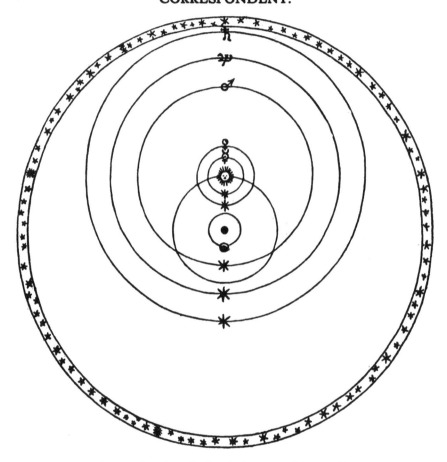

Figure 8.9. The Tychonic System of the world.

there are two more satisfying explanations available more or less directly from Tycho himself. The first is that Tycho stated explicitly that he did not derive his system merely by inverting Copernicus's.[39] He was undoubtedly an enthusiastic admirer of Copernicus. But his admiration was for the geometry (and perhaps the astronomical data processing) of Copernicus's planetary theory, not for his cosmology (or his observational prowess). Once he had done his thinking about cosmology and found himself unable to believe that the earth could be moving, he had little incentive to think about cosmology at all. Having once consciously adopted the geocentric worldview, he subsequently did all of his technical thinking within that framework.

Thus, although Tycho was capable of entertaining individual Copernican theories for the planets, when he was thinking technically, the framework within which he viewed them was a Ptolemaic one. This situation rendered it unlikely that Tycho would ever contemplate the Copernican cosmology as a whole long enough to see that it could be stood on one ear and converted instantly into a geostatic system. Although he certainly recognized that Copernicus's revolution of the earth provided a geometrical equivalent to Ptolemy's epicycles, even that knowledge seems to have been at a formal, academic level, for Tycho made too many references – even in his mature years – to "proving" the rectitude of his system for the equivalence to have been second nature.[40] But somehow and sometime, presumably during his contemplation of Mars's parallax, and sometime before the fall of 1584, Tycho struck on the conversion, inversion, or whatever means he used that produced his system. Unfortunately, however, getting a glimpse of the system was only half the battle. When he actually set out the scheme in detail, with the proper ratios of planetary orbits to the sun's orbit, he found that his new system required that the orbit of Mars intersect the orbit of the sun.

---

[39] VI, 178.

[40] See VI, 178–9, 236, 239, and VII, 130, 230, 294–5 for the most extended of such statements. They start by implying an equivalence between the systems and recognizing that the distinction between them rests on disproving the motion of the earth but then come around to testing astronomically instead of physically for the motion of the earth – the failure of comets in opposition to display any motion reflecting the earth's. In III, 175, Tycho speaks of testing the planets at their stations to see whether they show any parallax to indicate that the earth is moving! Derek Price, "Contra-Copernicus: A Critical Re-estimation of the Mathematical Planetary Theory of Ptolemy, Copernicus, and Kepler," in Marshall Clagett, ed., *Critical Problems in the History of Science* (Madison: University of Wisconsin Press, 1959), pp. 212–13, pointed out that Copernicus was guilty of the same kind of lapse, and Robert S. Westman likewise for Mästlin, "The Comet and the Cosmos: Kepler, Mästlin, and the Copernican Hypothesis," in Jerzy Dobrzycki, ed., *The Reception of Copernicus' Heliocentric Theory* (Dordrecht, Netherlands: Nijhoff, 1973), p. 24.

Although it is difficult to resurrect the ontology encapsulated in medieval and Renaissance references to the celestial spheres, there can be no doubt that in the second half of the sixteenth century at least, intellectuals in general and Tycho Brahe in particular believed that something real existed in the heavens to carry the planets through their appointed rounds.[41] The objections raised by such well-known contemporaries as Mästlin, Magini, and Praetorius[42] to the notion that two such spheres might be conceived to intersect could be cited as strong suggestions of Tycho's reaction to the proposition, if there were no other evidence. In fact, however, Tycho himself confessed that he initially "could not bring myself to allow this ridiculous penetration of the orbs, so that for some time, this, my own discovery, was suspect to me."[43]

What was at stake, for all of them, was the fact that the Tychonic system required an intersection of the orbits of Mars and the sun. The fact that the orbits of Mercury and Venus would have to intersect the sun's in exactly the same way seems to have escaped everyone. But whatever the illogicalities involved, the one required intersection was enough to ensure that it would take more than the flash of insight that inspired the system to make Tycho a believer in it. How many hours he must have stared at the relationships between the geometry and the astronomy, looking for a means of escape.

For if Tycho was not prepared to go so far as Copernicans did to retain the aesthetic sense of unity and simplicity that in the sixteenth century constituted the sole grounds for taking seriously the motion of the earth, neither was he willing to abandon what he had labored so long to achieve. One can well imagine, therefore, that the only way Tycho could cope with those intersections at the time would simply have been to refuse to acknowledge them – to draw the system with the orbit of Mars arbitrarily enlarged so as to encompass the orbit of the sun completely[44] – and then hope for some kind of inspiration that would solve the technical problem of accounting for Mars's stations and retrogradations. Complicating the whole issue

[41] See William H. Donohue, "The Solid Planetary Spheres in Post-Copernican Natural Philosophy," *Westman* (1975); and Eric Aiton, "Celestial Spheres and Circles," *History of Science* 19 (1981): 75–113.

[42] IV, 474–6; VIII, 206; Westman, "Three Responses," 299–301.

[43] VII, 130. Noel Swerdlow, "The Derivation and First Draft of Copernicus' Planetary Theory: A Translation of the Commentariolus with Commentary," *Proceedings of the American Philosophical Society* 117 (1973): 471–8 argued that Copernicus arrived at the same intellectual dilemma and (lacking the "evidence" found by Tycho) chose to put the earth in motion. For more argument on the subject, see *Archives Internationale d'Histoire des Sciences* (1975): 82–92, and (1976): 108–58. Tycho certainly thought Copernicus subscribed to literal spheres (III, 173).

[44] Indeed, this is exactly what *Gassendi* (78) says Tycho did.

was the problem of Mars's (null-) parallax, for any determination that ruled out the Copernican system also ruled out its inverse, the Tychonic system.

Exactly how long Tycho's system languished in this status is a matter of dispute. It has generally been assumed that Tycho had resolved his difficulties by the fall of 1584, when a mathematician by the name of Nicolai Reymers Ursus turned up at Uraniborg in the company of Tycho's friend Erik Lange. Ursus came from very deprived circumstances and is supposed even to have worked for a time as a swineherd while educating himself. By the time he arrived at Hven, however, he had published a Latin grammar (1580) and a work on surveying (1583), both dedicated to Heinrich Rantzov.[45] Later he was to work for a while at Cassel with the *landgrave* of Hesse and eventually go to Prague where he favorably impressed Hayek and even won appointment as imperial mathematician to Rudolph II (in which post he was succeeded by Tycho and Kepler).

Ursus should have been a prime candidate for employment at Hven, and indeed Tycho said that he paid Ursus for service of some kind. Things went well enough in the first week that Ursus composed a poem to Tycho thanking him for wining and dining and even giving money to a penniless scholar.[46] But Ursus seems already to have begun to rub Tycho the wrong way. Whether because he seemed just a bit too clever, a bit too ambitious, or a bit too presumptuous, Tycho became sufficiently distrustful of him, first, to exclude him from discussions of his system with Lange and the other guests and then to search and expel him from Hven when he learned that he had been snooping in the library. But his precautions were in vain. Four years later Ursus published Tycho's system – almost.

What is probably more important than the details of Tycho's reaction to the situation is the intensity of his conviction that he had been plagiarized. For although Ursus's system bore a general resemblance to Tycho's, it differed in two respects. The first was in using a rotating earth to account for the diurnal motion. This was a matter of taste as much as anything else. Tycho did not believe that the earth was rotating but admitted privately that such a phenomenon was conceivable.[47] The second difference was Ursus's orbit for Mars, which totally enclosed – did not intersect – the orbit of the sun. This was impossible. As Tycho told several correspondents in the next few years, publishing such a scheme was a confession of incompetence, because it simply could not reproduce the gross synodic

---

[45] *Dreyer*, 184.   [46] VIII, 204.

[47] VII, 80: VIII, 45–6. Tycho's principal disciple, Longomontanus, later adopted what has come to be called the semi-Tychonic system involving a rotating earth.

phenomena of Mars, its stations and retrogradations.[48] Yet in virtually the same dip of the pen, Tycho accused Ursus of having stolen the scheme from him, but, of course, from a diagram that had been misdrawn.

In fact, however, there are several reasons for doubting that the drawing Ursus found was defective, at least in the sense implied by Tycho. What Tycho had on his hands was a defective system – one that could not even be used "internally" on Hven for any serious purpose. A few months later, when Tycho computed some positions for Mars in the spring of 1585, the only diagram he could draw for it was one like that shown in Figure 8.3, which he labeled "Inversion of the Copernican Hypothesis."[49]

More curiously, as we have seen, Tycho had not yet made any plans to publish his system. In late 1585, he printed at Hven the astrological calendar that appeared under Elias Olsen's name and appended to it a brief discussion of the comet of 1585, which would have offered as plausible a context for announcing Tycho's system as his discussion of the comet of 1577 did. Shortly thereafter, he printed a title-quarto for what was to be an eight-chapter book on the comet of 1577, which included no mention at all of either Tycho's system or a table of distances that was eventually included.[50] What other conclusion can there be but that as of, say, the fall of 1586 Tycho's system was still in a form that he – as opposed to Ursus – knew was unpublishable?

In the summer of 1586, after entering into correspondence with the *landgrave* as a result of the comet of 1585, Tycho received Wilhelm's observations of the comet of 1577.[51] The opportunity to use these independent observations, made with instruments that were probably superior to those that Tycho had used in 1577, to make a new determination of parallax may well have been the spark that ignited Tycho's final drive to get his book finished. Because the results corroborated the supralunary findings that Tycho had obtained from his own data, Tycho eagerly placed them at the beginning of his chapter of critiques and went on from there.

Although the rest of the chapter could obviously have been written, altered, and rewritten any number of times, the fact remains that for several years this unfinished chapter had been the sole barrier to

[48] VII, 149, 200–1, 388: VIII, 47.
[49] X, 284. C. Jones, "The Geoheliocentric Planetary System: Its Development and Influence in the Late Sixteenth and Seventeenth Centuries" (Ph.D. diss., Cambridge University, 1964), pp. 35–6; and K. P. Moesgaard "Copernican Influence on Tycho Brahe," in Dobrzycki, "The Reception of Copernicus," pp. 311–55, both commented on this apparently anomalous diagram.
[50] IV, 496–7.   [51] IV, 207–38.

publishing the book. So, even though Tycho had surely looked over the literature on the comet as it came into him in 1578 and 1579 and had probably noted significant points and even made marginal comments or extensive notes for the day when he would eventually compose his critiques, it seems very unlikely that he actually had them written. The fact that references to the *landgrave*'s observations appear in the critiques of several other commentators and in particular in that of Michael Mästlin, argues in the same direction.

Although Mästlin is best known to history as the teacher of Johannes Kepler, he was a prominent astronomer in his own right. The fact that he had produced one of the few competent treatments of the new star had induced Tycho to search eagerly for his findings on the comet,[52] and when Tycho finally obtained them in 1579, he was not disappointed. Indeed, he must have been gratified to see how closely Mästlin's printed results agreed with his own. But in addition to the various angular coordinates Tycho had provided for his orbit, Mästlin had computed and plotted the comet's daily distances.[53] These distances ran from 155 earth radii (about three times the distance of the moon) upward past the sun (situated at about 1,150 earth radii) and out to 1,495 earth radii. Perhaps they should have meant as much to Tycho – to say nothing of Mästlin – when he first saw them, as they did later.

Indeed, the implication that the comet had gone right through what Tycho and everyone else regarded as the Ptolemaic spheres of Mercury and Venus, and thus that those spheres could not be the solid objects everyone seems to have thought them to be, was worthy of publication in and of itself. Presumably Tycho had either not noticed this table on first reading or, more likely, had not attached any special significance to it, because he had not yet formulated the system that made it an issue for him. Seven years later, however, the pieces appear to have fallen into place quickly, probably because Tycho was also just reading a manuscript by Christoph Rothmann that asserted (without any reference to numbers) that "the very motion of the comets is the strongest argument that the

---

[52] *Observatio & demonstratio cometae aetherei, qui anno 1577 et 1578. Constitutio in sphaera veneris, apparuit,* ... (1578), pp. 52–3. Hayek did not like Mästlin's conclusions and attacked them in letters to Tycho. When Tycho finally received the letters (they had gone astray), he replied to Hayek that he did not defend Mästlin out of friendship, because he did not even know him, but if he had to name the foremost astronomer in Germany, Mästlin would be one of his two choices. VII, 205–6, 48, 50, 52.

[53] IV, 177–9. Tycho later praised Mästlin's discussion by telling Hayek that "no one has produced so erudite and ingenious a work, and no one has provided anything so agreeable and probable." VII, 20. In his analysis of Mästlin's tract, Tycho said that the chart of distances was not perfect but that Mästlin was the only one in Germany to have come up with it: IV, 209.

planetary spheres cannot be solid bodies."[54] Mästlin's figures seemed to provide exactly the demonstration needed for Rothmann's theory. If both were correct, if the comet had really moved in this fashion, there could be no solid spheres – and therefore no reason that the orbit of Mars could not intersect the orbit of the sun. When similar computations from his own data yielded similar distances (173 to 1,733 earth radii), Tycho registered, in letters written in mid-January 1587, his first doubts concerning the existence of solid spheres.[55] All that now stood between him and his system was a reevaluation of the question of parallax.

With the dissolution of the planetary spheres, the $4\frac{1}{2}'$ of parallax that earlier had been so threatening to Tycho's conception of things were now indispensable to it. Not surprisingly, Tycho's view of the results of earlier trials for parallax underwent a similarly rapid and profound transformation. Perhaps he made at this time the confused calculations that so mystified Kepler and various later commentators.[56] But Tycho may also have simply rationalized slightly different values for one or more of the uncertainties in the calculation (refraction, the position of a reference star, or the proper motion of the planet) to raise his parallax to $4\frac{1}{2}'$ from whatever he first thought he detected.

Already in his letters of mid-January 1587, conveying the first statements of his disbelief in the existence of solid spheres, Tycho expressed his conviction that Mars, in opposition, showed more parallax than the sun did. Any residual qualms he may have had about this startling reversal of opinion were resolved a month later, with "live" but obviously none-too-tidy parallax checks on both Venus and Mars.[57] If this remarkable achievement is not exemplary of Tycho's best work, the fact that he thought he found something, sometime, must be explained away in any account of his discovery and is at least as comprehensible (if not defensible) as the product of excitement and haste in 1587 as it can be for any other circumstances.

To literally his dying day, Tycho regarded his system of the world as the most significant achievement of his career. Clearly, then, there

[54] Edward Rosen, "The Dissolution of the Celestial Spheres," *Journal of the History of Ideas* 45 (1985): 28–9. In his return letter to Rothmann, Tycho said he was wonderfully pleased with the treatise: VI, 85.

[55] VI, 88, 70.

[56] What is involved is an apparently circular computation (X, 283–6) that first assumes parallax and then derives it. Kepler's explanation (*Dreyer*, 179) of the parallax as a mix-up between Tycho and one of his calculators on observations of the opposition of 1582–3 is deficient in two respects. First, the calculations could have been done anytime after the event, and second, publication of the *Opera omnia* showed that the anomalous calculation was done by Tycho himself.

[57] XI, 181–7, 195–8.

could scarcely be any question about publishing it, particularly as he happened to have at that very moment a book going through his own press, dealing with a subject closely related to his discovery. Quickly, therefore, Tycho wrote up an exposition of his system, had his engraver cut the illustration shown in Figure 8.9, and labeled it Chapter VIII. Then, for no other apparent reason than to avoid having a nine-chapter book, he added what had previously been an appendix to Chapter VII, dealing with the sizes and distances of the head and tail of the comet, to his new table of distances of the comet and called that material Chapter IX. All that then remained was to finish writing the extremely lengthy final chapter, now labeled X, and print up a new title-quarto describing ten chapters instead of eight.[58] Even with some problems in obtaining the quantities of paper he needed, the printing was completed by the beginning of 1588,[59] under the title *De mundi aetherei recentioribus phaenomenis* . . ., or Concerning the more recent phenomena of the ethereal world.

In marked contrast with the misfortune that was to attend Tycho's decision to make a similar insertion into the *Progymnasmata* a few years later, the consequences of publishing the Tychonic system in *De mundi* were uniformly happy. In fact, it is doubtful that the system would now be called Tychonic if Tycho had been just a bit faster in getting his book printed,[60] or a bit slower in developing his

---

[58] In September 1588, after *De mundi* was out and Tycho was just beginning to hear responses to it, he wrote a lengthy letter to Caspar Peucer, in which he provided the most complete account of the genesis of his system that is available (VII, 129–30). Allowing for a bit of poetic license on the timing of his parallax computations, it outlines the one just given *in extenso*. First, the parallax test on Mars to decide between the accounts of Ptolemy and Copernicus (lines 8–26): According to Tycho, exhaustive trials showed a greater parallax for Mars and thus tended to favor the Copernican model. Although he referred these tests to the opposition of 1582–3, he was candid enough to mention that he confirmed the fit of the phenomena with Copernicus's numbers by a special trial on Venus, around 24 February 1587. So if his chronology were not perfectly consistent with the available evidence, it at least showed that Tycho was uncertain about the parallax issue right up to 1587 (lines 26–34). Given the tendency of the appearances to deny Ptolemy, Tycho continued, and the manifest absurdity of imagining the earth to move, there was nothing else to do but to find an alternative hypothesis. At first it seemed impossible, but finally it came to him (lines 34–40). And yet his inspiration had one flaw in it, that the orb of Mars had to intersect the orb of the sun in two places. Because he still subscribed to the notion that there were real orbs in the sky, his system remained suspect to him for some time (lines 6–13). Only when he had examined the motions and parallaxes of comets did he realize that they ruled out the existence of those orbs (lines 13–21).

[59] The earliest inscribed copy is dated 20 March 88 (*Norlind*, 122), but a correspondent of Kepler related having visited Hven on 6 January 1588 and seeing the book completely printed (G. W., XIII, 101–2).

[60] Thus the problem of obtaining paper for the book, which figures prominently in Tycho's correspondence in 1586–7 and ultimately induced him to construct his own paper mill on Hven, turned out to be a blessing in disguise.

thoughts, or a bit more cautious about altering his manuscript. Even as it was, the priority question was a ticklish one.[61]

Only a month or so after Tycho began to circulate *De mundi*, one of the several people to whom Rantzov sent the seven copies that Tycho gave to him reported that someone had shown him the same system a couple of years earlier. The correspondent, a rather prominent German astrologer named Rollenhagen, could not remember the name of his informant, for when Tycho alluded to the problem in September 1588, he was able to refer to the culprit only as "some run-away servant of mine."[62] But by the time Rothmann wrote saying that the *landgrave* had already had his instrument maker construct a model of that system a few years earlier,[63] Tycho had the solution to the problem in his hands: It was a book bearing the pompous title *Fundamentum astronomicum*, published in 1588 by the person Tycho had ejected from Hven four years earlier, Nicolai Reymers Ursus.

Because what Ursus published was the form of the system that Tycho had been using at Hven in 1584, Tycho could be morally certain that Ursus had indeed plagiarized it. But the only way a charge of intellectual theft could make sense was for Tycho to admit that he himself had been using through 1584 the nonintersecting (i.e., Ursus's) form of the system. And because in that form, the system could not account for the stations and retrogradations of Mars, such a claim would have to be accompanied by an elaborate explanation, indeed, if it were to be any more than a confession of his own incompetence. Whatever Tycho said about his conversion from one form to the other might even have revealed the role of Mästlin's table of distances in the conversion.

It cannot be surprising, then, that Tycho should have been unwill-

---

[61] In more ways than one. By Tycho's own testimony (IV, 159; VII, 130), his system depended on the destruction of the solid spheres, which, in turn, rested on the distances derived by Tycho from his orbit for the comet of 1577. Both internal evidence and Tycho's own testimony show that those distances were obtained after Tycho saw Mästlin's scheme and after he had written his own analysis of the comet's path. We therefore have the spectacle of Tycho's borrowing Mästlin's spatial conception of the orbit, without giving any credit to him, but then attacking Ursus for borrowing his (admittedly more significant) system of the world. Both Delambre and Westman suggested that Tycho borrowed from Mästlin, but for different reasons. Delambre, *Histoire de l'astronomie moderne*, p. 223, thought the similarity of Tycho's and Mästlin's inclinations (29°13′ and 28°58′) and nodes (8°21°) was "un hasard assez remarquable," because he did not note that the results were standard – that Schultz got an inclination of 29°36′ (IV, 296) and that he, Roeslin, and Hayek all got 8°21° for the node (IV, 251, 266). Westman, "The Comet and the Cosmos," p. 25, saw the heliocentric orbit as too similar for coincidence, although it, likewise, was surely written into the German tract and presented to King Frederick long before Tycho saw Mästlin's treatise.

[62] VII, 135, 387–9.   [63] VI, 157.

ing to risk this approach. He did, however, have two circumstances in his favor. One was the technical untenability of the nonintersecting system – the fact that it failed to represent the gross synodic phenomena of Mars and still involved intersections of spheres (Mercury's and Venus's with the sun).[64] The second was the fact that Tycho himself had tended – even before he knew that he had to defend his priority – to refer the genesis of his system not to the stage at which he finally decided that the celestial spheres were not a problem but to the time when he first saw how all of the planetary theories could be put together in an inverted Copernican scheme. Thus in his publication of the system, he mentioned having discovered it "four years ago."[65] So when Tycho poured out his tale of woe in letters to Brucaeus, Rantzov, Rothmann, and Hayek, he was at least able to point, with some plausibility, to the parallax investigations of 1583 and thus antedate Ursus's claim of having conceived the system in Pomerania during the winter of 1585–6.[66] All that remained was to patch the final hole by claiming that Ursus had found a faulty diagram.

Although Tycho's pride in his system may seem inordinate from a modern perspective, it is important to realize that in its day it represented the best of both worlds. Until the advent of the telescope, at the very earliest, the available evidence did not render belief in the mobility of the earth even plausible, let alone convincing. It is not necessary to deny either that the specter of a moving earth may have been the crucial incentive to a new physics of motion or that the traumatic seventeenth-century battle with the Roman Catholic church may have had the life-and-death implications for early modern science that its practitioners saw in the struggle. It is reasonable to insist, however, that neither of these issues had much to do with the progress of astronomy, at least during the generation or two after Tycho. The geometry of the Copernican system, on the other hand, represented a significant astronomical advance. By extricating it from the controversial, and perhaps even unrespectable, company of Copernicus's moving earth, Tycho provided an important technical service to his discipline and almost surely also hastened acceptance of the motion of the earth.

Between the excitement induced by his discovery and the haste involved in keeping the manuscript flowing to his printer, Tycho produced a description of his discovery that was considerably less than perfect in both organization and composition. His exposition of

[64] VII, 149, 200, 338, and VIII, 47. [65] IV, 155.
[66] Nicolai Raymari Ursi Dithmarsi, *Fundamentum Astronomicum* (1597), p. 37.

the system itself was so sketchy that at least one reader (Rollenhagen) envisioned actual collisions between Mars and the sun.[67] Tycho's discussion of the intersections was sufficiently detailed to remove any excuse for mistaking the nature of the problem in this way, but the method he adopted for dealing with the intersections was completely inadequate. All Tycho said was that the intersections were not a problem, because he could prove (and would in a later work) that there were no solid spheres.[68] How Tycho can have expected such an unsupported assertion to have resolved, for any appreciable fraction of his readers, a difficulty that he himself had found sufficiently fundamental to dictate an a priori rejection of the system, is hard to imagine.

Tycho's readers were indoctrinated in a physical and metaphysical worldview in which comets were terrestrial bodies. Only for those who had been convinced by the first seven chapters of *De mundi* that the comet of 1577 was actually celestial would Tycho's claim that it had penetrated the sphere of Venus not be a definitional impossibility. And even for those few, the rest of the "proof" would have to come from one or two delicate observations of one comet, and an intricate chain of trigonometric calculations.

Apparently, Tycho decided that his argument was premature at best, and that however convinced he might be (by the beauty of his system, if nothing else) that the celestial machinery that everyone else seems to have taken for granted could not exist, he would at least require an induction from several comets to establish reasonable proof. The result was that he left his crucial table of distances buried in the other new chapter, some twenty pages behind the presentation of his system,[69] and provided so little connecting commentary in either place that even historians, let alone his contemporaries, were unable to perceive its role in the development of his system.

Whatever purpose Tycho envisioned for his final chapter, he must have accomplished it. Longer than the preceding nine chapters combined, it consisted of point-by-point analyses of the results of eight authors, and concise but reasoned dismissals of the efforts of eleven others.[70] The most positive aspect of the enormous amount of work that went into these critiques appeared in the first discussion, devoted to the observations of the *landgrave* of Hesse.

Although the *landgrave* had hired an astronomer after Tycho's departure in 1575, he no longer enjoyed his services when the comet appeared. When Tycho inquired in 1586 about his conclusions concerning the comet, therefore, the *landgrave* could send only a few observations he had made himself and say that if the comet of 1577

---

[67] VII, 125.   [68] IV, 159.   [69] IV, 155-6, 177-9.   [70] IV, 180-377.

was anything like that of 1585, it would show too little parallax to be anywhere near the moon. Tycho converted those observations into a minor treatise.[71] He printed the observations, reduced them to usable data, and utilized them to conduct first a thorough investigation for parallax and then a concerted attack on the most frequently used method of finding parallax, the one developed by Regiomontanus a century earlier.[72]

The motivation for this attack was Tycho's conviction that the method was worthless from beginning to end but was so well known and ill used as to be responsible for most of the parallax that had been found for the comet.[73] The method consisted of noting two positions of the comet, and the time interval between the observations.[74] It included no provision for taking into account the comet's intrinsic motion during the interval. For the comet of 1577 this motion had been eight to ten minutes per hour during the first ten days, when most people conducted their checks for parallax. If uncompensated, it alone could produce "parallaxes" of several degrees from a long trial near the horizon and place the comet at a corresponding fraction of the moon's distance from the earth. Scarcely less palatable to Tycho was the dependence on clocks: An error of a few seconds in the timing of the interval could likewise produce a parallax of some degrees.[75]

In short, this method was a textbook example of Tycho's pet peeve: esoteric mathematical techniques with no practical felicity.[76] Of course, Tycho knew well the status that Regiomontanus enjoyed. Rather than alienate readers by seeming to denigrate him, or leave any room at all for the feeling that he disliked the method merely because it produced results that conflicted with his view of the comet's distance, Tycho decided to show that the method could work, provided – as he hastened to point out – that one used very good observations, such as the *landgrave*'s.[77] Tycho then proceeded to give seven examples, spaced from the beginning to the end of the *landgrave*'s observations, in which the method showed no parallax.[78]

From the *landgrave*, Tycho moved to Mästlin and then to two other writers, who had likewise found the comet to be above the

---

[71] IV, 182–207.

[72] "De cometae magnitudine longitudineque ac de loco ejus vero problemata XVI." Although written after the comet of 1472, it was not printed until 1531 (and thereafter).

[73] IV, 83, 129, 206, 440; VII, 107–8.   [74] IV, 194.

[75] IV, 441. Even for the new star, for which proper motion was not a problem, Tycho frowned on the method because of the difficulty of getting accurate timings (III, 202).

[76] III, 184; IV, 441.   [77] IV, 195.

[78] IV, 195–206. In fact, all the parallaxes were negative rather than null. But Tycho was not one to concern himself with philosophical technicalities and did not even bother to mention the difference of refractions uniformly responsible for the anomaly.

moon.[79] Their success, however, did not earn them immunity from Tycho's criticism of their instruments, data (e.g., Copernicus's star positions), computational errors, and the like. Neither did friendship, as Tycho began his attack on the mass of writers who had found sublunarity, by reviewing the efforts of his confidant Hayek and his old mentor Schultz. (Quoting what he took to be an ancient aphorism, Tycho said, "Friend of Plato, friend of Socrates, but more a friend of truth.")[80]

If Tycho's criticism was uncompromising, it at least was fairly gentle and thoroughly constructive. Although he felt obliged to point out that Hayek had committed three sins (observed with a cross staff, used existing star coordinates, and done his calculations on a small globe),[81] he was careful to praise his earlier results on the new star and to go through his work carefully enough to show that one of the grosser errors arose from a mix-up over his reference star.[82] After detailed critiques of two more parallax findings,[83] Tycho turned to more general criticisms of eleven writers whose efforts did not deserve anything more. He then concluded with a summary providing the first public description of any of his instruments.[84]

Although *De mundi* was a book characterized more by the perspiration than the inspiration that went into it, it was nevertheless a project that Tycho perceived as necessary to get the job done. Tycho did not expect *De mundi* to establish the celestial nature of comets, much less to overthrow the Aristotelian worldview. Fortunately, it did achieve the former task, so that Tycho's failure to complete a second work treating all the subsequent comets he observed at Hven cost nothing.[85] On the other hand, not even the fact that Tycho eventually published a similarly ponderous volume on the new star was enough to overthrow the Aristotelian cosmology. Only Galileo would achieve that, after Tycho's death. And he would do it with a completeness that would surely have left Tycho with doubts about the wisdom of having set the process in motion in the first place.

---

[79] IV, 207–58. The other writers were Cornelius Gemma and Helisaeus Roeslin. See descriptions of all writings on the comet in Hellman, *The Comet of 1577*.

[80] As Tycho said (IV, 336): "Amicus Plato, Amicus Socrates, sed magis Amica Veritas." Henry Guerlac, "Amicus Plato and Other Friends," *Journal of the History of Ideas*, 39 (1978): 627–33, showed that this passage was in Luther and that variants of it were used throughout the seventeenth century.

[81] IV, 263.   [82] IV, 265.

[83] IV, 337–55 (Andreas Nolthius and Nicolaus Wincklerus).

[84] IV, 368–77.

[85] See III, 26; V, 23; VII, 131, for examples. *Norlind* (142) described the order of printing projected for the volume, which was to concern comets observed in 1580, 1582, 1585, 1590, and 1593.

Chapter 9
High Tide: 1586–1591

A s of the end of 1585, Tycho's life and work on Hven had probably proceeded about as he had expected it would when he settled on the island. To be sure, the building of Uraniborg and its ancillary facilities, particularly his instruments, had taken longer than he had thought it would. But the finished product – as epitomized in his underground observatory, Stjerneborg – was better than the minute-of-arc accuracy he had originally sought,[1] and, in addition, Tycho had had the opportunity and hence the obligation to deal with more extraordinary astronomical phenomena than he could possibly have foreseen.

Each comet had contributed grist for Tycho's mill. The one in 1580 had gone around the sun, just as that of 1577 had. The comet of 1582 had shown a tail pointing away from Venus,[2] just as the great tail of the comet of 1577 had. That of 1585 had displayed no tail at all, which, as far as Tycho was concerned, merely meant that its tail was directed behind it away from the sun and was not visible because the comet was in opposition to the sun.[3] The pieces were beginning to fall in place: His observations of the comet of 1585 were in press; his manuscript on the comet of 1577 was going through right after it; and if Tycho had not yet decided to publish a work summarizing the appearances of all the comets he had observed, he soon would. But now that the years of preparation and investment were past, Tycho could look forward to an even more general period of harvest. With good instruments and a well-trained set of faithful assistants, Tycho was in a position to investigate almost any astronomical question he chose.

During the next half-dozen years Tycho did exactly that, chasing down a set of problems more or less associated with his work on the new star. While he was doing it, however, events conspired to provide new distractions that turned out to be almost as time-

---

[1] Although there was a sense in which Tycho's graduations to quarters and sixths of a degree were illusory, given the limitations of the human eye, we saw at the end of Chapter 6 that Tycho actually achieved accuracy to well under one minute of arc. He certainly thought he could measure to 10″ (II, 218), even if he seems to have seen its utility largely in terms of ensuring that such measurements gave him the correct minute.

[2] VII, 250, IV, 175. For a list of abbreviations of frequently used sources see Appendix 1.

[3] IV, 399, V, 170.

consuming as the construction activities had been that had complicated his life during his first nine years on Hven.

One of these events was undoubtedly the publication of Volume
IV of Braun's *Atlas*, with its pictures and descriptions of Uraniborg.
But it probably never would have been possible to make Uraniborg
the ivory tower that Tycho might have hoped it could be. If Queen
Sophie decided to come out to see what the royal disbursements had
wrought and was so impressed by her three-day visit that she soon
returned with her father (the duke of Mecklenburg) and mother
(King Frederick's sister),[4] what choice did Tycho have but to be
flattered? And as the word got into the travelers' "grapevine," Tycho
found himself facing an ever-increasing tide of summer tourists who
made the trip out to Hven and had to be entertained, or at least
tolerated, for a day or two.

Also contributing to the general visibility of his little empire and
leading to further demands on his time were the appearances of
Tycho's first publications from Hven – "Olsen's" *Calendar* with
Tycho's observations of the comet of 1585 and then (a few years
later) *De mundi* – which drew Tycho into correspondence and controversy that absorbed great amounts of time and energy.

The letter that brought Tycho once and for all out of the isolation
of his island environment was delivered to him in February 1586 by
Gert Rantzov,[5] the fiefholder of the king's recently completed castle,
Kronborg. As a neighbor of Tycho's across the Øresund and a son of
Tycho's fellow intellectual Heinrich Rantzov, Gert had already been
out to Hven to see Tycho's operation. Tycho, in fact, used that
occasion to mention that he had a book (Olsen's *Calendar*) all ready
for the press but lacked the paper to print it. Gert told Tycho (what
he had probably already known) that his father, an active publisher
himself, had his own paper mill and that Tycho should write to him
and see whether he could spare any paper.[6]

With this entrée, Tycho began corresponding with the elder
Rantzov and seems to have received paper from him for two
thousand pamphlets. However well acquainted Tycho and Rantzov
were before this (the first preserved) letter, Tycho apparently did not
feel that he could simply write to Rantzov and ask whether he could
purchase some paper. By the end of 1585, however, the *landgrave* of
Hesse seems to have thought they were reasonably close. For when
he wanted to suggest to Tycho that they exchange observations on
the comet of 1585, without making a request that Tycho might feel
he could not refuse, he chose to do it through Heinrich.

On the day after the *landgrave* himself first observed the comet,

⁴ IX, 45–7, VI, 64.   ⁵ VI, 33.   ⁶ VII, 89–90.

he wrote to Rantzov thanking him for a letter and book he had sent and proclaiming proudly that on the advice of Paul Wittich, he had had his instruments so improved that whereas he could previously scarcely observe to 2′ of arc, he could now observe to a half or even a quarter of a minute. He also had a sextant for observing interstellar distances and planned to use it to compile a catalogue of stars. After imparting his first couple of observations of the comet and a couple of thoughts about it, he concluded by asking whether Rantzov had observed it, whether Tycho Brahe had observations of it, and by what means he had observed it.[7]

Tycho received this letter at Kronborg Castle, probably after attending the funeral of his Uncle Steen (Bille) in Helsingborg.[8] To say that it seems almost to have made Tycho sick would be an understatement. For what the letter conveyed was the fact that the major – perhaps the only significant – accomplishments of his first ten years on Hven – the great engines he had built at such enormous expenditures of time, money, and intellectual energy to power his drive for the reformation of astronomy – were now in the hands of the only other person in Europe who had the inclination, the ability, and the means to do the same thing Tycho wanted to do.

Moreover, the candid *landgrave*, and now Rantzov, too – the two noble peers who shared Tycho's interests most intensely – had no reason to believe anything but that the ideas for improving the instruments were Wittich's. In fact, if Tycho read the in-other-respects nonaggressive letter correctly, the *landgrave*'s query concerning the means by which Tycho had observed the comet was a none-too-subtle hint that Wilhelm thought his instruments were now better than Tycho's. The shameless Wittich had obviously relayed all the ideas and techniques of Tycho's instruments without so much as mentioning Tycho's name.

Tycho had had a premonition that something like this might happen. Wittich seems to have been a convivial fellow who enjoyed, even more than most people do, being able to pass on the latest social and professional news as he traveled from one place to the next. While he was at Uraniborg, for example, he told Tycho that Hayek had lost his position at (and income from) the imperial court through "Machiavellian" intrigue.[9] Not knowing that Wittich had garbled (or at least exaggerated) the situation, a very concerned Tycho had included with his next letter to Hayek an invitation to move to

[7] VI, 31–2.
[8] The death and burial of Steen on 5 January and 17 February were recorded in Tycho's log: IX, 42.
[9] VII, 60, 73.

Denmark, where he could be sure that his own reputation and Tycho's influence would enable him to make a comfortable living.[10] Tycho then (as he wrote to Schultz a year later)

> gave to Wittich as he was departing from us, a letter to be sent to (Hayek), which he, however, attended to less than faithfully. And I have found that he acted likewise in many other things also, although he was received kindly here. For when he had secretly pilfered the things of mine that he was able to obtain and had seen the plans and methods for observing stars with our instruments (although he was not at all adept in using these and could not take even one observation either of a comet or of a star and still less find the parallax of a comet, as he wrongly persuaded Dudith), he finally, under the pretext that his uncle had died and claiming that an inheritance was owed to him, returned home from here.[11]

As it turned out, Wittich had delivered Tycho's letter. The reason that Tycho assumed he had not was that Hayek's return letter had miscarried. But actually, the misadventures of Tycho's correspondence with Hayek had little to do with the intemperance displayed in his letter. Some of Tycho's irritation was surely due to his disappointment over Wittich's failure to return to Hven: It seems clear from the expensive book he presented to Wittich at his departure and his several subsequent allusions to Wittich's mathematical prowess, that Tycho had relished having his help on Hven. But most of Tycho's ire seems to have stemmed from some kind of reference in the (unpreserved) letter he had just received from Schultz, to Wittich's having shown to members of the Breslau intellectual circle some observations of the comet of 1580. Although Tycho's reference to this transgression is a bit vague, a later letter to him from Hayek also refers to such revelations. Wittich's "generosity" was doubly upsetting to Tycho because Wittich had not seen fit to follow Tycho's routine of correcting the times of the observations for the errors of Tycho's clocks and was therefore circulating data that were substandard, as far as Tycho was concerned.[12]

Wittich, for his part, may well have thought that because he had helped make some of the observations, he was entitled to treat them as his own. It is more likely, however, that he just did not have the proprietary instincts that Tycho had, either for his own or other people's findings. Perhaps Tycho would have thought to speak to Wittich about the general issue of divulging information from Hven if he had known when Wittich left that he was not going to return. But he had not.

A year later, Tycho immediately concluded (which, however accurate it turned out to be, was certainly unwarranted at the time) that his other results were in jeopardy, too – that Wittich might "secretly have taken things prepared by me with much labor, and arrogate them to himself, which, as he did with the comet, I do not doubt that he will also do in the restitution of the course of the sun and emendation of the eighth sphere."[13]

For all of Tycho's concern about what Wittich might reveal in his perambulations, the one thing that seems never to have occurred to him was that a man as theoretically oriented as Wittich was might find any occasion to talk about the technical details of Uraniborg's instruments. In the letter from the *landgrave*, therefore, Tycho was brought face to face with the inconceivable. One can only guess how long it took Tycho to recover his composure and write the kind of restrained, tactful letter that had to be written to someone of the *landgrave*'s rank.

By the first of March, Tycho had gathered some observations on the comet, had one of his students copy out his solar ephemeris for the year, and had composed a diplomatic assertion of his priority in every aspect of Wittich's improvement of the *landgrave*'s instruments. Tycho welcomed news of the *landgrave*'s new instruments and observations, he said, and would be most interested in a detailed description of them, particularly a drawing of the sextant.[14] For he suspected (as well he might, as he had not only coined the term but had also had a sextant with him when he visited the *landgrave* in 1575) that it was very much like his own, especially in respect to the sights and divisions. Wittich had been on Hven with him five years previously and had seen every feature not only of the instruments he then owned but also of those that were then under construction. Since then, however, Tycho had built even better instruments and had some engravings to send to Cassel to help the *landgrave* imagine how they worked. After mentioning a few aspects of his own work, suggesting that they exchange observations, and apologizing for having "transcended the bounds of prolixity," Tycho then mentioned casually that he was sending his materials through Flemløse, who had been in his household for some years, was well qualified in astronomy as well as alchemy, and could answer any questions the *landgrave* might have about Tycho's operation on Hven.[15]

What Tycho really expected Flemløse to do, of course, was to testify in person to the *landgrave* that he had been on Hven when

---

[13] VII, 63. [14] VI, 35. [15] VI, 36–40.

Wittich was there, and that everything Wittich knew about instruments he had seen at Uraniborg. And when Flemløse returned two or three months later (after having continued on from Cassel to the spring book fair at Frankfurt to buy books for Tycho), he doubtless told Tycho that he had convinced everyone at Cassel of Tycho's priority.[16]

But Tycho was still very upset, and the news that the *landgrave* had given Wittich a gold chain in appreciation of his suggestions for modifying his instruments did not smooth his ruffled feathers a bit. On 1 July he wrote to Hayek:

> This overhaul done by the *landgrave* ... was patterned after some instruments of mine. For Wittich revealed to him the form, the division, and the sights developed here by me in no other way than if he himself were their inventor, making no mention of me, and for this he was given a gold chain by the *landgrave* himself. Nevertheless, what usually happens to those who act deceitfully happened to him: He did not show the right way of using four completely parallel pinholes, either because he forgot or because he presumed he could accomplish the business with fewer. As a result he cost the *landgrave's* observers considerable time and energy until they themselves noticed what was needed from experience itself, as you will understand more fully from the mathematician's letter. Indeed, Wittich (whom I held in high esteem both because of your recommendation and because of his own talents and whom I trusted with more things than I perhaps should have done) dealt with me less than faithfully because he makes my ideas public without my consent under his own name. It is not that I begrudge my inventions to others, because I myself have decided to treat the mechanical part of astronomy in a special work and share with posterity the construction of the many and various forms of instruments. But the thanks that I should have had from others (not undeservedly, wouldn't you think) for all the arduous labors and incredible expenses this person, as much as he could, tries to snatch away and attribute to himself, by adorning himself in the plumes of others like Aesop's crow.[17]

Tycho would find opportunities to allude (although usually a bit vaguely) to what he clearly regarded as Wittich's absolute and utter betrayal on a dozen more occasions extending right up to the end of his life.[18] The closest Tycho could come to consoling himself on the

---

[16] In the meantime, the *landgrave* had responded to Tycho's boasts concerning the size of his instruments with the subtle dig that his own were large enough to be accurate without being too large or heavy to use conveniently: VI, 51.

[17] VII, 108–9. See also IV, 453.

[18] III, 6, 29; V, 25, 79, 155; VI, 328; VII, 261, 323, 360; VIII, 201; *Norlind*, 377.

matter was that even at Cassel it was now understood that Wittich had bungled the duplication of Tycho's sights[19] and that in the meantime, Tycho had developed instruments that made obsolete all those Wittich had seen.

Flemløse also carried written news, in a letter from the *landgrave,* saying he was pleased that Rantzov had sent the Cassel observations on to Tycho and was delighted to have Tycho's in return. However, he found it difficult to comply adequately with Tycho's request for past observations, because it was only very recently, at the instigation of Wittich and through the diligence and industry of his clockmaker, Joost Bürgi (who was a veritable Archimedes),[20] that he had gotten his instruments functioning properly. But now he did, and he also had a mathematician in his employ, who, by very diligent observations, had confirmed the *landgrave*'s earlier discovery that the star places listed in the existing catalogues contained numerous errors. Whether these errors were merely typographical or whether the individual stars had actually moved, he would leave to his mathematician – who was not the ordinary "house" mathematician – to discuss with Tycho.[21] The mathematician, Christopher Rothmann,[22] duly did, in an exchange of a dozen long letters dealing with practically every technical and philosophical aspect of astronomy.

Tycho's correspondence with Rothmann provides the most intensive astronomical discussion found in any of his writing. It is the source of most of the descriptions already presented of Tycho's thought on such diverse matters as the nature of the celestial regions and the cause of astronomical refraction, the shortcomings of clocks as astronomical instruments, and the conceptual links between astronomy and alchemy. In addition, it offers an unusual view of Tycho's interaction with two people who, between them, were roughly his social and professional equals. Tycho's other correspondents were generally not in an intellectual position to challenge his findings, assertions, or judgments. But the *landgrave* and Rothmann were, on the few issues on which they did not see eye to eye with Tycho. Perhaps the most delicate of these was the question of a systematic difference of a few minutes in the longitudes of their stars.

The issue first arose in connection with the *landgrave*'s positions for the comet of 1585. The reason they differed from his, Tycho explained to the *landgrave,* was that the star coordinates from which they were measured were not completely accurate. The resulting

---

[19] VI, 55.
[20] VI, 49. Concerning Bürgi, on whom the verdict of history has been almost as enthusiastic as the *landgrave*'s, see *DSB* II, 602–3.
[21] VI, 49–50.   [22] *DSB* XI, 561–2.

errors, however, were only a few minutes of arc and had not prevented the *landgrave*, Tycho hastened to say, from arriving at the
correct conclusion, namely, that the comet showed no parallax.[23]
There were also a couple of minutes' difference in their meridian
altitudes, and because Tycho's were obtained with anywhere from
five to seven instruments of great size and exquisite construction,
it might be wise for Rothmann to check the orientation of their
quadrant.[24]

Although the *landgrave* replied that he was not at all concerned
about a discrepancy of five minutes (Tycho had mentioned six)
because nobody had previously achieved an accuracy better than ten
minutes, anyway,[25] Tycho could not let it go. He opened his next
missive by referring to "those five minutes concerning which there is
a disagreement between us in the longitudes of the fixed stars"[26] and
delivered a long discussion of the virtues of his instruments. He
realized he was making perhaps undue fuss over five minutes, but he
was convinced that the problem was in the *landgrave*'s observations,
whether in the irregularity of his clock, the suspension of his
quadrant, or some other little thing. It would be best if everything at
Cassel were rechecked.[27] Five pages later, after having explained the
whole foundation of his star catalogue, from the solar tables, to his
use of clocks, to his observations of Venus, Tycho reiterated that his
stellar longitudes could not possibly be responsible for any of the five
minutes.[28]

Unwilling to leave it even at that, Tycho told Rothmann in the
letter he wrote to him the next day that he was happy to learn that
the *landgrave* was going to compile a star catalogue and agreed that
the systematic five minutes between the *landgrave*'s longitudes and
his own was not important, as long as it was understood that the
problem was not in Tycho's longitudes.[29]

During his exchanges with Rothmann and the *landgrave*, Tycho
completed the printing of *De mundi*. As an almost inevitable result of
this, he was pressed into what was to become the highest-priority
element in his scientific work: the promotion and defense of his
system of the world. He could not start as vigorously as he would
have liked to, with the outright publication of his book, because
it was actually Volume II of a series, *On the More Recent Ethereal
Phenomena of the World*, and Tycho thought it inappropriate to release
it formally until Volume I, on the new star, was ready. But he could,
and did, start sending out copies – as people who could be drafted as

---

[23] VI, 65, 68, 86.   [24] VI, 66.   [25] VI, 106–7, 108.   [26] VI, 122.
[27] VI, 124. When Tycho was writing up his discussion of the comet of 1577 at about this time,
he described an uncertainty of 4′ in longitude and 2′ in latitude as "insensible": IV, 168.
[28] VI, 125–9, II, 282   [29] VI, 143.

messengers turned up on Hven – to everyone he thought might be interested in it. In virtually all cases, he sent covering letters[30] specifically mentioning the existence of his new system and encouraging the recipient to share the contents of the work with friends and return their comments to him. They soon did.

In a book as long as *De mundi*, it is only to be expected that almost everybody would be able to find something to object to, and with few exceptions, everybody did: Tycho's system. Concerning the hundreds of pages in which Tycho developed his conclusions on the comet and analyzed the findings of others, few respondents had anything to say (although Tycho's old friends Schultz and Hayek wrote to acknowledge his exposure of errors that had led them to believe they had found parallax). Only after a couple of years was Tycho challenged by the inevitable, obdurate Aristotelian who insisted on disputing the status of comets verbally, in Aristotelian terms.

Concerning the system, however, almost everyone had something to say, even if it sometimes took pointed inquiry by Tycho to elicit it. For the most part, the comments they made reflected one particular concern: the fact that Tycho's diagram showed the orbit of Mars intersecting the orbit of the sun.

A German astrologer named Rollenhagen, who received a copy of *De mundi* through Rantzov and then returned word of Nicolai Ursus's co-option of the system, actually thought Mars and the sun might collide.[31] Hayek wrote that he found the book most thorough and accurate generally but that he would have to ponder the new disposition of the heavenly orbs[32] (thus betraying by the very choice of his words what was troubling him).

Hayek had also shown the book to Rudolph's prochancellor, Jacob Kurtz, who likewise enjoyed the book but did not feel he could judge the hypothesis. Kurtz's title made it clear that he was a man of considerable prominence. It was he, in fact, who in 1590 wrote the copyright letter for the Hapsburg realms that Tycho printed at the beginning of *Progymnasmata*.[33] When Kurtz sent with that letter one in which he finally endorsed Tycho's system, and at the same time described Ursus as Tycho's plagiarizer, Tycho seized the opportunity to print that letter in his *Mechanica*.[34]

A few months later (in 1590) Tycho received a letter from the noted Italian astronomer G. A. Magini. Because it included a rather

[30] VI, 149 (Rothmann): VII, 120 (Schultz), 123 (Hayek), 131 (Peucer), 142 (Brucaeus).
[31] VII, 125. Duncan Liddell seems initially to have had the same misapprehension: *Norlind*, 367.
[32] VII, 145.   [33] II, 8–10.   [34] V, 120–1.

ambiguous line to the effect that Magini could not avoid giving his
strongest approval to Tycho's system,[35] Tycho decided that he had
to print it, too, in the *Mechanica*, immediately following Kurtz's
letter. But in fact Magini went on to say that he would have wished
that Tycho had not made the orbs of Mars and the sun intersect,
although he supposed that if, as Gellius related to him, Tycho's
instruments had really shown that Mars was closer at opposition
than the sun was, he had to acquiesce.

When Tycho sent to Hayek at least this part of Magini's letter
(even the most poorly prepared among his students could always
provide him with copies of documents), he received a statement
agreeing with Magini's reservations but urging him not to be
discouraged by lack of acceptance, because new ideas always provoked opposition. Tycho was certainly well aware of the latter
problem. In 1588, in fact, when Brucaeus sent an unqualified
approval of Tycho's "hypothesis,"[36] Tycho was unable to refrain
from expressing his astonishment that such an "ardent and steadfast
guardian and defender of the authority of the ancients" should have
accepted a new idea.[37] (Not surprisingly, Brucaeus took some
umbrage at Tycho's candor and answered his letters and handled his
business rather coolly for a while.)

As we have seen, some of Tycho's difficulty might have been
avoided with a better presentation of the system. But the majority of
Tycho's peers were probably unprepared to contemplate anything so
fundamental as a new system of the world, no matter how exhaus-
tively it was argued. And of those few who gave it open-minded
scrutiny, some would surely have remained troubled by the intersections, anyway. For the next question was how the planets
actually made their rounds of the heavens if there were no crystalline
spheres to carry them. Unless one could answer that question, it was
almost easier to be a Copernican.

As of Tycho's day, the motion of the earth was not yet a significant issue. Although Copernicus was known as a great astronomer
and his technical results were having considerable impact on the
professional literature, there is no indication whatsoever that his
cosmology was being taken seriously by a significant number of
astronomers.[38] It may have been 1587 before Tycho encountered
anyone who believed the earth was moving, for neither in his
lectures at Copenhagen in 1574 nor in the account of his system that

---

[35] V, 125–6.   [36] VII, 142.   [37] VII, 148.
[38] See Jerzy Dobrzycki, ed., *The Reception of Copernicus' Heliocentric Theory* (Dordrecht,
Netherlands: Nijhoff, 1973), passim.

was being printed in *De mundi* had he felt obliged to say anything about the proposition, other than that it was absurd.

But in 1587 Tycho learned that Rothmann was a Copernican. Motivated not only by the challenge of demonstrating to Rothmann the error of his ways but also by the task of defending his as-yet-unveiled system, Tycho set down in letters to Rothmann the arguments that, because they were later published, provided the framework for all seventeenth-century debate on the motion of the earth.[39]

Unless Tycho was radically different from everyone else who rejected the proposition that the earth was moving, his principal reason for doing so was the readily perceived fact that the earth is manifestly not moving. However, it was one thing to know something and quite something else to prove it, and Tycho was sufficiently imbued with the scholastic tradition to be as uncomfortable as all his peers would have been with any attempt to found intellectual arguments on the bouncing or rocking of everyday experience in travel on land or sea. In an effort to rationalize his instincts, Tycho used evidence and argument drawn from three different academic disciplines: theology, science, and mathematics.

Although theology was destined to assume a prominent role in guiding scientific belief in the generation after Tycho and to retain that role through the end of the nineteenth century, its explicit appearance in Tycho's day was relatively new. For the three hundred years prior to Tycho, science and religion had coexisted on terms under which science was regarded as merely a collection of "likely stories" – stories that could be interesting in their way, but from which it was completely inappropriate to expect any real picture of the physical world.[40] This understanding had relieved Christian theology of the embarrassment of having to cope with conflicting ideas from the realm of science and, conversely, in a culture in which Christian theology was preeminent, spared medieval science a struggle from which it could not possibly have emerged in any viable condition.

Just before Tycho's birth, this long truce was broken by Copernicus. However, the significance of Copernicus's assertion of the reality of his system was blunted by two circumstances: Osiander's preface reiterating the traditional fictionalist view of science, and the fact that over the intervening centuries, theological truth had come

---

[39] Ronald J. Overmann, "Theories of Gravity in the Seventeenth Century" (Ph.D. diss., Indiana University, 1974), pp. 15, 253.

[40] Edward Grant, "Late Medieval Thought, Copernicus, and the Scientific Revolution," *Journal of the History of Ideas* 23 (1962): 197–220.

to be as abstract in concept as physical reality was. As of Tycho's day, the curia that defined church doctrine had had no cause to render a (negative) judgment of Copernicanism.[41]

For members of the traditional church, therefore, Copernicanism posed no problems. Tycho, however, was a Protestant, and the essence of the Reformation that had created Protestantism was that church doctrine should be taken directly from the Bible rather than subjected to the academic – and sometimes political or economic – interpretations of the church hierarchy. Thus, the new broom that swept out the travesty of indulgences dragged in from the most unlikely corners of the Bible peripheral references to the natural world that now had presumptive status as divinely inspired statements of scientific fact.[42] Luther and Melanchthon both unhesitatingly ruled out the possibility that the earth could be in motion, on the basis of biblical references to the diurnal motion of the sun. Whether to avoid this problem or the even more serious ones latent in such terms as "the four corners of the earth," Osiander, as a Lutheran clergyman, felt obliged to impose his unauthorized preface on *De revolutionibus*. On the other hand, Rheticus seems to have been able to reconcile outright advocacy of Copernicanism with his Protestant calling.

Half a century later, Rothmann, Mästlin, and the young Kepler were also managing to circumvent the strictures of literal interpretation. Relatively early in his correspondence with Tycho, Rothmann expressed his conviction that the Bible had been written from the perspective and understanding of the common person and did not need to be taken literally.[43] Tycho's response was that the Holy Scriptures ought to be held in greater esteem and reverence than that and should not be used, as Rothmann was doing, as a stocking that could be pulled on by any and all.[44] (Interestingly enough, however, once Tycho decided that the comet of 1577 had ruled out the existence of solid spheres, he had no conpunction about quoting – to Peucer – Isaiah 40:22, "He stretcheth out the heavens like a curtain and spreadeth them out as a tent to dwell in," to justify his conclusion.)[45]

[41] But see Edward Rosen, *Copernicus and the Scientific Revolution* (Malabar, Fla.: Robert Krieger, 1984), p. 188, for an assertion that Pope Paul's personal theologian, Bartolomeo Spina, planned to condemn *De revolutionibus* but fell ill and died before he could do it.
[42] See J. R. Christianson, "Copernicus and the Lutherans," *Sixteenth Century Journal* 4 (1973): 1–10; and R. S. Westman, "The Melanchthon Circle, Rheticus and the Wittenberg Interpretation of the Copernican Theory," *Isis* 66 (1975):165–93.
[43] VI, 159.   [44] VI, 177.
[45] VII, 133–4. On the other hand, Tycho eventually published (III, 151) his view that in the face of the evidence of observation, even Scripture could not prove the existence of solid spheres.

Anyway, whether rightly or wrongly, Rothmann interpreted Tycho's admonition as an implicit charge of impiety and reacted very defensively (and very strongly) to the accusation.[46] It was then Tycho's turn to become defensive and soothe his correspondent by replying that Rothmann had read more into his words than Tycho had intended, that Tycho did not wish to sit in judgment on anyone's piety or impiety and certainly had not meant to accuse Rothmann of denigrating the Scriptures.[47]

Given Tycho's empirical orientation, there can be little doubt that he would have been able to rationalize any biblical references that challenged his instincts. At the same time, however, a considerable portion of Tycho's "instincts" rested on the quasi-theological conviction that God could not have been so inefficient as to have created a universe containing all the wasted space implicit in a Copernican system.[48] So in the end, all one can say is that whenever the subject came up, Tycho invariably mentioned his belief that the motion of the earth was contrary to Scripture.[49]

Just as the (as yet feeble) tide of Copernicanism enjoyed the benefit of ambiguity concerning its theological status, so, too, it should have reaped the benefit of at least a few doubts regarding its philosophical status, given Buridan's demonstration, some two hundred years earlier, of the impossibility of proving a priori that the earth was not rotating. And perhaps through familiarity with Buridan's analysis, Tycho felt obliged to hedge a bit on the question of rotation. Already in his letter to Brucaeus in 1584, Tycho professed to be unable to decide whether the diurnal motion belonged to the heavens or to the earth. But all he said about the issue was that despite the opinions of such distinguished ancients as the Pythagoreans and the followers of the "divine Plato," he was inclined to think it was the heavens that moved.[50]

Sometime during the next year, probably, Tycho received the posthumously published didactic poem of his old acquaintance George Buchanan, which elaborated the difficulties that would attend a moving earth. But if anything, Buchanan's recapitulation of the arguments seems to have unsettled Tycho's thought on the issue, for in referring to the book in his letter to Rothmann on 20 January 1587, Tycho stated that the possibility that the earth might have a natural perpetual motion around its own center was not as problematical as most people imagined – certainly not as ridiculous as the

---

[46] VI, 181.   [47] VI, 185.
[48] III, 63. One is reminded of the remark attributed to King Alfonso the Wise (ca. 1275, during the construction of the *Alfonsine Tables*) that if he had been present at the Creation he would have wanted to say something to God about the mechanism that moved Mercury.
[49] IV, 156; VI, 177; VII, 129.   [50] VII, 80.

Ptolemaic assumption that planets could circle equably around a noncentral equant point.[51] What he seems to have been alluding to was Buchanan's disbelief that the earth could rotate without creating some disturbances in the water and air surrounding it. However, in a somewhat belated return to the discussion on 24 November 1589, Tycho expressed grave doubts that a lead ball dropped from a high tower could land at the foot of the tower if, while the ball was falling, the earth was rotating out from under it.[52]

Rothmann's suggestion that the ball, being in essence part of the earth, might be able to continue to participate in the hypothetical circular motion of the earth's rotation while it was in vertical fall – and thus undergo simultaneously two of Aristotle's "simple" motions – was just too contrary to Aristotelian theory to be conceivable to Tycho. Before he could get his rebuttal (denying that a separated chunk of "earth" would retain the earth's circular motion) into a letter, however, Rothmann turned up at Hven for a visit, and the rest of the debate was conducted orally. We know the arguments Tycho used, because he published them in a postscript to his and Rothmann's letters.

Prominent among these arguments was the cannonball argument, according to which if the earth is rotating west to east, a cannon aimed toward the west should appear to propel a ball farther than when it is aimed toward the east.

Tycho was confident that that did not happen. In fact he implied strongly that he had actually tested an analogous situation:

> Some people think that if a missile were thrown upwards from the inside of a ship, it would fall in the same place whether or not the ship were moving. They offer these assumptions gratuitously, for things actually happen quite differently. In fact, the faster the motion of the ship, the more difference will be found.[53]

Needless to say, Tycho provided no details concerning his assertion. But his reputation for care in empirical matters was enough to make this "finding" an important datum in the early seventeenth-century arguments about inertia. Appropriately, it was Tycho's first biographer, Pierre Gassendi, who eventually laid the matter to rest in 1640 with a decisive disproof of Tycho's assertion.[54]

Although Tycho was so far from being a slavish follower of Aristotle as to be conducting in *De mundi* the first systematic campaign to show that one feature of the peripatetic worldview was

[51] VI, 102.   [52] VI, 197.
[53] VI, 219–20. Despite Tycho's apparent confidence, he was still very cautious in 1598 when discussing with Kepler the diurnal rotation: VIII, 45–6.
[54] *DSB* (*Gassendi*) V, 288.

simply wrong, he could not imagine that the entire web of Aristotelian physics, with its uniformly negative implications for the motion of the earth, might be wrong. With the weight of two thousand years of tradition and experience behind such axioms as that the element "earth" tended to seek the center of the universe and to move naturally in a straight line when doing so, it was almost impossible to contemplate both that the agglomeration of all the earth in the universe might not be at the center of the universe and that all that "earth" might actually be moving, not in a straight line, but with a circular motion that was natural to heavenly bodies. Nor does Tycho seem to have seen the need to try very hard.

For unlike the rotation,[55] the putative revolution was subject to empirical test. Copernicus himself had realized that if the earth revolved around the sun, there should be some kind of apparent displacement of the stars in consequence of and therefore in time with it. As the universe had been understood since Ptolemy, the swings involved should have reached 10°.[56] Undeterred by the fact that no such phenomenon manifested itself, Copernicus compounded his philosophical offense by suggesting that the stars were simply much farther away than anyone had until that time believed. With his rather limited concern for observation, Copernicus need not have been thinking of anything more than a factor of about forty. But by the time Tycho decided that he could not detect parallax with his instruments, the factor was seven hundred, making the volume of the universe over three hundred million times what it had to be with a stationary earth.[57]

As far as Tycho was concerned, this one absurdity by itself was sufficient to rule out the motion of the earth. But for those willing to suspend judgment even this far, Tycho posed another problem. Measurements of the angular diameters of stars showed them to range from about $\frac{1}{3}'$ to $2'$ of arc, considerably smaller than most previous estimates but appreciable nevertheless.[58] But for a star to have a parallax of $1'$ meant that from its distance the orbit of the earth subtended about $1'$ of arc. Thus, a star just out of reach of the $1'$ parallax Tycho thought his instruments were capable of detecting, but itself measuring $1'$ in angular diameter (third magnitude), had to

---

[55] Actually, the one test for the earth's motion that would have been feasible in the seventeenth century was the Foucault pendulum (*DSB*: V, 86), for detecting the rotation of the earth. But it was not conceived until the middle of the nineteenth century, after parallax had been found.

[56] Taking Tycho's estimate of 13,000 earth radii as the distance of the nearest stars and 1,150 e.r. as the radius of the earth's (putative) orbit, the half-angle of swing would be $\sin^{-1}0.0885$, or a bit more than 5°.

[57] II, 430.

[58] For a brief summary of previous efforts, see *Dreyer*, 191.

be inconceivably large – comparable in size not to the sun but to the entire orbit of the earth. And insofar as the star in question was either larger than third magnitude or hypothetically more distant than one parallax minute, it had to be correspondingly larger than the earth's orbit.[59]

It is regrettable that the extensive correspondence between Tycho and Rothmann includes no mention of one of the most significant topics of the day, the development of the prosthaphaeretic method. Their silence on the subject is not surprising, for Rothmann does not seem to have had much to do with it and Tycho appears to have had a positive interest in limiting the circulation of at least his particular computational scheme and probably also the ideas that might allow anyone else to duplicate it.

This interest dated from at least the fall of 1582, when Tycho was expressing to Schultz his concern over Wittich's possible disclosures. In that context, the thought struck Tycho that if Wittich had made and shown copies of the comet observations, he had probably done the same with the prosthaphaeretic scheme he had put together under Tycho's direction. Tycho could scarcely object to Wittich's sharing it, because the contents were largely Wittich's. But he could protect his share of the credit for it with a little devious ingenuity: Because Schultz's town of Görlitz was not far from Prague, Tycho was already planning to send a letter to Hayek with the traveler who was carrying his letter to Schultz. And because (as he thought) Hayek had not received last year's letter, he would also send a copy of that letter. Into that letter dated a year earlier he could append to his lines mentioning Wittich's arrival a statement that

> in the resolution of triangles, we have discovered some things, working together, that were not known before. For I have sweated on this business and, God willing, will sweat further, so that this method, which proceeds by prosthaphaereses without tedious multiplication and division, can be worked out more fully. He has not yet attained the goal, but he is getting there little by little. He seems, originally, as he once volunteered, to have been put onto this idea by something he heard me say when I was at Wittenberg while he was a student there, and we talked about his studies, but I cannot remember what it was.[60]

Because Tycho posted this claim with a man he knew had been working rather closely with Wittich and one he must, therefore, have presumed would be privy to all of Wittich's results, his belief that he was entitled to a share in the credit for his manual must have

[59] VI, 197. According to modern definition, it is half a minute of parallax that is at stake in this example.
[60] VII, 58–9.

been very strong.[61] Unfortunately, the basis for Tycho's claim is as vague as his claim. His allusion to having possibly influenced Wittich at Wittenberg[62] is pathetic, even if it somehow represents a germ of truth that we cannot readily imagine. And given the fact that Wittich arrived on Hven with some kind of booklet containing both a prosthaphaeretic identity and sample calculations made with it, it seems unlikely that Tycho could have played any role in either the recovery (or rediscovery) of Werner's identity or the recognition that it could be something more than just one more trigonometric identity.[63]

In regard to applying the method of prosthaphaereses to Tycho's data reduction scheme, Tycho's candid reference to the fact that it was Wittich who had not yet attained the goal, but was gradually getting there, makes it unlikely that Tycho was really sharing much in that process, either. What we seem to be left with then is some kind of quasi-institutional claim analogous to the one Tycho had on his instruments.

Institutional claims have a long and reasonably meritorious tradition. Even today, many coauthorships tend to be bestowed on people whose major contribution is having provided the facilities or milieu in which an idea has been developed. Thus, even though one cannot imagine Tycho's having ever sweated over a forge, the instruments that were made in his shop are regarded not just partly but virtually entirely as his because he directed the efforts of his artisans and paid for their bed and board. In general, those artisans surely did more things that Tycho did not know how to do himself than Tycho's observatory assistants did. Thus it is very unlikely that he regarded the products of his researches as any less "his" than his instruments were.

---

[61] Tycho later made even vaguer claims in 1590 (VII, 281), 1592 (VII, 323), and 1600 (*Norlind*, 377).

[62] The issues in this controversy were reviewed briefly by V. E. Thoren, "Prosthaphaereses Re-visited," *Historia Mathematica* 10 (1988): 32–9. Exactly when Tycho and Wittich met is simply unknown. But Tycho would not have needed the prompting of an introduction from Hayek if he had met Wittich only five years earlier. At the other extreme, if the meeting occurred in 1566, when Wittich enrolled at Wittenberg, Tycho (and Wittich) would have been less than twenty. There is no evidence that Tycho was even at Wittenberg in 1568. The citations of that date by Gassendi and Dreyer rest on misinterpretation of a statement by Tycho (*Norlind*, 24). On the other hand, we know that Tycho passed through Wittenberg on his trip home in 1570 (*Norlind*, 26). An encounter with Wittich on that occasion would have occurred when both principals were older but would necessarily have been brief enough (and remote enough from 1580) that Wittich could not have been sure that Tycho would remember him.

[63] Owen Gingerich and Robert Westman, "The Wittich Connection: Conflict and Priority in Late Sixteenth-Century Cosmology," *Transactions of the American Philosophical Society* 78 (1988), pt. 7, p. 12.

The manual of trigonometry was clearly in a different category, because of the unique input provided by Wittich. But it was obviously not sufficiently different to persuade Tycho to resign all the credit for it to Wittich. As a man who contemplated the trigonometric reduction of thousands of observations, Tycho may well have been the only person to whom Wittich showed his method, who saw it as a tool of marvelous utility rather than as simply a clever toy. So there is no reason to doubt that Tycho did sweat over it, both from his own exertions and from his exhortations of Wittich.

As a man who made his career by performing trigonometric (parallax) calculations more frequently and more successfully than any of his contemporaries did, Tycho was surely more familiar with the steps involved in his data reduction scheme than Wittich was and must in some sense have been leading the effort to apply the prosthaphaeretic method to it. However, because he thought that Wittich had discovered the one formula that he brought to Hven, he naturally thought that Wittich might be able to discover others, given proper encouragement and support. This expectation may well have played a role in the prized assistant's departure from Hven, for Wittich must sooner or later have come to realize that he would be doing trigonometry there until he discovered the identities necessary to complete Tycho's manual.

How completely Wittich was able to convert Tycho's data reduction scheme to prosthaphaeretic computation in 1580 is open to debate. Since the discovery in the 1880s of a copy of the manual of trigonometry[64] used on Hven, no one has assumed anything but that Wittich left Tycho with a reasonably complete scheme. Unfortunately, the evidence for this view – the fact that the manual uses not only Werner's formula for the product of two sines but also the practically identical one for the product of two cosines – is seriously flawed. It rests on the presumption that the manual known to us is the manual that was left by Wittich (or, more likely – because Wittich's work was incomplete and his absence was supposed to be a temporary one – written up by Tycho when Wittich failed to return). In fact, the extant manual is dated in a way that leaves little alternative to Dreyer's interpretation "that in the form we have it the book was put together on the first of January 1591, while the only copy now existing was written in 1595."[65] This not

[64] F. I. Studnicka, ed., *Tychonis Brahe Triangulorum planorum et sphaericorum praxis arith metica* (Prague, 1886).

[65] J. L. E. Dreyer, "On Tycho Brahe's Manual of Trigonometry," *The Observatory* 39 (1916): 127. According to M. Zeller, *History of Trigonometry from Regiomontanus to Pitiscus* (Ann Arbor, Mich.: Edwards Bros., 1946), the treatise displays a familiarity with the work of Viete. Because Viete published during the 1580s, this is more proof that the manual was developed further after Wittich left in 1580.

only suggests that the original manual was revised in 1591 but also invites speculation that because the result was not simply labeled "revised" or "new," there may have been an earlier revision.

Indeed, if Wittich did not know the identity for the product of two cosines when he was at Hven, then there must have been a revision that introduced that identity before 1588, when Ursus first published it, for Tycho printed a vague reference to it in *De mundi* in 1587,[66] and Gellius, during his travels, was able to interpret and expound on it when Magini queried him about it.[67] This probably means that Tycho himself made at least the first revision. But it would be straining the bounds of probability to assume that Tycho made an independent discovery of that second equation or any of the other innovations that entered the revision with it. Most likely, Tycho learned of the new equation from the same place Ursus got it – Cassel.

Both Ursus and Flemløse visited Cassel in the first half of 1586, and if Ursus left with the identity, there is every reason to believe that Flemløse did too. Ursus published both formulas in his *Fundamentum astronomicum*, without indicating his indebtedness for either.[68] Only in a later publication of 1597, after having been accused by Tycho of plagiarizing the formulas, did Ursus reveal his source. Wittich, he said, had come to Cassel with one formula and no proof for it: It had been Joost Bürgi, the *landgrave*'s clockmaker, who had devised a geometrical proof for the one identity and had then seen an analogous construction that gave the formula for the product of two cosines.[69]

Strangely – or perhaps not so strangely, given Ursus's record – no one seems to have taken Ursus's statement seriously. But whatever the problems with the credibility of Ursus's other statements, whatever the possibility that he may just have been making mischief or deliberately trying to transfer to Bürgi credit that he thought would otherwise accrue at least partially to Tycho (who, as far as he knew, had collaborated on the original discovery), the fact is that Tycho had four years after the publication of Ursus's statement to rebut the implied diminution of whatever contribution he could claim, and he never did.

While Tycho was corresponding with Rothmann and the *landgrave* at sufficient length to comprise what was eventually to be a book of three hundred pages, he was also researching and writing his book

---

[66] IV, 233.   [67] VII, 281.
[68] 16v–17. The diagram was dedicated to Wittich, but Dreyer, for one, certainly thought the context implied that the formulas were Ursus's.
[69] Ursus, *De astronomicis hypothesibus* I (1597), p. 3: "ex qua visa Demonstratione foecundum quoddam lucrum pariter elucescebat: vz. cum casus alter, una cum sua etiam Demonstratione, tum ratio solvendi quaecunque Triangula."

on the new star. In fact, well before *De mundi* cleared his presses, Tycho began to think actively about printing that book. Planned as Volume I of a series designed to show that the remarkable celestial phenomena of the 1570s and 1580s unanimously contradicted the Aristotelian worldview, it had been left to follow *De mundi* (designated chronologically as Volume II) through the press, because Tycho naturally wished to declare himself publicly on the comet before he issued a second book on the new star.

But if we can believe a casual remark by Kepler, Tycho had already begun writing or at least outlining his second work on the new star by 1582.[70] And certainly by the middle 1580s, he was well along on the composition of a book that at that time greatly resembled the structure of his volume on the comet. First, there would be a discussion of the original discovery of and arguments over the nova.[71] Then would follow a description of the instruments used, a reworking of the star's position from his own coordinates for the stars of Cassiopeia, various proofs of the supralunarity of the new star, and finally criticisms of other writers on the subject.[72]

Because these chapters follow so closely the material of *De nova stella* and the form of *De mundi*, it will suffice to mention their nearly five-hundred-page length and the fact that Tycho's sense of fair play[73] induced him to reprint his own technical discussion of the nova along with his critiques of other writings, even though it obligated him to recant his earlier statements about comets and spheres,[74] and revealed to the careful reader that his final treatment rested on slightly different data and arrived at somewhat different results than his original work had.[75] Although Tycho made minor additions to the manuscript as it went to press later, it must have been nearly complete by March 1588, when Tycho wrote in his diary, "End of Volume I."[76]

Even though the book's principal subject was written, the manu-

---

[70] Kepler's statement (III, 320) may well be merely an inference from the observations of Venus of 1582 referred to in this chapter, rather than direct information from Tycho. But that Tycho had plans at that time can be seen from a letter to Hayek: VII, 73, and III, 42–3.

[71] For reasons that will eventually become clear, this became Chapter III (II, 307–29).

[72] See Chapters IV (II, 330–52), V (II, 353–71), VI (II, 372–414), and VIII–X (III, 5–299), respectively. Most of Chapter VII (II, 415–35) was probably also written during this period.

[73] III, 112.   [74] III, 107, 111.

[75] Tycho pointed out (III, 108) the changes he made in three observed distances (compare III, 100 with II, 336) and defended the change by referring to problems with ocular parallax on his instruments. He also corrected a minor computational error and its chain of minor consequences but attributed them to misprints in the original (III, 108). He did not mention a difference of 12′ (II, 341; III, 102; see also I, 23) in the resulting declinations.

[76] IX, 62. See somewhat different views in Dreyer (186) and Norlind (144–6).

script for the entire volume turned out to be far from complete, for it now was scheduled to include presentations of Tycho's solar theory and star catalogue as well: At some stage in the writing of the volume Tycho decided that the positions of the stars of Cassiopeia (relative to which he had observed the nova) had to be established accurately not only with respect to the sun but even within the stellar system as a whole. How much of this decision arose from an extraordinary demand for perfection and how much from a very ordinary desire to print up his researches simply because they seemed to be ready for publication is difficult to determine. Presumably, Tycho rationalized the additional materials as enhancements of his claims for the superiority of his observations, but the fact that both his solar theory and his star catalogue were "essentially" finished by 1588, when he began to consider printing the book, cannot have been irrelevant to the decision. Because of the addition of these and several other subjects to his tome, Tycho soon decided that it required a more comprehensive title: hence, *Astronomiae instauratae progymnasmata*, or *Introductory Exercises Toward a Restored Astronomy*.

Given the logic of their presence in Volume I, the solar and stellar research had to come at the beginning of the book. After one last observation of the autumnal equinox in September 1588,[77] therefore, Tycho made the final adjustments to his solar theory and began to write up the justification of his various results. By the end of the year he had probably completed the hundred pages on the solar theory that would be inserted at the front of his magnum opus on the new star.[78]

Despite the essential similarity of Tycho's endeavor to those of Ptolemy and Copernicus before him, his discussion is remarkably different. The source of the difference is Tycho's introduction into technical astronomy of the concept of error: error not just in the individual discrepant observations that almost every astronomer in history must have had to learn to rationalize and reject, but error that might be so fundamental to some aspect of the collection and use of astronomical information as to invalidate, at least for Tycho's purposes, the result of the whole process. Although Tycho was not bashful about naming names and did not hesitate to devote a section to pointing out even Copernicus's errors in his latitude and obliquity, due to his (already-mentioned) neglect of refraction,[79] he was

---

[77] II, 15. The arithmetical tidiness of Tycho's table of equinoxes (six of the eight intervals between the successive equinoxes are $365^d5^h49^m$, and the other two are one minute shorter) makes it clear that one need not adhere slavishly to the chronology here.

[78] "Hoc ipso anno, quo haec scribimus 1588." II, 87–8.

[79] II, 29–32. In general, Tycho was willing to give the benefit of a doubt by ascribing discrepancies to errors of transcription.

really much more interested in avoiding the consequences of the errors of his predecessors than in enumerating them. The only way Tycho could do this was to depend on them as little as possible. For some things, such as the length of the year, the longer baseline seemed more important than the uncertainty of ancient observations. But in fact, when Tycho used an equinox determined by Ptolemy to reckon the length of the sidereal year ($365^d6^h9^m26^s43\frac{1}{2}^t$), he got a result 16^s too large; his determination of the length of the tropical year ($365^d5^h48\frac{3}{4}^m$), made from an equinox determined by Walther in 1488 was less than three seconds too small.[80] On the other hand, when Tycho established his rate for the progression of the line of apsides by referring to a determination of the apse made by Walther, he arrived at a figure of $45''$ per year; whereas referring to Hipparchus's apse would have brought him perilously near the modern value of $61''$ per year.[81]

The major problem in all of these decisions was what credence to give to the traditional assumptions of variability by which previous astronomers had accounted for the wide range of results obtained by different observers for the same parameters. Although Tycho obviously did not accord to the determinations of his predecessors the respect he gave his own, he stopped well short of dismissing them altogether. If he was sure that none of them had achieved the accuracy "to the minute" that he presumed for his own results, he was nevertheless willing to believe that the obliquity, eccentricity, apse, and even the length of the tropical year all varied in the long term.[82]

What Tycho did not believe was that the determinations of the ancients were sufficiently accurate to permit the establishment of any general theory of secular change, of the kind Copernicus had tried to construct.[83] He realized that failing to construct such theories would limit the useful life of his tables, but he figured that it was better to have the best parameters possible for the short run than to fall back on less trustworthy observations in order to try for longer-term results. Accordingly, he contented himself with computing four

---

[80] II, 34–7, 42–3, and *Norlind*, 397, note 5.

[81] II, 45. Using Hipparchus's longitude of apogee (65°30′) and his own (95°30′) would have had him dividing 1,800′ by about 1,750 years.

[82] II, 86, 28, 33, 38. The last is the notorious trepidation, concerning which Tycho took a sterner view on II, 255. A hundred years later, Flamsteed denied that the observations of the ancients were accurate enough to justify belief in changes even of the obliquity or apse (*DSB* V, 24).

[83] See this theme developed by K.-P. Moesgaard, "Success and Failure in Copernicus' Planetary Theories," *Archives Internationales d'Histoire des Sciences* 24 (1974): 73–111, 243–318.

hundred years' worth of tables based on fixed parameters and took refuge in the knowledge that if there were minor variabilities in some of them, they could not amount to much in that period of time.[84]

Concerning his tables, little need be said except that Tycho essentially tabulated the epoch values of the sun and apse for every single year and thereby somewhat reduced the labor – and the chance of error – in calculating positions. Tycho, however, said a great deal in the following fifty pages about various features of the foundation, construction, use, and, above all, accuracy of his solar theory.

Most of his comments concerned auxiliary tables, for refraction, parallax, conversion of coordinates, and entrance of the sun into the various astrological signs of the zodiac. All that requires description here is his obvious concern that his fellow astronomers – for he was not writing for tyros, he said[85] – should be convinced of the accuracy of which he was so proud. Not only did he show that his own observations corroborated his tables and disagreed with predictions from the Alphonsine and Copernican tables, but he also produced observations from Hainzel and the *landgrave* to back up his claims.[86] Moreover, he included his own six-page table for converting longitudes to declinations (at 10′ intervals, with differences for interpolation) so that his readers could readily check the longitudes of his theory with the meridian altitudes given by their instruments.[87]

By February 1589, Tycho was finished with the solar theory[88] and ready for the major effort on his star catalogue. Like most of his undertakings, this too, was a ten-year project, begun in the fall of 1581 and still far from completion. But because the methodological aspects of the research were complete and were the ones that required most of the discussion in his write-up, there was no reason not to keep writing.

In general, Tycho's discussion paralleled his original work. First had come the attempt, already reported, to use his clocks and quadrants to obtain stellar coordinates directly from meridian transits. Tycho duly reported its failure.[89] Toward the end of the same winter (1581–2), Tycho had started looking into another question: the possibility of using Venus instead of the moon as the intermediary by which he could establish the longitudes of his stars relative to the sun, whose observable crossing of the equator defined the celestial reference point.

---

[84] II, 32, 37, 44–5. [85] II, 58. [86] II, 60–3. [87] II, 58, 66–71.
[88] See the final checks of the theory and his table of solar refractions on XI, 311–16, and his transmission of both to the *landgrave* (VI, 165).
[89] See Chapter 5 and II, 156–8.

For six weeks, Venus was bright enough to be located with respect to the sun in the afternoon and then observed with the stars in the evening before it set.[90] Because its own motion in the interim, both real and apparent (parallax), was considerably less than the moon's, it proved to be as suitable a substitute as Walther had thought it would be.[91] Accordingly, Tycho had followed through on it at similar opportunities from the fall of 1585 to the spring of 1588, to establish as the second stage of his work the fundamental "skeleton" of his catalogue.

Relating those efforts in a way that would make convincing his claim to "scrupulous" – right down to the minute – accuracy required 120 pages of prose and computations, as well as pictures of the sextant and armillary used for the work.[92]

The foundation of his work, and therefore of his whole catalogue, was twenty-seven observations linking α Arietis to the sun by means of Venus and various second intermediary stars. Copernicus had used γ Arietis because of its position at the "front" of the constellation. Tycho opted for α because its brightness made observation easier.[93] And because he had learned to respect the potential for error inherent in refraction and was also concerned about parallax, he chose observations in which the altitudes of Venus and the sun were as similar as possible. When this measure still left results with a scatter of $16\frac{1}{2}'$, however, Tycho discovered the tactic of pairing up observations made on opposite sides of the sun. This reduced the scatter to 40″ for fifteen values and led Tycho to an adopted longitude that disagrees by only 15″ with modern understanding of where the star should have been.[94] This step also justified Tycho's confidence that the five-minute difference between his longtitudes and the *landgrave*'s involved no error on his part.

With his catalogue thus securely anchored, Tycho proceeded to establish positions for twenty other reference stars around the celestial sphere, checking the coordinates as he added stars to ascertain that the differences in right ascension for various groups of them always came within a few seconds of totaling 360°.[95]

Part of the task of fixing the framework for Tycho's star catalo-

---

[90] See especially X, 158–63.

[91] See Tycho's description in II, 159–61, and the discussion in *Dreyer* (348–51).

[92] II, 150–257 (137–257 in original pagination). Note that the pictures were the ones sent to Hayek and Rantzov in 1585, so that Tycho already foresaw the "imminent" publication of his star catalogue.

[93] II, 160–1. The observations and calculations are presented on II, 162–97.

[94] See II, 162–97, passim, and *Dreyer* (351).

[95] II, 198–233. For the illusory nature of Tycho's checks (200, 203, 206), see *Dreyer* (351–2), who found a probable error of ±25″ in the nine basic stars.

guing was establishing the rate at which the stellar sphere moved, for Hipparchus had already shown in the second century B.C. that the positions of the stars changed slowly over the years. In antiquity, it was assumed that the rate, found to be at least a degree per century by Hipparchus and so adopted by Ptolemy, was uniform. But through the years, as various Islamic astronomers contributed various findings on the position of the equinoxes, the notion that precession was nonuniform gained more and more credence, until Copernicus enshrined this so-called trepidation in an ingenious theory that simultaneously accounted for the diminution of the obliquity of the ecliptic from ancient times.[96]

Observationally, the question was primarily a solar matter, depending on the determination of the equinoxes. Tycho used the difference between the sidereal and tropical years as his fundamental datum for the parameter. But because precession affected the position of every star in the sky, any longitude that could be trusted could be used to investigate the phenomenon. Already in 1580–1, therefore, Tycho was finding a longitude for Spica that caused him to note in his log: "The hypothesis for the motion of the eighth sphere must be much different from that speculated by N.C."[97] The source of his discontent was Copernicus's assumption that the changes in the rate of precession must have returned to the low rate of 1° per hundred years that Ptolemy had chosen.

By 1587, after scores of additional observations of Spica (and the expedition to Frauenberg by Elias Olsen in 1584[98] to see whether Copernicus's ignorance of refraction might not have led him to incorrect determinations of his latitude and obliquity), Tycho was certain that Copernicus's theory was seriously in error. Utilizing an observation of Spica made by Copernicus in 1525 but correcting it by 7′ to compensate for the shortcomings in Copernicus's fundamental constants, Tycho reported to the *landgrave* a finding of 1° in seventy-three years.[99]

In his write-up for the star catalogue in the *Progymnasmata*, Tycho started with the results given by his solar data, believing, no doubt, that ancient observations of equinoxes were less unreliable than were those of individual stars. The 20^m42^s difference in time between the

[96] *De revolutionibus*: III, 3. On Copernicus's theory and its background, see K.-P. Moesgaard, "The 1717 Egyptian Years and the Copernican Theory of Precession," *Centaurus* 13 (1968): 120–8; and N. M. Swerdlow, "On Copernicus's Theory of Precession," Westman, 49–98. A. Pannekoek, "Ptolemy's Precession," *Vistas* 1, 60–6, showed how Ptolemy could have selected his data to derive the one-degree-per-century value for precession.
[97] X, 90, 108. "N.C." is Copernicus.
[98] Tycho's reference in this context is on II, 30.
[99] VI, 73. The determination is in XI, 125.

sun's return to an equinox and its return to a star implied an annual relative movement of nearly 51″, which accumulated to a whole degree after only seventy years and seven months.[100] Accordingly, Tycho chose Copernicus's other observation of Spica, made in 1515, to corroborate the result, because it gave almost exactly 51″.[101] Similar comparison with an observation of Regulus made by Hipparchus provided the same figure, as did one made with an observation of the same star by al-Battani in A.D. 880. This seems to have provoked Tycho to take a hard look at the entire question of trepidation, rather than merely at Copernicus's obviously inadequate treatment of it. The more he looked, the less impressed he was by the observations on which the supposed phenomenon was based.

In an allusion to trepidation written into Chapter I,[102] probably only a few months earlier, Tycho had expressed nothing but a vague agnosticism, listing it with the other secular mutations as phenomena that could be considered only in the short term because data sufficiently accurate for long-term resolutions did not exist. Even now, he was by no means aggressive in his rebuttal but merely cited a few ancient results and Copernicus's average result (50″12′) that tended to agree with his and promised to discuss the matter further in his anticipated general work on astronomy.[103] Still, Dreyer seems to have assessed things accurately in saying that "the mere fact of (Tycho's) having ignored the phenomenon of trepidation was sufficient to lay (to rest) this spectre which had haunted the precincts of Urania for a thousand years."[104]

Amidst the uncertainties and inaccuracies of the data pertaining to precession, one simplifying feature had emerged and endured through the years. Precession affected longitudes only: Latitudes remained constant. In his first letter from Rothmann, however, Tycho learned that Rothmann had found an explanation of some of the differences between ancient and modern star coordinates in a

---

[100] II, 253. The length of the sidereal year was $365^d6^h9^m27^s$, and that of the equinoctial year was $365^D5^h48^m45^s$. The sun's mean motion during the time interval between the two was 50″47′. The $71\frac{1}{2}$ years mentioned later in the *Mechanica* (V, 113) is apparently just a careless misstatement for the $70\frac{1}{2}$ cited in letter of 1599 (VIII, 198), because Tycho quotes the same 51″ and 70 years, 7 months in a letter to Scaliger in 1598: VIII, 106.

[101] II, 253. Swerdlow "On Copernicus' Theory of Precession," showed that Copernicus simply altered this longitude by 4′. Tycho, however, characteristically reprocessed Copernicus's original observation and thereby worked right through the "adjustment."

[102] II, 32–6.

[103] II, 254–6. Later, in writing the *Mechanica* (V, 113), Tycho was even more outspoken about attributing Copernicus's "erroneous" ideas on trepidation to "incorrect observations of the ancients." See also the doubts registered in 1590: VII, 268.

[104] *Dreyer*, 356. Tycho explicitly denies its existence in a letter of 16 November 1599: VIII, 198.

presumed systematic shift of stellar latitudes corresponding to the change in the obliquity of the ecliptic since Ptolemy's day.[105] It was an interesting new twist to an old idea. If the angle between the plane of the sun's orbit and the equator of the stellar sphere had decreased by nearly 20′ since antiquity, either the orb of the sun had shifted slightly or the axis of rotation of the eighth sphere had.[106]

The Islamic astronomers who had established the existence of the shift had settled on the latter explanation, apparently imagining that the poles around which the stellar vault rotated daily had slipped ever so slightly from the points at which they had contacted the eighth sphere in Hipparchus's day.

In the Copernican terms from which Rothmann was viewing the situation, it seemed more likely that the plane of the earth's orbit around the sun had shifted than that the tilt of the earth's axis had changed. Perhaps it was his prescient conception of the "inertial" stability of the axis that had conducted Rothmann to this (correct) alternative, for he later argued that Copernicus's third motion of the earth – a "countergyration" of the earth's axis to explain the cycling of the seasons – was unnecessary. But Rothmann did not present any theoretical argument, and Tycho probably would not have been impressed by it if he had. If Tycho gave any theoretical consideration to the question, it was probably only to the choice between imagining that the thousand and more stars comprising the eighth sphere had shifted in their courses, and assuming that only the sun had. What Rothmann seemed to be advancing, however, was an empirical proposition – even if he had not succeeded in confirming it empirically – and that is how Tycho approached it.

Distinguishing between the two possibilities should have been fairly easy. If the poles of the eighth sphere had shifted on the sphere, then the declinations – the angular distances from the equator – of all stars not on the hour circle of either equinox should have shifted slightly (and independently of the effects of precession) from the positions recorded for them in antiquity. If, on the other hand, it was the sun's orbit that had shifted, then the declinations of the stars would not have changed, but their latitudes above or below the sun's course would have, again everywhere except in the directions of the equinoxes. At the ecliptic, Tycho could expect the differences between ancient and current latitudes to follow a table of sines, being

---

[105] VI, 57.

[106] Tycho's operational understanding of the obliquity of the ecliptic [$(SIQ = WIQ)$, in Figure A.4.3 of Appendix 4] as half the angle between the solstices (SIW) may have been more important. If SIW had increased by 40′, it is not immediately obvious how any change in the position of Q (the equator) could account for the situation.

null in the equinoxes and reaching maxima of nearly 20' (plus and minus) in the solstices.

Finding such an amount with Tycho's instruments would have been child's play: Detecting it among the observations of the ancients would be difficult. And Tycho knew it would be. Already when establishing the positions of his reference stars (for the new star of 1572) in Cassiopeia, Tycho had found that the latitudes observed in antiquity (and passed through to Copernicus) differed by an average of 29' from what his instruments showed.[107] For fifty-two brighter reference stars, whose coordinates he had just sent to the *landgrave*, Tycho had found things somewhat less desperate, but their deviations still averaged 23'.[108]

If Tycho had believed that the observations of the ancients were that bad, he would have had to try to work with the catalogue latitudes and would no doubt have managed to find in them what he wanted to see.[109] But he was convinced that at least some of the error in the ancient positions had been injected into them in the process of copying them from catalogue to catalogue over the years and that if he could find observations that had not suffered this fate, he would be able to make meaningful comparisons. Such observations existed in the *Almagest*, in which Ptolemy reported declinations of a number of stars taken by Timocharis in the third century B.C., Hipparchus in the second century B.C., and Ptolemy himself in the second century A.D.[110] Declinations would not have been the coordinate of Tycho's choice, but with some extra trigonometrical computation, some judicious selection of his stars, and some wishful juggling of the resulting data, he succeeded in showing that three stars tolerably near the winter solstice – Castor, Pollux, and Altair – showed shifts in latitude of about 19', 18', and 18½', while four stars at intermediate points between the equinoxes and solstices – Bellatrix, Betelgeuse, Regulus, and Antares – displayed appropriately intermediate shifts.[111]

As a result of his labors, Tycho was able to write confidently to Rothmann, in his first return letter, that he, too, had found systematic changes in the latitudes of stars, corresponding to the change in the obliquity: He did not feel obliged to mention that the ink was

---

[107] II, 354.   [108] VI, 82–3.

[109] Tycho had access to hundreds of latitudes covering all longitudes. On the average, these latitudes were in error by about 20'. H. Vogt, "Versuch einer Wiederherstellung von Hipparchs Fixsternverzeichnis," *Astronomische Nachrichten*, no. 5354–55 (1925): 23, found 22' average for a group of 122 Ptolemaic latitudes. As the simulation in Figure A.4.4 of Appendix 4 illustrates, Tycho would have had to pick out the particular stars that would demonstrate the existence of the sine wave.

[110] VI, 94; II, 234. Ptolemy listed the stars in III, 7.

[111] VI, 237–43.

barely dry on his computations.[112] And he was kind enough, when he wrote up fourteen pages of argument for the *Progymnasmata* a few months later, to note that the *"landgrave's* mathematician" had confirmed his findings. But ten years after that, when he summarized his work on Hven for the *Mechanica*, he explicitly claimed priority for the discovery.[113]

About all that one can say concerning Tycho's appropriation of Rothmann's idea (besides falling back on the impossibility of proving that Tycho did not conceive it independently) is that scientific ethics were loose, at best, in the sixteenth century and that class ethics had undoubtedly conditioned Tycho to expect as his due the right to harvest the fruits of the labors of his social inferiors. At the same time, however, it is fair to point out that the ingenuity and effort required to prove the existence of a phenomenon that was not generally acknowledged and theoretically accounted for until the eighteenth century[114] – the shift of the plane of the earth's orbit, owing to long-term perturbations by the rest of the planets – were Tycho's.

Moreover, we can be sure there was more work than appears in Tycho's write-up. The stars whose latitudes could be made to confirm the shift represented fewer than half of the declinations that Ptolemy listed. Tycho also mentioned Aldebaran and went to some lengths to explain away the fact that its latitude did not fit the pattern.[115] He said nothing about the other nine sets of declinations provided in the *Almagest*, presumably because he could not get them to fit, no matter how heroically he manipulated the data. Part of the problem was undoubtedly observational. Ptolemy's declination of the star Tycho chose as the reference for his entire series of calculations, Spica, is simply and unaccountably 27' off from modern understanding of what it should have been.[116] But there was one

---

[112] VI, 93. The first reference to the work in his logs is in (or after) October, when he wrote (XI, 126): "Note that if Spica had a latitude of 2° at the time of Ptolemy, it ought now, because of the change of the obliquity, to have one of 2°6″. He apparently communicated his "findings" to Brucaeus before May 1587, for Brucaeus wrote him at that time (VII, 114) expressing his usual objections to the likelihood that Tycho could really have found anything so contrary to the opinions of so many for so long.

[113] II, 247; V, 113. In a letter to Scaliger in 1598, Tycho described the phenomenon without mentioning Rothmann: VIII. 106.

[114] The early history of the development of this ramification of gravitational theory is related briefly in Curtis Wilson, "Perturbations and Solar Tables from Lacaille to Delambre," *Archive for History of Exact Sciences* 22 (1980): 131–3.

[115] Faced with hopeless disagreements in the ancient latitudes of Aldebaran, Tycho simply rejected Ptolemy's result as an error of transcription and averaged the values of Hipparchus and Timocharis: VI, 244–6.

[116] VI, 354, note to p. 96.

other more interesting factor: the individual motions of stars that Tycho explicitly ruled out of consideration. Aldebaran was one of the three stars for which Halley was to discover the first cases of so-called proper motion 125 years later, largely because he could be reasonably certain that Tycho had not erred by 2′ in placing the star for his catalogue.[117] Sirius and Procyon, the only two other stars with which Tycho mentions having trouble, also have pronounced proper motions: Sirius was the second of Halley's three moving stars.[118] So if Tycho did not have the purest of reasons for discounting those stars, he had the best reasons. And by going on to establish the first independent catalogue in Europe, he enabled others to divine the right reason, the last one he would have guessed: that the fixed stars were not fixed at all.

Once the formal stages of his work were done, Tycho could proceed to the actual cataloguing. Until the failure of his second attempt to use clocks (in 1583), Tycho had hoped to obtain right ascensions directly from time intervals recorded between meridian transits. Combined with declinations, determined either directly from the great armillary or almost as easily from meridian altitudes taken with one of his large quadrants, they could have been reduced quickly to longitudes and latitudes with the special conversion tables he was then using for such purposes.[119] Abandonment of the clocks meant that his second datum had to be the distance to another star, an oblique "coordinate" that would entail messy, time-consuming, mistake-engendering trigonometry for every position.[120]

For the most part, Tycho's cataloguing would also require special observations, for although he had recorded hundreds of observations every winter since 1581,[121] he seems to have preferred fresh data, presumably because he was unwilling to work with anything but the best results of perfectly oriented instruments and experienced observers. So although some constellations (Ursa major and minor, Cepheus, and, of course, Cassiopeia) were charted from observations made during the two or three winters before 1588–9, most of

---

[117] E. Halley, "Considerations of the Change of the Latitudes of Some of the Principal Fixt Stars," *Philosophical Transactions of the Royal Society* 29 (1714–16): 456–64. The *DSB* (VI; 71) article on Halley errs in assuming that "palalicium" is Procyon rather than Aldebaran.

[118] II, 247; VI, 99. Halley's (ibid.) third star was Arcturus.

[119] II, 227.

[120] Although a right ascension (α, in Figure A.4.5 of Appendix 4, from clocks) would have given Tycho orthogonal coordinates directly (SU, the declination, was observed), the measurement of the distance from the reference (R) to the star (S) necessitated solving for α in a spherical triangle of three known sides ($90°$-SU, $90°$-RT, and RS).

[121] There were over seven hundred in 1581 and 1585, about six hundred in 1584, and perhaps half that number in 1582 and 1583. Tycho probably viewed most of these as test data, designed to check the performance of instruments or procedures.

the catalogue was done from observations made in 1589, 1590, and 1591.

The project does not appear to have been a model of organization. In general, Tycho did the winter zodiacal constellations (Pisces to Virgo) in 1589, the summer ones in the early fall of 1590, the northern constellations in the early and late months of 1590 and 1591, and the (very few) southern stars in the winter of 1590–1.[122] But rarely was any constellation done all at once: The usual pattern, if it can be called that, was to do somewhat more than half in a first pass and then to follow through on the rest – with some duplications – in a second pass.

The observational procedure seems to have been to take distance measurements one night with a sextant, from some convenient bright star (often not one of the twenty-one reference stars), and declinations (or altitudes) on another night, although this generalization, like the previous ones, probably suggests more organization for the enterprise than it actually had. The vast majority of the measurements was taken twice and provided readings that agreed within a minute. Some differences were as great as two minutes, but probably twenty times as many agreed within a sixth of a minute.

How accurate the observations for Tycho's catalogue were has never been investigated systematically. Given that the catalogue is the ultimate symbol of Tycho's image as an observer, it is ironic that the most qualified appraiser of it, Dreyer, "confessed to a slight feeling of disappointment" with it and obviously felt a great deal more.[123] The best he could say was that the brightest stars "rarely differ more than a minute or two from Flamsteed's positions – often less" but that the fainter stars were considerably less accurately placed.[124] Of course, there were some good excuses for these comparatively poor results. In order to see the fainter stars, and many of those in the lower regions of the summer sky, the observers would have had to open up the slits on the sights that were responsible for much of the accuracy of Tycho's alignments.[125] No doubt, also, the effects of refraction, aberration, and nutation combined to significant amounts in some observations.

But Tycho found ways of coping with these problems in all his other work. The obvious place to start would have been with taking

---

[122] These generalizations can be checked only by reference to the logs for each year. Although all the summer constellations are entered in a zodiacal catalogue compiled at the end of 1589 (XI, 383–412), many changes were made after observations in the fall of 1590 (XII, 77–80).

[123] J. L. E. Dreyer, "On Tycho Brahe's Catalogue of Stars," *The Observatory* 40 (1917): 233.

[124] Ibid., pp. 231–2.

[125] Tycho alluded to the necessity of observing faint stars on moonless nights (V, 112) and complained of the bad weather that accompanies the new moon.

more observations. It is hard to imagine that another reading or two for each position would have left Tycho's calculators bewildered by a hopeless redundancy of data. Nor would such an expansion have been an intolerable burden on the observers, for observing was far from being a full-time occupation for anyone on Hven. Any month with a combination of twenty noon and night observing sessions was a very busy one, and the average (for nonsummer months) was closer to ten. Sessions of more than two hours were rare, especially during cold weather, and the average was probably closer to one hour. Weather was clearly a problem, but not one that Tycho regarded as insurmountable.[126]

The fact is that when Tycho wanted to get observations of something, he got them. At the beginning of March 1590, for example, Tycho decided to check Venus for parallax. In five days, involving sessions of five to eight hours (but with occasional breaks), he took nearly a thousand observations,[127] more than three times as many as he had collected for the star cataloguing in the previous two months. During two weeks the following September, in a comparatively intense period of stellar observation, he accumulated nearly four hundred observations.[128] It seems clear that he could not take distances – which involved orienting the plane of a sextant to the line connecting two stars before the alidades could be pointed – at the rate of the forty or fifty per hour that his team could achieve for declinations or altitudes.[129] But even assuming five minutes for each distance and two minutes for each declination, Tycho could have had an extra measurement for each coordinate for perhaps one hundred hours of observing time.

Observing, of course, was only half the task. Calculators then had to compare the readings and make some kind of decision about them, reckon positions for each star, and do whatever checking was going to be done. What Dreyer saw as a complete breakdown in this last aspect was what he found most bothersome. Perhaps 6 percent of the final star positions have errors in them[130] that could have arisen

---

[126] Although the logs are filled with references to conditions of wind and weather that prevented observing (see X, 231 and XII, 275 for extended examples), Tycho seems not to have regarded his weather as any worse than elsewhere. In fact, he believed that the cold of the north offered an advantage (V, 135), in that it purified the air so that it was "often perfectly clear for several days running."

[127] XII, 48–54.

[128] XII, 77–9, 90–4.

[129] For the five days' observations of Venus (plus some for other planets visible at the time) cited in n. 127, the ratio of observations to hours (excluding breaks) was approximately 150/5, 225/3.75, 150/3, 175/4.25, and 200/4, respectively.

[130] Dreyer, "On Tycho Brahe's Catalogue of Stars," reported only that he traced the causes of fifty-five errors out of a thousand stars. Only a few more than that, however, are listed in

only from a failure to check each stage of the entire operation. It is probable that nothing short of constructing two independent catalogues could have ensured that the distances and declinations were copied accurately and paired up appropriately, the trigonometry done correctly, the conversions made without error, and the subsequent compilations and recopyings performed faultlessly.[131]

On the other hand, mere "care" with the details of such a complex activity would ordinarily have left many more bad coordinates (that is, disagreeing with modern expectations by more than 2′ to 3′) than Tycho's catalogue has, so there probably was some checking. In the procedure that Tycho used, it would have been natural to check by comparing "paper" distances between derived positions, with observations of the actual distance.[132] Some of the untidiness of the observing patterns may well represent either checks of this kind or reobservations to resolve disagreements among distances.

There can be no question but that cataloguing was a tedious business from beginning to end and that Longomontanus (who later said that he supervised it), Tycho, and the staff must have heaved a collective sigh of relief in 1592, when, after a few widely scattered nights of mop-up observation (including the assignment of magnitudes to each star), and no doubt a rash of computing, transcribing, comparing, and the like, the essence of Tycho's catalogue of 777 stars, reckoned to epoch 1589, was entered into his log.[133]

But how much time Tycho himself had invested in this first independent catalogue prepared in postclassical Europe[134] must be

his footnotes (to III, 344–73). Most are for the fourth- to sixth-magnitude stars, many of which were added in haste in the mid-1590s (see n. 138). Any serious investigation of the accuracy of Tycho's star places should distinguish between the two sets of coordinates. One recent attempt shows that Tycho placed his brighter stars with average absolute errors of 1.9 and 1.2 minutes in longitude and latitude, whereas for the fainter stars, the errors mounted to 2.8 and 2.6 minutes. See P. Rybka, *Katalog Swiazdowy Meweliusza* (Warsaw: 1984).

[131] A century later, Flamsteed used, in fact, different calculators who did not know one another. Even aside from calculations, errors of transcription are almost impossible to avoid in tasks of this magnitude. When editing Kepler's reprint of Tycho's catalogue in the *Rudolphine Tables*, Franz Hammer found – (*G. W.* X, 122) – twenty-five copying errors (that is, 2.5 percent) and missed at least five more.

[132] That is, if both stars *A* and *B* have been positioned by means of their distances from *C* (and their declinations), then computing the theoretical distance between *A* and *B* and comparing it with observation will check a great deal of work.

[133] XII, 231–65. The collection actually contains 723. The final catalogue (II, 258–80) included twenty-six stars (already catalogued) in Cassiopeia and a few others precessed to epoch 1600. Longomontanus's claim is in his *Astronomia danica* (1622), p. 202.

[134] A "John of London" established independent positions for forty stars in 1246: Paul Kunitzsch, "An Unknown Arabia Source for Star Names," in G. Swarup, Bag, Shukla, eds., *History of Oriental Astronomy* (Cambridge, England: Cambridge University Press, 1987), pp. 155–63.

seriously questioned. Observing of any kind was something he did so infrequently in later years that he had taken to identifying in the log those observations for which he had actually made the alignments.[135] Even the annotations in his hand, which show that he was supervising many of the concurrent planetary observations, are absent for the stellar observations. In regard to the data reductions, it would be straining the bounds of probability to presume that he performed more than an exemplary handful of them. So, when all is said and done, it appears to come down to Tycho's having obtained only as much as he was willing to pay for personally. In all probability, he turned the task over to his assistants once he had established the reference stars and perhaps the outlines of some of the constellations.[136] And the plans seem never to have included a star map, even though they had become sufficiently commonplace since Dürer's pioneering effort to inspire the publication of a collection of them by Galucci in 1588.[137]

There is also some evidence that in marked contrast with the rest of his work, for which only the very best was good enough, Tycho himself was willing to settle for something less in his star catalogue. He could certainly see from the logs that the stars were being observed twice, instead of the several times that the planets were routinely observed. He surely knew that the tables used for converting equatorial coordinates into ecliptic were never recomputed when he decided that the obliquity that connected them was $23°31\frac{1}{2}'$ instead of $23°31'$.[138] But on the other hand, much of the point in observing the planets was to determine how they were moving, whereas the stars were fixed. And if one wanted star positions accurate to the minute, then an outdated obliquity that could introduce maximum errors of half a minute could reasonably be viewed as being below the level of observational error. It is worth pointing out

---

[135] X, 127, 210, 243, 287; XI, 58, 103, 163, 279; XII, 18, 50, 72, 121, 211.

[136] As an indication of the semiautonomy of the cataloguers, we might cite (again) the situation in 1593 (XII, 275), in which the assistant entered five sightings and stated that further observation was prevented by cloudiness, and Tycho added "and indolence!!" (*pigritia*).

[137] Many "maps" were actually on globes, and Tycho also had the figures for the constellations engraved on his great globe. These figures may have had some historical impact, because Blaeu later introduced what have been called the "modern" figures for the constellations. Deborah J. Warner, *The Sky Explored: Celestial Cartography 1500–1800* (New York's Alan Liss, 1979), p. 28.

[138] Tycho mentioned the tables already at the beginning of 1586 (VI, 37). In 1587 and 1589 he wrote that he actually found an obliquity of $23°31\frac{1}{2}'$ (VI, 87) but used $23°31'$ for many years (II, 26). That he continued to use the $23°31'$ for the star catalogue (Chapter II in the *Progymnasmata*) after officially adopting the higher figure for the solar theory was revealed in an unpublished statistical analysis by Moesgaard. The reason is clear from his reference to his conversion tables on II, 227.

that Tycho certainly thought his final positions were good to the minute and that at least one modern commentator has estimated that they were.[139]

When Tycho decided in the mid-1590s that he should obtain enough additional positions to be able to say that he had catalogued a thousand stars, just as the ancients had, he recognized that the faintness of the stars and the haste in which they had been observed militated against their being up to standard and therefore did not actually place them in his catalogue.[140] If those standards were not as high as Tycho thought they were, they were nevertheless very high. Even with the half-measures described, Tycho was able to reduce the errors of Ptolemy and Copernicus[141] by fully an order of magnitude, that is, a factor of at least ten. When the time came to boast about his star catalogue, he could point to comparisons of his own positions with Copernicus's for thirty-six of the brighter stars placed strategically around the heavens that showed differences averaging 18' in latitude and 25' in longitude.[142] And when Johannes Bayer published the celestial atlas (*Uranometria*, 1603) that inspired modern

---

[139] V, 112–14; VII, 281, 361; VIII, 106, 209. A. Pannekoek, *History of Astronomy* (London: Allen & Unwin, 1961), p. 215, stated that "their mean error, by comparison with modern values, appears to be about 1'."

[140] The goal is expressed clearly in the final observations, made in 1597 and labeled "pro complendo millenario" (XIII, 98). The previously observed additions were catalogued at the end of the observations for 1596 (XIII, 61–77). The entire thousand was published by Kepler (with a few additions) in the *Rudolphine Tables* (105–41) in 1627. In 1598 and 1599, Tycho "presented a few beautifully made MS. copies (probably less than twenty) to princes and other influential people who might be of use to him in his self-imposed exile." J. L. E. Dreyer, "On Tycho Brahe's Catalogue of Stars," p. 231, n. 123.

[141] It is interesting to see that Tycho (II, 281) anticipated the argument that is usually credited to Delambre, namely, that Ptolemy's star catalogue was merely the appropriately precessed positions from Hipparchus's catalogue. For the classic "disproof," see Vogt, "Versuch einer Wiederherstellung von Hipparchs Fixsternverzeichnis," p. 23. For what appears to be the modern proof, see Robert R. Newton, *The Crime of Claudius Ptolemy* (Baltimore: Johns Hopkins University Press, 1977), pp. 245–54, and Y. Maeyama, "Ancient stellar observations (of) Timocharis, Aristyllus, Hipparchus, Ptolemy – the dates and accuracies," *Centaurus* 27 (1984), 280–310. For more recent consideration of the subject, see James Evans, "On the Origin of the Ptolemaic Star Catalogue," *Journal for the History of Astronomy* 18 (1987): 155–72, 233–78. Copernicus, of course, did exactly that with Ptolemy's catalogue. But because he used a different theory of precession than Alfonso's astronomers had, their longitudes still differed.

[142] The comparisons are on III, 386–7. The differences cited are actually low because in order to counter the selection effects that probably exist, one grossly erroneous (3°25') longitude and one latitude (1°54') were arbitrarily excluded from the averaging (which, of course, is not the way in which Tycho conceptualized the data). In fact, a quick check of the remaining twenty-nine second-magnitude stars shows longitude differences averaging the same 25' (again, rejecting one problematical position: γ Andromedae). The differences would be even greater for fainter stars, if Ptolemy had as little success with them as Tycho did.

stellar denominations, it was Tycho's positions that constituted the basis of his catalogue.[143]

Not least of the circumstances that extenuate the various sins of omission associated with Tycho's catalogue is the fact that during the entire period of its preparation, Tycho saw it as the last obstacle to the publication of his book on the new star. As of the beginning of 1589, he had been telling himself for two years that the book was essentially finished.[144] Wishing to get it into print, he decided that he simply could not be held up any longer by the mere fact that the manuscript was not completed. Thus, even before the long discussion of the star catalogue was written, Tycho had started the printing. But this necessitated other writing, for in the process of seeing the 850-page book through the press, Tycho found that it was not as finished as he had thought it was.

In nearly every chapter he found reasons to interpolate at least references to more recent findings, insights, or events.[145] Some activities, such as compiling his table of refractions for stars[146] and computing six examples of lunar parallaxes (to show that parallax was observationally detectable at the moon)[147] amounted to minor research projects in themselves, and Chapter VII was completely rewritten. And because this twenty-four-page section on the size of the nova necessitated Tycho's expounding the details of the astronomical world as he saw them during the later years of his life, it demands closer examination than does the rest of the book.

Like all of his contemporaries, Tycho viewed the physical world as an articulated unit created by God. For the most part he subscribed to the account of its workings synthesized by Aristotle and adapted by the medieval church. He was, however, strongly attracted to certain alternatives presented by Paracelsus (probably because they linked humanity to the physical world even more closely than Aristotle's doctrines did). They concerned primarily the terrestrial sciences of astrology and alchemy (medicine, for Tycho). But the Paracelsan concept of a dynamic universe, capable of spawning new creations or

---

[143] Noel M. Swerdlow, "A Star Catalogue Used by Johannes Bayer," *Journal for the History of Astronomy* 17 (1986): 189–97.

[144] Already in January 1587 (VI, 112), Tycho had told Hayek that he was preparing Volume I for the press, but only in March 1588 (IX, 62) did he write in his diary, "End of Vol. I." By August, however, he was telling Hayek (VII, 122) that the book would be ready for distribution at the next spring book fair at Frankfurt. Even in February 1589, Tycho was blaming his lack of progress on inadequate supplies of paper (VI, 180).

[145] The view expressed here is a compromise (based on Tycho's logs) between *Norlind's* (144–50) argument that everything was done except for interpolations and *Dreyer's* belief (186) that everything was written when the references to time were made. The references are on II, 301, 331, 383, 389, 397–9, and III, 17, 119, 130, 258.

[146] II, 287.   [147] II, 411–13.

extinguishing old ones, was obviously conducive to Tycho's interpretation of the new star. The view that the heavens consisted of a celestial element of fire – as opposed to the Aristotelian idea of a terrestrial sphere of fire immediately below the moon – could almost be taken as a prediction that fiery objects such as comets would be found to exist above the moon.[148] It may also have been the basis of his belief (following al-Bitrugi) that Mercury (at least) was self-luminous.

In general, however, Tycho developed his view of the astronomical world by altering the Aristotelian picture, under minimal influence from Paracelsus. Unable to believe that there could be any kind of sphere of fire above him, either Paracelsan or Aristotelian, he decided that air must extend essentially up to the moon and then give way to an ether with all the ideal properties postulated by Aristotle.[149] This division was not an a priori one. Tycho was willing to believe that the transition from one to the other was gradual. But he did believe that there had to be a transition, for if the celestial regions were also filled with air, it would unbalance the system of elements, resist the movement of the planets, and destroy the celestial harmony, and probably even produce perceptible sound.[150] Besides such a boundary was necessary to provide the optical interface required to account for astronomical refraction.[151] Unhappily for Tycho, who did not like to occupy himself in quarrels that could not be settled by an appeal to numbers and felt that physics was not the province of astronomers anyway, Rothmann challenged this rather standard explanation by pointing out that it would produce refractions all the way up to the zenith.[152] Tycho certainly agreed that refractions disappeared long before the zenith; but the best he could do to account for the evidence was to adapt Rothmann's idea of a lower atmosphere consisting of impure, watery, and therefore denser vapors.[153]

---

[148] IV, 282–3.

[149] II, 76–7, 377; III, 246, 305; VI, 93, 135–9; VII, 230. Mary Ellen Bowden, who is engaged in a general study of early seventeenth century "astrometeorologies" is of the opinion that the discarding of the Aristotelian fiery sphere was one of the most critical breakthroughs of the sixteenth century.

[150] VI, 135.

[151] This was Tycho's view (VI, 39) taken from "the opticians" Alhazen and Witelo, until Rothmann challenged him.

[152] VI, 56, 92–3, 111.

[153] Neither Rothmann nor Tycho seems to have understood fully that any theory that did not simply postulate some kind of "threshold angle" (say 45°) of incidence above which there was no refraction (both assumed a point of refraction) would imply refractions all the way to the zenith. Rothmann seems never to have realized that merely moving the interface down from the moon to some vaporous layer much nearer the earth solved nothing.

Above the moon, Tycho's view was traditional, except, of course, for his intrusion of the new star and comets and his rejection of spheres. Naturally, however, he felt entitled and even obligated to convey his understanding of things so as to provide a complete context for his placing of the new star. This he presented in Chapter VII of the *Progymnasmata*. Presumably his presentation of Ptolemy's and Copernicus's figures for the sizes of the sun and moon, and al-Battani's and al-Fragani's distances and sizes for the other five planets and the stars,[154] was written up with the main body of the text in the mid-1580s. Because the only appropriate observations in Tycho's logs are from January 1590, however, it seems clear that the bulk of the chapter, listing Tycho's values, was written around that time.[155]

Following the order of his predecessors, Tycho started with the "luminaries," the sun and moon. Observations with a *camera obscura* some eight meters long showed that the apparent diameters of the sun varied from just under 30' at apogee to just over 32' at perigee. The values were about 2' lower than Ptolemy's but, surprisingly, no better, as the mean value is actually 32' (instead of Tycho's 31') and as Tycho, like everyone before him, felt obligated to make the ratio between his high and low values conform to the ratio of distances implied by the eccentricity of his solar theory.[156]

In order to convert these angular results into linear size, Tycho needed to know the distance to the sun. For several reasons, 1,150 earth radii seemed a good mean figure. It was between Ptolemy's (1,165) and Copernicus's (1,142); it agreed well enough with his own determinations from precisely observed eclipses; and it was a nice approximation (Tycho did not want to be thought too mystical) to the 1,152 e.r., or $(24)^2$ earth *diameters*, derived plausibly some years earlier from Pythagorean–Platonic numerological considerations,

Tycho tried to come to grips with the issue by appealing to a ratio of path lengths in the vaporous and nonvaporous (diaphanous) portions of the air, but he left implicit, at best, the idea that below a certain ratio there was no refraction. Tycho was either not sufficiently proud of or sufficiently interested in these ideas to give them the same exposition in the *Progymnasmata* (II, 76–7) that he did to Rothmann (VI, 135–9, 167–9).

[154] II, 415–20.
[155] XII, 30, 39, 44, 46, 47, 72, 105.
[156] II, 420–1. Observations (XII, 108–13, 117–18) made in 1591 with a "new large wood channel" may have been made only to confirm what was already done, for Tycho said that Chapter VII was printed by October 1590 (VII, 276). Naturally, it is impossible to know how he evaluated his various results. But the ratio of his raw ranges (e.g., $24\frac{5}{12}$ to $25\frac{1}{6}$) seems much closer to actuality than it does to Tycho's reported range (30 to 32), which suggests that he relied heavily on his solar theory – with an unbisected eccentricity – to derive his "empirical" results.

by a writer named Offusius.[157] When applied to the 31' mean diameter, this distance made the sun $5\frac{14}{75}$ times the radius of the earth and 139 times its volume.

The numbers for the moon Tycho found much more complicated – too complicated, in fact, to go into in detail for the purpose at hand. What he really meant was that although he could read, almost without realizing it, the $\frac{1}{2}'$ adjustments into observations of the sun's diameters that were necessary to confirm its eccentricity, there was no way to do so for the moon. Ptolemy's model implied distance variations of nearly two to one, and Copernicus's alteration, although it was the most enthusiastically received feature of *De revolutionibus* during the sixteenth century, still implied that the lunar distances should vary from 68 to 52 e.r. Although Tycho eventually constructed a model that reduced the extremes to 60 and 52[158] and hence to a ratio much nearer that of the observed diameters, he had not yet done much on the theory of the moon. Accordingly, he contented himself with using median values of 33' (diameter) and 60 e.r. to obtain a very reasonable $\frac{2}{7}$ e.r. for the radius of the moon.[159]

For the rest of the planetary distances, Tycho's numbers departed considerably from the ancients', because the latter of course, had not known the true system of the world. Again, the actual numbers had to be taken from the existing literature (via Copernicus), as Tycho had not yet made the technical determinations that, with his adopted distance of the sun, would give the sizes of the planetary orbits. The angular diameters, however, were matters mainly of observation, and so Tycho assigned them to his sharp-eyed observer, the instrument maker Hans Crol.[160] When the data were in, Tycho derived the angular diameters, distances, and sizes listed in Table 9.1.[161]

The crucial elements in his computations were the (essentially

---

[157] II, 421–2, and editorial note. Both Tycho and his principal assistant of later years, Longomontanus, liked the idea of perfect numbers and celestial harmony. See VIII, 46; and K.-P. Moesgaard, "Tychonian Observations, Perfect Numbers, and the Date of Creation: Longomontanus' Solar and Precessional Theories," *Journal for the History of Astronomy* 6 (1975): 84–99.

[158] Tycho's chart showing all values is on II, 131. [159] II, 423.

[160] XII, 30, 39, 44, 46, 47, 72, 105. See Tycho's reference to Crol's skill on XI, 292. A few observations were also made by Flemløse, and a couple by Tycho himself.

[161] For his representative distances Tycho chose the average of each planet's two extremes from the earth. In the case of the inferior planets, this meant using the sun's mean distance. For the superiors, Tycho's results are consistent with the ratios given by Copernicus in Book V: Chapters 9, 14, and 19 except that the 3,990 for Jupiter seems to be a misprint for 5,990. (However, see XII, 280, for the *use* of this value in a calculation of 1593.)

The Lord of Uraniborg

Table 9.1. *Tycho's Estimates of the Sizes and Distances of the Planets*

	Diameter (mins.)	Distance (e.r.)	Radius (relative to earth's)	Volume (relative to earth's)
Mercury	$2\frac{1}{2}$	1150	$\frac{3}{8}$	$\frac{1}{19}$
Venus	$3\frac{1}{4}$	1150	$\frac{6}{11}$	$\frac{1}{6}$
Mars	$1\frac{2}{3}$	1745	$\frac{12}{5}$	$\frac{1}{13}$
Jupiter	$2\frac{3}{4}$	3990[a]	$\frac{12}{5}$	14
Saturn	$1\frac{5}{6}$	10550	$\frac{31}{11}$	22

[a] See footnote 161.

Ptolemaic) distance of the sun and the orbit geometry of the Tycho-nic system. The former ensured that none of the distances that Tycho derived would differ greatly from the traditional values taught to every university student. The latter had the effect of generally compressing the planetary system, by pushing the inferior planets outward to the mean distance of the sun and pulling the superior planets inward to the distances computed by Copernicus. Tycho's only comment was that these changes had reciprocal effects on the physical sizes previously understood for the planets.[162]

Because Tycho could not find any diurnal parallax for the stars, he had no direct way of getting information on their distances. He had no doubt, however, that they were situated just beyond Saturn. Much of his unwillingness to consider Copernicanism stemmed, as we have said, from his inability to imagine that God would have created a universe containing as much wasted space as the Coperni-can scheme and the absence of annual parallax required.[163] In order to establish the minimum distance of the stars, therefore, it was necessary to determine the maximum distance of Saturn. Here, again, as Tycho had derived no parameters for Saturn, it was necessary to fall back on Copernicus's. And in order to render the numbers more intelligible, it seemed desirable to present a diagram of the theory, which Tycho could do because he had settled on the double-epicycle form of Copernicus's mechanism.

Accordingly, Tycho provided a picture (Figure 9.1) of the most remote parts of the universe as he envisioned them. The technical

---

[162] II, 425. For a general discussion of the history of this problem, see A. van Helden, *Measuring the Universe* (Chicago: University of Chicago Press, 1985). For the traditional distances of the planetary spheres, see note 176.

[163] The most outspoken expression of this view is in III, 63, where Tycho said that if there were no other absurdities in the Copernican system, the necessity of assuming such a vast distance to the stars would be enough to rule it out.

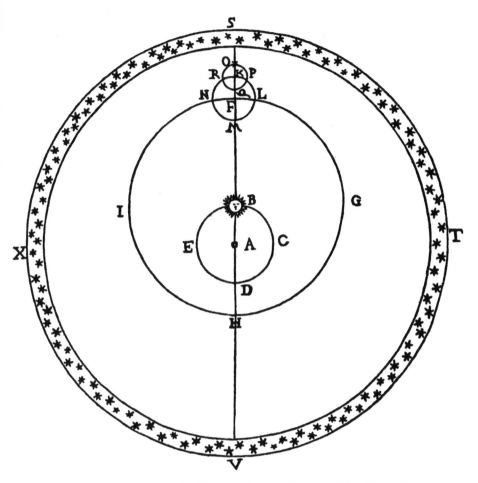

Figure 9.1. Copernican double epicycle model proposed by Tycho for his planetary theories.

aspects of the planetary theory need not detain us here, once we have noted that Tycho's engraver depicted the planet (O) in an impossible position.[164] Tycho described the technical aspects well enough (and correctly, his diagram notwithstanding) to show that Saturn should reach a maximum distance of about 12,300 e.r. and then suggested 13,000 e.r.[165] as a reasonable minimum and 14,000 as an average

[164] II, 426–9. Tycho gave the correct, purely Copernican account (although *Dreyer*, 180–1, described it erroneously) and even mentioned explicitly that the planet could never be at O.

[165] After mentioning that the planet could never be at O, Tycho stated that the apogee point on the epicycle is at a distance of 12,900, so that the stars could not be closer than 13,000.

distance for the stars (for there was no reason to assume that all had to be equally distant). Although these distances seemed inconceivably immense, the traditional view placed the stars around 20,000 e.r. away, and if the Copernican speculations were true, they would multiply Tycho's distances beyond credibility.

By overestimating (as Tycho saw it) the distances of the stars, the ancients had also exaggerated their sizes. Moreover, they had assumed that all stars of a given rank were the same size, which, as Tycho pointed out, was clearly not the case, since there were observable differences in brightness and unknowable variations in the distances of stars of the first magnitude. With these problems noted, however, Tycho presented his estimates of the sizes of the stars: First-magnitude stars were $2'$ and 4 earth radii; second- through sixth-magnitude stars were successively $1\frac{1}{2}'$ and 3 e.r., $1'$ and $2\frac{1}{5}$ e.r., $\frac{3}{4}'$ and $1\frac{1}{2}$ e.r., $\frac{1}{2}'$ and 1 e.r., and $\frac{1}{3}'$ and $\frac{2}{5}$ e.r.[166] Was it possible, instead, Tycho wondered, that all stars might be intrinsically the same size and differed in apparent magnitude only because they differed inversely in distance? It seemed unlikely. Given the ratio of six between the apparent diameters of first- and sixth-magnitude stars, the eighth sphere would have to have a diameter of 155,000 e.r.,[167] and Tycho simply could not conceive of such vastness.

Although it may seem anomalous to find Tycho speaking of the eighth sphere after he had renounced the other seven, and in a book in which he mentions that renunciation at every opportunity, the anomaly lies as much in Tycho's vocabulary as in his universe. Scientific terminology is notoriously tenacious. However Tycho conceived of the stellar cosmos, he was likely to retain the term *sphere* as a reference, even if he did not mean it as a description.[168] And there were indeed some senses in which he did mean it as a description. Certainly he thought of (and depicted) the stars as ordered into a sphere (or spherical shell). As a believer in a geocentric universe, he almost had to believe that the stars were interconnected in some way, for imagining that thousands of stars rotated in unison around the earth daily, in complete independence of one another, would have been absurd.

Thus when the *landgrave* suggested that the gross differences between some of the ancient and modern star places might be due to individual motions, Tycho had responded that it seemed unlikely,

---

[166] II, 431. These sizes were entirely due to the optical limitations of the human eye.
[167] II, 432. The figure is rounded from a radius of 6 (that is, $2'$ divided by $\frac{1}{3}'$) times 13,000.
[168] In his several references in the *Progymnasmata*, Tycho stated almost interchangeably that there were no "real" spheres and no "solid" spheres for the planets: I, 378, 398–9; III, 90–1, 111.

that both the constellations described qualitatively and the colineari-
ties mentioned specifically by the ancients were still intact after all the
intervening years.[169] But he did not picture a traditional crystalline
vault. He was willing to assume not only differing distances for the
stars, and rotations to account for their scintillation,[170] but even
differing densities of material in the heavens. When concentrated, it
formed a star.[171] Otherwise it was so well dispersed as to be invisible
and unresistant to the daily motions of the stars, except in nebulous
areas such as the Milky Way. There it was thick enough to be seen
and to allow the very occasional formation of something like the
new star, even if the result was imperfect and therefore only
temporary.[172]

The new star had been in the eighth sphere. During its appearance,
Tycho had not had an instrument capable of measuring diameters,
but he and others had made enough comparisons with Jupiter and
Venus to be able to adopt a maximum value of $3\frac{1}{2}'$ with some
confidence. Assuming an average stellar distance of 14,000 e.r. – for
having shown no parallax at all in numerous trials, it could scarcely
be any closer – it would have been $7\frac{1}{8}$ times the diameter of the
earth and approximately 360 times its volume.[173]

This enormous size was another proof that the Ptolemaic estimate of
the distance of the stars, 20,000 e.r., was excessive, for the star would
then have to have been more than a thousand times the bulk of the earth.
With age, of course, it shrank, so that by the beginning of 1575 it was the
size of a fifth-magnitude star (and the earth) and after a few more
months was too small to be seen at all. How some people could follow
Seneca in suggesting that its apparent decay was actually the manifestation of a motion straight up from the earth, Tycho could not
understand, for in addition to the impossibility of such rectilinear
motions in the heavens, the star would have to have been at least 300,000

[169] VI, 49, 52, 71. Tycho told Rothmann (VI, 94) that "all are in one orb and are moved
uniformly around one and the same pole." See also II, 234–6.

[170] The conventional explanation was that the stars scintillated because they were farther
away. Tycho could not believe either at the beginning of his career (*De nova stella*: I, 29) or
at the end (II, 375–7) that the difference in distance between the stars and Saturn could
be enough to be crucial. He inclined toward a view he ascribed to Plato, that the stars
revolved on their axes, which caused intermittent problems in the transmission of light
through the intervening medium (I, 29).

[171] VII, 235–6. Tycho believed that stars were self-luminous and probably intrinsically
brighter than the moon (VI, 171).

[172] III, 305. Tycho alluded to a gap in the Milky Way about the size of the half-full moon,
from which he thought the new star took its material. Similarly, he thought the comet of
1585 arose from the "great conjunction" of 1583 that occurred at about the same place. He
thought that comets were not in the eighth sphere because they moved: VII, 267.

[173] II, 433.

e.r. higher when it disappeared than it was when it was first sighted.[174] One might almost as well be a Copernican if he were to credit such a distance.

In view of the extent to which Tycho occupied himself with the question of astronomical distance during his career, and indeed, he could almost be said to have built his career on it, it is rather appalling to note the casualness with which he actually handled the subject. The distances he sought – and quoted – could be obtained (rigorously) only by the measurement of parallax. Yet when we examine his "determinations" for the comet of 1577, the sun, and Mars at opposition, we see that all of them were flawed. In fact, only for the extreme cases of the new star and the moon can Tycho be credited with unexceptionable treatment of the issue, and in the strictest sense, the list begins and ends with six examples for the moon compiled around 1590.[175]

To be sure, the other objects of Tycho's attention were sufficiently remote, and their parallaxes (reciprocally) sufficiently small that Tycho could not possibly have made convincing detections of them, anyway. But this does not mean that attempts on them were bound to be pointless. On the contrary, Tycho's proof that the new star had no parallax at all was much more significant to his contemporaries than any number of "successful" measurements of parallax for the moon could have been. And a similar null determination for Mars, Venus, or the sun would have been almost as devastating for the (non-Copernican) sixteenth-century view of the cosmos. However, such an achievement was not as accessible to Tycho as one might assume. The very fact that a null result would have been so contrary to Tycho's expectations was certainly a significant barrier in itself.

In the case of the new star, the fact that it neither moved like nor resembled the other "meteorological" phenomena, with which Aristotelian theory identified any object so obviously a product of change, at least prepared Tycho for the possibility that it might be a star and hence show no parallax. For the planets, there was no such alternative to hint that the 3′ solar parallax and all the planetary parallaxes derived from it might be equally and completely fictitious. When Tycho was faced with observations of the comet of 1577 that showed it to have too little parallax to be below the moon and too much motion to be at the other (stellar) boundary of the cosmos, he instinctively strained his observations until he found one that yielded the few stray minutes necessary to give the comet a definite "place" in what he considered to be the well-known geography of the Ptolemaic universe.[176]

[174] II, 435.   [175] II, 411-13.
[176] See Chapter 8 for the lower limits (in earth radii) of the successive spheres. Tycho's final version of the list is in II, 417.

Even though Tycho could never have made a convincing discovery of the amounts of parallax actually available to him,[177] he certainly could have shown that the amounts assumed by his predecessors were wrong, if parallax had been a more straightforward issue. The problems lay not in detecting the several minutes (variously) that should have been accessible to Tycho's instruments but in deciding how the contributions of the several small parameters associated with the observations should affect the result. For the sun, we saw that Tycho removed the possibility of detecting horizon-to-horizon parallax by building the parallax corrections into his table of refractions. Because the low-altitude refractions necessitated corrections (on the order of 20' to 30') that would have swamped any other parallax that he tried to find, Tycho never tried to use the entire baseline that was theoretically available to him. Instead, he tended to use about half of the horizon-to-horizon swing, by observing planets at positions three to five hours on each side of the nonagesimal, under which condition he thought that he could generally ignore refraction. His usual technique was then to compare the actual change in position with the expected change due to the planet's proper motion,[178] just as he had done for his studies of comets. The expected motion could be estimated from theory and checked by interpolating from positions taken before and after the night of the trial. But the quantities were small and the operations were tricky. When Tycho used the method for comets, he convinced himself that the comet's motions in opposition showed no decrease reflecting the motion of the earth and therefore constituted disproof of the Copernican system. When he used the method for Mars (in 1583) whatever he achieved was so far from being convincing that he was able to reverse his conclusion (in 1587, as we have seen) with no apparent struggle at all.

In the midst of several days of intense observations in 1590, Tycho stumbled over something. He noted that some observations seemed to show almost no parallax and that the matter should be looked into further.[179] But there is no record that he ever did. The series of tests recorded a few years later seems to have been made with the idea of deciding whether special tables of refraction were needed for the planets. Such a situation would not have been preposterous for

---

[177] The maximum is about 25" (for a "swing" of 50") displayed by Mars at (perihelion) opposition. It could not be detected convincingly until the late eighteenth century. See Tycho's putative version of the trial in VII, 263. But see also VII, 294.

[178] For a trial made at one of Mars's stations, the proper motion would not have been a problem. But at that time, even the Copernican geometry places Mars farther from the earth than the sun is, so that there would be no point in making such a test.

[179] XII, 52.

Tycho, because he was willing to comprehend refraction as the effect of a given medium on a given ray and to accept that rays of differing intensities might be affected differently by the same medium.[180] As we have seen, he had already found that starlight suffered less refraction than sunlight did (because it did not require any parallax corrections) and thus constructed sometime in 1590 a special table for the stars.[181] Sometime after 1595, he drew up an analogous table for the moon.[182]

In his observations, Tycho treated the parallax corrections as known quantities, routinely using 4′ values for Venus and, for a few horizon situations, using 10′.[183] Yet, not even these figures created enough problems to attract Tycho's attention or even stimulate any compensatory corrections for refraction. Influenced, on the one hand, by observations of more distant planets with smaller parallax corrections and, on the other hand, by the theoretical assumption that all planetary light would be affected in the same way, Tycho decided that its refractions were similar enough to the refractions of starlight – which had been tabulated with the aid of no parallax corrections at all – to permit use of the same table.[184]

So, despite all the difficulties involved, Tycho had his chance to raise some serious doubts about parallax if he had been willing to open his mind to them.[185] At the same time, however, Tycho would

---

[180] See IX, 145, where Tycho assumed that the moon must be suffering greater refraction than Aldebaran is at the same altitude and vowed to try to determine whether the crucial variable was the intensity of the source or the distance the light has traveled. Tycho regarded impurities in the air as the cause of refraction.

[181] At the end of 1589, after some rather extensive testing of stellar refractions, Tycho constructed a partial table (XI, 380) that does not agree with the one he printed in the *Progymnasmata* (II, 287). Virtually all tests thereafter, however, were for the planets (XII, 153–5, 224–6).

[182] The table is in II, 136. Its values differ only slightly from those for the sun. As late as 1595 (XII, 388) Tycho was simply using solar refractions for the moon.

[183] XII, 70–1, 224–5, and especially 463, 468, in which latter place Tycho assumed a distance of 319 e.r. and derived a theoretical parallax correction of $10\frac{3}{4}′$.

[184] Tycho was rather tentative about the equality in II, 287. That he had not made up his mind at the beginning of 1595 (XII, 463) is indicated by a remark interpolated into his logs: "I later found the refractions of Venus not to be greater than for the fixed stars at similar latitudes."

[185] Tycho's mind was never quite open on the subject. Already in 1578 (X, 82–3) he computed tables of parallax based on Copernicus's distance to the sun and changed them only to take account of his own very slight modification of the eccentricity of the sun's orbit (II, 389). He was, however, in distinguished company. N. S. Swerdlow, "Al-Battani's Determination of the Solar Distance," *Centaurus* 17 (1972): 97–105, showed that al-Battani backed off from a logic for deriving parallax that would have produced a much greater distance than Ptolemy had adopted, and Janice Henderson, "Erasmus Reinhold's Determination of the Distance of the Sun from the Earth," *Westman*, 129, stated that both Copernicus and Reinhold consciously sought a solar distance that would agree with Ptolemy's.

have been astonished by the suggestion that he had never seriously checked the solar and planetary parallaxes. In his view, the numerous successful reckonings of solar and planetary positions – particularly those of the sun in precisely predicted eclipse[186] positions – that he made during his lifetime all involved theoretical parallax corrections, and all, therefore, constituted implicit tests of the values used. The result was that it was only for the moon, where the parallaxes reached proportions comfortably above the "noise" of Tycho's observing conditions, that Tycho really came to grips with parallax. There he fought through all his uncertainties, including those arising from the moon's motion in latitude, to achieve the first real measurements of astronomical parallax.

[186] Tycho's reference to eclipses reinforces the supposition that in all considerations of parallax, Tycho thought operationally – that is, looked at the total apparent motion of the planet as a sum of various theoretical components. The apparently anomalous fact that he specifically cited lunar eclipses (II, 383, 413) suggests that he was thinking of the role of the size of the earth's shadow cone in eclipse timings.

Chapter 10
The Theory of the Motion of the Moon

I F the five-year delay imposed by the initial requirements of getting established on Hven had gotten Tycho off to a slow start in his work, that situation was only a memory by 1590. By this time, Tycho was well under way in his long-envisioned renovation of the whole science of astronomy. Already in 1588 he mentioned plans for presenting it in the form of a mighty *Opus astronomicum*, a sort of "New Almagest" that would begin with a discussion of his instruments and proceed from there in much the same way that Ptolemy and Copernicus had done. Even then, Tycho recognized that he was still a long time – five or six years, at least – from this achievement, because he still lacked some of the requisite observations for the slower moving (superior) planets.[1]

Each year, however, there was more competition for his research time, generally from projects he had not originally conceived as part of his "redintegration." Of longest standing was his cosmological trilogy on the "more recent celestial phenomena," designed to prove that the prevailing Aristotelian view of the world was simply untenable, that there were no comets below the moon, no solid spheres above the moon, and, by implication, no reason to adhere to any other aspect of traditional, largely a priori, dogma.

At the beginning of 1590, Tycho was still heavily mired in this undertaking. Volume II (*De mundi*) had been printed and was in the hands of various friends, dignitaries, and acquaintances. But even its status became uncertain when Tycho composed an addition to it in the fall of 1589 to answer rebuttals, made by an obscure Scots physician (John Craig), that Tycho apparently feared might be sufficiently representative and/or credible to undermine his arguments.[2] Volume III, on the comets that had followed the great one of 1577, was little advanced beyond the stage of raw observation, and Tycho actually lost ground on this project when another comet appeared for twelve days in February and March 1590.[3] Volume I had been under final

---

[1] Tycho's fullest discussion of his plans is in a letter to an old teacher, Caspar Peucer: VII, 132.

[2] Craig's original letter (VII, 175–82) is dated 1 May 1589. Tycho's lengthy response, published from manuscript in IV, 417–76, was returned in the fall and sent in duplicate to Hayek in November (VII, 216). Between the plans announced for printing it (VII, 225) and numerous later references to it (especially II, 439), *Dreyer* (IV, 416) thought it was actually printed in 1591, but *Norlind* (134–42) argued convincingly against it.

[3] See III, 27, and V, 23 (among many) for references to this third volume. So much

editing for some time but was expanding in scope about as fast as Tycho was completing chapters for it.

Tycho had already begun printing the bulk of the completed manuscript, as we have seen, even while he was still researching and composing – or rewriting – Chapter VII. In order to hasten the task, he had started through his press all three parts of the book concurrently.[4] By estimating the printed lengths of the various sections of manuscript, he could paginate Parts II (ending with the incomplete Chapter VII) and III. If he could have contented himself with this plan, the project would have gone more smoothly for him than such a stunt deserved, but for some reason he introduced one more irregularity. While he was printing Chapter I (the solar theory) and before he could have finished even writing Chapter II, let alone compiling the star cataloguing described in it, he decided to start printing Chapter II.

This last split turned out to leave him with too much space for the solar theory and too little space for the star catalogue. The latter problem he circumvented relatively easily, by simply using recto and verso letterings for each of sixteen pages.[5] The former could be solved equally simply, Tycho decided, by adding a concise review of the virtues of the Gregorian calendar reform. The calendar was, after all, sun based, and had been a very controversial issue since the introduction of the reformed version into Catholic – but not Protestant – Europe on 5/15 October 1582. Reverberations had reached Hven by the spring of 1584, when Tycho asked Brucaeus to try to obtain for him Mästlin's tract against the new calendar, and shortly thereafter received a letter from Johannes Major describing the controversy that had accompanied its introduction at Augsburg.[6]

Unlike such Protestant colleagues as Mästlin (whose fulminations earned all of his writings a place on the church's Index of Prohibited Books in 1590),[7] Tycho could not see why the calendar should be a matter of concern to theologians or, conversely, why people should

   manuscript material has survived Tycho that the absence of any part of this book makes it very doubtful that Tycho ever actually started it. All the observations (and some calculations), however, are printed in XIII, 305–93.

[4] The printing had begun by February 1589 (VI, 165). That it was concurrent is related by Kepler (III, 320–1) in his appendix to the work and borne out by a few references to time that Tycho made as he went over each section before sending it to the press, for example, the beginning of Part III (III, 74), where Tycho refers to "circa finem huius 1589 Anni." Everything except Chapter X was apparently printed by October 1590 (VII, 276). See also VII, 274, 281, 302.

[5] The doubled pages are 257–72 (II, 257–80). Other adventures of the composition and printing of the *Progymnasmata* are related in *Norlind* (144–50).

[6] VII, 82–3. Tycho's response is on VII, 86–7.

[7] *DSB* IX, 168. For the abbreviations of frequently used sources, see Appendix 1.

care now if they used a papal calendar, when they had been using one all along that had been established by the popes, anyway. As he told Major by return messenger, if Regiomontanus had worked out a papally sponsored reform before the Reformation, Luther would not have dreamed of altering it, because it contained nothing contrary to his teachings. So, why, Tycho wondered, was there so much fuss from his followers now? It was technically imperfect, but the church could not be blamed for the shortcomings of the astronomers. As far as Tycho was concerned order was the most important thing, and because the pope had more authority than anyone else did, he was the person who could most easily establish a universal convention.

In addition to his cosmological trilogy, Tycho was also occupied with a suggestion made by Rantzov three years earlier,[8] that he publish his scientific correspondence. Because one of the benefits of this undertaking would be the registering of his conviction that Ursus had plagiarized his world system, Tycho was eager to press ahead and had not only started soliciting letters but even started printing some of them.[9]

Not content with these manifold pursuits, Tycho also began in 1590 the research phase of his work on the planetary theories. And when he quickly made what he thought was a breakthrough that augured well (see Chapter 13), he was induced to plan a further augmentation of his publications. Volume I was, by February 1592, so close to completion that Tycho printed up the title page and sent it to friends. An extant copy reveals that the already-printed Volume II would have an appendix (in addition to the *Apologia* against Craig already mentioned) treating the three superior planets, Volume III – which was still no further advanced than it had been in 1590 and in fact was never even partially composed, let alone printed – would contain the theories of the inferior planets.[10] A letter sent with the title page to Hayek reveals that the last of the planets, the moon, would also be included in the set: It would go into Volume I in the three-quarto lacuna at the end of Chapter I,[11] instead of the formerly planned disquisition on the Gregorian calendar.

As Tycho's projects multiplied, so did his other activities. In order to keep his printing press staffed and supplied with paper, Tycho had to circulate correspondents such as Rantzov, Brucaeus, and Vedel almost continuously. Thus, when he initially offered to print materials at his own expense for Vedel and Hayek, and Hayek took him up on his offer, Tycho had to beat a hasty retreat and give his friend

[8] VI, 29–30.   [9] VI, 190; VII, 216–18, 223, 230.
[10] II, 439–40. See also VI, 307, and VII, 327.   [11] VII, 351.

all the reasons – heavy social schedule, backlog of his own printing, shortness of paper – why he could not print things for him. Tycho subsequently received no letter from Hayek for over two years and clearly feared that reneging on his offer had alienated him. When a very anxious letter praising his friend's writing and promising to cite his work elicited no response, he wrote a second one asking specifically for a reply. Finally he learned that the problem was only that two successive letters from Hayek had miscarried.[12]

Building and maintaining a professional library required even greater exertions, even though Tycho was frequently able to send people such as Vedel, Aalborg, or Flemløse to Frankfurt for the book fairs. Brucaeus, Hayek, and the Augsburg correspondents had standing orders – probably of the kind Tycho had given Camerarius early on, for "anything mathematical or chemical, old or new"[13] – to buy books for him.

As Tycho's printing press began to turn out finished material, Tycho began to worry that someone might pirate his works. In 1585, he asked Hayek to request from Rudolph II a blanket copyright (for the Holy Roman Empire) on all books printed at Uraniborg. He assured him that none of the books would contain any material that was either religiously or politically objectionable and left Hayek to do whatever was necessary.[14]

Even with Tycho's status and Hayek's connections, however, this turned out to be difficult. Because so many more books were projected than actually completed, Hayek said, the court was accustomed to having a copy of a book in hand before issuing an imperial "privilege." But by May 1586 Tycho had his ten-year copyright[15] and was pressing on to other realms. His negotiations with the French court were conducted by Dançey. They bore fruit in the spring of 1588, only a year and a half before Tycho went to Copenhagen to attend the funeral of his old friend.[16] Dançey must have given the request everything he had for when the privilege arrived it was penned in such style and length, on real parchment and signed by the king, as to make Tycho dissatisfied with the one he had received from the emperor. He therefore wrote again to Hayek, asking if it were possible to get something equally imposing from Rudolph, who was, after all, "first among the Kings of Europe."[17] It

The Lord of Uraniborg

was. Hayek, as we have seen, had already been sharing Tycho's letters with the emperor's vice-chancellor, Jacob Kurtz, and Tycho soon received through the latter a more aesthetically pleasing privilege.[18] Most of the copyrights were more difficult to obtain. Tycho entrusted his hopes for an English copyright to Daniel Rogers, who came to Hven in 1588 after attending Frederick II's funeral as Queen Elizabeth's representative. When Rogers died in 1590, however, Tycho still had nothing and eventually had to resort in 1596 to asking one of his former students, Pontanus, to check into the matter while he was in England.[19] Whatever channels Tycho used to approach the king of Spain produced a document dated (6 July 1596) but not signed.[20] And a letter to the Italian astronomer Antonio Magini requesting a copyright from the Venetian Republic brought the news that such privileges were bestowed only on books printed in Venice but that it would be easy to get one from the duke of Mantua and Ferrera, because Magini had friends and relatives there.[21]

Tycho's enthusiastic reorganization of the *Opus astronomicum*[22] had left each of the three volumes he envisioned dependent on the completion of one or more planetary theories, and Tycho did not have a single one done. He had been observing the planets more or less regularly since his student days. But he had done relatively little serious work with the data from the observations. Only for the moon did he have anything that could be cited as a "result" from his efforts, and that had emerged almost accidentally from rather routine and certainly very leisurely observations.

From the inception of natural philosophy in pre-Socratic Greece, eclipses had constituted its most spectacular challenge. Generation after generation had struggled with them in attempts first to explain and then to predict them: Various aspects of the problem were solved at regular intervals through the fifth, fourth, and third centuries B.C.[23] But it had apparently been Hipparchus, in the second century B.C., who first quantified all of the elements required for successful solar and lunar theories and synthesized them into a coherent theory of eclipses.

---

[18] The privilege was used for the *Progymnasmata* in 1602 (II, 8–10). Some kind of emendation was made in 1592 (VII, 347).

[19] VII, 141, 377–8.   [20] *Norlind*, 165.   [21] VII, 304.

[22] Tycho still planned to collect the results of his researches into a comprehensive opus, which is mentioned on the preliminary title page of the *Progymnasmata* (II, 439), in the *Mechanica* (V, 112), and in personal correspondence (VIII, 199).

[23] Victor E. Thoren, "Anaxagoras, Eudoxus, and the Regression of the Lunar Nodes." *Journal for the History of Astronomy* 2 (1971): 23–8.

For the purpose for which it had been devised, Hipparchus's lunar theory was reasonably good. From it, the extent and duration of eclipses could be predicted about as closely as they could be recorded. The timing, though less accurate, was generally good within an hour and, at any rate, was not improved appreciably until Tycho did it. Despite his achievement, however, Hipparchus does not seem to have been satisfied. Having successfully accounted for the moon's appearances in syzygies, he then examined the performance of his lunar theory at other places in the orbit.

Nothing is known about this undertaking except that the results were available to Ptolemy[24] three hundred years later. We are certain that Hipparchus noted discrepancies, but whether he (or anyone else) managed to reduce them to rule can only be conjectured. As far as we know, it was Ptolemy who first modified the theory with an eye to representing the moon's positions in noneclipse situations. By examining the errors of the Hipparchan account, Ptolemy succeeded in generalizing them as (1) functions of the angular distances of the moon from syzygy and (2) proportional to the existing Hipparchan corrections themselves.[25] Recognizing that this implied a periodic "swelling" of the epicycle, from $5°$ at new or full moon to about $7\frac{2}{3}°$ at first and last quarters (Figure 10.1, left half), Ptolemy utilized what was, in effect, a highly elliptical orbit[26] to obtain purely optical magnifications of the traditional $5°$ epicycle.

Unfortunately, although Ptolemy's industry and ingenuity produced an almost perfect accounting for the basic departures of the moon from circular motion in longitude, they failed spectacularly to represent the moon's distances. For, as was obvious from the most cursory observation of the moon's angular diameters, the moon's distances did not vary by anything approaching the extremes ($65^P15'$ to $34^P7'$)[27] implied by the theory.[28] Until the Middle Ages, however, none of Ptolemy's successors registered any complaint about this

---

[24] *Almagest*, V, Chaps. 1 and 2.   [25] *Almagest*, V, 2.

[26] However, the ellipse would have been synodic rather that anomalistic, that is, centered on the earth and oriented with its major axis along the (swiftly moving) line of syzygies. (*Almagest*, V, 3). Neither Ptolemy nor anyone else referred to the orbit as an ellipse, but Copernicus may have known that a body moving in such a way would trace out an ellipse. See J. L. E. Dreyer, *History of Planetary Systems from Thales to Kepler* (Cambridge, England: Cambridge University Press, 1906), pp. 330–1.

[27] These "parts" refer to a unit deferent of 60^P. The equivalent figures in earth radii were $64^P10'$ and $33^P33'$: *Almagest* V, 13. For a glossary of technical terms, see Appendix 2.

[28] For a discussion of the situation, see G. Hon, "Is There a Concept of Experimental Error in Greek Astronomy?" *British Journal for the History of Science* 22 (1989): 1129–50.

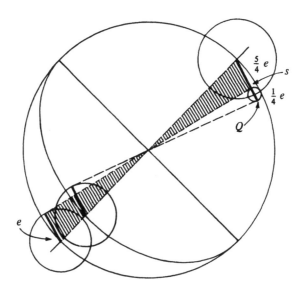

Figure 10.1. Equivalence diagram for the Ptolemaic and Copernican
Lunar Theories. The Copernican Theory (right) uses a small epicycle to
provide the correction (required in quadrature) achieved by the "elliptical" orbit of Ptolemaic Theory.

shortcoming.[29] Only as a rather incidental part of the general over-
haul of Ptolemaic astronomy did Copernicus rectify it.[30]
   Copernicus accomplished the task through essentially the means
rejected by Ptolemy – by arranging to increase the size of the
epicycle as the moon moved from syzygy to quadrature. He first
enlarged the basic epicycle by half the value of the required mag-
nification (see Figure 10.1) so that it produced a mean correction of
$6\frac{1}{3}°$. He then mounted on that enlarged epicycle a second epicycle
whose radius was the other half of the required additional correction.
By making the moon move in the second epicycle, in such a manner
that it was always at *S* in syzygies (subtracting from the contribution
of the major epicycle) but at *Q* in quadratures (adding to the
corrections by 50 percent), Copernicus achieved virtually identical

[29] Except for the recently recognized fourteenth-century anticipator of Copernicus's theories,
      Ibn ash-Shatir: Victor Roberts, "The Solar and Lunar Theory of Ibn al-Shater," *Isis* 48
      (1957): 428–32; Regiomontanus seems to have been the first to make an issue of it: J. B. J.
      Delambre, *Histoire de l'astronomie au moyén age* (Paris, 1819), p. 359.
[30] Because Ptolemy's lunar theory utilized an equant, albeit one different from those used in
      the planets, Copernicus felt obliged to redo it. Once the transformation was complete, he
      was able to make an issue also of the conflict between parallaxes and diameters: *De
      revolutionibus*, IV, 2.

longitude effects[31] while reducing the distance variation from four times to about two times its actual range.

Given the problems that the planetary (and even solar) parallaxes posed for Tycho, one would expect that those arising from the much larger and more erroneously represented parallaxes of the moon would have been insurmountable. In fact, however, they did not prove inordinately troublesome, perhaps because the fact that they could exceed 100′ was a warning that they had to be handled with extreme caution. Traditionally, the uncertainties of parallax had been circumvented by the use of lunar eclipses, through which the moon could be placed by its geometrical position opposite the sun, in the earth's shadow. And for positions outside syzygy, nonagesimal observations had been invented, in which the sightings were planned for the time at which the diurnal rotation brought the observer and the moon to the same celestial longitude. Under such circumstances (Figure 10.2) only the latitude coordinate would show parallax. At that point, however, there was no further recourse short of moving to a lower geographical latitude. Finally, the component in latitude, which could still be considerable, had to be removed theoretically, using the weakest aspect of lunar theory, the representation of the moon's distances.

Over the years, Tycho performed several such computations, always using the corrections generated by the Copernican model and hence recognizing, as virtually everyone else did in the sixteenth century, the obvious superiority of its representations.[32] He even occasionally attempted to check the accuracy of the corrections by comparing them with the difference between the true and apparent (observed)[33] latitudes of the moon. Of course, the true latitude had to come from theory, but because Copernicus had agreed with Ptolemy on that aspect of the theory,[34] Tycho thought he was on safe ground.

In fact, however, the latitudes predicted by this theory were almost perpetually erroneous, by amounts that could range up to nearly 20′ but that varied independently of the errors of the Coper-

---

[31] *De revolutionibus*, IV, 3.

[32] Note that already in his write-up of the new star in 1573, when Tycho wanted to show how much parallax the moon would have at the altitude of the new star, he used the Copernican minimum (I, 26) for the moon's distance, evidently believing that the more spectacular figure obtained from the Ptolemaic value would have no credibility.

[33] The observed correction, however, had to be corrected for refraction, which, because Tycho regarded it as requiring empirical determination, injected another unknown into the situation.

[34] That is, they agreed except insofar as their theories of longitude provided different arguments of latitude as inputs to the theory of latitude.

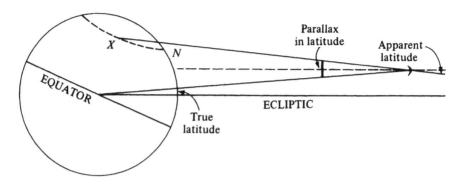

Figure 10.2. Diurnal parallax of the moon. The parallax in latitude is minimized and the parallax in longitude is null when the earth rotates the observer (X) into the nonagesimal (N).

nican theory of parallaxes. The result, needless to say, was a complicated pattern of errors compounded from the error curves of the theories of latitude and "altitude" (parallaxes). So although the problems would become apparent to anyone who scrutinized the moon's latitudes throughout the orbit, they would not be readily soluble in theoretical terms and would not be soluble at all in purely empirical terms.

As of the beginning of 1587, there was little likelihood that Tycho would solve on any basis any of the problems in the motion of the moon. For although he had made observations of it on perhaps 150 occasions since early 1582, he had reduced few of those observations to usable positions and had compared even fewer with predictions computed from theory. And although even the few comparisons he made had provided some potentially interesting results, Tycho had not analyzed them sufficiently to detect any inconsistencies.

In April 1585, for example, Tycho simply rejected an observation that gave a latitude of $5°$ $14\frac{1}{2}'$ because the maximum theoretical value of the moon's latitude was supposed to be $5°$.[35] Only on 9 January 1587, did he encounter a situation in which he decided that something was definitely wrong: that he was unable to account for $7'$.[36] At the time, he did not resolve the contradiction. By August of that year, however, he had, through observations that indicated that the inclination of the moon's orbit was $5\frac{1}{4}°$ rather than the $5°$ found by Ptolemy and adopted by Copernicus.[37]

[35] X, 379.   [36] XI, 146.

[37] The circumstances of Tycho's discovery were laid out in detail by Victor Thoren in "An Early Instance of Deductive Discovery," *Isis* 58 (1967): 19–38.

Although this discovery, which Tycho interpreted as evidence of secular change analogous to that in the obliquity of the ecliptic (rather than as an error in Ptolemy's determination), was important enough to merit the pains of a last-minute (but not particularly prominent) announcement in *De mundi* as the book went to press,[38] it was scarcely something that could stand by itself as a restitution of the lunar theory.

Yet when Tycho decided to include such a restitution at the end of Book I of his *Progymnasmata*, that was still all he had to report. During 1590 and 1591 he made some observations that were labeled as being suitable for examining theoretically critical points of the moon's orbit and even reduced his observations and compared them with theory, but if he actually reached any conclusions, his logs do not show it. He made his most determined effort in August 1592. A detailed examination of the moon's position on the seventh left him unable to account for a 33' discrepancy and grasping at the possibility that a change in the size of the second Copernican epicycle might ameliorate the situation.[39] Eventually, Tycho changed the sizes of both epicycles, but what he needed for the problem at hand was what he later discovered as the *variation*. In any case, it seems clear that he made no progress on either alternative in 1592. By the end of the month, when he had concluded his investigations, all he had to show for his efforts was some minor emendations in the reference values for the mean longitude, mean anomaly, and ascending node.[40]

After these investigations, directed specifically at checking the moon's parallaxes but involving all the rest of the lunar theory because of the need to know the moon's "true" position, Tycho essentially abandoned the moon for over two years. Presumably, the aforementioned changes in his epoch values would constitute his restitution of the moon, even though his two final observations had shown the moon agreeing with the ephemeris position on one day and then departing by over 20' on the next.[41] Alternatively, he may simply have decided that he had done all he could for the time being and so turned to the superior planets, which had recently been added to the moon as projects that had to be finished before Tycho could publish Volume I. Whatever the reason, Tycho made only three observations of the moon in the next two years. And when he returned to it in October 1594, it was not for general observations of

---

[38] IV, 42–3.   [39] XII, 199.

[40] Tycho subtracted $5\frac{1}{3}'$ and $50\frac{1}{2}'$ from the Prutenic mean longitude and mean anomaly, and averaged the Prutenic and Alfonsine nodes (XII, 198–207). For details, see Victor Thoren, "Tycho Brahe on the Lunar Theory" (Ph.D. diss., Indiana University, 1965), pp. 87–92.

[41] XII, 207–9.

the kind that had characterized his earlier work but for a specific investigation of something he had noticed in connection with an eclipse.

Ever since his student days, observation of eclipses had been virtually obligatory for Tycho. Both in their own right as phenomena that epitomized the demands on and skills of astronomers and as valuable checkpoints for the more general theories of the sun and moon, Tycho observed, scrutinized computationally, and logged them for ready use in a systematic way that was unique among all of his observations.[42] If there was one activity for which the whole staff at Uraniborg turned out, it was probably the recording of eclipses. Eclipses that occurred at reasonable hours and were blessed by good weather were observed independently by three crews: Tycho superintended one on the large instruments in Stjerneborg, and senior assistants oversaw operations in the north and south observatories on the second floor of the house (Gellius and Flemløse are named in the 1587 records).[43]

The goal of the observations was to generate a series of pictures like that shown in Figure 10.3, which is unusual only in that it depicts a longer (more central) eclipse than average and that it was drawn by a doodler (probably Flemløse) who had some extra time on his hands.[44] From such records, carefully timed, the details of the eclipse – time of first and last contacts, angle of traverse, maximum obscuration and so forth – were scrupulously determined and meticulously compared with predictions from various ephemerides.[45]

Preparations for such efforts took time. If first contact were to be recorded accurately, the clocks had to be checked, the instruments prepared for use, and the crews ready for work when first contact occurred. Because the earth's shadow is not itself an observable entity, the moon's approach to it cannot be traced, and so the only advanced warning available was ephemeris predictions.

Unfortunately, however, one of the major reasons for observing eclipses was precisely the fact that ephemeris predictions still left quite a bit to be desired. A collection of three or four could easily scatter their forecasts of first contact over a period of two hours.[46] If

[42] Tycho published a list of the thirty that he had observed as of mid-1600 in Book I of the *Progymnasmata*: II, 98.
[43] XI, 163–5. See also the eclipse of 1596 (XIII, 15–19) and those listed in note 44.
[44] 2 March 1588: XI, 257–64. For a series of twenty drawings, see the eclipse of 1590: XII, 21–4.
[45] For a completely explained, step-by-step example of a traditional eclipse computation, see Victor Thoren, "Extracts from the Alfonsine Tables and Rules for Their Use," in E. Grant, ed., *A Sourcebook in Medieval Science* (Cambridge, England: Cambridge University Press, 1974), pp. 465–87.
[46] For a range from 4:53 to 6:51, see XII, 24.

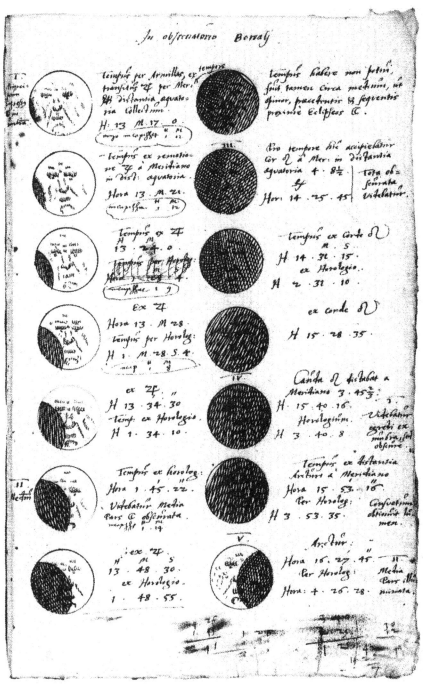

Figure 10.3. Tycho's log of the lunar eclipse of 2 March 1588.

one could not be certain that the earliest of them were systematically premature, one would have to be looking at the moon even before the time of the first prediction and so might wind up waiting two hours for the beginning of a three- or four-hour eclipse. This, of course, was an astronomer's lot, and probably not the worst of it, although it is hard to believe that anyone ever really accustomed himself to winter observation on Hven. Clearly it would have been better to eliminate as much uncertainty as possible, and by 1590 Tycho seems to have conceived a means of doing it.

This involved a last-minute check on the position of the moon before the eclipse. That is, instead of starting with a long-range approximation of mean opposition (and correcting by theory to get true opposition and the resulting eclipse conditions),[47] Tycho apparently decided to use an empirical observation of the moon's position two days beforehand (in case the night before the eclipse was cloudy?) and then to estimate its arrival at true opposition from the theoretical velocity of the moon in the interim.

As straightforward as Tycho's calculation was, however, any application of it was bound to disappoint his expectations, for the anomalistic acceleration of the moon,[48] which was the only one he would have accounted for in determining the moon's average velocity, was not the only velocity phenomenon involved. The *variation*, which presents itself in each octant as a 40′ departure from the position predicted by the two traditional inequalities, shows up at syzygy (and quadrature) as a departure from the velocity predicted by those inequalities. Through syzygies, therefore, the moon was always traveling faster than Tycho's tables stated. Within rather narrow limits, its extra progress during the day before opposition was about 15′ of longitude. In the nearly two days intervening between Tycho's preliminary observation and the eclipse, it would have picked up almost 30′, essentially an hour's progress.

Tycho observed the moon on 28 December 1590 "because of the coming eclipse" and decided that the eclipse would begin at 6:24 P.M. on the thirtieth – nice information to have, when ephemeris predictions ranged from 4:53 to 6:51.[49] Unfortunately, therefore, everyone seems to have been inside eating supper when the eclipse began. At least, when the moon was first observed at 6:05 (and at an altitude of

---

[47] That is, although eclipse tables could provide the time at which the mean moon would be opposite the mean sun, the theories had to be used further to determine when the individual epicyclical motions of the true sun and true moon would find them physically opposite each other. See Figure A.4.6 in Appendix 4, and note 45.

[48] That is, the departure from the uniform motion posited for the center of the epicycle, due to the moon's motion in the epicycle.

[49] XII, 20–5.

20°) it was already over halfway into the shadow:[50] Instead of being twenty minutes away, the eclipse was an hour in progress. Later, when there was time for puzzling over the situation, Tycho vowed to look into it further.[51] But as of the eclipse of 1592, he did not have it explained.

Tycho's next chance occurred in October 1594. This time he made the most of it.[52] From one pre-eclipse observation and the eclipse observation itself, he developed the suspicion that the moon must be traveling faster through full moon (position 4 in Figure 10.4) than theory allowed. Moreover, he must have followed through on his suspicion by assuming a symmetrical acceleration[53] through new moon (8) and then deducing a compensating reduction in the moon's velocities between syzygies, in the quadrants (2, 6). Finally, he seems to have perceived that the most noticeable effects of such a *variation* in velocity would be a displacement, which should accumulate so that the moon would lead theory by its greatest amount about halfway from syzygy to quadrature (5, 1) after accelerating through syzygy and lag by an equal amount midway from quadrature to syzygy (7, 3) after decelerating[54] through quadrature. Although Tycho did not present this analysis anywhere, and it undoubtedly will seem esoteric for pre-Galilean concepts of motion, his actions leave little alternative to the proposition that they were directed by some such thoughts.

On the morning of 28 October 1594, nine days after the eclipse

---

[50] The moon was already seven digits (out of twelve) eclipsed: XII, 22.

[51] This account is inferred from Tycho's summary: "It is necessary to redo the examination of the lunar observation made two days ago. For the place of the moon deduced at that time by calculation from the observation, is seen not to agree reasonably with the motion of the moon in the eclipse unless perhaps (as I suspect) a blatant error is contained in the time given here by the clock; then again the preceding calculation from the lunar observation of two days ago might not have been correctly carried out, of which it is necessary to make a test" (XII, 22). Tycho found fifteen minutes of error in his clocks but never did account for the other missing hour.

[52] All of the argument outlined here is presented in full detail in Thoren, "Tycho Brahe," pp. 97–109; and Victor Thoren, "Tycho Brahe's Discovery of the Variation," *Centaurus* 12 (1967): 151–66.

[53] There is undoubtedly a sense in which it is anachronistic to speak of accelerations in connection with it. But the evidence available – and laid out more fully (see ibid.) requires attributing to Tycho some kind of grasp of the notion of "change of velocity." In fact, Tycho did make some tabulations showing both first and second differences, which involve precisely such a conception. Kepler described Tycho as having found that the moon's motion "was irregular – that the moon moves exactly as much faster in the syzygies as it moves slower in the quadratures." Hence the name *variation*. Tycho eventually generalized the phenomenon (correctly) in the form of a displacement, equaling approximately $40'\sin 2\alpha$, where alpha is the angular distance of the moon from the sun.

[54] Ibid.

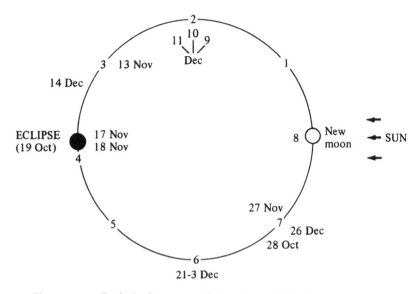

Figure 10.4. Tycho's discovery of the effects of the Variation, 1594.

and eight months after his last previous attention to the moon, Tycho observed the moon in the seventh octant (7), noting explicitly that it was "halfway between quadrature and conjunction,"[55] – the first reference to such a position in thirteen years of log entries. Two weeks later he checked again, and again noted that the moon was in "the mean place between first quadrature and opposition."[56] Four days after that, he instituted a two-day investigation of the velocity of the full moon. Observing at opposition, he reduced his readings and found that his result agreed reasonably well with both theories.

Repeating his procedure the next day (18 November) as the moon left opposition, he found that it had traveled half a degree farther during the interval than the theories predicted.[57] After completing the first circuit with another observation in the seventh octant and tracing the moon through first quadrature on 9, 10, and 11 December, Tycho was able to generalize that he would make his next observation "between quadrature and opposition ... in that place, namely, where both the Prussian and Alfonsine tables differ most from the appearances."[58]

In December, February, and March, Tycho conducted investigations designed to establish more precisely the qualitative effects and quantitative limits of the phenomenon.[59] Exactly what he concluded

[55] XII, 319.   [56] XII, 320.   [57] XII, 322.   [58] XII, 324.
[59] XII, 324–31, 378–88, but especially the tables on pp. 330 and 385.

at that time is somewhat doubtful,[60] but whatever it was, it concerned the first new astronomical phenomenon to be discovered
since Ptolemy's era.[61] Justly proud of his achievement, Tycho found
an opportunity to mention it to Hayek in his next piece of scientific
correspondence:

> And even though he (Copernicus) has arrived at a much more
> agreeable and probable theory for the moon than the Ptolemaic was,
> nevertheless not even this theory is sufficient to explain the lunar
> cycles in every case. In syzygies and quadratures he deserves praise,
> although not even in these places does he set forth everything with the
> necessary precision. But in the four places already mentioned, which
> are intermediate, he by no means saves the appearances, unless we are
> prepared to consider as nothing the throwing away of half a degree
> when the moon is situated near the mean distances of the larger
> epicycle, and nearly a whole degree when she revolves near apogee or
> perigee of the same.[62]

Inspired by his discovery of a generalization that removed perhaps
three-quarters of the error formerly inherent in predictions of the
moon's longitudes, Tycho turned to other aspects of the theory,
Chief among them was the old problem of parallaxes. As with the
sun, the question was bound up empirically with refraction: All that
could be observed was the sum of the conflicting effects. Thus the
only way to solve either problem empirically was to resolve the
other one theoretically first. As of early 1595, Tycho still had not
made up his mind how to handle the issue.[63]

Complicating matters was the fact that the moon itself, like the
planets, followed a different course across the celestial sphere each
time around the earth. Of course, Tycho thought he had this part of
the problem under control through his discovery that the inclination
of the moon's orbit had increased since Ptolemy's day. But actually,

---

[60] See the description sent to Hayek and the discussion of Tycho's restitutions of 1596–7, in
Appendix 4.

[61] One might cite as counterexamples to this claim the progression of the sun's apse
(al-Battani), the progression of the planetary apsides (Copernicus), and trepidation (Thabit
ibn Querra); but all of them require some rather extended defense in a way that attribution
of Tycho's discovery to him does not. Actually, the *variation* seems to be the best candidate.
An Indian named Bhaskara apparently discovered it and established its maximum at 34′ by
about A.D. 1150. News of his discovery was not transmitted to the West, however, and
remained unknown until the twentieth century. See "The Evection and the Variation of the
Moon in Hindu Astronomy," by Dhirendranath Mukhopadhyaya, *Bulletin of the Calcutta
Mathematical Society* 22 (1930), pp. 121–32.

[62] VII, 370.

[63] See XII, 388, for a note Tycho was taking lunar refractions *"comm. refract. solari sunt."* By
the time his restitution of the lunar theory was complete, he had a special table of lunar
refractions similar to those for the sun: II, 136.

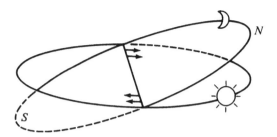

Figure 10.5. Inclination of the moon's orbit to the sun's, and the retrograde motion of the line of intersection (nodes) of the two orbits.

the latitudes he had been reckoning "from our own tables"[64] since 1589 were no more correct than the corresponding Ptolemaic or Copernican accounts.[65] Discovery and resolution of this situation were prerequisites to success with either the parallaxes or the refractions of the moon. In 1595 the time was ripe for this feat, and Tycho achieved it, in a style that makes it stand out as the jewel among the efforts and results of his entire career.

Since their earliest efforts to cope with eclipses, astronomers had known that they were not confined to special seasons but could occur at any time of the year. Any attempt to work seriously with Anaxagoras's (ca. 530 B.C.) explanation of eclipses – as arising from intersections between the apparent paths of the sun and moon – therefore, required a provision for allowing those points of intersection to move.[66] Centuries before Ptolemy, this motion had been reduced to rule, in terms of a uniform retrograde motion of about $1\frac{1}{2}°$ per month which carried the nodes backward along the ecliptic in a period of about eighteen years (see Figure 10.5). Obviously as the nodes moved, so did the limits: the positions (N and S) in the orbit where the moon departed farthest from the ecliptic and where it was, accordingly, easiest to determine the angle between the two orbit planes.

---

[64] XII, 123, 189.

[65] The latitudes predicted by the accepted theory were almost perpetually erroneous, differing from actuality by an amount varying with the distance of the moon from the node, and the distance of the node from quadrature. The ancient representation of the latitude was effectively $\delta = 5°\sin\beta$; the modern is essentially $\delta = 5°9'\sin\beta + 9'\sin(\beta - 2\phi)$, where β is the distance of the moon from the node and ϕ the distance of the node from the sun. The difference between the two is $\delta = 9'[\sin\beta(1 + \cos 2\phi) - \cos\beta\sin 2\phi] = 9'[2\sin\beta\cos^2\phi - 2\cos\beta\sin\phi\cos\phi]$. If we let $\theta = \phi - 90° =$ distance of the node from quadrature; then $\delta = 18'\sin\theta\cos(\beta + \theta)$.

[66] The chronology is reviewed and its implications are argued in Thoren, "Anaxagoras," pp. 23–8.

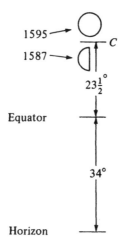

Figure 10.6. The moon's meridian transits at summer solstice.

Theoretically there were two opportunities a month to measure this angle, but practically, some occasions were considerably more auspicious than others. To minimize the uncertainties arising from parallax and refraction, it was desirable to observe when the moon was as high in the sky as possible, at the beginning of Cancer where the sun reached its summer solstice altitude.

Fortunately, that was precisely the region in which the moon had happened to be attaining its maximum latitude when Tycho made his determination in 1587. But the limit then involved had been the southern limit, rather than the northern one, so that the moon had been 5° ($+\frac{1}{4}$°) below C (Figure 10.6) instead of 5° above it. By 1595–6, the moon's orbital plane had slewed halfway around in its orientation, so that the northern limit was in Cancer. Conditions were therefore the best they would ever be for Tycho's purposes, and he determined to take advantage of them to test his results.

Already on 11 February 1595, Tycho took up the task, stating in the log explicitly what he was after.[67] Unfortunately the clouds closed in on him and stopped the work after one sighting. Thus, Tycho's long-standing policy of averaging pairs of sightings prevented him from reducing the data and seeing that the moon's latitude was over 10′ less than he expected it to be. Any of several observations made in succeeding months would have shown the

---

[67] XII, 378. The argument summarized in the following pages was presented in greater detail in Thoren, "Tycho Brahe," n. 40, pp. 54–66; and Thoren, "Deductive Discovery," n. 37, pp. 19–36.

same thing if Tycho had reduced them. Only on 9 July, in an observation made paradoxically at the southern limit where the least favorable conditions obtained, did he note a problem. At the time he was not prepared to place the blame for the discrepancy, but neither was he prepared to overlook it, and so he noted: "Differ 5', but to be examined again."[68]

Apparently, Tycho did reexamine the offending observation, but no direct record of his procedure has survived. It would be interesting to know how he managed, in an observation designed explicitly to test parallax and refraction, to wind up attributing the discrepancy to the theory of latitude. But however he did it, there can be no doubt that sometime in the following two months he broke the problem wide open, discovering not only one but two inequalities in latitude.

Until the summer of 1595 Tycho's conception of the moon's motion in latitude was the traditional one depicted in Figure 10.7. His discovery in 1587 had affected this conception only to the extent of indicating that the tilt of the moon's orbital plane was no longer 5°, as it had been in Ptolemy's day, but $5\frac{1}{4}°$ – that the pole of the lunar orbit had shifted from P_0 to P_1, about 15' farther from the pole of the ecliptic.

In his review of the discrepant calculation of 9 July, however, Tycho was forced to reconsider his interpretation and ultimately to conclude that the moon's maximum latitude actually varied throughout the synodic month, that the critical difference between his and Ptolemy's determinations had been not the epoch of the observation but the phase of the moon, and that the pole of the lunar orbit must be at P_0 when the moon was in syzygy and at P_1 when it was in quadrature.

From these data, the conclusion that the pole nutated in a circle of diameter P_0P_1 in a period of half the synodic month would have been, in Tycho's era, inescapable. By supposing the pole of the lunar orbit to describe twice each month the circle shown from a polar perspective in Figure 10.8, Tycho obtained a model that provided the accepted inclination ($EP_0 = 5°$) at conjunction and opposition, and his larger finding ($EP_1 = 5\frac{1}{4}°$) at the first and last quadratures. It also provided a smooth continuous variation of the inclination between those extremes, easily computed from a circular function: $i = 5° + 7\frac{1}{2}'$ $(1 - \cos^2\alpha) = 5° + 15' \sin^2\alpha$ in the best mathematical tradition.

Tycho could not have worked with his new model very long before noticing that the model itself implied another phenomenon.

[68] XII, 392–3.

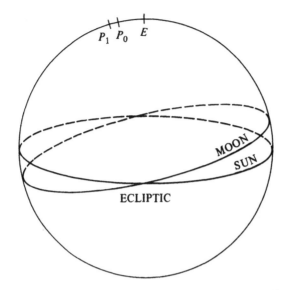

Figure 10.7. The moon's motion in latitude under a fixed inclination of the moon's orbit.

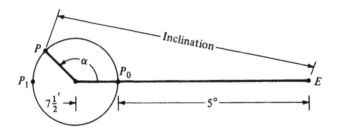

Figure 10.8. Detail of the nutation of the pole of the lunar orbit envisioned for Tycho's Theory of Latitude.

As the pole of the lunar orbit moves around its circle, the line determined by it and the pole of the solar orbit oscillates back and forth. In Figure 10.9 (which is a greatly exaggerated view), as P progresses away from P_0 in the small circle, the line *EPL* moves to the right, displacing L to L'. Continuing toward P_1 and beyond it, P moves the line back to the left through EP_0P_1L, displacing L in the other direction. Now the line in question is what determines the positions of the limits and nodes of the moon's motion in latitude. L is the southern limit. As it oscillates, the nodes, being 90° from the

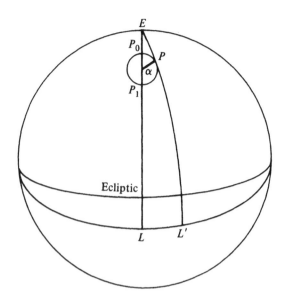

Figure 10.9. Tycho's deductive discovery of the oscillation of the lunar nodes.

limits, naturally oscillate in the same manner. Hence the model that Tycho adopted to explain the one inequality implied a second!

Such an implication might have been ignored, or deplored and eliminated, depending on one's attitude toward models,[69] but it could scarcely have been overlooked – particularly because Copernicus had used the same type of arrangement to connect and account for the changing obliquity of the ecliptic and the supposed oscillation of the equinoxes.[70] Tycho did not ignore it.

Following a routine observation of the moon on 11 September, he made some calculations headed "Examination of the node from the meridian altitude." These calculations, which involved "taking the angle of maximum latitude for this place to be 5°7′,"[71] implied that the node was 35′ away from where Tycho expected it to be. Such an error could not also obtain in syzygy without making an absolute shambles of eclipse predictions, and so it almost had to be a temporary situation. When similar trials with two old observations[72]

---

[69] Tycho could, for instance, have used a second circle to convert the circular motion into rectilinear morion, as Copernicus had done for his liberation of Mercury: *De revolutionibus*, V, 25.

[70] *De revolutionibus*, III, 3.

[71] "*Resp. Distantia a ☊ 39¼, posito angulo maximae latitudinis hoc loco 5⁸7′*" XII, 394.

[72] "Conferatur cum anno 92 *die* 12 Feb. [XII, 124] *Item anno* 95 *die* 17 Feb" (XII, 394).

showed similarly temporary displacements of the node, Tycho felt justified in concluding: "Whence it becomes clear that the nodes are changed unequally because of the small circle, and by the greatest difference in aspects of 60° and 120°, just as it [the small circle] changes the maximum latitude in syzygy and quadrature."[73]

Even in post-Newtonian science, such deductive discoveries have a pronounced mystical tinge. In Tycho's day, the theoretical prediction of the existence of a previously unnoticed phenomenon must have been completely unprecedented. Yet, because discovery of any kind was almost unprecedented in the sixteenth century, Tycho had no reason to view his theoretically derived finding on the lunar theory in any different light than he did his empirical studies of refraction or his investigations of the parallaxes of comets. Thus, he nowhere said anything at all concerning the nature of his discovery of the nodal inequality but merely tied it trigonometrically to the variation of the inclination by stating in his final (and sole) exposition of them that the one followed from the other "both by observation and by the doctrine of triangles."[74]

73 "*Unde palam fiet nodos inaequaliter mutari ratione parvi circelli & differentia maxima in* ⚹ & △, *quemadmodum in*(☍)& □ *mutat maximam latitudinem*" (XII, 394). Of course, *aspectus sextilis* (⚹) and *aspectus trigonus* (△) do not convey the situation accurately, but Tycho had no alternative to taking refuge in astrological terminology. Because the octants had never before been critical points in lunar theory, no shortcut reference to them had ever been needed. Tycho, it is true, had understood the effects of the third inequality for the better part of a year but could still describe its maximum points only as "four places that are intermediate." With the discovery of a second context in which these intermediate places figured, Tycho soon coined the term *octants*: V, 187.

74 II, 122.

Chapter 11
The Last Years at Uraniborg

THE spectacular discoveries in the lunar theory were the most remarkable results of Tycho's career, but they were also almost the entirety of his scientific work during the last decade of his life. Already by the spring of 1591, when he had finished writing and printing almost all of the ten projected chapters of his 850-page tome on the new star,[1] Tycho was facing the first of a series of distractions that was to plague him to the end of his days. As is so frequently the case in human affairs, most of these distractions were of Tycho's own making. But the most time-consuming – a steady stream of curious tourists whose comings and goings had already turned his summers into periods of negligible productivity – was one that was probably beyond his control.

Tycho's diary for 1590 shows what proportions this activity could assume. Typically for these years, the most frequent visitor was Tycho's sister Sophie, who made a dozen stops, ranging in length from a night to a week. Half a dozen other relatives and friends also made two or three visits each to the island, and several others who seem to be named as friends likewise put in appearances. In addition, numerous groups of unknown and usually unnamed travelers dropped in. During 1590 the diary mentions "Niels Krag and three others," "some Scots nobles," "two German students," "four German students," "Wilhelm Katt and three others" (plus a party of four from eastern Europe, on the same night), "a German noble," "some German nobles," "a noble youth with his preceptor and servant," "four Copenhagen students," "a learned Scotsman," and "a party of travelers with two other nobles."[2] Through much of the bustle, the son of a Dutch instrument maker, Jacob van Langgren, was at Uraniborg for seven weeks surveying Tycho's operation and copying out Tycho's new star places for use on a new generation of celestial globes.[3] Finally, this year, which seems to have been somewhat busier than usual, was crowned with visits from three people who, each in their own way, gave Tycho special cause to remember them.

On 20 March, the keeper of Tycho's diary recorded: "The king of Scotland came this morning at the eighth hour and left at three." It

---

[1] VII, 274, 290, 301.   [2] IX, 81–95.
[3] IX, 83, 86; II, 282; VIII, 322.

was the future James I of England, in Denmark to claim as his wife Frederick's adolescent younger daughter, Anne. James fancied himself an intellectual and must have felt at home when he saw a portrait of his old tutor, George Buchanan, hanging in Tycho's library.[4] Coincidentally, James had already been to Roskilde to see Tycho's old teacher Niels Hemmingsen. Whatever discussion James had with Tycho was probably on theological topics, because James had already developed the interests that would lead him to authorize the first English translation of the Bible. But Tycho seems to have managed both to work into the conversation an exposition of his world system and to extract from it a thirty-year copyright for his books,[5] so he cannot have felt that he wasted that day. James, on his part, was sufficiently pleased with the reception and companionship to present Tycho with two English mastiffs, which Tycho installed in the entrances to Uraniborg to announce the arrival of visitors.

James remained in Denmark for another month, during which time various members of his entourage made trips to Hven that were not recorded in Tycho's diary.[6] Just before James left Denmark, he participated in the wedding of his bride's older sister, Elizabeth, to the duke of Brunswick and mentioned to the groom the remarkable sights to be seen on the little island within their view out in the sound. Two weeks later the duke and his party appeared at Uraniborg.

Unfortunately, this second royal visit was much less pleasant than the first had been. Whereas James had left Tycho with two barking mementos of his visit, the duke departed so far from that example as to decide that his own castle would not be complete without the half-meter-high revolving statue of Mercury that adorned the roof of Stjerneborg's warming room.[7] Because he promised to have a copy made and returned to Hven, Tycho had to give it to him, but he was unable to do so graciously.

When the duke, after some kind of afternoon refreshment, remarked that it was getting late and time for his party to leave, Tycho half-jokingly reminded him that excusing people from the table was the host's prerogative. At this, the undercurrents that must have been flowing since the discussion over the statue came to the surface, and the duke simply left the house and began walking toward the boat that had brought his party to the island. After initially refusing to stir himself, Tycho apparently decided that he had better try to mend things and went after the duke, carrying a cup of beer for him. When he was rebuffed, however, Tycho returned to Uraniborg and coolly ignored the rest of the preparations for departure.[8] Needless

---

[4] VI, 269.   [5] II, 11.   [6] *Dreyer*, 204.   [7] *Gassendi*, 115.   [8] *Gassendi*, 80.

to say, whatever chance Tycho had had of getting his statue back vanished with this incident. Rather than accept the situation, however, an unrepentant Tycho sent several reminders to the duke and eventually even published a letter to the *landgrave* in which he had related the fate of his Mercury.[9]

Of all the people who might have come to Hven, the one whom Tycho would most have welcomed was the *landgrave* of Hesse. For a while, it looked as if Wilhelm would manage a visit in the spring of 1588, when he was going to make a trip to Hamburg for a meeting of the North German princes. But when the most powerful member of the group, King Frederick, died, the conference was canceled, and with it went the *landgrave*'s visit to Uraniborg.[10] Two years later, however, Christoph Rothmann, the man through whom the *landgrave* had been doing his astronomy vicariously for nearly six years, got restless, and was given a vacation by the *landgrave* for the purpose of visiting Hven. On 1 August 1590, he turned up at Uraniborg.

Although Tycho and Rothmann had never met, they had since 1586 exchanged several long letters comparing practices at Cassel and Uraniborg in almost every aspect of technical astronomy. In addition, Tycho had entertained Rothmann's brother at Uraniborg,[11] and Rothmann had worked with Wittich at Cassel for some months in 1584–5 and had spoken with Flemløse and Gellius on their respective trips through Cassel in 1586 and 1588. It would be remarkable, therefore, if Tycho's and Rothmann's astronomical discussions contained any surprises.

But there were other things to do besides talk about astronomy. Rothmann had theological interests that eventually induced him to publish on the subject, and the usual heavy quota of August visitors would have diluted the conversation further. Tycho took his guest on an excursion to Sophie's estate, where they saw exotic gardens that impressed Rothmann greatly.[12] On one occasion the keeper of Tycho's diary even recorded that "Master Tycho and Rothmann observed with us tonight." The latter event was probably precipitated by Rothmann's announcement that it was time for him to leave. That the announcement came only after a month's stay must be taken as some kind of tribute to Tycho's exertions as a host, for, unbeknownst to Tycho, Rothmann was in the process of deciding to abandon astronomy.

When Rothmann left Hven, he returned to his native town and simply dropped out of sight. Under the circumstances, it is ironic that Tycho apparently marred his role as host by making rather a

---

[9] VI, 256.   [10] VI, 202.   [11] *Norlind*, 213–14.   [12] VI, 315.

nuisance of himself with the already reported attempts to convert Rothmann from his Copernican convictions. But the enterprise seems to have been worth the trouble, from Tycho's standpoint. For in the end, he said, Rothmann, although generally an obstinate man, finally yielded to the force of his arguments and not only declared himself convinced of the folly of his former beliefs but also assured Tycho that he would never publish any of his Copernican writings. [13]

In addition to entertaining visitors, Tycho did his own share of traveling. The meteorological diary records nine departures from the island in 1590, without providing much information about the trips. Until King Frederick died in 1588, most of Tycho's excursions were probably directed toward maintaining good relations with his patron. And however much Tycho detested having continuously to protect his lucrative tap on the royal treasury from the machinations of those who envisioned alternative uses for the state revenues, he clearly gave a good account of himself in such exchanges. For he managed not only to retain but actually to increase the level of support promised him in the original granting of Hven, while compiling a record bordering on irresponsibility in the administration of his various estates.

We earlier saw the problems that Tycho experienced at Hven virtually from the beginning of his tenure and noted that conditions there might well have posed difficulties for anyone who tried to assert control. But Tycho seemed to encounter such problems with his other properties, too, and in this capacity he seems to have been emulating his foster father, Jørgen Brahe, who had established enough of a contemporary reputation for liking to keep underlings in their place to be described as a "peasant baiter."[14] Moreover, Tycho combined with this trait an epic disdain for administrative responsibilities, which created another set of problems. The result was a virtually annual "situation" for the crown to cope with in one way or another.

In the original establishment of Hven, Frederick provided Tycho with land for his estate, a one-time cash payment of four hundred dalers to defray the costs of hiring skilled labor to build a house (for which Tycho was to furnish materials from his own resources), and the goods and services of the farms on the island to supply the year-to-year necessities of the estate. In lieu of the long-promised (1568) but not yet available canonry at Roskilde, Frederick had added a five-hundred-daler annual pension to meet the special expenses of

---

[13] VI, 218–23.
[14] "Bondeplager": *DAA* III, 566. For a list of abbreviations of frequently cited sources, see Appendix 1.

founding the observatory. Already in 1577, Tycho was back for
more and received the manor of Kullagaard. Looking this gift horse
in the mouth, Tycho learned that the incomes of the manor carried
with them the responsibility for keeping a beacon fire going on
Kullen, to warn ships of a sharp promontory out into the sea-lane
north of Helsingborg. Finding this simple duty unworthy of his
organizational efforts, Tycho initially managed to gain special exemption from it, but after whatever alternative arrangements were
made turned out to be unsatisfactory, the king made him assume the
burden anyway.[15]

In the spring of 1578, Tycho had again applied for more money,
and the king had again responded in royal style, granting the use of
eleven farms outside Helsingborg and the incomes from an enormous crown estate in Norway, temporarily until Roskilde became
vacant.[16] In the next year, it finally did: Tycho was officially notified
that after the widow of the deceased incumbent received the customary extra year's rents from the chapter properties, the incomes would
be his for life. Apparently, however, Tycho could not see any reason
to wait out this year of grace. Even though he had been awarded two
other sets of properties in lieu of the Roskilde canonry, he began
collecting whatever rents he could easily gather at Roskilde and
badgering those tenants who were disposed to honor the traditional
widow's rights.[17] These latter activities seem to have interfered with
his supervision of the lighthouse keeper on Kullen, for by the fall of
1579, the king had received enough complaints to compel him to
reassign the duty – and the fief – to Otte Brahe's successor at Helsingborg Castle (who was a second cousin of Tycho's).

Perhaps Frederick meant only to teach Tycho a lesson, for Tycho's
plea that the manor constituted his only adequate source of firewood
for Hven won a reversal.[18] But if the king thought Tycho had
learned anything from the experience, he soon found otherwise.
Within three weeks of issuing the order to restore Kullagaard, he had
to command Tycho to surrender his illegally collected rents from
Roskilde.[19] Nor had he heard the last of either situation.

In 1582 repairs became necessary on the chapel at Roskilde.
Because the previous holder of the prebend had enjoyed its rents for
some forty years, Tycho decided – not entirely without logic but
certainly without precedent – that the cost of the depreciation would
be more appropriately borne by the heirs of his predecessor and
accordingly dunned them for the money. Both on this occasion and
later (1585) when the lighthouse tower at Kullen had to be rebuilt,

---

[15] XIV, 7.   [16] XIV, 6, 8.   [17] XIV, 12.   [18] XIV, 13.   [19] XIV, 13.

the long-suffering king stepped in with public money to avoid wrangles with his highly paid astronomer.[20]

In the summer of 1580, Tycho was given lawful possession of the rents from Roskilde. Because Kullagaard, the Helsingborg farms (two hundred dalers), and the Nordfjord estates (one thousand dalers) all had been given to Tycho as temporary substitutes for Roskilde (seven hundred dalers), the king naturally issued writs recalling them. Tycho, however, quickly pointed out that these incomes were by no means superfluous and found sufficiently persuasive arguments for additional support to get them back under his control within three months. At this stage, Frederick assumed that he had surely done his duty by now and canceled Tycho's five-hundred-daler pension. But again, Tycho rose to the occasion. Surmounting even the fact that the most visible expense of his establishment – building Uraniborg – was now behind him, he managed to gain indefinite extension of his pension.[21]

Nor were pensions and fiefs the limits of Frederick's generosity. When Tycho decided in 1583 that the requirements of freight (and state?) between Hven and the mainland(s) justified a larger boat than he owned, the king directed that Tycho be given a ship from the royal fleet that was "as good as new, and not too old." In addition, sometime in 1585 or 1586, Tycho was awarded the Order of the Elephant for distinguished service to the crown. Shortly before his death, Frederick mentioned to a correspondent what these services were: Tycho was "a faithful servant and friend (who) executed conscientiously that for which he had originally been employed. He cast dependable horoscopes for all my sons and gave me notice throughout my life as to all heavenly portents sent by God Almighty to warn me and my kingdoms."[22]

Although all of these demonstrations of favor appear to have raised Tycho's support to well over twice what his benefactor had originally envisioned, even they left Tycho less than completely satisfied. Tycho's greatest concern was precisely that his incomes were contingent on royal favor and thus perpetually vulnerable to the claims of people who had no better way to spend their time than to maintain almost constant attendance on the king. The foregoing catalogue of what can be described only as remarkable diplomatic successes was doubtless perceived by Tycho as a series of struggles

---

[20] XIV, 24–5, 28–9. See also documents overlooked for the *Opera omnia*, found by J. R. Christianson, in "Addenda to Tychonis Braha Opera Omnia," *Centaurus* 14 (1972): 231–4.

[21] XIV, 19, 20.

[22] XIV, 25. The elephant medallion is shown hanging from Tycho's neck in the frontispiece. It is described by Harold Mortensen in "Portaeter af Tycho Brahe," *Cassiopeia*, 1946, pp. 59–60. The king's testimonial is quoted (without reference) in *Gade*, 107.

that proved only that he could not turn his back on the king for a moment without losing crucial funding. The one consolation he could find from Frederick's death in 1588 was that he would henceforth be free of the politics and infighting of court life. But even this relief depended on the fortunate circumstance of Tycho's being a member of the party of nobles who would control power in Denmark until Frederick's ten-year-old son came of age.

Tycho's income was not only less secure than he could have wished but was apparently also less than sufficient to meet his expenditures. Thus, despite the sincere expressions of gratitude that Tycho never failed to include with any mention of his patron's name, he proceeded with almost indecent haste after Frederick's death to apply to the government for reimbursement of his excess outlays since 1576. The reason for this extraordinary action was doubtless Tycho's belief that the regency that would rule until the young king attained his majority would be especially receptive to his claims.

Tycho may well have submitted – or at least drawn up – his claim in the expectation that Queen Sophie would be the regent. If so, he was presumably depending on the enthusiasm with which she had reacted to his enterprise during her visit to Hven in 1586. But the fact that his stepmother, Inger Oxe, had been Sophie's head lady-in-waiting from 1572 to 1584 and that his mother, Beate Bille, had served in that capacity since 1584 (to retire only in 1592) must have been part of his hopes. Indeed, Tycho seems to have enjoyed the queen's favor. Sometime during the next year or so she lent him one thousand dalers, and at the beginning of 1590 she drew up – or at least signed – for Tycho a declaration stating that King Frederick had intended to install one of Tycho's sons on Hven if either of them displayed the capacity to follow in his footsteps.[23]

But whatever the degree to which the regency rights of the queen mother may have been institutionalized in Germany, it had been centuries since the Danish Rigsraad had let such an opportunity to exercise power slip from its grasp. Sophie was quickly and crisply informed that the Council of the Realm – through a protectorate consisting of four of its members – would "advise" the young king until he turned twenty. As it turned out, this situation worked as favorably for Tycho as any alternative he could have imagined.

In describing the happy circumstances of Tycho's birth, we alluded to his family's deep and extensive roots in the Council of the Realm. Beate Bille's father, her two grandfathers, and all four great-grandfathers had been *rigsraads* (see Figure 1.1). Otte's and

---

[23] XIV, 53, 51–2. See also the documents Sophie sent to Tycho asking him very firmly to repay the loan and then dunning one of the cosigners when Tycho apparently did not: Christianson, "Addenda," nos. 74a and 83a.

Jørgen's father and both grandfathers had also been *rigsraads*. Their father (and Tycho's namesake) had been killed when they were boys; his brother Axel Brahe and their mother's brother Knud Rud were soon taken into the council, and when their mother remarried (to her brother's wife's brother) she brought another *rigsraad*, Erik Bølle, into the family. This kind of inbreeding of wealth, power, and influence was as pervasive in sixteenth-century Denmark as it is in most other societies.

Although there were well over a hundred noble families, some of them were definitely more noble than others were. Of the approximately 120 men who were inducted into the Rigsraad during that century, the Bille (7), Brahe (7), Rosenkrantz (5), and Gyldenstierne (5) families accounted for 24, or 20 percent. Fifteen other families had 3 or 4 representatives each, for another 40 percent, and another fourteen families had 2 members each, for about 23 percent.[24] Thus, fewer than one-third of the noble families produced five-sixths of the *rigsraads*. And with few exceptions, the other one-sixth of the councillors came from families that either had produced *rigsraads* in the fifteenth century or would produce them in the seventeenth century. The conservative nature of the system was compounded by consistent intermarriage.

The full extent of the resulting interrelationships among these so-called conciliar families can be grasped only by a survey of the council's membership. For 1552, just after the death of Axel Brahe and the election of Peder Oxe, among the twenty-five members,[25] seven were first cousins of either Otte or Beate; another four were first cousins of one of their parents; and three were second or third cousins. Another three were married to cousins of Otte or Beate, and five more were married to first or second cousins one generation removed. Of the remaining three members with no familial connection to Otte or Beate, one was a brother (Peder), and another was a brother-in-law of Inger Oxe, Tycho's stepmother. Thus, only Johan Rantzov (the father of Heinrich) was in a position to be completely objective about the careers of young Otte and Jørgen Brahe.

In 1588, the situation of the Brahes was generally much the same.

---

[24] These figures come from my own scrutiny of the genealogies of about a hundred of the largest noble families, published in various (annual) volumes of the *Dansk Adels Aarbog* since 1884. The Lykke, Ulfstand, and Urne families had four *rigsraads*; the Bild , Bjorn, Bølle, Gøye, Lange, Lunge, Munk, Parsberg, Podebusk, Rud, Sparre, and Thott families each had three; the Brok, Brockenhuis, Friis, Hardenberg, Huitfeldt, Krabbe, Krognos, Marsvin, Oxe, Rantzov, Rosensparre, Skram, Trolle, Ulfeldt, Valkendorf, and Viffert families each provided two.

[25] The council was actually a bit smaller, because although Iver Krabbe, Niels Lange, Peder Oxe, and Holger Rosenkrantz were added that year, they were replacements for Esge Bille, Knud Gyldenstierne, and Niels Lunge, who died during the year.

For although the representation of the conciliar families in the council had been diluted a bit overall, this dilution had been sufficiently general to leave the Brahes as well represented, relative to the other conciliar families, as they had been in Tycho's youth. Among the *rigsraads* serving at the time of Frederick's death, Tycho had a brother (Steen), a brother-in-law (Christen Skeel), three of his mother's first cousins, two more remote cousins, and the husbands of a few of his near relatives.

In any issue in which nepotism was going to be a serious consideration, therefore, all Tycho lacked was a noble wife from one of the conciliar families to augment his connections. Brother Steen had set a good example in this regard. Although his wife, Birgitte Rosenkrantz, had been orphaned at an early age, she was descended from five generations of Rosenkrantz *rigsraads* and still had the connections to prove it. Seven of her uncles had sat in the council, and she had a sister and a cousin married to *rigsraads*, not to mention the cousin married to Steen's (and Tycho's) brother Axel (who was likewise to become a *rigsraad*). There was an element of circularity between marrying into power and becoming powerful: Power brokers with daughters to marry off were in a position to select the most able and eligible husbands for their daughters, and Steen Brahe was clearly a very able individual, for all that he happened also to be a Brahe and to have married a Rosenkrantz.

Because Tycho was (by all indications) on reasonably good terms with his brother, it cannot have been disadvantageous for him to have had Steen on the council. But if the Rigsraad with which Tycho had to deal had consisted of many men with Steen's "warrior" background, it seems at least questionable that Tycho would have received the support he did. By the late 1580s, however, the Rigsraad consisted primarily of men whose training had included formal university education and whose interests now included at least some things intellectual.

The three most important members were probably the triumvirate of administrators who kept the books through the years of Tycho's patronage. Old Niels Kaas, who had been in the chancellory since Tycho's youth, was a veteran of Niels Hemmingsen's household and the University of Wittenberg. If Christoffer Valkendorf seems not to have had a reputation for sustained patronage the equal of Kaas's, he certainly compensated for it in 1595 when he endowed scholarships at the University of Copenhagen for no fewer than sixteen students. In retrospect, the Rigsraad's most distinguished intellectual was their younger colleague and one-time fellow student of Tycho's, Arild Huitfeld. In 1588 he was still collecting materials, but from 1595 until his death (1609) he published a multivolume history of sixteenth-century Denmark (to the accession of Frederick II) that earned permanent regard as a trail-blazing effort.[26]

Important more because of their wealth and political power than because of their intellectual credentials were Peder Gyldenstierne, Erik Hardenberg, and Jørgen Rosenkrantz. Little is known about Gyldenstierne (who was raised in the warrior tradition and served eighteen years as marshal of the realm) except that he was interested in history and supported Vedel in a project or two. Similarly, all that is known of Hardenberg (besides an education at Wittenberg in the house of Melanchthon) is a statement by him that he waited on the court only to please his father and preferred to stay at home passing the time with his books and studies. Jørgen Rosenkrantz was a protégé of Niels Hemmingsen and had gone to the University of Wittenberg and had been raised (in the 1530s, after the death of his parents) by an aunt who happened to be married to the old *rigsraad* Axel Brahe. Although Rosenkrantz's own intellectual interests in history seem to have been more enthusiastic than profound, he was in the process of raising a teenaged son who became widely known – and was inducted into the council – as "Holger the Learned." In the late 1590s, Holger married a daughter of the younger Axel Brahe (Tycho's brother) and became an admirer and confidant of Tycho.

It would be hazardous to assume that all of these members – and even Tycho's close relatives – invariably endorsed his appropriations. The same caveat would surely apply to Breide Rantzov, the son of Tycho's friend and intellectual ally, Heinrich. However, Tycho may have had one automatic vote, that of his former partner in dueling exercises at Rostock. A *rigsraad* since 1580 (with a cousin married to Tycho's brother Jørgen since that same year), Manderup Parsberg seems to have been somewhat sensitive about the misadventure of their youth. At least, when an allusion to it was made in the (afterwards published) funeral oration for Tycho in 1601, Parsberg reacted by seeking redress through young King Christian. Claiming that no one had ever insinuated that Tycho's loss was anything but the accidental result of a fair fight and that he and Tycho had remained good friends for the rest of their lives, he petitioned to have the allusion to the duel and its consequences stricken from the oration, on the grounds that even though he was not named, the incident was so notorious that the mere mention of it constituted "a defamation and injury" to him.[27]

According to whatever books Tycho had kept (or reconstructed) up to 1588, the insufficiency of his incomes over the years had accumulated to six thousand dalers. It was a large sum, equal to the

[26] *DBL* X, 649.   [27] XIV, 244.

total of all the annual cash pensions he had previously received from Frederick and enough, if invested prudently, to provide an interest income equal to his annual cash pension. Tycho's request seems not to have gone to the full council but to have been handled instead by the four councillors who had been chosen as regents. At least, it was two of their number, Niels Kass and Jørgen Rosenkrantz, who turned up at Uraniborg for a tour of inspection on 8 July, and the other two, Christoffer Valkendorf and Admiral Peder Munk, who joined in signing warrants for the payment of the full six thousand dalers issued four days later.[28]

When all of this went smoothly, Tycho – not one to concern himself about avoiding an appearance of excess – pressed further the regency's manifest goodwill. In short order he received a document committing the regency to maintain Uraniborg during their tenure and promising to advise the young king (when he attained majority) to do the same. And because one aspect of these provisions was particularly important to Tycho, he succeeded in getting the same document issued a year later under the signatures of the entire council.[29]

In the meantime, the regency had ordered the mayor of Copenhagen to grant Tycho the use of part of the city wall for building an observatory tower and also had given him "and his heirs to have, enjoy, use, and hold forever" two houses next to the one he already owned in Copenhagen, provided only that he build at his own expense another house for the dyer who was currently using one of them.[30]

There was one last favor. In 1586, Frederick had taken the Nordfjord fief from Tycho in what appears to have been an almost purely administrative shuffle. He had given Tycho, in return, cash incomes that were supposedly equal to the net incomes from Nordfjord. Apparently they were not. On 26 June 1589, Tycho got Nordfjord back from the Regency.[31] He was, without doubt, riding high.

No matter how firmly entrenched Tycho was in the system, however, even he could not treat the system itself with impunity. There were at least two respects in which he seems to have been doing exactly that. In the early 1590s, the chickens began to come home to roost.

The first hint that Tycho himself perceived any problems in his circumstances appears in a letter written to the *landgrave* in the spring of 1591. With the departure of Rothmann from the scene and the fact that most of the scientific business between them had been settled, the letters had become largely personal. Thus, although the *land-*

---

[28] IX, 46; XIV, 39–40.   [29] XIV, 41–3, 49.   [30] XIV, 45–7.

[31] XIV, 32, 34–5, 38, 48–9. Tycho seems to have been able to get one thousand dalers out of Nordfjord, and was getting only three or four hundred dalers from the treasury.

grave's letter of February 1591 related that Rothmann had not returned to Cassel and asked Tycho to send his star places as soon as
possible, its main burden was an animal called a '*Rix*,' which the
landgrave had heard was taller than a deer and native to Norway.
Wilhelm had not seen one and hoped Tycho could have a picture
painted from one of the specimens in Copenhagen or in the royal
deer park at Fredricksborg.[32] Tycho thought the *landgrave* must be
thinking of a reindeer and said he was sending a picture drawn partly
from Gesner's etching of it (which did not err much) and partly from
the advice of a Norwegian student of his (Cort Axelsen, no doubt)
who had seen many of them.

At the same time, sensing that what Wilhelm really wanted was an
addition to his own zoo, Tycho said that he himself had an elk at
Knudstrup that he would be happy to ship to the *landgrave*.[33] The
landgrave already had an elk, as it turned out and he also had had a
reindeer sent from Sweden on two occasions. He would be grateful
to get an elk or two from Tycho, he said, but reindeer did not seem
to be able to survive the summers at Cassel.[34] Tycho's subsequent
arrangements to send his elk to the *landgrave* foundered at Landskrona. He got the animal to the estate of his niece's husband, but
before he could ship it, the elk walked up some steps to the manor
house and drank so much beer that it broke a leg going down the
steps again and died.[35]

The *landgrave* had also wanted to know how things were going in
Denmark after the king's death. Not surprisingly, after his dealings
with the regency, Tycho reported that the country was comparatively happy and tranquil:

> Our most gracious prince and king-elect grows daily, not only in
> years and physical strength, but also in virtue and knowledge, which I
> value even more. We have, thus, the firm hope that he will eventually
> follow in the heroic footsteps of his virtuous and greatly-to-be-praised
> father and will be suited to run the government very well. He seems
> born to it: It seems as if God intended him to replace his most
> praiseworthy father, who was taken from us much too soon. In the
> meantime, while the young prince is still under age, four of the senior
> *rigsraads* are in charge of the government.... If anything especially
> difficult or important arises, it is handled [either by the Rigsraad or an
> annual meeting of the nobility, held at midsummer]. Our government
> is, therefore, a kind of oligarchy, which works out well for us until
> our most gracious king-elect attains his majority.[36]

---

[32] VI, 225.   [33] VI, 230.   [34] VI, 233.
[35] VI, 235. Tycho subsequently got two from Axel Gyldenstierne and shipped them to Cassel
after Wilhelm's death: VI, 337–8.
[36] VI, 229.

Having expressed these pious hopes for the future, Tycho went on to say that things were going well for him and that he did not concern himself much with politics but did only what was absolutely necessary. However, he had already confessed to Wilhelm that there were certain obstacles that prevented him from achieving everything he planned to do. He hoped to free himself from them one way or another, but any soil was a country for the brave, and the heavens were overhead everywhere.[37] What seems to have prompted these prophetic remarks were some difficulties that Tycho was having with one of his tenants.

Tycho's relationship with Rasmus Pedersen seems to have begun in 1587, when the latter bought the life tenancy of a small manor that was part of Tycho's Roskilde prebend.[38] The estate, Gundsøgaard, was a small and old-fashioned one, consisting of nine cottages whose tenants still owed labor service to the manor, and a manor house that had just burned down. It was too modest a holding to be worthy of a nobleman but was certainly a comfortable country estate for a gentleman of middling background. Pedersen, however, does not seem to have been the usual middle-class retiree living the life of a country gentleman but was probably of peasant origins. He seems to have known more about farming than most country gentlemen did and almost surely recognized in Gundsøgaard a potential that others before him had not seen.

Pedersen paid what seems to have been a relatively high lease fee and then drew on the unpaid labor dues of his cottage tenants to rebuild a large and splendid manor house. From there he apparently relocated some of the cottages so that he could incorporate their tiny plots into the manor's broad fields. He may even have used his tenant labor to build (and sell) extra cottages, because Tycho's complaint alleged that Rasmus was selling the cottages and keeping the money. Such a policy would have intensified the burden on the commons, and this would surely have worked hardships on the other villagers of Gundsømagle. The records show that Pedersen claimed free fishing rights in Gundsø Lake and dragged with nets until a royal missive ordered him to desist on the grounds that only the crown enjoyed these rights.

Exactly how Pedersen ran afoul of Tycho is not clear. At worst, it may have been as simple as Tycho's coveting the fine manor house Pedersen had built: Tycho is known to have used it in 1594 and may well have realized soon after it was built that it would make a nice

---

[37] Ibid.
[38] The details of the background of Gundsøgaard come from an unpublished paper by John Christianson.

administrative center for the Roskilde fief. At best, Tycho may have been attempting to prevent an aggressive entrepreneur from exploiting the manor of Gundsøgaard by disrupting traditional relationships and reducing poor cottagers to the status of landless laborers. Most likely, the truth lay somewhere in between: Pedersen probably saw a loophole in the law (which, for example, allowed him to build extra cottages) and generated from it enough unanticipated (by Tycho) income to induce Tycho to try to renegotiate the lease.

Whatever Tycho's motivation and whatever his opening move, it brought a remonstrance at Hven from an emissary of Pedersen, the lord mayor of Roskilde. Tycho was so far from being impressed by the mayor or his arguments that he simply seized the manor in October 1590 and expelled Pedersen without refunding his lease fee.[39] Pedersen, however, did not give up. When Tycho's men plowed the fields at Gundsøgaard and sowed most of them, Pedersen had his men sow fifty-two and a half bushels of rye right behind them. Tycho now resorted to more direct action. His men seized Pedersen as he sat at the high seat of his table, clapped him in irons, and carried him off to Hven, where Tycho kept him incarcerated for six weeks until he signed an agreement capitulating to Tycho. In the same swoop, Tycho's agents inspected and registered the condition of the disputed estate and sealed and impounded all of Pedersen's business records.

High-handed action of this sort was by no means uncommon in that day, and usually it worked. In this case, Tycho had a second arrow for his bow, Pedersen's papers. In the spring of 1591, Tycho took the papers and his agents' survey of the estate first to the local court and then to the provincial court. Both of them found for the bloody but unbowed Pedersen. Tycho then appealed to the supreme court of the land, the king and the Rigsraad, and they not only upheld the lower courts on 10 July 1591 but ordered Tycho to return Pedersen's papers unexamined. At least one aspect of Tycho's complaint had been deferred (and thus delayed) to a provincial court. Presumably it too was decided against Tycho, for on 15 February 1592, on the way home from the hearing, Tycho composed an eight-line Latin epigram complaining of the unfair treatment he had just received from the judge.[40]

At this point, when even what appears to have been an attempt to link Pedersen to a man who had drowned in a well had failed, one would think that Tycho could have written off the issue as a lost

---

[39] What is known of the proceedings is found in the royal judgment of 10 July 1591: XIV, 57–60. See also IX, 88.
[40] IX, 192.

cause. However, the determination that was such a virtue when Tycho grappled with scientific and technical problems was in this matter a vice that would not allow him to leave Pedersen in peace. By November 1592, when Tycho had probably prevented Pedersen from farming the estate during the summer and was currently incarcerating a brother and a servant of his, Pedersen succeeded in getting a royal warrant asking Tycho to consider the situation of his subordinate and act toward him in a Christian, fair, and lawful manner so that the crown would not have to intervene again.[41]

To accommodate Tycho's concerns, the prince (or his advisers in the regency, which temporarily included Arild Huitfeldt and Tycho's brother Steen,[42] instead of Munk and Valkendorf) had ordered another hearing, at which all (and only) the allegedly unjudged issues would be discussed and for which Tycho would himself be able to nominate some impartial panelists to participate in the decision. Nothing is known of the final outcome, except that two years later, Tycho seems to have had the use of Gundsøgaard.

However strongly Tycho felt about his case in the Pedersen affair, the fact is that several different judges and juries, composed largely of his noble peers, found it unconvincing. Nobles did not often feel compelled to decide that right lay on the side of the commoner in such disputes, and some of them must have been as embarrassed about the matter as Tycho apparently should have been.

Meanwhile, in August 1591, just a month after the first round of the Pedersen fiasco, Tycho came up for official attention in another respect. The Chapel of the Three Wise Men in Roskilde Cathedral was in disrepair: Because its maintenance was Tycho's responsibility, he was commanded in the king's name to attend to the matter before winter.[43] Under such circumstances, most people would look to the repairs and then try to keep a low profile for a while. Not Tycho. In February 1592, he unblushingly (and successfully) appealed to the regency for an edict requesting rags for his newly built but still under-utilized paper mill.[44]

If none of these contacts with the government seems to have reached the fifteen-year-old prince himself, Christian nevertheless heard about Hven in the spring of 1592 from his chancellor, Niels Kaas. So impressed was he by Kaas's glowing account that he used his next exercise in Latin composition to request permission to visit Hven.[45] He was initially refused, apparently by his superintendent, *rigsraad* Hak Ulfstand, who had been one of the judges in the Pedersen case. But after an appeal to Kaas, over Ulfstand's head, and

[41] XIV, 63–4.   [42] XIV, 61.   [43] XIV, 60.   [44] XIV, 61.   [45] XIV, 62.

then a postponement due to what Tycho feared might be the appearance of plague on Hven, Christian was taken there on 3 July by Regents Munk and Rosenkrantz (and, of course, Superintendent Ulfstand).[46]

The prince seems to have generally enjoyed himself and to have been particularly fascinated with a wind-up globe that showed the revolutions of the sun and moon around the earth. Tycho naturally presented the globe to him and received in return a massive gold chain with Christian's portrait hanging from it.[47] The tour does not seem to have taken long, and there was conversation afterwards about fortification, shipbuilding, navigation, and the rest of the mechanical arts that were supposed to be of interest to kings and the special province of mathematical consultants like Tycho. Christian may even have promised to provide additional funds for a kind of school of navigation at Uraniborg, but nothing ever came of it.[48]

The experience seems to have been agreeable enough that the prince must have been quite shocked to travel to Roskilde the next summer and find that the chapel in which his father (Tycho's patron) was entombed, and on which Tycho had been ordered to make repairs two years earlier, was in a shambles. In view of the real possibility that the roof would collapse and because of Tycho's past (non-) performance, the king specified that the work be ordered done by an independent builder at Tycho's expense if it were not completed before the onset of winter. Perhaps Tycho was satisfied to have it attended to in this manner and was sufficiently naive to take the prince at his word. The next fall (1594), however, Tycho received an angry complaint "that the aforesaid chapel becomes daily more dilapidated and aggravating." After expressing his astonishment that Tycho had disregarded his command to repair the building, Christian ordered Tycho to focus his undivided attention on the matter and get it done by Christmas, without any excuses. If he did not, Christian said, "We will command our deputy in Roskilde to repair the chapel at your expense and immediately give the fief, with its incomes and perquisites, to someone else." This message did attract Tycho's attention. But even then, he was insensitive enough to request (and receive) permission to replace the arched chapel roof with a flat one, in order to save money on the repairs.[49]

Although his troubles at Roskilde with Pedersen and the Chapel of

---

[46] IX, 110, 112; XIV, 63. Whatever the situation in 1592, Hak had been on sufficiently good terms with Tycho three years earlier to agree to function as a guarantor of the loan that Tycho obtained from Queen Sophie. Perhaps the fact that Tycho was show in repaying the loan so that Sophie had dunned Hak for the money had caused some problems. See Sophie's letter in Christianson, "Addenda to Tychonis Brahe Opera Omnia," pp. 239–40.

[47] V, 30.　　[48] *Dreyer*, 216.　　[49] XIV, 68–9.

the Three Wise Men were surely the greatest factors in whatever loss of royal favor Tycho suffered, neither of those issues was anywhere near so personally threatening or time-consuming as what began to occupy his thoughts during the same period: the status of his children.

When Tycho had taken up with Kirsten, he was by no means engaging in an unprecedented activity. Mixed-class relationships, both long-term and short-term ones, had for centuries been a well-documented part of Danish history. So had common-law marriage. From custom predating the introduction of Christianity into Denmark – and for practical reasons dictated by a population that remained largely rural, impoverished, and illiterate – the services of a priest and the formality of church records had never been made compulsory.

From at least the thirteenth century, therefore, Danish law had recognized the act of living together openly for a period of three years as equivalent to matrimony. The Lutheran theology from the Reformation had likewise held that mutual consent – betrothal – and consummation by physical union were sufficient to establish a binding marriage. These customs proved especially convenient for situations in which formal ceremonies were not appropriate because the two parties involved were, in a legal sense, of different "species." It was understood, of course, that the relations could be asymmetric in only one direction: The male commoner who associated with a noblewoman was executed,[50] not just labeled *slegfred*. But in its proper context, the term *slegfred* was not dishonorable. In particular, *slegfred* children were legitimate.

However, it was one thing to be legitimate and quite another to be noble. In sixteenth-century Denmark, nobility was begotten by nobility. In the last generation before the Reformation came to Denmark, many aristocratic clergymen anticipated the event by living with women whom, under the circumstances, it was impractical for them to marry. Yet, in those cases in which the women were noble, the offspring were usually regarded as noble, even though they were not legitimate. For example, Tycho's mother had a distant cousin who was the son of one of the last Catholic bishops of Denmark. Because his mother had been noble, he had no trouble living – and marrying – as a noble.[51]

In Tycho's day, the situation was much the same. As a matter of custom, nobles celebrated a wedding. But living together openly without benefit of clergy was not particularly scandalous, if only for

[50] S. C. Bech, *Danmarks Historie* (Copenhagen: Politikens Forlag, 1963), vol. 6, p. 356.
[51] See Joachim Rønnow's son, Floris.

the eminently practical reason that so many rich and powerful men seemed to find themselves linked to a commoner whom they could not marry. (When Ebbe Andersen Galt went so far as actually to marry a commoner, however, his estate was declared forfeit in 1599, and his own status remained dubious for the rest of his life.)

Clandestine transitory adventures between nobles, however, were strongly discouraged. Marriageable daughters were a resource to be protected by the patriarchs of the realm. Tycho's brother Knud was fortunate, indeed: first, to escape the king's immediate wrath and, then, to find that two years in exile (and perhaps some kind of fine) was deemed to have expiated the long-term demands of justice. A similar peccadillo cost a third cousin of Knud's (and Tycho's) a promising career and an early grave.[52] But the offspring of that illicit union, though illegitimate, could be regarded as noble. When the son in question subsequently redeemed himself on the battlefield, he could petition for reinstatement as a Rosenkrantz and sire a son who would be a baron.

When the mother was a commoner, however, there was ordinarily no question but that the offspring would be commoners. For instance, Tycho's father had a cousin who was a son of the mighty Mogens Gøye. As the product of a common-law union, Falk the younger (as he was entered in the family genealogy) was entitled to use the surname Gøye, just as the twenty children of Mogens's first two (noble) wives were. But he was never recognized as a noble. To be sure, there were a few who achieved this latter feat. Erik Lange had a distant cousin (Niels Lange) who died in 1565 without any issue (after a marriage of thirty years) except an illegitimate son, presumably by a commoner. Lange, a *rigsraad* himself and the nephew, grandson, and great-grandson of *rigsraads*, had his son educated at Wittenberg and Paris and got him a job in the Danish chancellery. After Niels's death his son succeeded first in marrying a noble wife and then in being ennobled himself (although under a modified coat of arms), in 1572. The son of Tycho's Uncle Steen (Bille), *rigsraad* Anders Bille, died in 1633 leaving several children by a *slegfred* wife. After a half century that included many years of faithful service and the threatened extinction of the line, one of them was ennobled and given the right to use the Bille coat of arms.

When Tycho took up with Kirsten, he must have known that noble rank for any offspring they had would be unlikely, in even the best circumstances. Such considerations do not ordinarily trouble twenty-year-olds, however, and Tycho probably did not give it a great deal of thought in the early 1570s. But by 1576, when he buried

---

[52] Frederick Rosenkrantz. See Chapter 13.

his oldest daughter under a tombstone labeling her a "natural daughter,"[53] the distinction was clearly on his mind. By the early 1590s, when he had two sons approaching adolescence, it was a matter of great concern.

Although the status of nobility carried with it subtle or not-so-subtle perquisites affecting every aspect of life, the essence of rank was the feudal right to hold landed estates. It was a privilege for which the Danish nobility had fought determinedly a century earlier. By 1513, in the coronation charter of Christian II, the nobles had won guarantees against pressure on their lands by the crown and legal sanctions against any form of alienation to commoners. With those two safeguards, the nobility achieved virtually a system of entailment, which ensured the perpetuation of their landed wealth and hence their power and influence.

Of particular significance to Tycho's situation was the explicit statement that no commoner could inherit and continue to possess noble land but had to sell such an inheritance to a nobleman. It is hard to imagine that this article would have applied to many cases unless it was specifically intended to include the offspring of *slegfred* wives. However, any remaining ambiguity in the law was cleared up by two ordinances enacted in the early 1580s. The first, dated 11 June 1580, condemned the fact that many "nobles and commoners are said to be living evil scandalous lives with mistresses and loose women, whom they keep in their houses and with whom they openly associate without shame, just as if they were their good wives."[54] The ordinance commanded all parish clergy and rural provosts to put an end to such practices within their parishes and circuits, under the threat of punishment.

By eschewing the use of the word *slegfred*, the ordinance avoided direct conflict with Jutish law. But its intention was clearly to establish a greater degree of royal control over marriage, through the Lutheran church, and to promote a more formal concept of marriage, as a matter in which the state church, and not simply the families or individuals concerned, would be involved. At the same time, by describing *slegfred* wives as "mistresses and loose women," the ordinance must surely have acted to discourage future arrangements like Tycho's. But by 1580, the marriage of Tycho Brahe and Kirsten Jørgensdatter had been well established according to Jutish law and had produced some five or six offspring. There is no evidence that Tycho was admonished by his local parish clergymen or anybody else when the ordinance was first published.

53 IX, 174–5.
54 V. A. Secher, ed., *Forordninger, Recesser og andre Kongellge breve, Danmarks lovgivning vedkommende 1558–1660* (Copenhagen, 1890), vol. 2, pp. 160–1.

Two years later, however, a royal ordinance, dated 19 June 1582, suddenly put Tycho's situation into a new legal context that did not bode well for the future. The ordinance did not attempt to invalidate marriages under Jutish law, but it did provide that henceforth couples would be punished if they lived together without first having been betrothed in the presence of a clergyman and five other witnesses, then having had their banns read on three consecutive Sundays, and finally having been married by a clergyman. Tycho's marriage to Kirsten Jørgensdatter was not legally threatened or altered by this ordinance. But any remaining ambiguity in the legal status of their children was cleared up once and for all, by its stipulation that "when any nobleman has taken a commoner as wife, or hereafter shall do so, and has children with her, then shall these same children of theirs not be or be considered to be noble children or have noble privileges."[55]

If Tycho had previously harbored any illusions about the status of his children, the ordinance of 1582 effectively dispelled them. According to the law, they could not now even use his family name, let alone inherit his noble rank, his landed property, or any of the privileges that might attach to kinship to him or his family. Perhaps the births of sons Tyge and Jørgen in 1581 and 1583 gave the matter an urgency that it had not had as long as only daughters were involved. (Magdelene, Elizabeth, Sophie, and Cecily – at least one of the latter two probably born after little Tyge – survived to adulthood.) At the same time, sons provided – in a way that daughters did not – the opportunity to do something about the future of his family.

By 1584, Tycho had conceived, apparently in collaboration with Vedel,[56] a scheme for getting Hven granted to his heirs in perpetuity, not in the style of a noble fief, but as an institute for the pursuit of mathematical studies. Presumably Tycho approached the king with the proposition at some time. But obtaining a special extralegal dispensation from Frederick would have been difficult. Frederick himself had struggled for ten years with the problem of being in love with a woman of inferior status to his (Anne Hardenberg, a cousin of *rigsraad* Erik Hardenberg), who was, moreover, betrothed to *rigsraad* Olaf Krognos, before he finally resigned himself to setting a royal example by marrying his niece Sophie.[57] So it is not surprising that as of Frederick's death in April 1588, Tycho still had nothing to show for his efforts.

With the friendly regency that ruled Denmark during the minority

[55] Ibid., pp. 296–9.
[56] According to Friis (*Dreyer*, 141) the handwriting of the document (XIV, 26–7) seems to be Vedel's.
[57] Bech *Danmarks Historie* pp. 298–9.

of Christian IV, however, he obtained almost immediate satisfaction. In August 1588 the same men who granted his request of six thousand dalers in reimbursement for personal expenditures on Uraniborg issued a recommendation that Tycho's work be perpetuated

> so that good people could be moved to cultivate this [astronomical] art and that it might always be maintained and cultivated here in this realm. And if any of the aforenamed Tycho Brahe's own [family] could be found to be disposed and fit, then this one shall be recognized for this ahead of others, . . . and if none of his own in particular nor of his close relations nor his staff are found to be fitting, then others of the native born Danish nobility shall be appointed; and if there are no nobles who are fit, then other native born Danes shall be appointed.[58]

Included in the recommendation was the explicit suggestion that permanent funding be established for the institution from canonries and other ecclesiastical properties, the only royal fiefs that were customarily awarded to commoners as well as nobles in that era.

Tycho could scarely have written a better document himself. In one motion it provided for the perpetuation of his work on Hven and the honorable placement of whichever son might display interest and proficiency in astronomy. Its only shortcoming was that the regency did not have the authority to make it binding on the future kings of Denmark. In order to invest the project with as much authority as possible, Tycho obtained a similar document in 1589, sealed and signed by the members of both the regency and the Council of the Realm, twenty men in all.[59] And finally, early in 1590, he – or more likely his mother and/or stepmother – persuaded Queen Mother Sophie to record a statement testifying that King Frederick had intended before his death to install one of Tycho's own children as his successor at Uraniborg.

Having accomplished as much in this direction as he could ever have hoped to may have made Tycho feel better about the fact that when he took little Tyge to Denmark's elite school (Sørø Academy) in the fall of 1590, he had to enroll him there as one of the forty commoners, rather than among the forty noble youths in which Tyge's cousins were numbered.[60] But there still remained the more general task of establishing his children as his legal heirs.

Sometime before 1594, Tycho drafted a document declaring that his entire family recognized the children as his heirs and he appointed

[58] XIV, 41–3.   [59] XIV, 49.
[60] 16 July 90: IX, 88. V. Møllerup, Danmarks Riges Historie: 1536–1588 (Copenhagen: Det Nordisk Forlag, 1933), p. 282. Little Jørgen went off to school in May 1593: IX, 119.

two of his brothers (Jørgen and Knud) and two of his noble friends (Falk Gøye and Erik Lange) as wards in the event of his death.[61] Because under the terms of the ordinance of 1582, there was no question of any of the children's inheriting Tycho's half of Knudstrup, this document must be regarded as evidence of Tycho's concern that his children might not even be recognized as legitimate, that in the absence of such certification, even his personal fortune might revert to the Brahe family in the way that his half of Knudstrup would. If such were the case, then it must have been disconcerting to find that someone among his Brahe heirs – although it might have been a person as remote as the husband of one of his sisters or the ward of widow Sophie Brahe's infant son – would not sign the declaration.

If Tycho had not already entered into negotiations with Steen for the sale of his half of Knudstrup, he must have done so at this time. By mid-1594, the arrangements were complete. Steen obtained sole ownership of Knudstrup without having to wait for Tycho's death and to deal with a committee of family heirs: Tycho converted his share into a form that he could at least plan on passing on to his children, while retaining the lifetime right to refer to himself, in the style of the nobility, as being Tycho Brahe of Knudstrup.[62]

One last project remained: making suitable matches for the daughters, the eldest of whom turned eighteen in 1592. A husband from the offspring of Tycho's noble peers would have been ideal: The children of such a marriage would have had some kind of chance of being ennobled in later years. But noblemen willing to leave such a matter to chance cannot have been numerous. Tycho had two preferred alternatives. One was the so-called lesser nobility, whose pedigrees were short and undistinguished and whose wealth tended to be in urban commerce and buildings rather than rural fiefs and estates. This group tended to marry within itself but occasionally managed a match in the higher nobility and frequently married into the middle class.

The Mule family of Odense was one such that happened to have some dealings with Tycho. Patriarch Hans was some twenty years older than Tycho and entered the public record in 1561 when he killed a fellow citizen of Odense. When he subsequently assaulted and wounded nobleman Eiler Bryske in 1569, Mule had to go into exile for a while, but in 1571 he was given a fief by Frederick II. In 1581 Mule was able to buy a landed estate from the crown, and in 1599 he served as mayor of Odense. In 1573 a daughter, Else, married Master Iver Bertelsen, the controversial priest of Tycho's young

---

[61] *Norlind*, 368–9 (Bilaga 3).   [62] XIV, 68.

manhood. After Bertelsen died, Else in 1591 married Tycho's historian friend, Professor Niels Krag. Presumably through her, a younger brother of hers, Claus Mule, associated himself with Tycho for an unknown period in the late 1590s.

A second possibility, and one that was generally more accessible to Tycho, was the intellectual elite of professors and canons. Men of this class enjoyed high social status and exercised privileges that were similar in some respects to those of the nobility. Because their education at foreign universities inevitably thrust them into the company of Danish noble youths traveling for the same purpose, they formed social and intellectual ties that tended to solidify as the two groups rose to positions of control in their respective arms of the establishment.

From this class, a particularly suitable candidate turned up on Hven in the summer of 1593, Tycho's old student, Gellius Sascerides. Gellius's father had been a professor at the university since Tycho had attended. Gellius's younger brother had spent some time at Hven, too, and a sister who had died in 1589 had been married to a theologian by the name of Jon Jacob Venusin, a native of Hven who was on familiar terms with Tycho and his circle. Already in 1584 Tycho had secured one of the university's prestigious traveling stipends for Gellius, and when Gellius actually departed in 1587, he went with the additional plum of serving as preceptor to Tycho's nephew, Albert Skeel. While pursuing his medical studies in Italy, he acted as Tycho's emissary to the well-known astronomer Magini, even to the extent of supervising the construction of a model of Tycho's sextant for him. At one stage, Tycho contemplated outfitting Gellius for an astronomical expedition to Alexandria. Small wonder, then, that when his former student returned to Denmark with an M.D. degree from Basel, Tycho should have regarded him as a reasonably appropriate candidate for son-in-law.[63]

How his daughter Magdalene felt about it was not recorded and indeed was probably never even discussed. Marriages arranged by their fathers were the common fate of nearly all women in an age of hard economic circumstances. To be sure, family gossip reported an occasional girl with the ill grace to balk at the arrangements made for her, but the same records show many more situations in which a noble bride married a brother of the man to whom she was originally betrothed, after her first intended died. Most of the brides were a bit younger than the twenty years old Magdalene would have been at the time of her wedding, and the average noble groom was probably just about the thirty-two years old that Gellius was.

It seems not to have taken long for Tycho and Gellius to reach some kind of agreement in principle, during one of the latter's several trips to Hven in 1593–4. Working out the details, however, proved to be more difficult than might have been expected. There was probably some kind of quarrel at the end of July 1594, when Gellius left Hven after apparently having been there for much of the year. Tycho's sister Sophie left the island at the same time and probably assumed on the way back to the mainland the role of principal mediator that she seems to have played throughout the proceedings. Gellius was back overnight on 4 August, only to find Tycho off the island. He was back again overnight on 18 September and returned on 16 October long enough to be pressed into service on observations of the lunar eclipse of the nineteenth.[64]

Exactly what the problems were is far from clear. Gellius appears to have feared that Tycho wanted to keep him on Hven "working like a slave." But the root of the problem was clearly pecuniary. When Aunt Sophie got things patched together in October, her letter to Tycho stated that Gellius "desired only a small philosophical wedding."[65] Friis and Dreyer interpreted this as a concession to Tycho's demand, and it is certainly a plausible interpretation. Tycho would let out all the stops for the wedding of his daughter Elizabeth, six years later, but that was a marriage with a (Bohemian) nobleman: He might well not have been eager to make a great display of Magdalene's marrying a commoner.

However, it is just as easy to read the passage in the opposite way, as a final condition imposed by Gellius. Tycho's already-weary response – that it and Gellius's agreement to stay at Hven until the following Easter were fine with him, as long as Gellius did not keep coming up with new problems and conditions[66] – in many ways fits the opposite interpretation better. Noble weddings in Denmark lasted for three days. And although Tycho would have borne most of the expense, even the groom's relatively minor obligations would probably have looked enormous to anyone who did not have the rents from hundreds of farms with which to pay his bills.

Sophie's letter apparently led Tycho to believe that Gellius would come to Hven and sign a contract. When the reluctant bridegroom had not shown up by 6 November, Tycho interrupted the just-launched investigations that culminated in the discovery of the (lunar) *variation* to go to Copenhagen. Negotiations during the next ten days or so – for part of which Tycho may have been at Roskilde checking the work on the chapel roof – finally produced a wedding date of 15 December.[67] On 30 November, Tycho went to Copenhagen again, presumably to attend to the details of the wedding. On that evening or the next, something happened to cast a pall

---

[64] XII, 318; IX, 122–7.   [65] XIV, 81.   [66] Ibid.   [67] XIV, 85.

over the contract. There is no question that Gellius at some point demanded a dowry. The most charitable explanation of this action is that it was a response to his discovering, during what probably seemed like lavish preparations for even their small philosophical wedding, that Tycho expected him to continue to clothe his wife as befit a noblewoman. Whatever Gellius's excuse, one cannot imagine Tycho reacting with anything much less than apoplexy to the proposition that he pay a commoner to marry his daughter. That his excuse may not have been a very good one may be inferred from some of Gellius's other excuses: for example, that he did not know where and when the wedding was to be held, that his wedding clothes were not ready, and that Tycho demanded that he spend a year on Hven after the wedding, doing whatever Tycho wanted him to. What Tycho did admit to was having insisted that Gellius provide his wife with silk and damask clothes and agree to keep her thereafter in clothes as good as those in her wedding chest, and of course, to having refused Gellius a dowry in no uncertain terms.[68]

Whatever happened, the breach was irreparable. When Tycho went to the house of Niels Krag to pick up two Latin verses (of his composing, doubtless) that were to have been sung contrapuntally on the evening of the wedding, the two of them and Krag's wife (Else Mule) began discussing the problem.[69] When the Krags' efforts to mollify Tycho were reinforced by Sophie Brahe and Mogens Bertelsen (administrator of Copenhagen's hospital), Tycho apparently sent Krag and Bertelsen "to corner the loose and wobbly character wherever he might be and ask definitely: Did he or did he not want the girl?" When Gellius responded with an answer that was apparently neither yes nor no, Tycho and Sophie returned to Hven on 4 December.

A few days later, Gellius turned up at Niels Krag's house. Krag was on the way to the house of Jon Jacobsen Venusin, a theologian who was very close to Tycho. Because Venusin happened also to have been married to Gellius's sister until her death in 1589, Gellius was not bashful about inviting himself along. On the way, he told Krag that he had now made up his mind and wanted Krag to tell Tycho that he declined Magdalene's hand and that Tycho could do with her as he liked. Krag, however, said that he would do nothing of the sort, that he would have been obliged to do so earlier when he was acting as Tycho's emissary but that it was now Gellius's duty to talk to Tycho.

By this time they had arrived at Venusin's house, where Venusin, his second wife, and Niels Krag's brother, Professor Anders Krag,

[68] XIV, 77.   [69] XIV, 89.

had just finished lunch. In this company Gellius continued to press Krag to tell Tycho that he renounced the match. Krag again refused and was joined by Venusin in urging Gellius to go to Sophie and ask her help in getting released from the betrothal. They offered to accompany Gellius to visit her, but Gellius was apparently unwilling to take responsibility for breaking a betrothal contract that was for many people already tantamount to marriage. Although the reluctant bridegroom was later to claim that he must have been joking or drunk at the time, the brothers Krag swore that he was both serious and sober.[70]

Before Tycho heard about any of this, he had already decided that whatever period of grace Gellius deserved for making a decision had expired. On 12 December 1594, therefore, he wrote a formal declaration canceling the wedding contract and recorded a summary of Gellius's conduct since the betrothal. Magdalene wrote two similar documents, concluding her description with an expression of her happiness "that our Lord has let me become so mercifully and well freed and delivered."[71] In a fifth document Tycho summarized what had happened on the evening of the great falling-out. A day or two later he arranged to have Niels Krag, Jon Venusin, and Mogens Bertelsen meet him at Gundsøgaard, where they signed his fifth document as witnesses of the dispute. He then sent Krag and Bertelsen to read and present copies of the five documents to Gellius. As they set out, Krag asked what they should do if Gellius indicated that he would abide by his promises. Tycho replied that they should stop wherever they were in the reading and arrange a meeting between Tycho and Gellius in Copenhagen. But it was wishful thinking. Gellius listened and then refused to accept the copies or even acknowledge the reading with his signature.[72]

In January Tycho was back in Copenhagen and heard about the exchange between his friends and Gellius at Venusin's house a month earlier. Because all of Tycho's friends had scrupulously avoided saying anything to him about the incident before then, it seems likely that what Tycho heard was gossip that stemmed from Gellius. It was one thing to cancel the wedding contract himself; it was quite another to have Gellius running around Copenhagen bragging on the one hand that he had rejected Tycho's daughter and complaining on the other hand that he had rejected her because Tycho had set impossible conditions. Unwilling to put up with the situation, Tycho felt that he had to try to set the record straight and that the best way to do it was to sue Gellius for breach of contract.

Jurisdiction over marriage contracts (as well as manslaughter and

[70] XIV, 86–9.   [71] XIV, 82.   [72] XIV, 72.

sorcery) lay with the Senate of the University of Copenhagen. In a similar case, that august body had followed Niels Hemmingsen in declaring that "in this country persons who have betrothed themselves to each other and have witness and proof of this, and have given each other tokens of betrothal – those persons are spouses before God and man."[73] The problem in this case, however, was that Magdalene was on record as being relieved that the marriage was off, and Tycho was probably past the point also of being able to embrace Gellius as a son-in-law, so that the plaintiffs did not want a judgment by that precedent. Instead, what they wanted was a public verdict against Gellius's perfidy. As the distinguished son of a former member of the Senate, however, Gellius was far from an easy mark. And the academic Senate may well have been reluctant to be used in the way that Tycho wanted to use it. After an intensive discussion and even consultation with officials outside the university, the Senate refused to rule against either party but, rather, drew up a "contract of reconciliation" that was supposed to assuage the wounds on each side and close discussion of the matter once and for all.

What brought it up again is unclear. It is hard to believe that the appointment of Gellius as provincial physician of Skaane was not part of the problem. The award of such a plum in Tycho's backyard, a few weeks after Tycho's lawsuit against him, must have been widely perceived as a slap in Tycho's face. But Tycho did not file suit for nearly a year. The general grounds for the suit seem to have been that Gellius would not refrain from defending his side of the issue, in violation of the contract of conciliation. The specific offense occurred at a banquet in the brewers' hall in Copenhagen, when a half-drunk Gellius approached Else Mule and berated her, in front of witnesses, for her husband's support of Tycho.[74]

Tycho sought a formal airing of the whole business, through a legal deposition of all of the testimony relevant to the case. Because Gellius, through his new appointment, was now a resident of Skaane, the case would be heard by the cathedral chapter at Lund. With this action, Tycho seems to have gotten at least an ounce of flesh. Gellius first attempted to get the case delayed, but Tycho succeeded in blocking that move by appealing to Jørgen Rosenkrantz.

Gellius then submitted a rather frightened-sounding petition to the rector and professors of the university, complaining that Tycho was persecuting him, claiming that he had never said or done anything to

[73] J. Nellemann, "Retshistoriske bemarkninger om kirkelig vielse som betingelse for lovligt agteskab i Danmark," *Historisk tidsskrift*, 5th series (Copenhagen, 1879), vol. I, p. 389.
[74] XIV, 95.

harm or annoy Tycho or any of his family, humbly asking the Senate to send two people to persuade Tycho to abandon his vendetta, and promising to be Tycho's humble servant in word and deed for the rest of his life.[75] Whether the Senate rose to Gellius's defense once again was not recorded, but they did not have jurisdiction in these proceedings, anyway. And Tycho certainly knew better by now than to accept Gellius's promises.

The hearing was convened on 25 February 1596 in the consistory of the cathedral, with as little enthusiasm as had prevailed in the previous proceedings. Gellius continued to use every possible tactic for delay. Tycho deposited statement after statement from influential friends and acquaintances, recounting Gellius's misdeeds. Most of the documents have been lost, but each was logged into the minutes of the proceedings with a summary of the argument attending its insertion, and a quotation of its opening and closing lines. The hearings were conducted off and on until 1 July, when the consistory finally decided that the case was outside its jurisdiction and refused to issue a verdict.[76]

The worm had turned, with a vengeance. After years of trading on his favor in high places to win, or at least stymie, cases he probably would have lost if they had been decided on their merits, Tycho was now in the frustrating position of being unable to get a judgment in a case that was probably his to win if it could be decided on its merits. What he could have won, however, is another question. Any pecuniary damages that might have been awarded could scarcely have mattered to Tycho, and gag rules are notoriously difficult to enforce. Nor for that matter, is it even clear that Gellius was an appreciable part of the problem once the betrothal was broken. People talk, and the higher the subject of the gossip is, the more edification there is in bringing him or her down to the level of those who repeat the gossip. At the same time, the protection that Gellius seems to have enjoyed makes it clear that there were people in high places who were eager to see Tycho discredited – people with enough influence to get Gellius appointed provincial physician of Skaane over whatever opposition Tycho could muster.

Rightly or wrongly, Tycho seems to have assumed that Gellius was actively collaborating with them. When on 4 April 1597, a royal inquiry was launched into the theological practices of the parish priest on Hven, under the pretext that the peasants on Hven had complained to the king, Tycho immediately refiled his suit against Gellius, this time to a jury of Tycho's peers.[77] It is not unlikely that Gellius was the source of the information that was to be the basis of

---

[75] XIV, 73–4.   [76] XIV, 97.   [77] XIV, 102–4.

the final indignity heaped on Tycho. But before Tycho could follow through on his last attempt to punish his one-time student, he himself was on his way out of Denmark.

Exactly when or exactly why Tycho decided to leave Denmark will probably never be known. Even Tycho could probably not have singled out one reason for leaving. It would have been remarkable if the personal and political problems that had occupied him almost continuously since the spring of 1591 had not exacted some kind of psychic toll: They certainly seem to have affected his work. Except for his discoveries in the lunar theory, it is hard to see any solid accomplishment in his last six years on Hven.

The eight-hundred-page *Progymnasmata* that had in 1591 lacked only an introduction, a conclusion, and material to fill a three-quarto (twenty-four-page) gap created by Tycho's already-mentioned attempt to shortcut the printing was still unfinished in 1597. To a certain extent this result was bad luck. If Tycho had stuck with his original intention of using the three quartos for an exposition and defense of the Gregorian calendar, he would surely have completed the printing. He might also have made a considerable dent in the Protestant opposition to the papal reform, which prevented its being adopted in Scandinavia until 1700. Sometime in the early 1590s, however, Tycho decided that logic argued for a discussion of the lunar theory. At the time he made the decision, he doubtlessly envisioned nothing more elaborate than a presentation of the Copernican lunar theory, improved by new constants derived from his observations.[78] After the discovery of his new inequalities in 1594 and 1595, however, the composition of the insert became so far from routine that Tycho would not live to see his theory in print.

Before 1591, Tycho had also begun printing his astronomical correspondence. At that stage, his plan was to publish all of his professional letters, in roughly chronological order, grouped according to the degree of their astronomical interest.[79] Already in January 1590, he was hoping to have the first volume published by the following summer.[80] But by August, he could report only that a lack of paper was holding up the printing.[81] It was a five-year-old problem, and he had already started work on a solution to it, his own paper mill.

A diary entry for July 1589 and a letter to Rothmann five months later suggest that Tycho already had the mill running in 1589.[82]

---

[78] See notations suggesting Tycho's preference for Copernicus's lunar theory in XII, 199, 202, 195, 192, 119.
[79] VI, 190.   [80] VII, 223.   [81] VII, 274.   [82] VI, 198.

Either Tycho was exaggerating, however, or it was only the mode for grinding grain that was then in operation, for it was not until March 1590 that the crown gave Steen Bille's widow permission to let Tycho cut a large oak from the royal forest at Herrevad for the main beam of his paper mill, and Tycho's dedication plaque for the mill stated that it was begun in 1590 and completed in 1591.[83] Meanwhile, because the two papermakers he sent to Germany to buy paper did not return, the printing did not progress.[84] It turned out to be just as well, for the death of the *landgrave* in August 1592 suggested to Tycho an alternative way of organizing the correspondence for publication.

Instead of dividing his correspondence chronologically, with the first volume to include all letters up to about 1589, Tycho now decided to extract all the correspondence with the *landgrave* and Rothmann and print it as a memorial to the *landgrave*.[85] When Tycho had originally broached the project to Rothmann, the latter had expressed the conviction that the correspondence should be edited so that a writer did not have to see in print ideas that he had subsequently modified or abandoned. Tycho responded that it was too late for that, as the letters were already being printed (probably a bit of an exaggeration) but that he did not see anyway why the readers should not be made aware of the difficulties that had to be overcome in the quest for truth.[86] Having promised in this context that Rothmann's prestige would not suffer, Tycho followed up on it in a way that Rothmann could not have envisioned: In a letter (unknown to Rothmann) in which the *landgrave* had referred to Rothmann's suffering from "Morbum Gallicum" (syphilis), Tycho altered the phrase to the less indelicate "a serious and harmful disease."[87]

In the fall of 1594, four years after Rothmann had dropped out of sight from a very perplexed *landgrave*, he wrote to Tycho saying that he had been very ill but was now somewhat better and was wondering why there had been no announcement of Tycho's book on the comet of 1577 in any catalogue of the Frankfurt book fair.[88] It is not clear that Rothmann got a real answer to his query, but he certainly got a long one, one that filled twenty-two printed pages as the last letter in the Cassel correspondence.[89]

The letter was mostly a general review of the arguments for and against the supralunarity of the comet, and it was doubtless composed at least partly because Tycho was feeling combative in the

---

[83] XIV, 52–3; VI, 365. An English translation of the plaque is given in *Gade*, 110.
[84] VII, 216–18, 226.   [85] VI, 13.   [86] VI, 190.   [87] VI, 231.
[88] VI, 314–15.   [89] VI 315–17.

middle (14 January 1595) of his first hearing against Gellius. But it also seems clear that Tycho did not envision getting his *Apologia* against Craig into print in the near future and thought that he should use this opportunity to get the crux of his arguments into circulation as soon as possible.

At that time, he did not even know when the correspondence could be printed, for as he told Hayek eight months later, his publications were being held up not only "by some serious impediments from things foreign to philosophy," but also by a lack of paper, arising from the fact that he had been unable for three years now (virtually since his paper mill had been completed) to find a competent papermaker, even though he had been searching diligently. In the same letter of August 1595, however, Tycho was able to report that he had just hired a papermaker and hoped to have an abundant supply of paper soon.[90] Whether or not abundant, the mill did eventually turn out paper, bearing various watermarks depicting either the Brahe coat of arms or the view of Uraniborg shown in Figure 11.1.[91]

The letter conveying the final solution of Tycho's paper problems was the same one in which Tycho shared with Hayek his discovery of the moon's perturbations in latitude. Yet neither in this newsy letter nor in any other communication did Tycho report on another undertaking he was apparently sponsoring at the time: the stamping of a small (3.2 cm) silver medallion bearing his portrait on one side and the Brahe coat of arms on the other, surrounded by Tycho's motto, "to be rather than to be perceived," and the date 1595. A slightly larger one with two different mottoes on it is also preserved. About all that is known about them is that very few of them can have been made, for each survives in only one example. How and to whom (or even whether) they were intended to be displayed, how and where they were made, and what event they might have commemorated all are unknown. Friis thought one might have been struck by Heinrich Rantzov.[92] Indeed, it is not hard to imagine his having done the one with the two unfamiliar mottoes and Tycho's deciding to try his hand at it, perhaps somehow in his paper mill. However, Friis also suggested that the medals might have been meant to celebrate the completion of the star catalogue. This is much less likely.

---

[90] VII, 369. William Lengertz, "Boktryckeriet pa Uraniborg" *Cassiopeia*, 1946, pp. 104–11.
[91] The watermarks are depicted in Lengertz, "Boktryckeriet pa Uraniborg," p. 106. Ove K. Nordstrand, "Småstykker," *Fund oq Forskning*[5] (1958–9): 218–21, found two previously unknown types of watermark and one variant on one of the old types.
[92] *Friis*, 214–15; *Dreyer*, 228, gives all the particulars about both medals.

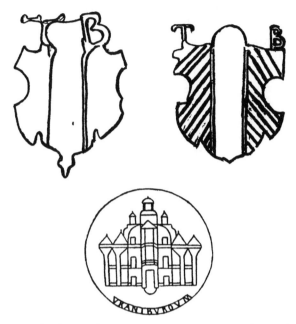

Figure 11.1. The three styles of watermarks used by Tycho on the paper made in his mill.

To be sure Tycho told Rothmann in his final letter in the corres-pondence that he now had a thousand stars catalogued, and he hinted at the same thing in a description of his great globe.[93] But Tycho had told Hayek just a few months earlier that he had 800 stars done,[94] and that seems to have been much nearer the truth. He did a few stars just after writing to Rothmann and about 150 in the winter of 1595–6, but at the head of a few observations made just before leaving Hven in 1597, an assistant was still noting that "60 are required for completing a thousand."[95]

The medal of 1595 was not Tycho's only vanity production. In 1586, when Tycho had had a figure of himself painted on the wall of his mural quadrant, he had Gemperlin prepare at least two other paintings of him. One was a (similarly) life-sized portrait that wherever it was hung at the time, eventually found its way to the

[93] In the letter to Rothmann (VI, 327), Tycho was using the statement to buttress an argument. He repeated it (VI, 337) in a more neutral context, expecting no doubt, that it would be true by the time the letter was published. See also V, 103.

[94] VII, 361.

[95] XII, 482–6, XIII, 98–100. They seem to have done only about half those needed at the time.

royal castle at Fredericksborg where it was hanging when it was destroyed in a great fire in 1859.[96] Another, most likely smaller, one was used as a model for a woodcut that looked very much like our frontispiece and was printed in a very few prepublication copies of the *Progymnasmata* that were sent out after 1592.[97]

Whether because there were a few "miscuts" in the family tree represented in the arch over his head or because Tycho felt he needed the finer detail obtainable from a copper plate, he had a similar picture etched in Holland in the late 1580s. For some reason, however, Tycho did not like the artist's first effort and filed it away where it was used only because his heirs printed it in a book published after Tycho's death.[98] However, he promptly commissioned a second plate, printed as our frontispiece, by the same well-known artist (Jacques de Gheyn) but had it done from a picture that seems to show him somewhat older – if less bald, because of the added beret – even though it carries the same legend claiming to depict Tycho in his fortieth year. Whatever Tycho had originally planned to do with the picture when he first had it executed as a woodcut, when the final version became available to him in the mid-1590s, he had decided that it should be used to embellish the front of his *Astronomical Letters* when it was printed.[99]

By the spring of 1596 Tycho's paper mill had produced enough paper for the fifteen hundred copies Tycho routinely printed of all his editions, and the printing of the *Astronomical Letters* was going smoothly enough to encourage him to compose (and date) a preface on the vernal equinox. As late as the end of May it was still possible to add a poem, composed impromptu by a visiting physician from Bergen,[100] to the prefatory praises Tycho had already collected over the past few years. By late summer, however, Tycho was distributing copies.[101]

[96] Harold Mortensen, "Portraeter af Tycho Brahe," *Cassiopeia* 8 (1946): 53–77. The author shows one copy that was painted from the painting and lists numerous others.
[97] Ibid., p. 62.
[98] Ibid, pp. 63–4. Mortensen speculated that Tycho regarded the first plate as too ostentatious because of all the gold chains he was wearing in it.
[99] Regarding this reconstruction, the print sent to Thomas Savile (VII, 285) must have been from the first plate, because if the second plate (which Tycho liked) had been ready by 1592 when Tycho printed the advance copies of his *Progymnasmata*, he would have used it instead of his woodcut.
[100] *Norlind*,223; IX, 139.
[101] See *Norlind*'s discussion (232–3) of the copy sent to archbishop Johan Adolf of Bremen. *Norlind* has an extensive catalogue of the existing copies of Tycho's circulation (232–7), a discussion of Tycho's various departures from strict chronological order in the printing of the letters (220–3), and a description of the contents of some of the letters (224–32) that differs in some of the subjects it presents from those discussed here (in previous chapters).

For a few presentation copies, Tycho now had ready an additional treat: woodcut prints of his eighteen principal instruments. Some of these had been available for circulation to friends (and printed in *De mundi*) for as long as eleven years, and two of them had been sent to the *landgrave* in Tycho's first letter to him. Over the years, however, Tycho had had his entire collection depicted, with an eye toward publishing a volume on the construction of astronomical instruments. Through a combination of circumstances arising from the cancellation of the *landgrave*'s trip to Denmark and the peculiar conditions of Rothmann's trip there, part of such a volume was already in the *Astronomical Letters*.

When it had become clear to Tycho that Rothmann was not going to return to the *landgrave* after the latter had sent him to Hven "for a little travel and a change of air"[102] and to let him see Tycho's instruments, Tycho had decided that he himself should give the *landgrave* the (thirty-three-page) description of the instruments and buildings that Wilhelm would not get from Rothmann.[103] It was therefore by no means unreasonable that he should have appended his pictures to those *Letters*, except, of course, that he was planning to publish them separately, anyway. It seems likely, however, that there was more to the situation than that, for things were not going well for Tycho.

On 12 April 1596, Prince Christian turned nineteen, and plans were made for his ascent to the throne. The coronation was held on 29 August 1596, and Tycho attended the week-long ceremonies and celebrations. Within a month he learned that Nordfjord, at an income of one thousand dalers his greatest single source of revenue, was being recalled by the crown.[104] A trip into Copenhagen four days after the signing of the transfer probably only confirmed Tycho's worst fears. After years of having staunch support from the likes of Niels Kaas (d. 1594) and Jørgen Rosenkrantz (d. 1596), in a government that was generally inclined to share out the crown revenues among the country's noble elite, Tycho now faced an inner circle that was at least somewhat hostile to him personally and had been selected for its philosophical commitment to a more centralized royal government.

The new chancellor was Christian Friis, who had been one of the judges in Tycho's unsuccessful suit against Pedersen.[105] In addition,

[102] VI, 212.
[103] VI, 250–62 in German, with a much fuller acccount (262–95) in Latin.
[104] *Norlind*, 249.
[105] XIV, 55–7. Friis had sufficient general concern for the underdog to have sponsored studies abroad for poor students. He had been somewhat of an outsider during the regency: *DBL* VII, 398–9.

the economical Christian had appointed someone, Christoffer Valkendorf, to the post of treasurer for the first time since the death of
Peder Oxe in 1575. It was these two that Tycho would subsequently
blame for his problems, and in the case of Friis, he may have been
partially correct.[106] But it is clear that in any case there would have
been changes under Christian. It was only a question of how many
and which ones.

In any such atmosphere of retrenchment, Tycho would have been
extremely vulnerable. Not only had he been funded to an almost
unprecedented extent, but he also had been funded on a completely
unprecedented and a largely temporary basis for twenty years. Even
without having compiled a record of administration that virtually
demanded some sign of official disapproval and having maintained a
style of life that was offensive to at least some of the nobility and
some of the clergy, Tycho's funding would surely have been the first
place the government looked once the decision had been made to
recover control of the royal budget. Nor was Tycho the only one
cut. Jørgen Rosenkrantz's son-in-law, Oluf Bille, lost his fief the
same year, and Tycho's brother Jørgen was relieved of his post as
governor of Varberg, after having had command of most of the
province of Halland since 1590.

Of course it was one thing to grasp the larger picture of government austerity (and it is by no means certain that Tycho ever did) and
quite another to accept the consequences for one's own situation.
By the end of the year, after having failed to achieve anything with
an approach to Valkendorf,[107] Tycho decided that he should make a
more or less complete presentation to the government of the case for
his support. In fact, it is more than likely that he made the decision in
September and was held up in the execution of it only by the time
it took to get his little pamphlet of *Illustrations* ready to append
to presentation copies of the *Astronomical Letters* destined for King
Christian and Chancellor Friis.

Tycho seems to have sent the king his copy first, probably with a
few well-chosen lines of dedication but with no covering letter. On
31 December 1596 he sent Friis his copy, accompanied by a letter
nominally appealing for Friis's help in presenting his case to the
king,[108] but probably really predicated on the assumption that Friis

---

[106] *Dreyer* has an extensive discussion of the imponderables including tales about Tycho and
Valkendorf falling out over a dog (pp. 232–4). Tradition accords Tycho's former friend
Peder Sørensen a role in his downfall supposedly because of anger over Tycho's dispensation of free medical advice and preparations. As royal physician, Severinus had access to
court, but it is extremely unlikely that a commoner would have had much influence in
political matters.

[107] XIII, 101.   [108] XIV, 99–101.

was the one actually making individual budgetary decisions. Frederick II had not devoted a great deal of time or energy to such decisions, and Tycho doubtless thought that the young king would feel he had better ways to occupy himself, too: Certainly when things eventually worked out badly for Tycho, he took the time-honored tack of blaming the king only for having received bad advice.

After some opening compliments to Friis,[109] Tycho reviewed for him the progress of his work at Uraniborg, starting with the observatory and instruments and continuing with his observations of the stars and his studies of the motions of the sun, the moon, and the rest of the planets. It was a work, Tycho reminded Friis, that Frederick had planned to perpetuate but had been prevented from doing so by his unfortunately premature death. The Rigsraad had thought the work worthy of perpetuation, too, and Tycho enclosed a copy of its endorsement (which had been signed by Friis). He then called Friis's attention to the evidence of international repute manifested in the *Letters* and expressed his concern that the young king might not have had time to study the book given to him and might not appreciate how important royal support was to Tycho's work. He then concluded with the hope that Friis would intercede for him generally, although he asked "at this time nothing but that the fief in Norway taken away from me might be mercifully restored, if not for the long term, at least until the first of May" when the annual rents were due.[110]

At first sight, Tycho's letter represents nothing more than simple recognition of the realities of the new governmental order. From it he appears to be accepting the necessity of getting along without the income from Nordfjord but hoping, by means of a thorough presentation of his contributions, to show cause for its restoration. However, the fact that he chose this occasion to raise the issue of long-term funding, by introducing the Rigsraad's resolution of 1589, clouds the picture. One has to wonder whether Tycho might have been so optimistic as to imagine that he could convert a short-term loss into a long-term gain by negotiating perpetuation of his other revenues in exchange for Nordfjord. The alternative to that interpretation almost has to be one at the opposite extreme, that Tycho regarded the status of his other funding as so desperate that he had to

---

[109] For whatever reasons, Friis had been to Hven with his wife and daughter in October 1592 and had stayed for a few days (IX, 115). He was generally interested in things intellectual, as his uncle, Johan Friis, had been. After Tycho's death, Friis sponsored an extraordinary appointment at the university for Longomontanus and paid most of his salary until a regular post opened: *DBL* VII, 401–2.

[110] XIV, 101.

invoke the authority of Frederick and the Rigsraad in an attempt to save it. Whatever the case and however long Tycho worked at analyzing the situation and formulating his response to it, his efforts came to naught.

Three weeks later, Friis wrote a short, cold note saying that he had presented Tycho's request[111] to the king but that, first, Christian had no interest in providing funds from the treasury to maintain Tycho's instruments and, second, the prefect of Bergen could not spare the money from Nordfjord without great inconvenience. Furthermore, although Tycho might not yet have realized that his annual pension also was in jeopardy, Friis had given him the strongest possible hint that it, too, was a thing of the past.

It took Tycho less than a month to react, spectacularly. Even though it was winter, he prepared to leave Hven. As he wrote to Rantzov on 22 February 1597 in a letter that seems to have been overlooked by previous biographers,

> I am in the process of moving my family and part of my household to [Copenhagen]. I think I will stay here for a while, for a certain reason. I believe I will also set up some of my astronomical instruments here, in a tower that lies near my house and that is well suited to observations, and establish here a chemical laboratory with a printing press. I am heavily occupied with all these things. Besides, I have decided to go over to my island today (Friday) to have some of my things brought here and to sort out the rest of the stuff I have left there. Until all this is taken care of, I am so busy that I scarcely have a free hour to write.[112]

What Tycho had in mind is very unclear. Can Dreyer have been right in speculating that there may have been "some disturbances at Hven, which perhaps, had more to do with Tycho's departure than we are aware of"?[113] Tycho later was to state that if he ever did go back to Denmark, he would never return to Hven. Is it conceivable, alternatively, that Tycho thought he could mend his fences by spending time at court? If he had either that illusion or the hope that a partial withdrawal from Hven would exert pressure on Christian, he misjudged, for on 18 March 1597, Valkendorf received an official letter stating that "although His Majesty has hitherto had Tycho Brahe given five hundred dalers annually from the treasury, he now gives a counter order that the aforesaid sum of money will no longer be dispensed, and His Majesty graciously commands that the aforesaid Tycho Brahe be informed of this."[114] On 21 April one of

---

[111] XIV, 101. The word Friis used, *postulata*, ordinarily means "demand." It is hard to know whether or nor Friis was using a pejorative phrase.
[112] VII, 380.   [113] *Dreyer*, 241.   [114] XIV, 102.

Tycho's assistants penned into the meteorological diary: "We catalogued all of the *junker*'s books."[115] Sometime afterwards Tycho
wrote that he and his family left Hven for the last time, shortly after
Easter, on the twenty-ninth of March.[116]

It is hard to believe that Tycho could not have held on. If he had
lost something on the more or less forced sale of his half of Knudstrup, he still had the payments in cash and kind from Hven and his
other small holdings to augment the interest on the money he had
received from Steen as well as, apparently, a share from Inger Oxe's
estate after her death in 1591 and a share in the estate of an aunt,
Birgitte Bølle, in 1595.[117] Presumably these incomes could have
maintained Tycho's family in something approximating a Brahe
style. In addition, he held his seven-hundred daler income from
Roskilde for life and should have been able to maintain Uraniborg
with it if he had ever envisioned being able to build anything from it.

But Tycho had been spending huge sums of money at Uraniborg –
75,000 dalers during his twenty-one years there, according to a
memorandum he drew up after his departure. Even assuming considerable exaggeration (Tycho had mentioned 25,000 dalers in 1590),
he could well have been spending 3,000 dalers annually on all expenses;[118] and although the one-time costs for the establishment of
Hven must have been considerable in the early years, continuing
expenses to maintain an institution have a way of rising over the
years, too. Tycho's subsidies had paid about 2,400 of those dalers.
Fifteen hundred of them were now gone. However frivolously some
of Tycho's money may have been spent, cutbacks to accommodate
such a loss of income would have had to be drastic. There certainly would have been no more printings of large books in fifteen-
hundred-copy editions.

Although someone bothered to continue the meteorological diary
– probably in Copenhagen – until 22 April, the last observation on
Hven was made on 15 March. If the move was all a bluff, undertaken
in the hope of getting at least some of his funding restored,[119] Tycho
played it to the hilt. Except for a few alchemical furnaces that must

---

[115] IX, 146.   [116] IX, 146.

[117] I infer this inheritance from records showing that Tycho's brother Axel inherited an estate
from her. Presumably, Tycho got an equal share in the total estate of Otte's and Jørgen's
half sister, who was known as "the mad aunt of the Brahe's," and was renowned for her
wealth.

[118] XIV, 128. In Prague, Tycho stated that his income from Roskilde – 700 dalers and all he
had left after the revocation of Nordfjord and his annual stipend – did not meet a quarter of
his expenses: VIII, 175–6. The 25,000 figure (VII, 201) had risen to 100,000, a *tun* of gold
by 1598 (VIII, 48).

[119] This speculation was registered by Hugo Yrwing (cited in *Norlind*, 250). I do not find it
convincing.

have been part of the masonry in the basement and his four largest instruments, he took everything, even his printing press. Exactly where he initially took them is not clear, because it was only on 11 April that he wrote in his log that they had just moved into their "new museum," in Copenhagen.[120]

There he apparently made some effort to resume his observations, until he was forbidden by the city authorities (of whom the most influential was Valkendorf) from setting up his instruments on the section of the city wall adjacent to his house. Meanwhile, his world continued to crumble. A few days after Tycho left Hven, the king issued a warrant commanding the chancellor and another nobleman to look into the charge that Tycho had mistreated the peasants on Hven and also to see "if he has dared to act against the ritual, as you, Christian Friis, are aware."[121]

Tycho's previous biographers have unanimously taken this document at face value and assumed that the peasants on Hven, at a time when they had just spent many days removing all of Tycho's possessions from Uraniborg and when all they could hope to reap from further agitation was a reversal of the decision that seemed to promise an end to their troubles, actually lodged a complaint. In fact, the reference to Friis in the warrant makes it highly likely that it was the chancellor himself who lodged the complaint and that he used the peasants as a smoke screen to cover his harassment of Tycho.

If Tycho's refiling of his lawsuit against Gellius on the next day was connected with the warrant, he apparently thought that Gellius must have supplied some of the information on which Friis based the warrant. The task of accompanying Friis on this fact-finding mission fell to the holder of the capital fief of the area, Helsingborg, which holder happened to be Tycho's brother, Axel. The consequences of an adverse judgment were not trivial: Ludvig Munk had lost his fief the previous year over just such a complaint. In Tycho's case, Axel's influence seems to have been equal to whatever merit (or influence) there might have been behind Friis's complaint, so that the civil component of the investigation came to nothing. On the religious component, however, the prefect of Landskrona was commanded to take the priest of Hven before the bishop of Seeland, Peder Winstrup, to be judged by the two of them "according to what is Christian and right."[122] The specific charge against the priest, Jens Jensen Wensøsil, was that he had been omitting exorcism from the ritual of baptism.

This was an issue that had been contended in the Danish church for over thirty years. Even Luther had had some doubts about retaining

<hr>

[120] IX, 146.   [121] XIV, 102–3.   [122] XIV, 104.

the traditional words adjuring evil spirits in the baptismal ceremony, and Melanchthon had openly characterized them as irrelevant. Naturally, therefore, Niels Hemmingsen had been unenthusiastic about them, and some Philippists in Denmark had actually campaigned for their abolition. However, the exorcism was not only part of the authorized ritual of the Danish church but had the additional virtue of having been omitted by the great revisionist, Calvin. It had, therefore, come to be seen by many as a symbol of ecclesiastical authority and discipline and also as a touchstone against the crypto-Calvinism that had to be defended against with vigilance.

In 1567, a pastor with ties to the intellectual circle in which Tycho was raised, Iver Bertelsen, had been removed from his post and put into prison for omitting exorcism from the rite of baptism. In 1588, Jon Jacobsen Venusin had been temporarily deprived of his call for the same offense.[123] Now, in April 1597, Tycho's appointee on Hven, Jens Jensen Wensøsil, was tried for the same offense. According to Tycho, Jens could have been beheaded if powerful friends had not intervened.[124]

In fact, however, it was obviously Tycho himself who was the target of these proceedings. The result of the inquiry was a foregone conclusion. The prefect of Landskrona was Ditlev Holck, who had actually visited Hven once in 1594 to notarize the sale of Tycho's half of Knudstrup to Steen. That had been business. This visit was probably pleasure, for Holck was a son of Tycho's predecessor in the prebend of Roskilde and, of course, a son of the wife to whom Tycho had tried to deny the rights to a year of grace in the revenues of the prebend. The learned bishop Dr. Winstrup had played a heavy-handed role in protecting Gellius through both of Tycho's proceedings against him. Within a few years he himself was to baptize a daughter of King Christian without uttering the exorcism.[125] But he and Holck found Wensøsil guilty and the subject of "extreme royal disfavor," not only for omitting the words of exorcism, but also for failing to "punish and admonish Tycho Brahe on Hven, who for eighteen years has not taken Holy Communion but lived an evil life with a concubine."[126]

In a few months after Christian IV's coronation, then, all of the building and planning Tycho had done to secure the future of his work and his family had come unraveled. Nor could that have been Tycho's only concern. The implication that his labors had not been

---

[123] *DBL* XXV, 327–9.   [124] VIII. 7.

[125] *DBL* XXVI, 114–15. Winstrup seems to have omitted the exorcism, contrary to his own beliefs and only at Christian's explicit command.

[126] XIV, 105.

worth the money expended on them would have been upsetting to
anyone and all the more so to Tycho, who knew, on the one hand,
how hard he had worked and how much he had accomplished over
the years and, on the other, how little of any value was being
produced by the various other expenditures of crown revenues.
Already in 1581, in describing to Schultz the capitalization of his
enterprise, Tycho had acknowledged that his expenditures were
considerable but maintained that such things had to be done right
and that considering how much money was spent foolishly on other
things, it was certainly appropriate to spend some on science. [127]

Could Tycho really now stay and work in a country that had just
officially repudiated everything he stood for? Could he even be sure
that the catalogue of his misfortunes was yet complete? Some two
years earlier, Vedel had been replaced in his post as royal historiographer. Was Tycho in danger of suffering the ignominy of being
replaced at Uraniborg? When Vedel had been turned out, he had
been forced to give up the materials he had gathered for his project to
Professor Niels Krag, on the grounds that they had been gathered at
public expense. [128] With such a threatening precedent, how could
Tycho be sure that he would not be deprived of his observations or
stripped of his instruments? Above all, if his wife of twenty-five
years could now be officially described as a mistress, what future
could there possibly be for his children in Denmark? Would he be
able to pass on to them even the liquid assets he had been at such
pains to accumulate in the last few years? Indeed, the *slegfred* children
of his cousin Torbern Bille were finding their situation sufficiently
threatening that they moved to Sweden.

Yet emigrating was no bargain, either. The seriousness of that
undertaking had been envisioned by Hayek over fifteen years earlier,
in response to a suggestion by Tycho that he could escape the
intrigue of court life at Prague by joining Tycho for work in
Denmark:

> As far as your invitation to move there is concerned, it is a harsh and
> difficult question. I ask you first, my good Tycho, to consider it
> yourself more closely, and ponder all the reasons which make it so
> difficult for me. I am at an advanced age, fifty-six years, and the father
> of many dependents. To migrate to a strange land and, as it were,
> choose exile freely, where both the customs and the language of the
> people are foreign is obviously difficult, grim, and full of misery.
> Nor, will I add, would it bring honor to me and mine to move into
> alien lands without an honest and decent vocation. For how can I

---

[127] VII, 62.   [128] *Dreyer*, 225, 252-3.

think other than that your countrymen will say, or at least suspect, when they see me come uninvited to your country, that I have sought foreign residence because I have been either deported or compelled by penury. And what about mine? Won't they think that I am restless and capricious and insufficiently prudent to allow myself to be so easily convinced by the friendly invitation of one person that I would choose uncertainty for certainty and accept words of honor in place of things of substance?[129]

If the cast was different, the haunting scenario was all too similar. Tycho was only fifty. But his father had died at fifty-three, and he himself would soon be characterized by Kepler as being enfeebled by old age and would die without reaching fifty-five. Although Magdalene was twenty-two, she was still dependent and had five younger siblings down to about twelve years of age. If Tycho's prospects were potentially – and eventually – greater than Hayek's would have been, they still were only prospects and did not materialize for over two years. Nor would they, even then, ever reproduce the security and stability Tycho had known on Hven.

How carefully Tycho thought out the implications of emigrating before he made the decision cannot be known. He had certainly begun to recognize the magnitude of the step by July, when he made a last-ditch attempt to work out a compromise with King Christian: "It is by no means from any fickleness that I now leave my native land and relations and friends, particularly at my age ... and burdened with a not inconsiderable household."[130] But that was five weeks after Tycho, following a few unpleasant weeks in Copenhagen, had taken his household of two dozen persons and set sail for Germany at the beginning of June.

[129] VII, 70. [130] XIV, 109, as translated in *Dreyer*, 245.

Chapter 12
Exile

THE ship carrying Tycho's entourage out of Denmark stopped at Köge and Lübeck before discharging them at Rostock sometime before mid-June.[1] It was a city that Tycho had come to know fairly well from the travels and studies of his youth. In particular, it was the place in which, thirty years earlier, he had fought and recuperated from the ill-starred duel with Parsberg. As Tycho stayed there through the summer of 1597, figuratively licking the wounds to purse and pride suffered in the previous year and renewing his acquaintance with the wise old men who had seen him through his earlier crisis, he must have sensed more than one melancholy reverberation from the past. His old friend and agent, Brucaeus, had died in 1593.[2] But the theologian Chytraeus, and the clergyman who may have been Tycho's former landlord, Lucas Bacmeister, were still alive, and there was still a considerable colony of Danes in the city, which Tycho quickly sought out for news and opinion.

The news was that literally on the day Tycho had left Copenhagen, 2 June, the king had ordered a third commission to go out to Hven for some kind of investigation which Tycho could only interpret (but probably wrongly) as further harassment.[3] And on the tenth, an executive order had been issued transferring Tycho's canonry at Roskilde from him to the chancellor, Christian Friis.[4] The

---

[1] On 16 June (VIII, 3) David Chytraeus wrote a short letter welcoming Tycho to the "City of Roses."

[2] *Allegemeine Deutsch Biographie* III (1876), 374–5.

[3] The inspectors were Professor (of Hebrew) Ivar Stubbe and Professor Thomas Fincke. The latter was a mathematician of some note who in 1583 had published the trigonometry text in which the words tangent and secant were introduced as useful definitions. By 1597 he was a practicing physician (he became professor of medicine in 1603). The report attributed to the inspectors is that the instruments were useless. But since three of the four instruments left there were among Tycho's best, and he subsequently reclaimed and used them all without complaint, this judgment may have been clouded by the disinterest of a mathematical theoretician in things observational. Possibly the reference was directed primarily at the alchemical laboratory, which had doubtless been left with a fair amount of debris in it, and, besides, represented an enterprise to which Fincke – a dedicated anti-Paracelsan – was notably antagonistic. Finally, the fact that Fincke was a cousin of Peder Sørensen's wife, and an avowed enemy of Tycho may also have entered into the assessment: See XIV, 105. Tycho was not the only one whose interpretation (XIV, 102) was colored by third- or fourth-hand reports of the mission: See XIV, 144. The report seems not to have survived, and no one has offered a convincing suggestion as to the nature of the mission.

[4] XIV, 106.

opinion that counted, as far as Tycho was concerned, was that all was not lost, that Tycho should give the king one last chance before burning his bridges by applying abroad.[5]

In exactly what state of mind Tycho moved first from Hven and then from Denmark, no one has yet managed to argue convincingly. To all appearances, he left in a huff, insulted, confused, and perhaps a bit frightened by the speed with which his empire had crumbled. But there were other considerations, too. The flurry of investigations on Hven, particularly the last one, undertaken after Tycho had left and with the apparently benevolent intention of inhibiting vandalism of Uraniborg, may indicate that conditions there had deteriorated even further than the records suggest. Nine months after his departure, Tycho was to say that he was not willing to return to his island under any circumstances.[6] At the same time, he had strong reasons for believing that a return to any location in Denmark would limit the futures of his children in a way that he obviously found hard to accept.

Thus, it is not clear that Tycho should have been particularly eager to win reinstatement. Yet, ordinary prudence dictated that he communicate with Christian in some way, to explain his position and justify his "resignation," in order to get a benediction from his king, if nothing else. There were still several ways in which Christian could harass or even harm Tycho, who had property to sell in Copenhagen, rents to collect from Hven, instruments to fetch from Uraniborg, and enough uncertainty already associated with his future abroad to want no complications from antagonism in his home country. If his appeal proved unexpectedly effective, he might salvage an institution for one of his sons or one of his assistants. Even if he had to run it himself for a while, it would be a small price to pay for the immortality that would be conveyed by the perpetuation of his observatory. And if it came to nothing, it would not have cost him anything, either: He did not feel that he could apply to the emperor, anyway, until he could send with the inquiry a printed copy of his *Mechanica*.

However strongly Tycho may have felt about returning to his position, he did not let it show in the letter he finally wrote to Christian on 10 July. The problem, he explained, was money. It had been his ambition since youth to devote his life to the pursuit of astronomy, and as long as it had been possible to do this in Denmark, he had been happy to live there. In fact, in addition to the revenues he had received from the crown, he had spent considerable sums of his own money, particularly after being given reason to believe, first by the late king and then by the regency and the council, that Uraniborg would be endowed permanently. Unfortunately, after twenty-one years of loyal service to king and country, he had found his expectations unfulfilled – to say nothing of all the other things that had happened to him through no fault of his own. And because he was so far from being able to carry out the work from his own resources that he had been obliged to part with his hereditary estate, he had seen no alternative to seeking support elsewhere. Tycho hoped that Christian would not resent this: It was strictly a last resort to avoid being a burden to the king and the kingdom. In fact, if he were given an opportunity to continue his work in Denmark under fair conditions, he would not refuse to do so. If not – if it were ordained that he work abroad – he would remain a loyal subject of the king, wherever the Almighty might send him.[7]

Not until nearly three weeks after writing this noncommittal letter did Tycho do anything to reveal how much he actually hoped to achieve with it. The occasion seems to have arisen from an investment opportunity that had presented itself in the meantime – a decision by the duke of Mecklenburg to float a large loan on behalf of his two young nephews and wards. Duke Ulrich was the sovereign of the Baltic region of Germany of which Rostock was the principal city. More importantly, from Tycho's standpoint, he was the father of Queen Sophie of Denmark and, hence, the grandfather of King Christian. Some ten years earlier, in fact, he and his wife had accompanied their daughter on a visit to Hven. Apparently Tycho had not thought that this brief introduction – or a subsequent opportunity to provide the duke with some chemicals[8] – was sufficient to entitle him to appeal to Ulrich for influence on his grandson, for it seems unlikely that he would voluntarily have delayed his appeal to him until the end of July, when his letter to Christian of 10 July may already have been acted on.

But now, presumably, Tycho saw the possibility of realizing something besides mere annual interest on an investment. If so, he probably first sought preliminary contact with the duke's financial agents, to establish the fact that he was in a position to provide the entire requisite, ten thousand dalers. Indeed, he would eventually do so, after negotiations in August that culminated in terms of 12 percent annual interest, the security of ten noble cosigners (and their heirs), and a mortgage on an entire country[9] (Tycho's extreme caution would turn out to be justified when he set about recalling the loan).

[7] XIV, 108–11. The entire letter is translated into English in *Dreyer*, 243–5.
[8] VII, 106; XIV, 62.
[9] The documents revealing the negotiations are printed in XIV, 115–18.

Before committing himself to anything, however, Tycho wrote a polite letter to Ulrich, announcing that he was in the duke's domain after having had to leave Denmark for reasons that he would rather not set down on paper and was now seeking ways to continue his work somewhere else.[10] He did not explicitly ask the duke to intercede for him with Christian, but he certainly implied that restoration of his funding in Denmark was one way of ensuring the continuity of his work. And lest the duke not take the hint, Tycho wrote to his chancellor, asking him to bring the matter up with Ulrich.[11] One of them rose to the occasion quickly. On 4 August Tycho received, for his approval, a draft letter from Ulrich to Christian,[12] asking the new king not to allow the work that had brought such credit to his father and his country to remain unfinished.

If Tycho's letter to the duke of Mecklenburg is taken as evidence that Tycho was actively seeking reinstatement, it leaves only one interpretation of Tycho's letter to Christian. Tycho was too proud to bend his knee in humility, even to his king, even at a time when, after three months of living in temporary circumstances, the enormity of "leaving his native land and relations and friends, and going abroad at the age of fifty, burdened with a not inconsiderable household,"[13] may just have begun to strike home to him.

Whether Tycho's two letters were merely the final stages of a calculated bluff to win reinstatement or whether they represented the last-minute erosion of confidence in a conscious decision to relocate the family abroad, they failed in their objective. Nor can there be any doubt as to the fundamental reason for their failure. It was the obvious lack of humility in Tycho's declaration of his position. This is not to say that Christian saw no substantive problems with Tycho's letter. On the contrary, he presented an itemized rebuttal to Tycho's various claims:[14] So far was Tycho from being faultless as a landlord that he had for some years taken the incomes and tithes of the church on Hven without bothering to maintain it in any way or even to pay the parson an appropriate wage. And it was common knowledge that he had not only sanctioned religious unorthodoxy but also permitted economic abuses of his tenants that had given the king ample reason to transfer the fiefs "to others who would keep them under law, right, and established custom." As for Tycho's being so reduced by his expenses that he had had to sell his right in Knudstrup, "it is said here that you have (money) to lend in thousands of dalers to lords and princes, for the good of your

---

[10] XIV, 111–12.   [11] VIII, 5.   [12] XIV, 113–14.   [13] XIV, 109.
[14] XIV, 121–3. Again, *Dreyer* translated the entire Danish letter into English: 248–52.

children and not for the honor of the kingdom or the promotion of science."

What disturbed Christian most, however, was Tycho's effrontery in challenging the withdrawal of royal support. Fiefs and benefices had ebbed and flowed from the crown for as long as it had existed as an institution, without anyone's having the temerity to think – or to express the opinion aloud, at least – that he had a right to be funded by the king. Unable to understand Tycho's view that he was providing unique services that were both costly and had to be paid for by someone and were just as significant as the traditional contributions of the crown's noble servants, Christian saw only insufferable insubordination.

After recounting the points of Tycho's letter in the style of the day, Christian let Tycho know in no uncertain terms that he greatly resented Tycho's "not blush(ing) to (act) as if you were our equal" and warned him "that we expect from this day to be respected by you in a different manner, if you are to find in us a gracious lord and king." In ending his letter, Christian repeated his astonishment that "your letter is somewhat peculiarly styled and not without great audacity and want of sense, as if we were to account to you why and for what reason we made any change about the crown estates."[15] Thus, whereas Tycho had "humbly stated" that he "would not refuse" to continue his work in Denmark if he were treated justly, the king now "graciously answer(s) you that if you (wish to) serve as a mathematicus and do what he ought to do, then you should first humbly offer your service and ask about it as a servant ought to do. . . . When that is done, we shall afterwards know how to declare our will." In the meantime, although professing to be mortified that Tycho was soliciting support abroad, "as if we or the kingdom were so poor that we could not afford it unless you went out with women and children to beg from others," Christian was unwilling "to trouble ourselves (further) whether you leave the country or stay in it."

Christian's response did not find Tycho in Rostock. At the end of August some kind of pestilence had struck the city, and Tycho had been happy to accept an offer by Heinrich Rantzov to move into one of his castles in Holstein until his course of action became clearer.[16] The first estate to which Tycho had gone, however, had turned out to be too remote for his taste. As he told Rantzov in a letter dated 17 September, the castle was beautiful, but he had left Hven in order to relocate, not retire. He had a book to finish and needed access to skilled people who could engrave pictures and do the printing for him. Accordingly, he wondered whether Rantzov had something near either Hamburg or Lübeck that might better serve Tycho's needs.[17] Rantzov did (as Tycho doubtless well knew). Around the middle of October,[18] therefore, Tycho settled into a nicely renovated old castle on the outskirts of Hamburg, where the long-awaited reply from the king soon caught up with him.

Christian's response had not been chasing Tycho very long. Apparently everything else that had crossed the royal desk through August and September had been regarded by either the king or his ministers as being more significant than Tycho's letter, for according to Christian, it had just been presented to him during the week before the 8 October date of his response. For Tycho, this delay was doubtless just one more piece of evidence that it was really Valkendorf and Friis who were responsible for all his troubles.

But whatever the case, it no longer mattered. Whatever Tycho might have expected, the opportunity to return to work and then see what bounty might fall from the hand of a gracious and benevolent majesty was not much of an offer. Indeed, for a man of Tycho's pride, it was probably impossible even to recognize such an offer as a realistic option. Still, as he wrote to a friend in Denmark, it was about what he had expected, except that he had not thought it would be so harsh and undeserved:

> It is regrettable, extremely regrettable, that the young king who is otherwise noble and sensitive by nature should be so blinded that he cannot see and understand what is true and right, and probably of greater value to his realm than other useless things. No doubt the time will come when experience and circumstances will render him more clear-eyed and sensitive, but it will be too late for me and my researches.[19]

So it was to be exile, after all. On 20 October, a defiant Tycho entered into his log a prose account "concerning the cause of the interruption of the observations and my going into exile" and poured out his soul in a poetic *Elegy to Denmark*.[20] On the next day he resumed his observations of the planets and began the final push toward publication of his *Mechanica*.

The circulation of a systematic presentation documenting the

---

[17] VIII, 5.

[18] Tycho seems to have been at Rantzov's estate of Bramstedt on 8 October talking over the various alternatives when he sent a letter to Basel: VIII, 8. On the twentieth, Tycho composed his *Elegy to Denmark* at Wandesbeck.

[19] VIII, 10. Norlind (261) assumed that Christian probably did little more than sign a letter prepared by his advisers.

[20] XIII, 101–2, 102–4. Dreyer (254) gives a précis of the *Elegy*.

superiority of his instruments was a project that Tycho had envisioned almost from the initial successes of his workshop.[21] Already in 1584 he was mentioning the possibility to Brucaeus, and by mid-1585, he had prints of his most advanced sextant and small equatorial armillary available to send to Hayek.[22] Two others, depicting the instruments used to observe the comet of 1577, were ready when De mundi went to press in 1586–7,[23] and the several used in various aspects of work on the new star were printed for the almost-completed Progymnasmata in 1592–3.[24]

In 1591, meanwhile, Tycho had sent to the landgrave of Hesse a lengthy letter containing descriptions of his various instruments. In 1596, this letter, which included six plates of Hven, Uraniborg, and Stjerneborg, was published as part of Tycho's Epistolarum astronomicarum liber primus.[25] As an afterthought, which was probably associated with the first of the cutbacks in his revenues from the crown, Tycho printed the eighteen woodcuts of his instruments that were then at hand and sent them in the form of a supplement to his Astronomical letters to King Christian and a few other dignitaries and friends.[26] Although the illustrations of Tycho's instruments did not influence Christian (who, after all, had seen the instruments themselves a few years earlier), Tycho remained convinced that they were crucial to his hopes of finding a new patron.

Once he settled into Rantzov's estate of Wandesbeck, therefore, Tycho quickly produced his Mechanica. Deciding that if eighteen illustrations were good, twenty-two would be better, he augmented the woodcuts of his "working" instruments with copper-plate engravings of the great Augsburg quadrant, his first sextant, the old Q. max., and his giant globe.[27] While they were being produced, Tycho composed long descriptions to accompany each of his twenty-two plates, shorter descriptions of a few lesser instruments, several thousand words on his life and work to date, and an extensive account of the facilities on Hven where he had achieved his results. By the end of 1597, he was writing a dedication to Emperor Rudolph II[28] and proofreading pages from the press.

[21] VII, 79. [22] VII, 96. See also VI, 38, and VII, 109.
[23] IV, 370, 375. [24] II, 249, 251, 332, 340, 345. [25] VI, 243–95.
[26] "Icones instrumentorum quorundom astronomiae instaurandae gratia." Surviving copies are very rare, and Dreyer estimated that fewer than twenty copies were made. III, 409–10.
[27] Tycho had hoped (in 1586) to do all the plates as engravings (VII, 103), because he thought they were artistically superior (VII, 330). Presumably he could not get an engraver out to Hven and did not get the woodcuts redone at Hamburg because he did not want to delay the publication of the Mechanica.
[28] V, 10.

While Tycho was working on the *Mechanica*, the few assistants who had followed him into exile were finishing up another project he had conceived some years earlier and now wanted to enlist in the same cause as the *Mechanica* – his star catalogue. We saw earlier that the star catalogue, which has survived as Tycho's definitive work on the subject, was printed as part of Tycho's exhaustive study on the new star, at a time when Tycho thought that publication of the *Progymnasmata* was imminent. As it turned out, problems with the lunar theory had kept the book unfinished and would continue to do so until 1601. In the meantime, however, Tycho had resumed his cataloguing, vigorously enough to obtain positions for another 150 stars during the winter of 1595–6. There seems to be no doubt that this activity was motivated by the fact that Tycho's catalogue contained 777 stars instead of the thousand-odd featured in the traditional ones.

But what did Tycho intend to do with the additional places when he got them? Given that he was concerned enough to make the observations, it is hard to imagine that he did not plan to publish them, perhaps even by reprinting an expanded catalogue of one thousand stars to replace the "short" version in the *Progymnasmata*. Yet the extra positions were never printed by either Tycho or his scientific heir, Longomontanus. These contradictions can be resolved by assuming that although Tycho obtained the number of positions he wanted, he did not obtain the quality he demanded. Presumably, Tycho originally intended to collect two independent sets of data and average the results, but his decision to leave Hven and his failure ever to get reestablished well enough for serious observations left him without time to do the things that remained to be done.

In February 1597, suspecting that he had to be prepared to impress a new patron, Tycho chose to devote his last weeks of observing time on Hven to "filling out the thousand"[29] – doing enough new places, however roughly, to bring his total count up over one thousand. If the results were not good enough to allow either Tycho or Longomontanus to print them, they were good enough for advertising purposes (and Kepler). At Wandesbeck, Tycho's assistants quickly incorporated them into the primary catalogue, to produce an elegant manuscript version suitable for presentation to the emperor. On 2 January 1598, Tycho dedicated the catalogue to Rudolph as a New Year's gift, composed a preface outlining its virtues relative to the existing catalogues, and made plans to send it

[29] XIII, 98.

to Prague with Tycho junior,[30] who was now nearly seventeen and beginning to travel in his own education.

While Tycho was devoting his energies to seeking support abroad, he seems not yet to have abandoned hope of salvaging something in Denmark. Tycho was certainly aware that he was in serious disfavor. When he learned in December that Rantzov was about to go to court, he wrote a solicitous letter urging him not to say or do anything that would get him into trouble on Tycho's account. If the issue came up, he suggested that Rantzov say only that he had merely extended the hand of friendship to people threatened by plague at Rostock, who, as far as he knew, had done no harm to either king or country and who in any case would find shelter somewhere else if not from him. In the meantime, Rantzov's hospitality kept Tycho near Denmark, in case the king should wish to recall him.[31] Given the tone of the king's reply to Tycho's first letter, Tycho could scarcely attempt any more overt negotiation, all the more so as any second letter he sent would have to run the same gauntlet – Friis and Valkendorf – that had been so devastating (as Tycho was convinced) to his first appeal.

Christian's wedding at the end of November, however, presented Tycho with an opportunity that he did not waste. Just before Christmas, the bride's parents, the (soon-to-be) elector of Brandenburg and his wife, stayed with Rantzov on their way home from the wedding. After meeting them at some kind of reception and finding them generally interested in his work, Tycho seized the occasion to enlist them in his aid – not to win reinstatement, ostensibly, but merely to ask their son-in-law not to think ill of Tycho for having been forced by economic circumstances to seek patronage elsewhere.

In order to document his situation and increase the likelihood that the elector would follow through on his promise, Tycho submitted a letter to him on 22 December, saying that he regretted the king's displeasure over his departure but that, under the circumstances of his age and encumbrance, one could scarcely think that he had done so capriciously. At any rate, it was now done, and he had no wish to be reinstated on Hven. What he would appreciate from the elector was a letter assuring Christian that wherever Tycho might find himself, he would prove to be a loyal son of Denmark. Moreover, if Christian should decide after all to endow Uraniborg permanently, Tycho would leave his four large instruments in it and attend personally to

---

[30] The preface to the catalogue and the catalogue itself are printed in III, 335–77. Tycho's covering letter is in VIII, 18–20. For some reason, however, Tycho did not send the younger Tycho (or the documents) to Prague at this time .

[31] XIV, 124–5.

the selection of a suitable director. Otherwise, Tycho hoped it would be possible to remove his four instruments and perhaps even obtain some compensation for the expenses of relocation![32]

Tycho got his letter: four of them, in fact, as both the elector and his wife sent letters to both Christian and their daughter.[33] Unfortunately, the appeals were sent through Friis and Valkendorf, where, again, they seem to have languished for some time. Whether they would have had any effect under any circumstances is doubtful.[34] Whatever the politics of the situation the overwhelming probability is that Christian simply had different priorities for the money.

Christian's priorities led to the construction of many of the grand palaces of Denmark, but they also cost Denmark its unique historic observatory. If Christian had any concern for the existence of Uraniborg, he no doubt felt that Tycho should now be able to maintain it himself after the extraordinary funding he had received in the past. If he ever thought that Tycho might pack up and emigrate, he certainly could not have imagined that it would become an international incident. But as regrettable as Christian found it to have Tycho "seek(ing) for help from other princes," he probably saw himself as engaged in a contest of wills whose outcome might establish the balance of power between the crown and the aristocracy for the whole of his reign. Therefore, once he had more or less promised Tycho some kind of support – if he were to return to the kingdom and go about his business – Christian undoubtedly felt that he had gone as far as he could go and that the next move had to be Tycho's. That Tycho's move might be an attempt to enlist public opinion on his side rather than to accept the king's pardon, either Christian or his advisers had anticipated, for the concluding thought of Christian's letter to Tycho had been an injunction "forbid(ding) you to issue in print the letter you wrote to us, if you will not be charged and punished by us as is proper."[35]

Tycho barely adhered to the letter of this prohibition, let alone the spirit. The *Elegy to Denmark* written into his log on 20 October immediately found its way to Rantzov, who displayed it prominently enough in his castle that not only visiting nobility but even King Christian himself saw it within a year.[36] At the same time, Tycho expressed some of the sentiments concerning the ingratitude of his homeland in the introduction to his *Mechanica*, which was being printed even as Tycho pressed his case through Christian's in-laws. Finally, although the elector was now, in a sense, part of the

---

[32] XIV, 126–7.   [33] XIV, 131–4.
[34] Tycho expressed his doubts in VIII, 88, 98.   [35] *Dreyer*, 251.
[36] VIII, 63–4. See also *Dreyer*, 255.

"family," Tycho was soliciting influence from a foreign prince. And whether or not Christian knew it, Tycho, in order to document his situation for the elector, gave him a German version of his "Reasons for leaving Denmark," which presented and even amplified the substance of his original letter to Christian.[37]

Nor was Tycho yet done. As he began to envision sending out copies of his *Mechanica* and star catalogue to the dignitaries of Europe,[38] he decided that there was no reason to omit Christian from the list. On February, therefore, he drafted a letter to the king, congratulating him on his marriage, musing philosophically that perhaps the difficulties Tycho had encountered had been ordained by fate, thanking Christian for not interfering with his departure from the country but expressing regret that his explanation of his action had not proved satisfactory, and presenting to him the latest fruits of his research.[39] It is not clear that this letter was actually sent to Christian, for the books whose presentation the letter was intended to cover were only delivered after a delay of a year and a half.[40]

Tycho turned his wiles on Valkendorf in the spring. He began by thanking the treasurer for all of his past kindness and particularly for the support he had given Tycho's steward in various dealings with the tenants on Hven. Events since his departure had only borne out "how contrary and disobedient the peasants on that little land are, and what I have suffered from them all the time I lived there, and yet had patience with them and been more kind to them than they deserved." Tycho then slid neatly through his point – requesting that Valkendorf see why the first half of the 1597 rents at Nordfjord had not been paid to his agent in Denmark – to an apology for having bothered Valkendorf and an expression of confidence that the treasurer would see that justice was done.[41]

Life at Wandesbeck was not all business; in fact, Tycho and his family seem to have enjoyed living in the style of a German provincial governor. A letter written by an obscure Danish nobleman named Stygge, who was clearly no admirer of Tycho's, corroborates the inclination toward ostentation suggested by other bits and snatches of information about such things as the furnishings, decorations, instruments, and even the coach used by Tycho on Hven. When Tycho went to Hamburg, according to Stygge, he drove with so many horses and attendants and such pomp and ceremony that one would think he was more than a nobleman. In general, Tycho had so little to do with his countrymen that he could scarcely even be

---

[37] XIV, 127–30.
[38] Lists of the known recipients of each are provided in *Norlind* in 286–93 and 295–6.
[39] XIV, 133–4.   [40] XIV, 156.   [41] XIV, 139.

regarded as a Dane. In church (on the rare occasions when he went), he sat in a place reserved for persons of special distinction and was shown other honors, too, that were reserved for high-ranking dignitaries.[42] All in all, it must have been a heady experience for a family whose status had been a subject of public gossip, or worse, for years.

And yet, as is so often the case, Tycho's status is revealed more truly in the abuse of his enemy than in the praise of his friends. People assume, Stygge wonderingly told Tycho's former preceptor and friend, Hans Aalborg, that with Tycho's departure Denmark had been robbed of both eyes. Some regarded it as a barbarian outrage, others as an example of Machiavellian intrigue, or envy. Others reproached the Danes' ignorance and crudity, as if, like the rooster in Aesop's fables, they could not distinguish between a pearl and a kernel of wheat. How could one refute the judgment of so many distinguished men? Practically everyone valued Tycho so highly – as if no other nobleman in Denmark were more learned – and looked at the rest of the Danes as illiterates. If only it could be shown in print how insignificant Tycho was and how useless he had been, so that one could vindicate the homeland from all the accusations by irresponsible people![43]

If Tycho wished to avoid the company of those Danes who wanted to be able to say that they had seen their notorious countryman in exile, he was at the same time in sufficient need of moral support to be urging various friends and relatives to come to visit him. In February 1598, one of them, Holger Rosenkrantz, did. Holger was a son of Tycho's old protector Jørgen Rosenkrantz and was thus a remote kinsman of Tycho's and about to become more closely tied by marriage to brother Axel's daughter Sophie.

Educated abroad in much the way Tycho had been, Rosenkrantz had acquired enough interest in things intellectual to become a kindred spirit, even though he was almost thirty years younger than Tycho. He had already contributed a poem for the introduction of Tycho's *Mechanica* (in which he diplomatically exhorted Tycho to investigate the deepest secrets of the heavens and smile at the foolishness of the world)[44] and was to continue to exchange letters with Tycho which contain some of the most interesting details (certain sensitive aspects of it in a code they worked out) about Tycho's exile. In later life, Holger was to establish a modest reputation as a writer on religious subjects and the collector of a great number of books and manuscripts, while participating in national

---

[42] XIV, 142–5.   [43] XIV, 142.
[44] V, 11. Rosenkrantz was employed in the Danish chancery at the time!

politics with enough effect to be made a councillor of the realm.[45]
At Wandesbeck, Rosenkrantz provided a great boost for Tycho's
morale, probably most of all by simply answering numerous questions about the reaction in Denmark to Tycho's plight, which was a
subject of considerable morbid fascination for Tycho. But he also
pleased his host by staying with him through the night observing a
lunar eclipse (12 February 1598) from beginning to end.[46] If Rosenkrantz lent Tycho some money, his visit would have been all the
more timely, for Tycho was embarrassingly short of money in the
early months of 1598.

Most likely, in his zeal to accommodate King Christian's grandfather, Tycho had overextended himself somewhat with his loans in
Mecklenburg. He had certainly been counting on some funds from
Denmark – such as the aforementioned half-year's rents from Nordfjord – which did not flow as smoothly as he might have wished. It is
also clear that living abroad on the scale to which Tycho had always
been accustomed was very expensive. A "household" of two dozen
people, living in temporary circumstances, would be a burden in any
case.

But as Tycho complained to Holger when he asked him if it would
be convenient to bring along an extra five hundred to six hundred
dalers when he came, the cost of living in Hamburg was frightful,
three times that of Rostock. He had run through seven hundred
dalers in three months. Rosenkrantz would be able to get his money
back right in Denmark, from two thousand dalers that Esge Bille
owed Tycho, or from a mortgage for one thousand dalers Tycho
held on Schyrmann's house in Copenhagen, or from the sale of
Tycho's own house there. If it were not convenient, however,
Tycho did not want him to go to a lot of trouble, and he certainly did
not want him to delay his trip to wait for it.[47] Whatever Tycho got
from Rosenkrantz, it was not enough, for similar requests went to
his sister Margrethe (that we know of) and even to Vedel.[48] In the
end, as the period through which Tycho was "between jobs"
dragged out to two years, he had to fall back on the good offices of
his host, Heinrich Rantzov, for loans that eventually mounted to at
least five thousand dalers.[49]

Vedel, too, had promised to visit Tycho at Wandesbeck but had
finally decided that he would not be able to come. Feeling disappointed and generally forsaken, Tycho wrote a very harsh letter

[45] *DBL* XX, 81–90.   [46] XIII, 126.   [47] VIII, 24.   [48] VIII, 48.
[49] XIV, 174. Whether Tycho borrowed 10,000 and repaid half of it relatively soon, or
borrowed 5,000 and still owed it after Rantzov's death is not clear. See letter 1 in Appendix 6.

to his old friend which provides an interesting commentary on the relations between the two men. From Tycho's defensive responses, it is obvious that Vedel thought not only that Tycho had overreacted to the loss of his support but also that he, Vedel, was entitled to volunteer that opinion to Tycho.[50] Of course, Vedel had some credentials in this respect himself, having lost his post as royal historiographer in 1595, possibly at least partly because of his association with Tycho. Vedel had even been forced to turn over to his successor, Niels Krag, all the materials he had gathered for his long-projected history of Denmark.[51] But it is clear that the basis of what, in any other circumstance would have been intolerable presumption on Vedel's part, was not his own similar experience but his confidence in their nearly forty years of friendship.

Tycho responded hotly enough, but it was a response based on the merits of the point at issue, not on rank. (Vedel should be wishing him well in his new position, instead of reproaching him for having left the old one.) And Tycho registered enough other complaints with Vedel to worry, when he did not get a reply, that he had offended him. He therefore wrote to Vedel again, asking for his help in getting victuals out of Denmark and asking him, once again, to come to visit him. Winter no longer (10 April 1598) posed problems, and if he were not feeling well, the excitement of travel and the change of air would be good for him. His wife could surely look after things while he was gone, and if the problem was money, purses were more easily healed than old friendships.[52]

From the time that Tycho first contemplated leaving Denmark, to the moment at Wandesbeck when he finally realized that nothing miraculous was going to happen (in the near future at least) to make it feasible to return home, the alternative source from which he envisioned obtaining patronage was Rudolph II, ruler of the Holy Roman Empire. It was not just a matter of having connections at the imperial court through Hayek or even of knowing that the Hapsburg monarchy drew its resources from the productions of a quarter of Europe's population. Overriding both of these considerations was the fact that Rudolph was known to be greatly interested in intellectual esoterica of all sorts and indeed had attracted to Prague and supported through the years a considerable, if motley, throng of delvers into arcane subjects.[53] Even Rudolph's former vice-chan-

---

[50] VIII, 25. See a further defense of his decision to emigrate in VIII, 175–6. Many of the letters between Tycho and Vedel survived a couple of centuries before being lost: *Norlind*, 174.

[51] *DBL* XXV, 183–92.   [52] VIII, 49.

[53] See R. J. W. Evans, *Rudolf II and His World* (Oxford, England: Oxford University Press, 1973).

cellor, Jacob Kurtz, had been a mathematician of some accomplishment, who had read Tycho's work and gone beyond the call of duty to provide an imperial copyright for Tycho's books.

As fall turned into winter, therefore, Tycho began to consider the mechanics of finding his place at the imperial trough. As of 2 January 1598, his plan was simply to outline his work and his situation for the emperor and beg – implicitly, of course – for support. It is unlikely, however, that the letter written to initiate the plan was ever sent. Certainly the copies of the printed *Mechanica* and the manuscript star catalogue, which were cited in the letter, were not sent at that time. Perhaps the baldness of the appeal was too much for Tycho, and he temporized by sounding out Hayek instead. More likely, however, the delivery was first postponed because of delays in the printing of the *Mechanica* and then canceled entirely, when a completely unexpected and very embarrassing complication emerged from the past, in the form of Tycho's plagiarist of long before, Ursus.

Although nearly ten years had passed since Ursus's infuriating deed, it seems safe to say that the offense had never slipped far out of Tycho's mind. The short complaints injected into a couple of letters that provided a context[54] and the extended tale of woe described to a couple of correspondents who questioned him on the issue[55] had doubtless registered Tycho's grievance in the intellectual "grapevine," but they had by no means vented his anger. Already in 1592, after hearing from Hayek (in the summer of 1591)[56] that Ursus had been appointed as court mathematician to Rudolph II – and on the recommendation of Kurtz – Tycho mentioned the existence of a cache of evidence against Ursus and the possibility of filing a lawsuit charging plagiarism.[57]

In actuality, however, Tycho had contented himself with the ingenious device of publishing his correspondence. Of course, there were good reasons independently of this issue for publishing what was in effect the first "journal" of scientific research and opinion. But the fact that doing so would carry his case against Ursus to the public and essentially co-opt the eminent recipients of his letters (Rantzov, Peucer, Rothmann, Hayek, and Brucaeus) as implicit witnesses for his prosecution cannot have been an insignificant aspect of the decision made in 1589–90 to proceed with a project that had been suggested by Rantzov in 1586. Despite the ancillary incentive, however, Tycho did not succeed in getting out the first volume until 1596. And even then, that volume was the correspondence with the

---

[54] VII, 135–6, 149, 200–1, 281.
[55] VI, 157, 179; VII, 305, 321–6, 387–8.
[56] VII, 304–5.   [57] VII, 325.

landgrave, which, with its almost exclusively technical focus, contained only one exchange (with Rothmann) on Ursus's activities. The one statement, however, was enough. Ursus, who already in 1588 had apparently been expressing the opinion orally that Tycho was more an astrologer than an astronomer and would not achieve what people expected from him,[58] retaliated in a way that must have made Tycho wish, early in 1598, that he had left well (or bad) enough alone.

The burden of Ursus's 1588 *Fundamentum astronomicum* had been a desperate appeal for patronage. To enlist recommendations for his work, Ursus had dedicated each of twenty-one diagrams to people he thought might be able to help him. Along with Clavius, Mästlin, Brucaeus, Schultz, Mercator, Dee, Bürgi, Wittich, and the *landgrave* were many lesser figures who are today barely identifiable.[59] Tycho was conspicuous by his absence: The diagram of Ursus's "new hypothesis" was dedicated to the *landgrave*. And Ursus had had the foresight to defend his priority in advance, by documenting his discovery of the system. He had worked it out during the winter of 1585–6, while he was employed in Pomerania as a tutor to the children of Georg Schwauen and had displayed it at Cassel in May 1586.

Although it doubtless took longer than Ursus had hoped it would, his appeal for patronage had borne fruit in Prague. The swineherd whom Tycho had once called "Erik's (Lange's) boy" had now, for some six years, styled himself "mathematician to Emperor Rudolph II."[60] Early in 1597, Ursus published a book entitled *Chronotheatrum*. Like his *Fundamentum astronomicum*, it contained an elaborate documentation of its history: "It had been conceived by Ursus in 1581; thought out in Holstein in 1582; begun in Denmark in 1584; sketched out in Hesse in 1586; corrected in Alsace in 1589; and finished in Bohemia in 1597."[61] And like his *Fundamentum astronomicum*, it was immediately greeted by charges – at least two different ones, this time – of plagiarism.

Independently of this problem, Ursus's reputation was suffering at court. For although he had correctly forecast the death of Rudolph's younger brother two years earlier, "it was discovered that he did not [cast the horoscope] himself but took it from the books of Elias of Prussia and pretended that it was his.... Hence it happened that he was examined by His Majesty with regard to the horoscope of the

---

[58] *Rosen*, 53, 58. For a list of abbreviations of sources, see Appendix 1.

[59] Biographical sketches, when enough is known to make them possible, are provided in *Rosen*, 66–80.

[60] *De hypothesibus astronomicis tractatus* (Prague, 1597), p. F, recto.

[61] *Rosen*, 88.

Prince of Siebenbürgen, in which he knew neither the beginning nor the end."[62]

In the midst of these threatening developments, Ursus probably learned that negotiations were under way to bring Tycho to Prague. Tycho had probably not even considered that the post into which he was hoping to move was, if not actually the one Ursus held, certainly in Ursus's territory. He certainly could never even have contemplated that a peasant could offer any kind of challenge to him. But Ursus, probably feeling that he had absolutely nothing to lose, met him with a fury that he himself likened to the biblical (Hos. 13:8) "bear bereaved of her whelps" in the form of a book with the innocuous title *De astronomicis hypothesibus*.

This book may well have been conceived simply as a response to the publication of Tycho's *Astronomical Letters*, for its format – a half-column-width reprinting of the lines in which Tycho had (1) reported Ursus's appearance on Hven in 1584, (2) related his inability to explain to Rollenhagen the nonintersection of Mars's orbit with the sun's, and (3) accused Ursus of plagiarism[63] – was dictated by Ursus's desire to issue rebuttals in the adjacent half-column. But Ursus finished the book late enough to include a reference to Tycho's having fled Denmark after some great act of villainy. And by that time, Ursus's basic strategy was to counterattack.

So far was the situation from being as Tycho represented it, Ursus said, that it must have been Tycho who stole the system from him, through the betrayal of that "[Judas] Iscariot" Rothmann. And Tycho himself was so ignorant of the technical details of the problem that when he modified the theory, he unwittingly arrived at one already proposed by Apollonius of Perga. For this reason Tycho should have stuck to the observing at which he undeniably excelled and left the higher pursuits of the discipline to more exalted intellects such as Ursus's. As he warmed to his subject, Ursus abandoned all pretense to anything but ridicule. Tycho was an "island Polyphemus" [the Cyclops who imprisoned Ulysses and was subsequently killed by him] who wished to be made "leader, King, and even God among astronomers." What point could there possibly be in having such a large and varied collection of instruments as Tycho had? How could anyone plagiarize the name *sextant*? Ursus had, indeed, visited Rollenhagen in Magdeburg, but Rollenhagen was simply wrong in "thinking that I was a runaway employee of Tycho which I never was nor wanted to be. Let Tycho therefore say who that runaway employee was, or how he could have escaped from him and that tiny

---

[62] Report of Rollenhagen's brother-in-law from Prague (VIII, 73) as quoted and translated in *Rosen*, 128.

[63] VI, 179–80.

island."[64] At one point, Ursus got so carried away with his cleverness that he even admitted to Tycho's basic charge. In the margin next to the accusation of theft, he noted, "Let it be a theft, but a philosophical one. It will teach you to look after your things more carefully in the future."[65]

But if in his vengeful hacking Ursus destroyed what remained of his own reputation, he seriously damaged Tycho's honor, too. For in addition to casting aspersions on every aspect of Tycho's work, he descended to mocking Tycho's nose, crediting Tycho with the ability to "discern double-stars through the triple holes in (his) nose."[66] Nor were Ursus's hatred and vulgarity to be satisfied with mere insult. Seizing on the fact that Tycho had accused him of plagiary (of his hypothesis), he noted:

> The word *plagium* applies to persons, and strictly to a wife or a daughter. But Tycho never married or had a wife. And his daughter, though the most nobly-born of girls, was not yet nubile at the time I was there and so not of much use to me for the usual purpose. But I don't know whether or not the merry crew of friends who were with me had dealings with Tycho's concubine or his kitchen-maid.[67]

Tycho was not Ursus's only target. Ursus felt he had to share his abuse with Rothmann, too. This he did by consistently misspelling the name as Rotzmann (Snivelmann)[68] and describing him as having died of debauchery. At first sight the major offense of Rothmann, who was still very much alive but was indeed suffering from the so-called French pox, had been to receive without comment Tycho's account of Ursus's nefarious activities and refer to him gratuitously as a dirty scoundrel. And even that one reference – at least in respect to the mention of the name – was later claimed by Tycho to have been a proofreading oversight[69] and, moreover, is now known to have been a piece of creative editing by Tycho.

Apparently, however, Ursus and Rothmann had had the proverbial "frank exchange of opinion" at Cassel in 1586, as a cause or

[64] *De hypothesibus astronomicis tractatus*, p. F2r, as reported and translated in *Rosen*, 195.

[65] Ibid., p. F1r. The translation is from Nicholas Jardine, *The Birth of History and Philosophy of Science* (Cambridge, England: Cambridge University Press, 1984), p. 30. Although Rosen (195) followed Kepler in assuming that this was a confession, Jardine regarded it as "clear that Ursus is merely using, somewhat ineptly, the standard defensive tactic of *concessio*, (an)argument of the form 'even if your claims are conceded, your case is still not proven.'"

[66] Ibid., p. B3 verso, as translated by Jardine, p. 35.

[67] Ibid., p. F2r, as translated by Jardine, pp. 35–6.

[68] Dreyer, 273.

[69] VIII, 204: "In my writings I do not deliberately assail anybody's life and character." And indeed, when Tycho published a letter by Kurtz naming Ursus as his plagiarist, Tycho edited the name to N. N. (Noman Nescio): V, 121. For a more complete account and the translation of Tycho's letter, see *Rosen*, 226–7.

effect of the fact that Ursus left the *landgrave*'s observatory without recommendation.[70] For a man who had as much trouble as Ursus did in finding and keeping employment, one can well imagine that both the consequences and the memory of the interaction may have been sufficiently bitter to sustain an eleven-year grudge. Part of the problem might well have been an unwillingness by the very feisty Ursus to accept the kind of high-handed "pecking" that Rothmann seems habitually to have inflicted on Bürgi. Rothmann, on the other hand, was clearly no admirer of Ursus. In a letter dated 22 August 1589, Rothmann concluded, "I would also write about Ursus of Dithmarschen and his theft of your system of the world, but the courier is in a hurry, so that I shall postpone that subject for another time."[71] He then went to the trouble of continuing his thoughts in an extensive discussion (which was never mailed to Tycho) denigrating the competence of Ursus (and, even more so, Bürgi) and referring to him as "Thersites" (the prototypical purveyor of intemperate and scurrilous speech in the *Iliad*).[72]

Perhaps Rothmann transmitted enough of this attitude during his month on Hven to leave Tycho with the feeling that it was appropriate to use the "dirty scoundrel" appellation.[73] But even in the best interpretation, Tycho's action ill accords with his assurances to the *landgrave*'s son (when proposing to publish the correspondence) "that I have interpolated nothing in [the letters] of my own (an action which otherwise would itself had been a crime").[74]

Ursus appears to have thought of responding to Tycho before August 1596, when he applied for and received an imperial privilege for an astronomical book. With privileges came responsibilities, however, and by the time Ursus finished his manuscript he had so far exceeded the bounds of responsibility that the script approval that constituted the final hurdle for copyright was out of the question. The title page, therefore, explicitly stated that it was printed without permission and conspicuously omitted any reference to the printer.[75] One scholar labeled it "a work of quite extraordinary scurrility and virulence" and judged that a satirical work "to which Kepler compared it on the strength of Tycho's description of its contents, pales by comparison."[76]

Tycho got his first look at Ursus's libels when a copy of the book

---

[70] *Rosen*, 56. [71] VI, 183, as translated in *Rosen*, 225.
[72] VI, 361. [73] VI, 351–2.
[74] VI, 14, as translated in *Rosen*, 225. [75] *Rosen*, 114.
[76] Jardine, *The Birth of History*, p. 34. Elsewhere (11), this defender of Ursus's priority calls his work "savage and scurrilous even by the ferocious standards of sixteenth-century polemic."

arrived "about the beginning of March by a courier coming from Helmstadt." By a coincidence that is remarkable in many ways, the same courier presented him with a package from young Johannes Kepler, containing a letter of self-introduction and a copy of his first book. Serious books on astronomy did not appear every year, and Tycho had, in fact, already seen Kepler's *Mysterium cosmographicum* and even "read it, as much as my other occupations permitted."[77] But now, as he incredulously recoiled from one after another of Ursus's fulminations, he found in the middle of *De astronomicis hypothesibus*, seeming to be a glowing testimonial to Ursus's work, a letter from Kepler.[78]

Tycho presciently chose this occasion, when he must have been as upset as he had ever been about anything in his life, to be almost the only instance in his life when he did not cater to anger by losing his temper. Perhaps this very fact is the best measure of the stunning impact of Ursus's vilifications. How does one react to the public suggestion that one's wife and daughter are available to consort with the kind of person who would make such a suggestion? How does one subsequently broach the issue in communicating privately with friends? Above all, how does one mitigate the damages? Is there any reasonable action other than retribution? If there is, it is doubtful that Tycho ever seriously looked for it. He had been right about Ursus from the start.

In 1591, when Hayek had informed him that "Nicholas Reimers, whom you accused of plagiarism" had been appointed as imperial mathematician, Tycho had complained about Hayek's withholding of judgment "as though you doubted whether he should deservedly and flatly be called a plagiarist, perhaps because he so cleared himself in your eyes with his crafty and cunning words."[79] He had told Hayek that Ursus had not only "copied most of my papers here without my knowledge" but also had "tried to impair my reputation and honor at Cassel in the court of the Landgrave and perhaps elsewhere by his insolent and malicious lies.... Yet, not even with a little word [had Tycho] deemed it worthwhile to reply to this nonsense, because by themselves, insults, falsehoods, scurrility in the end lead to absurdities." (So Tycho had probably heard some of Ursus's gutter talk in 1584 or, secondhand, in 1586 before he saw it in print in 1598.) He had come to know the man more than is necessary, he told Hayek, and had warned (using Virgil's phrase)

[77] VIII, 46, 44.
[78] *De astronomicis hypothesibus tractatus*, pp. Dlr, Dlv. The letter is translated in its entirety in Jardine, *The Birth of History*, pp. 53–4.
[79] VII, 305, 321. Translation in *Rosen*, 266–7.

that "a snake lurks in the grass" and "that that shrewmouse will soon betray himself by such a proof as to leave no difficulty in his being understood by you too."[80]

Just a few days after reading Ursus's book, Tycho got a letter from Hayek that would have included a second copy if the book had not been lost or stolen in transit. Kurtz, it turned out, had died three years earlier, but the new vice-chancellor, Corraduc, was equally well disposed toward Tycho. However, appointments at the level Tycho would require were made by Rudolph himself. Tycho would have to apply in person. In his reply, Tycho asked only that Hayek try to determine whether the emperor would frown on Tycho's filing a lawsuit against Ursus for plagiarism and defamation of character.[81]

In Tycho's circumstances – being between patrons, being low on ready cash, spending a great deal of money on temporary accommodations, and being eager to launch a lawsuit against Ursus – one would think that he might have gone to Prague immediately. Hayek's letter had already been delayed considerably through having been sent to Denmark first, because either Tycho had sent his inquiry to Hayek (about the possibility of getting patronage) before he left Denmark or had failed to mention that he had left Denmark. Yet Tycho not only did not make haste, but actually seemed to find one thing after another to do instead of embarking on the journey.

The problem appears to have been Tycho's meticulous nature. As with the careful foundation he laid for all his scientific work, Tycho approached the task at hand with the expectation of leaving no stone unturned. If he were about to beg for something for the first time in his life – perhaps he had learned a little bit from Christian, after all – he would do his best to ensure that it was also the last time he would have to do it. If his preparations were excessive, so be it: The more objective consideration he could provide, the less pride he would have to swallow. He could start with a recommendation from Rantzov. Rantzov suggested soliciting one from his friend the influential elector of Cologne, Archbishop Ernst. (Rudolph was, after all, Catholic.) Once Tycho's student, the Westphalian nobleman Franz Ganzneb Tengnagel, had delivered Rantzov's request to the archbishop at Cologne, he could just as easily go through Holland on the way back and deliver a few more inquiries as well. Tycho himself would request one from the duke of Mecklenburg.

The person on whom Tycho most wanted to rely for the impression he needed to make was himself. He had enough projects in the final stages of development that completing and presenting them to the emperor should cultivate his favor if anything would. By May the *Mechanica* was printed and ready for distribution, starting with copies going with Tengnagel on his errands. With each of the books went a manuscript copy of his star catalogue,[82] as the slow, tedious process of reproducing them by hand turned out copies for distribution.

What appears to have been the decisive hurdle for Tycho was the *Progymnasmata*, which had been almost done since 1592. The stumbling block for it had been the lunar theory. References in Tycho's log to a "hypothesis of '96" and "our restitution of the moon done in '97"[83] suggest that Longomontanus had probably finished the substantive work by the time he parted with Tycho when they left Uraniborg. All that should have remained, then, was the verbal exposition of the theory and what may have been considerable numerical work on the various tables customarily provided to facilitate computation of positions from the theory. Even with the still-unwritten preface and conclusion, however, Tycho should have been able to finish the book quickly. If there was any particular problem, no allusion to it has survived.

That there was a general problem, however, stemming from the onset of age, seems obvious. To be sure, Tycho must have been hindered by the temporary conditions under which he was working, and oppressed by the uncertainties clouding his future. But he was also free, for the first time in years, from the distractions of power politics, lawsuits, and the social obligations attached to Uraniborg. Yet the burst of energy that produced a hundred pages of prose for the *Mechanica* in November seems to have been the last such effort that Tycho could muster. And even the *Mechanica* was so largely descriptive as to bear no intellectual comparison at all with the sustained thrust of the 850-page *Progymnasmata* written between 1585 and 1592 or the brilliant flashes of insight that produced the remarkable discoveries in the lunar theory during 1594–5. Whatever was the general case for veterans of five decades of Renaissance life, the temporary lord of Wandesbeck seems to have been past either the capacity or the inclination for serious scientific work.

Under the circumstances, the help of collaborators became a consideration second only to that of finding support for his work. Yet Tycho had apparently had no assistant of any real ability since the departure of Longomontanus at the closing of Uraniborg. Not surprisingly, a certain "Johannes Eriksen from Hamburg,"[84] accompanied him out of Denmark to Germany, and then to Prague.

[82] Norlind, 286–93, 295–6. [83] XII, 205, 10, 115.
[84] Eriksen arrived on Hven on 15 August 1596; he appears to have been a Dane: *Norlind*, 322.

Because he was eventually kept on by Kepler, he must have been reasonably competent, but he would still have been in training in 1598.

In the aforementioned letter to Vedel dated 10 April 1598, Tycho touched tangentially on his lack of competent help by alluding to a student who aspired to become an astronomer but had no talent for the work, and expressing his uncertainty about how to deal with the situation. The subject of Tycho's concern was probably Claus Mule,[85] a member of Denmark's lesser nobility and a younger brother of Else Mule, Niels Krag's wife. But Tycho could equally well have been referring to Tengnagel, who had been with him since February 1595 and was still of more value as a courier of Tycho's letters to Archbishop Ernst and the Dutch Estates, than for whatever astronomical work needed to be done in the spring of 1598.

Sometime around the end of February, Tycho began to receive some skilled help. When he had spoken with the elector of Brandenburg in December 1597, he had apparently had the presence of mind to inquire after the elector's mathematician in residence, Johannes Müller, who had visited Hven briefly in 1596 when the elector had come to Denmark for Christian's coronation. In February, Müller turned up at Wandesbeck, with a note from the electress asking Tycho to instruct him in the art of preparing medicines, as well as other chemical exercises.[86] That Tycho also used Müller in some astronomical capacity is almost certain, for when Müller later left Wandesbeck, he was given Tycho's working copy of the almost-finished *Progymnasmata*[87] and left behind him some kind of agreement to try to join Tycho in Prague.

The person Tycho really wanted to have working for him, however, was Longomontanus. So when he learned in March that his disciple's tour of a few northern universities had brought him back within range of Wandesbeck (Rostock), he wrote saying that he was in urgent need of his services and asking that he do whatever he had to do to get to Tycho quickly.[88] Apparently Longomontanus turned up sometime during the summer. When he did, he was probably put to work on the lunar theory. For although Tycho received a letter from Hayek toward the middle of June[89] expressing his and the vice-chancellor's judgment that it was time for Tycho to come to Prague, Tycho delayed his departure for over three months. If there is a reasonable explanation of Tycho's reticence, it must be that he was determined to have the lunar theory printed for the

[85] VIII, 50.   [86] XIV, 135. VIII, 65.
[87] *Norlind*, 164.   [88] VIII, 34–5.
[89] Hayek's letter was dated 30 May: VIII, 68.

Progymnasmata by the time he was interviewed by Rudolph. And the fact that when Tycho left Wandesbeck, he left the lunar theory in the hands of a printer in Hamburg, speaks to the same conclusion, all the more so because Longomontanus then again effectively left Tycho's employ.

Exactly what Tycho himself accomplished during the summer is not clear. In May he managed to locate Erik Lange's former secretary, Michael Walther, and obtain a deposition (from Cassel) asserting that Ursus had actually confessed to sneaking peeks at Tycho's instruments and private papers.[90] When Tycho was warned about Ursus's activities, Walther continued, he had arranged to have the student who shared Ursus's garret room search him while he slept. The one pocket Tycho's counterspy managed to search (he had not searched the other for fear of waking Ursus) had yielded four whole handfuls of notes. When Ursus had discovered his loss in the morning, he had run around screaming like a madman, even though the notes that did not pertain to Uraniborg were returned to him. Moreover, his behavior had worsened in the following weeks. When he had finally taken to screaming out fears that his employer might hang him, Lange had dismissed him.

The letters written as Tycho began his second year away from Hven continue to show the moods — melancholy, defensiveness, defiance — already displayed in his *Elegy to Denmark*. From his viewpoint, Tycho was purely and simply a victim of malice and ingratitude. He had built ponds on Hven that stored water in a volume that had never previously been available during the summers. Yet the thanks he had received had been rebellion from below and accusations from above. Roskilde had been taken from him, even though it had been granted for life and carried a year of grace for his heirs.[91] His only crime had been to develop his research to such a high level that its value could not be understood by those who might have taken his side against the machinations of his enemies. His regrets, he insisted, were solely for the loss of honor that had been suffered by his beloved homeland because of his departure from it. He himself looked to the future without fear. The whole world would be his country, henceforth, because he was appreciated everywhere else, if not in Denmark. Even a year later when Tycho arrived in Prague, his attitude was much the same. When he was told by a high official that if he had come ten years earlier, arrangements for

---

[90] The deposition was preserved with Kepler's papers and is in the Pulkova Observatory collection of Kepleriana (Vol. 5, p. 302; as cited in G.W. 14, 469). Dreyer apparently regarded it as unworthy of being reprinted with Tycho's collected works.

[91] XIV, 128–9. On the year of grace see Chapter 11.

funding him would have gone much more smoothly and freely, Tycho observed that he too would have spared himself much grief if he had come to Bohemia when Frederick II died.[92]

Even though Tycho had been implicitly enjoined from publicizing his side of the dispute, he took considerable and obvious satisfaction from learning in the spring of 1598 that King Christian had seen – and been uncharacteristically affected by – a copy of his *Elegy*.[93] At about the same time and, no doubt, in much the same mood, Tycho decided when his *Mechanica* came off the press that he had to send a copy, along with his star catalogue, to Christian. Unhappily, however, the authorized channels to the court were closed to him: Tycho seems to have known that the four letters written to the king and queen by the queen's mother and father at the beginning of the year had still not made it through Friis and Valkendorf by mid summer.[94] Similarly, Tycho had applied the previous September to Eric Sparre, the chancellor of the king of Sweden, with the suggestion that Sweden might have an interest in reestablishing Tycho's enterprise on one of its numerous islands in the Baltic.[95] Despite the fact that Sparre had promised to present in person Tycho's written petition to the king, nothing had come of the venture. Tycho's conclusion, he told Rosenkrantz in a letter dated 8 May, was that diplomats could not be depended on.[96] Therefore he wanted Rosenkrantz himself to find a way to put the books into Christian's hands.

Four months later, Tycho learned that Rosenkrantz still had not found an appropriate occasion. Upset with poor Rosenkrantz on other grounds, anyway, Tycho heatedly denounced his unwilling agent's suggestion that for now they forget about presenting the books. It was not in his character, he said, to withdraw from a course of action once he had set his mind to it. He had sent those books for delivery to the king, and he wanted them delivered no matter how the king reacted to them. If Rosenkrantz did not have the spine for it, there was surely someone else in Denmark willing to stand up – or at least mutter – to "that Philistine king." Councillor Oluf Rosenspaar would do it, or if not he, Councillor Henrik Ramel would.[97] It was the latter who finally literally handed the books past Friis to Christian, but only in the following September: The king received his surprise with the statement that he certainly would not understand the books but that because they were gifts from Tycho he would accept them.[98]

By this time – the end of the summer – Tycho was resigned to

[92] VIII, 168–9.
[93] This was, of course, Tycho's interpretation of a report from Rantzov: VIII, 63–4.
[94] VIII, 88.   [95] XIV, 119–20.   [96] VIII, 59.
[97] VIII, 66, 111–12.   [98] XIV, 156.

exile. He still harbored much bitterness. People to whom he had done no harm had stabbed him in the back, and people he called friends had looked the other way or even criticized Tycho for his reaction. The behavior of some had been outright betrayal. To one of them (most likely Hans Aalborg, whose letter from Stygge had apparently reached Tycho not long before)[99] he had many years ago in Augsburg lent eighteen dalers when he had been broke and drunk. That person would have to be a lot worse off before Tycho would bail him out again.[100]

At the same time, Denmark was a very small country, as he told Rosenkrantz. Even on a two-cubit globe, it was just a little speck: In comparison with the universe as a whole, it was nothing. And from the princes in the rest of Europe, he had had nothing but goodwill. Tengnagel's mission to the archbishop of Cologne had produced not only strong letters of recommendation to the emperor and his secretary but also a letter to Rantzov assuring him that the archbishop himself would do what he could for Tycho if Prague turned out to be disappointing.[101] Nor was Tycho even telling Rosenkrantz the full extent of Ernst's favor. When Tengnagel had estimated Tycho's lost incomes at four thousand dalers or less, Ernst had not been able to believe that such a man could be had for so little money. What a disgrace! One nobleman could lose that much in an evening. Gold one could always get; such men one could not.[102] So great was Ernst's enthusiasm for Tycho that he had even presented his young emissary with a gold medallion and a riding horse.[103]

Tycho's inquiry to Holland had elicited some interest too. Both the civil and military leaders of the country, Olden Barneveld and Maurice of Orange, had promised to seek support for him from the Dutch government.[104] The preeminent classical scholar of the day, J. J. Scaliger, offered his regrets that the deliberations of the Dutch Estates would probably be too slow to give him the pleasure of getting so illustrious a colleague.[105] Therefore, by the time Tycho replied to Scaliger, at the end of the summer, he was his old optimistic, egotistical self: If his trip to Prague came to naught, he would be happy to come to Holland, and because the kings of both England and France knew his work, there was a good chance of getting support in those countries, too.[106]

---

[99] XIV. 142–5; VIII, 90.   [100] VIII, 94.
[101] VIII, 111–12. The letters from the archbishop are printed in XIV, 140–1.
[102] Tengnagel's words, but in German, presumably Ernst's originally: VIII, 81.
[103] VIII, 80.   [104] VIII, 82, 95, 98.
[105] Tycho's former student, Christian Johannsen of Ribe, was studying with Scaliger.
[106] VIII, 109. In a more candid moment two weeks earlier, Tycho had told Krag (VIII, 98) that Cologne would be his first choice and that England would not be tenable because the customs there were alien.

Preparations for the move were made at a leisurely pace. As of mid-August Tycho's plan was to travel light — to take only his sons, a few students and servants, and some of his instruments. With them, he would be able to show the emperor what Tychonic astronomy was, without the fuss of moving and accommodating his whole family. He might well have to return to Wandesbeck for the winter, he told Krag, even if he got satisfactory terms from Rudolph.[107] He did not mention another good reason for leaving the bulk of his household behind: Having a plausible reason for leaving Prague, under any circumstances, would take quite a bit of the strain out of his dealings with the emperor. However much negotiating Tycho thought he might be able to do, it would be nice to know that he would never be in the position of having to choose between refusing the emperor's offer outright and sneaking his bulky entourage out of the country: If either the conditions of life in Bohemia or the conditions of employment with Rudolph did not meet Tycho's expectations, he could simply announce that he was going to fetch his family and then concoct reasons, from Wandesbeck or wherever, for being unable to return to Prague.

Tycho's decision to move in two stages may also have been influenced by the logistics of moving. This seems a bit unlikely, because both Hamburg and Prague were on rivers, and water transport was the most efficient way of getting sizable loads from one place to another. But the possibility of boat or barge travel seems never to have figured in Tycho's plans. From as early as January, when he made arrangements through Rosenkrantz to buy a used travel coach in Denmark, Tycho had obviously decided to move his entourage not on the river itself but, rather, on the roads alongside the river.

When Hayek's signal to come arrived in June, the coach was still in Aarhus awaiting transport. Tycho, a lifelong resident of Skaane, in "Swedish" Denmark, twitted Rosenkrantz about the stereotypical inefficiency of the Jutland Danes and expressed his surprise that now that Rosenkrantz was about to become (in a triple wedding of Axel Brahe's daughters on 13 August) a "Skaansk" by marriage, he had not been able to rise above the handicaps of his heritage.[108] Three months later, the situation was no longer humorous. In letters written on 5, 9, and 14 September, Tycho inquired urgently after the coach, stressing that the need for it was all that stood between him and departure for Prague.[109] On the twenty-fifth, finally, he received word that it had arrived at Lübeck. Four days later, on Michaelmas,

---

[107] VIII, 98.   [108] VIII, 76.   [109] VIII, 110, 111, 113.

Tycho left the last quarters in which he would ever be settled for even as long as six months.[110]

In what might have been an impromptu change of plans, the whole Brahe entourage started the journey. It must have been a formidable enterprise. In his impatience over the delay of his coach from Aarhus, Tycho had finally purchased another one in Hamburg. Including the one he had brought with him from Hven, as an indispensable badge of rank, there were three coaches in the train that must have included several baggage wagons and at least a few people on foot and horseback. For the first leg of the journey they traveled some two hundred kilometers to Magdeburg.

They arrived no later than 5 October, on which date Tycho paid his respects to Duke Otto II of Brunswick by presenting to him an autographed copy of the *Mechanica*.[111] Whether this gesture was merely polite custom when one traversed the domain of a ruler of some rank, whether the duke had a special connection with the emperor, or whether Tycho himself had a special connection with the duke, he left with a letter of recommendation from the duke that he would present to Rudolph — along with those from the elector of Cologne and the duke of Mecklenburg — at their first meeting.

There was other important business to conduct in Magdeburg, too, for the city was the home of Georg Rollenhagen and the temporary refuge of Erik Lange.[112] Lange himself had been abroad since 1591 and was never to satisfy his creditors sufficiently to return to Denmark. In 1602 he was to meet Tycho's sister Sophie long enough to marry her, but he subsequently wandered alone to Prague in 1608 and died there in 1613.[113] The visit of the two exiles — if Tycho could yet think of himself in those terms — may have been largely social. But Lange had also been Ursus's patron at the time of the excitement at Uraniborg, and Tycho wanted a notarized endorsement from him of the document, provided by Michael Walther, describing Ursus's conduct on Hven.

Rollenhagen's testimony held even more promise. He was the rector of the local academy and a poet of some repute but was important to Tycho as the man who in 1586 had witnessed Ursus's inability to give a satisfactory explanation of his system. For the first time, now, Tycho had the opportunity to pump him for every potentially useful detail of the encounter, and there can be no doubt that he did so *ad nauseum*.

There were surely also discussions of strategy. Rollenhagen had sent Tycho a copy of *De astronomicus hypothesibus* the previous

[110] VIII, 114, 125.   [111] *Norlind*, 288.
[112] VIII, 52, 128, 210.   [113] DBL XIII, 549–50.

spring, along with a letter in which he had counseled caution in Tycho's response to Ursus. Even though at his age, he said, he had read many philosophical debates in which the arguments were vehemently disputed, and in all his experience he had never seen a more insolent or impertinent work: "This I surely know, if I wrestle with dung, win or lose, I am always defiled."[114] During the summer, Rollenhagen had given to Tycho inside information sent from his brother-in-law, who was a gem cutter for Rudolph and thus had access to the gossip at court. If Ursus were not exactly in disfavor, his already-related misadventures with two recent horoscopes had certainly removed him from the ranks of Rudolph's favorites. More significantly, there was a rumor "that Tycho Brahe is coming here to our court, and His Majesty has a strong desire for him and will give him a castle three miles from Prague to establish his residence there, a very fine place called Brandeis."[115]

This information probably reached Tycho before he left Wandesbeck. Whenever he received it, a rumor of such specificity should have quelled whatever apprehensions Tycho still had about his reception in Prague. But it clearly did not. When the time came to leave Magdeburg, after eight days,[116] Tycho left his instruments in storage and sent his household back to Wandesbeck. Accompanying them was Longomontanus, who, for some reason had come as far as Magdeburg but could not now be persuaded to abandon his intention of going home for a visit.[117] Tycho and his sons headed for Prague.

The Brahe men, accompanied at least by Tengnagel and a retainer by the name of Andreas, continued south along the Elbe until they reached Dresden. At that point Tycho sent a messenger to Prague to inquire about the protocol for approaching the emperor. Whether this was a standard procedure or a special precaution, it was well advised; for the dirty crowded capital of the Holy Roman Empire turned out to be in the grip of some kind of epidemic, and Rudolph had scurried to the countryside to seek safety in isolation. So effectively had he sealed himself off from possible contamination that it took a month even to get Tycho's communication to him an answer back out.

Given the extremes to which Rudolph had gone to protect himself, his response was a foregone conclusion: Tycho should stay a safe distance away and wait for the emperor to call him when the

[114] VIII, 50–1.
[115] VIII, 72, as translated in *Rosen*, 128.
[116] VIII, 125, 128.
[117] VIII, 132, 133. Longomontanus observed the eclipse of the moon on 31 January 1599 in his home village.

plague abated.[118] With a few words, therefore, it was ordained that Tycho would spend the winter in neither of the places in which he could reasonably expect to find comfortable lodgings for the entire family: Wandesbeck was too far away from the emperor, and Prague was too close.

Fortunately, however, during the several weeks Tycho sat waiting in Dresden, wondering why the messenger was so long on his errand, he seems to have stumbled into a satisfactory alternative. On the way down through Wittenberg, he had met a professor of medicine named Jesensky. Jessenius, as he latinized his name, lived in a house that Tycho had probably known during his student days at Wittenberg as the former house of Melanchthon. The house had passed from the revered theologian to his son-in-law, Tycho's old teacher Caspar Peucer, who may well have been the means by which Tycho met Jesensky in 1598. Whatever arrangements Jesensky and Peucer had, Tycho must have seen some possibilities for himself, for within a week of receiving Rudolph's temporizing message, he had retreated one hundred kilometers back to Wittenberg, moved into the Melanchthon house, and sent Andreas and Tycho junior to Wandesbeck to fetch the family.[119]

As Tycho headed north out of Dresden, Tengnagel had gone south, on another mission of some delicacy. He carried with him the routine presentation package, with a covering letter, for the grand duke (Ferdinand) of Tuscany, in Florence, and similar gifts for the doge of Venice and the noted mathematician and astronomer, Antonio Magini.[120] Tycho's dealings with Magini dated back to the beginning of the decade, when Gellius Sascerides, in Italy pursuing his medical studies, had given to Magini a copy of *De mundi*. Magini's letters of thanks and inquiry to Tycho and Gellius had provoked a long technical response from Tycho and a three-year correspondence with Gellius. The subjects covered included Tycho's instruments, prosthaphaeretic computing methods, world system, and star catalogue; Tycho's and Magini's struggles with the orbit of Mars; the question of whether the geographical pole and hence the geographical latitudes had changed since antiquity; and the possibility of establishing in Africa an auxiliary observatory to catalogue the stars of the Southern Hemisphere.[121]

As of 1598, however, the only contact they had had in seven years had been Magini's dedication of his *Tabula tetragonica* to Tycho, and

[118] VIII, 128–9, 131–2.   [119] VIII, 129, 132.
[120] VIII, 118–19; Norlind, 374–5; VIII, 120–5.
[121] V, 125–8; VII, 289–99, 303–4. Magini's correspondence with Sascerides was printed by Antonio Favaro in *Carteggio inedito di Tichone Brahe, Giovanni Keplero e di altri celebri astronomi e matematici dei secoli XVI. e XVII. con G.A. Magini* (Bologna, 1886).

because Magini had apparently not bothered to send a copy of the book to Tycho, Tycho had discovered Magini's gesture only through a routine acquisition of it. In the foreword to his book, Magini had exhorted Tycho to complete his imminent restoration of astronomy, and Tycho now felt he should explain why nothing had appeared.

First, Tycho still lacked some of the observations necessary for completing the planetary theories. Second, circumstances beyond his control had forced him to leave Denmark and now dictated that he send Tengnagel on a trip with a secret request of Magini, which turned out to be that Tycho wished to have some kind of tribute to his life and work written by an Italian. Whether Tycho simply wanted more front material for his soon-to-be-published *Progymnasmata* or something more extensive is not clear. Most likely, Tycho hoped that Magini himself would assume the task, because of the goodwill demonstrated in the dedication of his book to Tycho. To help persuade Magini, Tycho sent along to him a copy of the new lunar theory that he and Longomontanus had left with the printer in Hamburg. Magini, however, foisted the job onto someone else, or at least told Tycho that Bernardino Baldi would do it. In any case, Tycho did not appear in either Baldi's "Lives of the Mathematicians" or his "History of Mathematics."[122]

Little is known about the six months Tycho spent in Wittenberg. The family left Wandesbeck on 6 January,[123] and must have been happy to arrive at Wittenberg no matter how crowded they were when they all moved into whatever space Jesensky could make available to them. But Uraniborg had not been a large building, and Tycho mentioned nothing about their living conditions other than that there was room for observing and a place for his laboratory. In addition to living in close proximity with Tycho, Jesensky was a physician and recently appointed (1596) professor, so it is not surprising that he should have become sufficiently well acquainted with Tycho through the winter and spring to be chosen to conduct the funeral oration when Tycho was encrypted at Prague less than three years later. Almost the only date recorded in connection with the stay in Wittenberg is 1 February 1599, when Tycho, his sons, and one of his assistants matriculated at the university. Tycho, however, had clearly met at least one of the other professors well before that time and already enjoyed a profitable collaboration with him.

It was probably inevitable that Tycho would establish contact rather quickly with any professor of mathematics whose path he

[122] Norlind, 195–6.
[123] Frobenius reported that they left the place a mess, with neither payment for cleaning or repairs nor tips for the servants. See Friedrich L. Hoffmann, *Der gelehrte Buchhändler Georg Ludwig Frobenius in Hamburg*, Hamburg, 1867, p. 8.

crossed, but Melchior Jöstel was special. Not only does he seem to have satisfied all of Tycho's criteria for the post of astronomical assistant, but he seems to have gotten on very well with Tycho. Presumably it was his astronomical expertise that commended him to Tycho, for things proceeded so quickly on that front that it is hard to believe they were not discussing intricate details of Tycho's new lunar theory before Tycho had been in Wittenberg a month. In any case, by the time of the lunar eclipse of 31 January 1599, "Joestelius" was in print, with a short tract depicting the geometry of Tycho's theory and predicting the phenomena of the eclipse according to the "hypothesis and tables of TB."[124]

What earned Jöstel the honor of prepublishing Tycho's theory is clear: He had found the missing link to a complete solution of triangles by means of prosthaphaereses. Whatever the details of the understanding might have been, some kind of informal swap must have been made. Jöstel got into print – probably at Tycho's expense – and Tycho recorded in his observation log the final steps in the computing scheme he had envisioned when Wittich had broached the concept of prosthaphaereses to him nearly twenty years earlier.[125] A few months later he told one of his correspondents that he and a distinguished professor of mathematics by the name of Melchior Joestelius (for two heads are better than one) had discovered the final steps of the prosthaphaeretic process, so that almost nothing was now lacking for the easy solution of every conceivable problem.[126]

By no means the least interesting aspect of the intellectual barter between Tycho and Jöstel was the fact that each conveyed property worth somewhat less than either party had reason to believe it was. Jöstel's computational scheme was simply an alternative to methods already published by Christoph Clavius in 1593,[127] and Tycho's lunar theory produced a prediction for Jöstel that was so wide of the mark – in Tycho's view, anyway – that he soon had Longomontanus revamp the entire theory.

Even before Tycho knew of this problem, however, he was trying to talk his reluctant assistant back out of Denmark. On 31 December 1598, Tycho was able to respond with equanimity to the news that Longomontanus had inspected the quartos of the lunar theory printed at Hamburg during Tycho's journey to Dresden and had found them deficient in some respect. They could be reprinted at Wittenberg, he replied, while thanking him both for his attendance on the ladies during their trip back to Wandesbeck and for his

[124] Victor E. Thoren, "An 'Unpublished' Version of Tycho Brahe's Lunar Theory," *Centaurus* 16, (1972): 207.
[125] I, 297.   [126] VIII, 201.
[127] Christoph Clavius, *Astroloabium* (Rome, 1593).

promise to write a refutation of John Craig's Aristotelian arguments on the comet of 1577. But if Longomontanus returned and rejoined Tycho, his tract against Craig could be printed at Wittenberg as an appendix to *De mundi*, and Tycho would be happy to support him in studies at some German university (Wittenberg, no doubt) until Tycho was settled in Prague.[128]

Only twelve days later, Tycho appealed more directly and definitely. If Longomontanus were to come to Wittenberg at Tycho's expense, Tycho would get him the post held by Ursus in Prague. And if he did not wish to be so far away from Denmark, Tycho knew of a professor at Wittenberg (Jöstel, no doubt) who was not averse to going to Prague and whose chair Longomontanus might well expect to get.[129]

A year later, when Tycho finally recovered the services of his valued associate, he indeed assigned him to the lunar theory. But that subject became an issue only after Joestelius' printed prediction went awry. It must therefore have been something else that inspired Tycho's recall of Longomontanus: perhaps the feeling, awakened by his letter to Magini from Dresden, that it really was time for the final push on the last remaining task in his restoration of astronomy, the theories of the planets. Tycho had certainly tried to enlist other assistants. He had extended a veiled invitation to young Johannes Kepler some nine months earlier but had heard nothing from it. He had had Johannes Müller's services in Wandesbeck for perhaps even a few months but now had from him only a promise to try to return when Tycho was settled in Prague.

The one thing that looked as if it might work out was the acquisition of Wittich's many diagrams of planetary mechanisms. Tycho had learned in March 1598 that Wittich's library was still more or less intact, when an old friend from his (that is, Laurentius the Dane's) rooming house in Augsburg, Jacob Monau, had written from Breslau saying that Longomontanus had been through there.[130]

In his published correspondence, Tycho, misled apparently by a report about Monau's brother, had written to Rothmann that Jacob had died. The latter had naturally confided to Tycho that he felt very much alive but that he wanted to thank Tycho for his beautiful obituary, anyway, and that Tycho's troubles reminded him of an inscription he had once seen above the entrance to a house in Magdeburg: "Ingratitude has been, is, and ever will be one's reward from the world." Tycho was so impressed by this evidence that someone else had experienced, or at least imagined, the kind of treatment he had received, that he quoted it in his next letter to Vedel.[131]

---

[128] VIII, 133–4.   [129] VIII, 136.   [130] VIII, 28–31.   [131] VIII, 49.

Tycho also sent a letter to Longomontanus directing him to see whether Wittich's library included any manuscripts and whether any or all of it was for sale.[132] Apparently his letter arrived after Longomontanus had left Breslau, for Tycho repeated the inquiry to Monau at the beginning of 1600, asking specifically "if you could get the three copies of Copernicus's *On the Celestial Revolution* that had certain annotations by himself, what he composed about triangles, as well as *On the Celestial Hypotheses*."[133]

At the end of January, the emperor's secretary, Corraduc, wrote from Prague for Tycho to come.[134] This time, too, Tycho was unprepared to accept the call. At the end of February he wrote Hayek indicating that it would be nice to be able to come immediately but that the roads were still covered with snow and he had recently sold all of his horses in Leipzig, anyway. However, he would buy some more horses shortly and hoped to be on his way in a couple of weeks. Any hints Hayek could give on bad roads or plague-infested areas to avoid would be greatly appreciated. Meanwhile, he would determine the latitude of Wittenberg and make observations of Saturn's maximum latitude to complete the twenty-nine-year cycle he had begun at Augsburg.[135] On 7 May, however, Tycho was still at Wittenberg. He had not left, even though he had just received another independent testimony of the emperor's goodwill toward him, he told Rosenkrantz, because the snow on the roads had been followed by heavy spring rains, and then his daughter Magdalene had been gravely ill. There had also been a problem with the completion of some books he wanted to present to the emperor.[136]

Until at least the time of the lunar eclipse of 31 January 1599, the book Tycho had been counting on to accompany the *Mechanica* and the manuscript star catalogue was the *Progymnasmata*. He had probably stayed at Wandesbeck in order to get the lunar theory ready for printing and had, no doubt, assumed that when it was printed he would compose a preface dedicating his magnum opus to Rudolph, have it and a conclusion run off quickly, and present everything to the emperor at his interview. But after he decided to have the lunar theory altered in some respect, Tycho apparently concluded that he could not present an incomplete *Progymnasmata* and therefore had to find an alternative production to impress his prospective patron.

Tycho settled on an ephemeris of daily positions of the sun and moon for the year 1599.[137] It involved a considerable amount of

[132] VIII, 34.
[133] VIII, 237, as translated by Owen Gingerich and Robert Westman, "The Wittich Connection: Conflict and Priority in Late Sixteenth-Century Cosmology," *Transactions of the American Philosophical Society* 78 (1988): pt. 7, pp. 21–2.
[134] VIII, 136–7.   [135] VIII, 145–7.   [136] VIII, 154, 163.   [137] VIII, 163.

computing and may well have been the real cause of the delay in
Tycho's departure, for it is not clear that Johannes Eriksen could
have received any significant help from the likes of Franz Tengnagel
or Claus Mule.[138] Money was also a problem, because selling his
horses to avoid boarding costs over the winter is the first consideration of economy recorded in Tycho's life. However, the provisions
and/or funds from Denmark, for which Tycho told Hayek he was
waiting in February, had still not come in June; yet, Tycho was able
to leave almost immediately when he received a letter from Hayek
indicating that the emperor inquired almost daily as to whether
Tycho had arrived.[139]

They left "right after summer solstice,"[140] traveling en masse to
Dresden. Perhaps humbled by his recent financial woes, a grateful
Tycho recorded for Rosenkrantz the fact that the company had been
treated to overnight hospitality in a castle along the way, by a
nobleman named Loeser. In Dresden two days later, they stopped
long enough to leave the household with an (unnamed) acquaintance
who was a high official of the city,[141] and then they raced to Prague.
On arriving there "about the beginning of July," Tycho discovered
that his reputation, or at least word of his standing with Rudolph,
had preceded him. Some weeks later he was still basking in the
memory of his reception. He recounted the events for Rosenkrantz:

> Immediately on my arrival (at court), I was received in a respectful
> and friendly way by many distinguished men, above all by the em-
> peror's private secretary, Lord Johannes Barwitz, who handles every-
> thing for the emperor. He greeted me with a hearty welcome in the
> name of the emperor, the Senate, and himself, in a garden not far from
> the palace (for he does not stray far from the emperor). He reported
> that the emperor had just heard of my arrival and said many times
> how well disposed the emperor was toward me, even though he had
> not yet met me personally. When I most humbly showed Barwitz the
> three writings I wanted to present to the emperor, and the letters of
> recommendation from the elector of Cologne and the duke of Meck-
> lenburg, Barwitz said that he would speak with the emperor about

---

[138] Tengnagel was in Italy, but Jöstel could have helped with the enterprise.
[139] VIII, 163.
[140] VIII, 187. It was actually 14 June, according to a notation by Tycho (*Norlind*, 338), and it
must have been 14 June in the "old style" (as opposed to *Norlind*, 300), for the "new style"
would have been before the solstice, and anyway, the notation was made at Wittenberg, in
Protestant Germany, where the Gregorian reform had not been adopted (and from which
Tycho provided only one date – old style – on all of his letters).
[141] VIII, 163.

them, because he did not know from whom the emperor wanted to receive them. The following day he related that the emperor did not want to receive them from anyone other than me, myself, and that he would call me to the palace in a short time. I then showed Barwitz a letter written in my behalf by the most illustrious Duke Otto of Brunswick, near Hamburg, to the illustrious noble Lord Rumpf, the most powerful man after the emperor. Barwitz advised me to show the emperor this letter, also, in a day or two (which I had otherwise thought I would have my son do). At an appropriate time, therefore, I took my carriage to the castle and was straightaway admitted to the same Lord Rumpf, who also received me very warmly and behaved in a very friendly manner. Among other things, he said that he had previously heard of me from many high-ranking persons and rejoiced that he was now able to meet me in person. He and others, however, could not get over their astonishment that the king [Christian] had been willing to let me leave Denmark. This he ascribed to the king's youth, as the young are generally more inclined toward martial than peaceful enterprises. But I countered by defending the king and cited his capacity and ability. Whereupon Rumpf responded that those who acted for the king and wielded his authority must either be completely ignorant of learned things or be very ill disposed and envious, to have so completely disregarded the honor of king and country. To this I replied only that I did not doubt that if every aspect of my case had been rightly presented to the king, and he had been advised appropriately by those closest to him, he would have protected my astronomical research in his realm, no less than his father, King Frederick of most laudable memory, had done before him. But perhaps (I said) God has acted by some special providence in order that the astronomical investigations with which I have been so long and so thoroughly occupied should now come elsewhere and redound to the credit of the emperor himself. About this and similar things I conversed with the aforementioned Lord Rumpf (who spoke the most excellent Latin) and left Duke Otto's letter with him. Rumpf then drew me aside and promised to take on my cause himself and to press it with the emperor.... When I came down from the castle, Barwitz met me and drove me, following the emperor's instructions, to a splendid and magnificent palace (which the former prochancellor, Jacob Kurtz, had built in the Italian style, with beautiful private grounds, at a cost of more than twenty thousand dalers); whereupon he showed me all the amenities there and said that the emperor would purchase the whole estate for me from Kurtz's widow if I were pleased with it. I saw that a tower had been built by Kurtz for astronomical observations and that the house was situated near the castle where the emperor lived and worked, so that the resident could readily get there. When Barwitz deduced from what I said and did not say, that the tower would scarcely suffice for a single one of my instruments, much less for many of them, and that I was not really interested in that situation, he mentioned another option: If I did not want to live in Prague, the emperor would gladly turn over to me one of his castles located a day or two outside Prague, where I would be more undisturbed, away from the envy of others. When he noticed that this attracted me more, especially when I said that I had chosen in Denmark to inhabit an island just to enjoy peace and not be disturbed too much, he said he would mention this to the emperor and that he understood that the emperor was already inclined to grant such an alternative proposal. The emperor had also in his graciousness determined to give me an annual grant. I would hear about it more fully as soon as I received an audience (which he thought would happen in two or three days).

After a few days the emperor had me called to the castle. It had been determined in the council, beforehand, that the chancellor, Rumpf, should formally introduce me, as this would be more honorific. But the emperor chose another way at this time, either because some of the populace of the city had died from the plague in the past few days and some of Rumpf's staff had been infected or because the emperor preferred to talk to me in private without witnesses, which is, indeed, what happened. For when Barwitz escorted me to the door entering the chamber where the emperor was, he stayed outside and indicated that he would not go in unless he was called by the emperor. I therefore went in to the emperor alone and saw him sitting in the room on a bench with his back against a table, completely alone in the whole room without even an attending page. After the customary gestures of civility, the emperor immediately called me over to him with a nod, and when I approached him he graciously reached out his hand to me. I then drew back a bit and gave a little speech in Latin, in which I said that I had been called at his gracious command by a letter from Prochancellor Corraducius and was now here – and would have been quite a while ago if it had been possible – in order that if he might graciously condescend to excuse the delay, he might support and patronize with his imperial favor me and the research I had conducted long and well. Therefore I wanted humbly to leave with him, himself, some documents from the archbishop of Cologne and the duke of Mecklenburg, which he also benevolently received and opened in my presence. But he laid them on the table without reading them and immediately responded to me graciously with a more detailed speech than the one I had delivered to him, saying, among other things, how agreeable my arrival was for him and that he promised to support me and my research, all the while smiling in the most kindly way so that his whole face beamed with benevolence. I could not take in everything he said because he naturally speaks very softly. I then thanked him humbly for this proof of his grace and mentioned the three books I had brought with me to give him with the utmost deference. When he graciously responded that he would accept them I immediately fetched them from my son Tycho, who had them where he was waiting in the antechamber. I then entered again and left them with the emperor, who was still sitting at the table. When he took them and laid them out on the table, I reviewed the contents of each briefly. Then the Emperor again responded with a splendid speech, saying most graciously that they would please him greatly. I then removed myself according to the proper courtesies.

I found Barwitz still waiting outside. And when he was called in to the emperor by a signal, he told me not to leave the room before he returned. He came back immediately and said that the emperor wished to have the mechanical device on my carriage shown to him (for he had seen it from the window when I drove up to the castle). I therefore had my son fetch the mechanism from the carriage, gave it to Barwitz, and showed him how to explain its construction to the emperor. When Barwitz had done this and returned to me, he said that the emperor said he had one or two similar devices but none so large or made in exactly the same way. He did not want to accept mine from me but would have one made for himself by his astronomer, from the pattern of mine. Barwitz added that the emperor was very favorably disposed toward me and that after he referred the case to his council, in a short time, he would settle with me the matter of an annual grant and suitable quarters. In the meantime, they told me to have the rest of my family, which he understood I had left at Dresden, come to me, and the emperor himself would do everything necessary to make sure that we lacked nothing needed to live comfortably. Wherefore, on the following day I immediately sent my son Tycho and Claus Mule to bring them from Dresden, and they arrived eight days later.[142]

Although Tycho could not regard the Kurtz house as the answer to his long-term institutional needs, it was the emperor's first offering to him, and the widow had already "voluntarily" moved out of it. It seems, therefore, that Tycho was probably domiciled there well before the family arrived from Dresden.[143] To the inevit-

[142] Letter to Rosenkrantz (30 August 1599, old style): VIII, 163–6.
[143] VIII, 158–9.

able excited inquiries about the family's future circumstances, Tycho probably still had few answers.

The court calendar was very crowded, and the stipend suggested by Rudolph as a result of Tycho's itemization of his expenses (and, no doubt, some subtle mention of the fact that Tycho's support in Denmark had been in excess of 2,400 dalers) was very large, so the matter of Tycho's salary took some time. Because Rudolph's figure (3,000 guldens, each roughly equal to the more or less debased but standard-weight gold pieces used in every other realm of Europe) was higher than the salaries enjoyed by even the counts and barons of high standing and long tenure in the emperor's service, some of the councillors dared to speak against the proposal. But because neither Rumpf nor the chamberlain was so incautious and Rudolph was insistent, the motion carried.[144] The star-struck emperor also agreed to pay certain incidental expenses, which Tycho thought "might amount to some thousands," and even ordered that Tycho's salary start from 1 May 1598, when he had first been called to Prague, in as much as Tycho had not accepted any other patronage in the meantime.

By the end of July, Tycho may have had in hand the two thousand guldens of relocation allowance he said he received almost immediately.[145] By that time, too, the matter of housing had surely been reexamined. The emperor had been as good as Barwitz's word. When he learned that Tycho wished to live outside the city, he offered him a choice among three estates lying a few hours from Prague. Although one of them, Brandeis,[146] was Rudolph's favorite hunting lodge, another had recently been purchased from heirs who had been unable to agree on how to divide it up and had the additional advantage of being on a hill about sixty meters above the floodplain of the river adjacent to it.[147] Because of the tendency of the area to flood and surround the castle with water (and possibly because of the general beauty of the surroundings, in Tycho's opinion) it was alternatively called "the Venice of Bohemia."[148]

The proper name of the estate that Tycho finally chose was Benatky, and it was six hours (38 km) by level road from Prague (near a small village of Protestants with Calvinistic tendencies, Tycho reported approvingly).[149] After seeing Tycho's pictures in the *Mechanica*, Rudolph was now inspired to remodel the estate into something superior to Uraniborg in every respect. To top every-

---

[144] VIII, 166–7, 177–8.   [145] VIII, 167, 178, 187.
[146] For a discussion of Benatky and other places occupied by Tycho in Bohemia, see Appendix 5.
[147] VIII, 159.   [148] VIII, 168, 228.   [149] VIII, 168.

thing, he promised Tycho a hereditary fief as soon as one became vacant.[150] By 28 August Tycho had his first instrument set up at Benatky[151] and, after two and a half years of uncertainty and anxiety, could at last report to both Rosenkrantz and Vedel that he was squarely on his feet again.[152]

[150] VIII, 167, 178, 188.　[151] VIII, 219.

[152] *Gassendi* (49) melodramatically attributed to Tycho the thoughts of Plutarch's Themistocles, "*perieramus, nisi periissemus*" (we were near to going under if we had not already gone under).

Chapter 13
A Home Away from Home?

ALTHOUGH Benatky had been Tycho's choice among three manor houses, it was far from being the new Uraniborg that Rudolph had agreed to provide for Tycho. Its best feature was its location. It was the farthest of the three from Prague; it was situated on high ground, offering (as Tycho wrote Longomontanus) a clear view of the horizon in all directions; and it was splendid and commodious.[1] Compared with Uraniborg – to say nothing of the temporary quarters in which Tycho had been living during the year since his departure from Wandesbeck – all three of the places probably looked commodious. But although Benatky had room for people, it had no obvious place to set up instruments or perform chemical distillations.

Taking Rudolph's promises literally, Tycho began to make repairs, modify rooms, and even plan a completely new wooden building[2] – whether as an alchemical laboratory or some kind of analogue to Stjerneborg is not clear. Within a few weeks the administrator of the estate, Caspar von Mühlstein, complained to Barwitz about the cost of Tycho's projects. By late November, when the original estimates had doubled and Tycho had shown Mühlstein a letter from Barwitz stating that Rudolph had granted Tycho a salary that far exceeded the proceeds of the estate, Mühlstein had had enough. He wrote a letter to the chamber of deputies outlining the problem and formally refusing to authorize further expenditures without an official order – and the money to pay for it – from the treasury.[3]

Beneath the confrontation between Tycho and Mühlstein ran at least four undercurrents. The first of these was that although Mühlstein had been told by Barwitz that Tycho was in great favor, Tycho had, purely and simply, made a greater impression on Rudolph than Mühlstein could possibly have imagined. Tycho was not exaggerating when he wrote to Niels Krag the following March that "the most benign favor of the emperor increases daily."[4] And the situation was all the more remarkable in that it was a case of absence having made the heart grow fonder.

Less than a month after Tycho had moved to Benatky, another

---

[1] VIII, 159. For a description of modern Benatky and Tycho's other haunts in Bohemia, see Appendix 5.
[2] XIV, 168; VIII, 178, 193.
[3] XIV, 165–6. The letter is in Czech. I have used the Latin summary on XIV, 164.
[4] VIII, 273. See also letter 3 of Appendix 6 for Tycho's descriptions of Rudolph's solicitude.

plague had struck Prague. Even though Tycho had armed his fainthearted patron with his exclusive prescription against epidemic disease and had even turned down Rudolph's offer to pay relocation expenses for Tycho and his household (Tycho preferred to depend on the protection of God, he wrote his mother),[5] the emperor had taken to his heels. Thenceforth, Rudolph had consulted Tycho by letter. But just as Tycho's oral interviews had been completely and almost unprecedentedly private, so too did Tycho's messages go straight to Rudolph without being seen, either before or after, by any of the Imperial Council.[6] It is not surprising then that Tycho should have had the confidence, when Mühlstein balked, to threaten him with the emperor's displeasure and even to suggest that he might leave Bohemia and let the world know why.

A second source of misunderstanding was the fact that Tycho was accustomed to an administrative system in which the sovereign's word was his – and everybody else's – bond. Orders on the Danish treasury had produced instant results. The possibility that promises might rest on practically nothing more than good intentions was something that it took Tycho a while to learn. In the meantime, Mühlstein was saying that he would not and/or could not honor the demands inherent in Rudolph's promises, and six months after Tycho's salary was supposed to have started, he still had not seen a pfennig of it.

But Tycho's naïveté regarding the conditions of civil employment in Bohemia was nothing compared with the incredulous Mühlstein's inability to grasp what Tycho conceived for his new Uraniborg. Even Rudolph probably did not understand exactly what Tycho had in mind, and the changes that were being made and the costs that were being incurred were obviously beyond the ken of poor Mühlstein. So he may well have applied to the council and reported Tycho's threat even if he had fully appreciated the extent of Tycho's standing with the emperor. Nor can one doubt that his fellow administrators would have agreed unanimously with his opinion that the sums being expended would have been much more efficiently invested in the construction of fish ponds or the rebuilding and restocking of the manorial farms.[7]

Finally, even if Tycho had been naive when he reached Prague, he seems to have learned very quickly the realities of life as a foreigner dependent on imperial patronage. Foremost of these was that the Bohemian aristocracy was not going to allow anyone, let alone a foreigner, to come in and simply shoulder them aside, without

---

[5] XIV, 175–6. See a shorter description in letter 4 of Appendix 6, and also VIII, 273.
[6] VIII, 273.   [7] XIV, 164.

putting up some kind of opposition. Tycho had been warned by Hayek during the initial euphoria of his reception by the emperor that there were just as many envious opportunists and detractors in Bohemia as there were in Denmark: It was just that in Bohemia there were also people who greatly valued his work.

Already when Tycho wrote to Scaliger in October, he referred to the ill will that had manifested itself among those jealous of his position with the emperor.[8] Rudolph may have promised Tycho an hereditary fief as soon as one became vacant. But Tycho soon learned that it would take a while to apply for and obtain citizenship so that he could even become eligible to obtain a fief. So, although he certainly retained the strongest possible hopes for the fief that would secure the futures of himself and his family,[9] as he approached the age at which his father had died, common ordinary prudence would have dictated that he proceed on the assumption that his new Uraniborg would have to be established at Benatky, rather than at some splendid hereditary estate of his own.

But if Tycho was never going to prevail against this fourth and most fundamental undercurrent – to many of his peers he would have remained an expensive, presumptuous outsider, whatever the legal outcome might have been on the specific issues of citizenship, nobility, and fief – he at least won this particular battle. Although the chamber told Mühlstein that he should await a written clarification of the authorization from Rudolph, its members were sufficiently aware (and wary) of the situation to advise him to continue construction of Tycho's wooden house (but as cheaply as possible) if Tycho absolutely had to have it.[10] A few days later, on 10 December, Rudolph responded to the chamber. Tycho could have his wooden house and the little rooms that were being built (to house various instruments, no doubt), but no more.[11] And because there was no response to Mühlstein's indignant objection to supplying Tycho weekly with enormous quantities of wood and charcoal from the imperial forest, the long-suffering superintendent probably had to continue with that too.

The arrangements for the payment of Tycho's salary, however, did not work out quite so well. Mühlstein was ordered to pay Tycho one thousand gulden a year from the rents of Benatky and Brandeis[12] and must have done so reasonably regularly, inasmuch as we hear nothing more of him except that he attended the wedding of Tycho's

8 VII, 169, 188.
9 See references to it in VIII, 167, letter 6 of Appendix 6, and the application on XIV, 217.
10 XIV, 166–7. The response is in Czech. I rely on the Latin summary.
11 XIV, 168.
12 XIV, 164. There was produce from the estate besides: See letter 4 in Appendix 6.

daughter a year and a half later. This money probably provided most of what was needed in cash and kind to run the Brahe household. It seems to have been considerably more than Tycho's share of Knudstrup had been and was, at any rate, more than the seven hundred dalers from Roskilde that Tycho had forfeited by leaving Denmark.[13] The remaining two thousand gulden for each fiscal year was ordered to be paid from general funds. But the first year's sum was not disbursed until 13 July, and things dragged on so long the following year that Tycho still had not seen any money when he died, six months into his third year.[14]

As of the holiday season at the end of 1599, the remodeling effort at Benatky had probably subtracted more space than it had added. Indeed, it may have been partially responsible for Tycho's deciding in the first half of November, after two thousand people had died in the district, to take the family thirty kilometers down river, to a castle called Girsitz, to escape the plague (the women were frightened, Tycho told his correspondents).[15] At any rate, when the family moved back to Benatky in the new year,[16] and Tycho junior arrived from Denmark with Longomontanus,[17] there apparently was no room for the veteran assistant amidst the construction at Benatky. On 17 January 1600, therefore, Tycho sent him to Prague with a note explaining the circumstances to the rector of the university and asking if he could find a place for Longomontanus to live and work at Tycho's expense.[18]

This was an unfortunate precedent from Tycho's standpoint, for when Johannes Kepler turned up a month later, the fact that Tycho did not give him an analogous option (at least for some future time when his wife and child would join him) seems to have been a bone of contention between them. Whatever had been done to the residential portion of the house must have been finished by the end of March, for by that time not only Longomontanus and Kepler but also Johannes Müller and his wife were ensconced there. However,

---

[13] J. R. Christianson has estimated that the income from half of Knudstrup must have been about 650 dalers: "Tycho Brahe's German Treatise on the Comet of 1577," *Isis* 70 (1979): 116.

[14] XIV, 193, 207–8; XIV, 232, 242–3, 246–7.

[15] VIII, 193, 273, and letter 3 of Appendix 6. Tycho said (XIII, 236, and XIV, 185) that they were there for six or seven weeks.

[16] Tycho wrote his first letter from Girsitz on 16 November. Subsequent letters (e.g., to Kepler on 9 December) were occasionally addressed from Benatky just to avoid complicating things, and so it is hard to know when Tycho actually returned. The last one officially sent from Girsitz was dated 31 December (VIII, 226), with the comment that Tycho expected to go back to Benatky soon.

[17] XIV, 175, 176, and letters 4 and 5 of Appendix 6.   [18] VIII, 239–40.

work on the observatory and the laboratory was still in progress, and even the private entrance to the premises being built to connect with a nearby house reserved for the emperor was still unfinished.[19] And by the beginning of May, Tycho was sufficiently impatient with the delay to ask Corraduc to prod Mühlstein on the matter, so that when his instruments arrived from Hamburg and Magdeburg they would have protection from the wind and rain.[20]

By almost any standards, Tycho's first "working season" in Bohemia was an exercise in futility. The distractions associated with the funding and renovation of the new Uraniborg were a serious part of the problem. But they were compounded by a conspiracy of events that frustrated Tycho in everything else he tried to do, as well. One of the first things on Tycho's agenda was the task of cleansing himself and his family from the stigma of Ursus's calumnies. As long as Ursus went unpunished for his insolence, Tycho apparently felt that he could not hold his head up in Bohemian society.

Already by 18 September, Tycho had investigated the situation enough to know that Ursus had fled Prague shortly after his arrival. He had also consulted (secretly) with the archbishop of Prague, who was the official censor of the city, and had learned that because Ursus had published his book without submitting it to censorship, the archbishop was willing to summon him to court and punish him according to the law. Accordingly (Tycho told Vedel), when he got a bit less busy, "the beast" would have to be tracked down and dragged out from wherever he was hiding.[21] Unhappily, however, with Rudolph in hiding from the plague, there was no court into which to drag Ursus, and Tycho had to wait.

Right behind Ursus on Tycho's list was the much more difficult task of publishing at least one or two of his almost-completed-but-long-delayed books. Most likely Tycho thought that such an achievement would justify, and perhaps solidify, his standing with Rudolph. But he also regarded 1600 as a "jubilee" year, during which it would be especially appropriate to get his new findings before the public.

The easy one would have been *De mundi*, which Tycho had distributed in the form of presentation copies to two or three dozen people but had never actually published. All that stood between it and publication was Tycho's desire to append to it a rebuttal of John Craig's Aristotelian arguments against Tycho's conclusions, and the fact that the fifteen hundred uncirculated copies printed in 1587 were still in storage at Magdeburg and remained so until almost the end of 1600. The rebuttal (*Apologia*) was so close to being ready that Tycho

---

[19] See letter 4 of Appendix 6.   [20] VIII, 315.   [21] VIII, 181.

knew how many pages it contained and even led Schultz to believe that he had had it printed in Hamburg.[22] But it was never completed, and *De mundi* remained unpublished until Tycho's heirs sold it to a bookseller in 1602.

A second, relatively easy publication would have been Volume II of Tycho's astronomical correspondence. If we can believe another statement by Tycho to Schultz (notwithstanding that no printed *Apologia* has ever been found) he had actually had printed about one-fifth of the book before he left Denmark. Even if some of the editing remained to be done, therefore, finishing that book should have been relatively easy. Unfortunately, however, that project, too, aborted. The immediate cause was that the already-printed sheets, along with Tycho's press and type font and the paper from Tycho's mill, were incarcerated in Magdeburg with such other incidentals as Tycho's professional library.[23] But as the winter – and then the following winter – dragged on and Tycho's salary failed to appear, it became ever less reasonable to consider parting with hard cash to pay the costs of printing.

Although the work remaining on the *Progymnasmata* made it a more difficult project, there can be no doubt that it was the book that Tycho most wanted to complete. Seven hundred fifty of its 850 pages (in fifteen hundred copies) were already printed, and so Tycho could seriously consider paying a printer to do the rest. In fact, when he wrote Schultz on 7 January 1600, it was to tell him that he had searched Prague in vain for a printer who was good with numbers and diagrams and to ask if it might be feasible to have the printing done at Görlitz.

Schultz's reply (he had already failed to reply to one letter from Tycho) has not survived. But it is clear that the weak link in the chain (aside from the fact that the printed portions were in Magdeburg with everything else) was the writing. Longomontanus would finish the technical work on the lunar theory by mid-summer, but the preface and conclusion would remain unwritten until Kepler composed them when the book was finally published in 1602.

Aggravating the difficulty of establishing a new Uraniborg under the conditions imposed by Bohemian politics was that doing astronomy had ceased to be the most important thing in Tycho's life. It was at least partly a matter of age, not just the infirmities of age but also the concomitant, more or less normal, onset of the inclination to talk science, rather than to do science, and even to talk about past achievements rather than current problems. Thus, Tycho seems to have found it more rewarding to bask in the fact that Mercator's

---

[22] VIII, 235. See also note 2 in Chapter 10.   [23] VIII, 147, 234.

The Lord of Uraniborg

celestial globes were now obsolete, thanks to the appearance of two
different Dutch globes made from his star positions,[24] than to do
what had to be done to get his own version of those positions (in the
Progymnasmata) into the public domain.

As long as Tycho's inclination was manifested only in such
activity as an autobiographical account for the *Mechanica* that went
well beyond the requirements of advertising his qualifications to the
patrons of Europe, it was neither unusual nor particularly time-
consuming. Indeed, the value to science of a role model of Tycho's
stature many well have been greater than whatever else Tycho might
have achieved with the time he spent publicizing his accomplishments. In an age when science was not a widely chosen – or even
completely respectable – means of occupying one's time, who can
say what effect the chance to visit with Tycho in the winter of
1599–1600 may have had on the decision of a young scholar like
Willibrod Snel to devote his life to it.[25]

As for the autobiography itself, it is hard to exaggerate just how
extraordinary a production it was. Tycho can scarcely have been the
first man of the Renaissance whose family heritage had instilled in
him an expectation that he would make his mark in some way – or
even the first one to judge in his declining years that he had, indeed,
made his mark. But he certainly was the first one to imagine that
achievements outside the traditional bounds of military or civil
service could constitute grounds for such distinction. And he may
also have been the first one to have spent enough time in the
company of historians to ponder what his place in history might later
be perceived to have been.

Until his exile, such considerations seem not to have had much of
an impact on Tycho's life. He may have used the ideas of Rothmann
and Mästlin without acknowledgment, exaggerated his contribution to Wittich's development of the prosthaphaeretic method, and
zealously defended his own priorities against the real or imagined
claims of others, but he did not devote much time to considering
what he was doing. However, once he had left his native land, under
the circumstances that he did, he became and remained extremely
sensitive to what was being said about him and his situation, no

---

[24] The globes were made by Willem Janzsoon Blaeu (who was with Tycho on Hven for six
months in 1595–6) and Jacob van Langren, whose son Arnold visited Tycho for seven
weeks in the spring of 1590. See VIII, 259, 309, 322, 332. The Dutch instrument maker
Adrian Metius was also at Hven sometime before 1595: *DSB* IX, 335–7.

[25] Snel (another Dutchman, whose father taught Metius; see note 24) became the eponymous
discoverer of the law of optical refraction: *DSB* XII, 499–502.

doubt because his noble upbringing had accustomed him to assuming that the world revolved around him.

One way that Tycho chose to respond to this concern was to attack those such as Ursus who threatened his status. Another was his rather strained attempts to promote recognition of his work, such as the solicitation through Magini, in late 1598, of a laudatory poem written by an Italian. Even this relatively safe approach required a certain amount of discretion, however, and on one occasion, at least, Tycho was not able to exercise it. In early 1600, he heard that David Chytraeus was planning to publish a chronology of events at the century's end. Deciding that his exodus from Denmark and reestablishment in Bohemia were as newsworthy as any other happenings were, he submitted a description of his travails.[26] Unfortunately, Chytraeus died suddenly, and the project went with him. Unable to accept the loss of an opportunity to air his story, Tycho wrote to several other people in Rostock in a futile attempt to ensure that the project would be completed in some form and that he himself would figure in it.[27]

But if Tycho left behind him a substantial part of his self-image when he fled Denmark, that was only part of his problem. Most of what ailed him was that whether because of the loss of familiar home surroundings and the company of old friends or because of some objective isolation in his new environment, he was profoundly and irremediably homesick. To be sure, he had many problems to deal with, arising from both unfinished business in Denmark and the difference between the ideal new circumstances he described in his letters and the real situation he gradually came to know. But whether even the most complete resolution of these difficulties would have made Tycho happy is doubtful.

Kepler's remark that Tycho spent much of his time writing letters to Denmark may well have been based on a specific two-day period in March 1600, not long after Kepler joined Tycho, when the availability of a courier prompted Tycho to send off nine letters.[28] Similarly, a report by the visiting Flemish astronomer Adrian van Roomen, that Tycho seemed to be doing more drinking than science at Prague,[29] certainly referred to Tycho's first few weeks there, in

---

[26] VIII, 283–6.

[27] See the many letters to Bacmeister, Brasch, and Sturtz in VIII, 375–9, 389–91, 402–3, 410–16. During this period, Herwart von Hohenburg sent Tycho a copy of Rothmann's star catalogue but then expressed qualms about the propriety of having done so. Tycho responded that his own catalogue was effectively printed and that he would not say anything to Rothmann about the situation: VIII, 347–8.

[28] XIV, 175–91. They include letters 3 through 6 of Appendix 6.

[29] *Norlind*, 313; VIII, 350.

temporary quarters, and may illustrate nothing more than unfami-
liarity with the life-style of the Danish nobility. But it is hard to
dismiss Tycho's letters to Denmark. Written primarily to convey the
news that Tycho had landed on his feet and to make an issue of the
fact that he had indeed chosen exile – even though his work had not
been appreciated in Denmark, he said, it was highly valued elsewhere – what they show most clearly is that behind the proud
rhetoric about his prestigious appointment, Tycho sorely missed his
native land.

As is by no means unusual even for people who are happily
expatriated, the most important thing in Tycho's life was contact with
his homeland. Unfortunately, there was little of it. Tycho appears
to have enjoyed truly warm relations only with Sophie, among all of
his siblings. He does not seem to have been close even to his mother
(or his stepmother, before her death). Beate was on Hven a couple of
times, but there is no record that Inger Oxe ever was. Nor does
Tycho's log mention the death or burial of Inger in the fall of 1591 –
or any departure by Tycho from the island at that time. Tycho was
on Hven (with Sophie visiting him) when brother Axel remarried in
1589 and when Steen followed suit in nearby Helsingborg in 1590.
And Tycho does not seem to have left the island when sister
Margaret's husband, Rigsraad Christen Skeel, died in 1595.[30]

It cannot be surprising, then, that Tycho should complain in 1598
that he had written to Margaret two or three times without getting
any response and that his brothers seemed to vacillate between caring
and not caring about what happened to him and his affairs.[31] Sadly,
this assessment – though one of Tycho's all-too-typical petulant
reactions to minor adversity – probably sums up his relations with
his family.

There is no evidence of bad blood between them: Tycho could still
call on them for the help that they owed him as "kin," and Steen, in
particular, seems to have done what he could to mitigate the troubles
that drove Tycho out of Denmark. Furthermore, there were very
few summer months when Tycho was not visited at Hven by one or
more of his cousins, nephews, or nieces, and even in Prague he
received a number of letters and visits from the next generation of his
family. But from those people who would ordinarily be most able to
provide it, he received little emotional support in exile.[32]

There were letters from home. Niels Krag and Holger Rosenkrantz continued to advise Tycho of events in Denmark that might

[30] See the appropriate dates in Tycho's meteorological diary, in Volume IX.
[31] VIII, 48, 68.
[32] XIV, 160, 167 (letter 2 of Appendix 6). Steen, among other things , used two of Tycho's
students as preceptors for his sons. See XIV, 182, 190, 204–5.

concern him, such as the letter from Stygge to Aalborg and rumors that Hven might be taken from Tycho. To Rosenkrantz's suggestion that it might be worth resigning his lifetime fief just to recoup some goodwill and make himself judicially independent of the king, Tycho gave short shrift: There was no legal way the island could be taken from him, he told Krag, and he would not give it up without compensation for what he had spent there.[33] Whether Tycho actually cherished hopes of selling his interest there or whether he just could not bring himself to give up his connection with home is not clear. Certainly the income from Hven seems to have been negligible, at least compared with the grief.

Although an obscure warrant from the regency awarding Tycho two hundred dalers from the tolls collected in the Øresund seems to have been paid annually into Tycho's account right up through 1601,[34] what Tycho was obtaining from Hven seems to have been grazing rights for some horses he still had there, the few barrels of fish that could be harvested from the ponds and salted for transport to Bohemia, and whatever cash the landlord's share of the island's crops would bring on the mainland. As for the memories, they all were bad. When Tycho decided to sell the horses that remained on the island, he told Esge Bille to do with them as he pleased but that he "would prefer that (he) sell them to others than to the peasants there on the island."[35]

Perhaps part of the problem was that Tycho wondered whether his agent on the island (who would have handled any sales to the peasants) was dealing honestly with him.[36] If so, it would not have been particularly unusual, for to judge from similar comments about his long-time agent at Nordfjord and a person who was handling affairs for him in Copenhagen, Tycho seems to have assumed that most of the people who served him were trying to sneak some kind of extracontractual advantage from the relationship.[37] Whether this was enlightened sophistication or undue suspicion, a typical sample of Tycho's attitudes and the problems he left behind him in Denmark can be found in a letter he wrote to his remote cousin, Esge Bille,[38]

[33] VIII, 127–8. Tycho was a bit more diplomatic in his response to Rosenkrantz (VIII, 148, 170).

[34] J. R. Christianson, ed., "Addenda to Tychonis Brahe Opera Omnia," *Centaurus* 16 (1972): 241–2. Tycho apparently also still held the fief of a dozen farms on Kullen: VIII, 175.

[35] See letter 6 of Appendix 6.

[36] VIII, 127.

[37] See also the references to Dirick Farver and Di(de)rich Dyer in letters 1 and 6 of Appendix 6. Note that "Farver" is the Danish equivalent of Dyer and may refer to the "dyer" who had a vested interest in a shop next to (or in: see the end of letter 1 of appendix 6) Tycho's house in Copenhagen.

[38] Esge was a third cousin of Tycho's and was about five years younger. Their friendship dated at least from the early 1580s, when Esge helped Tycho observe on Hven: X, 385.

who was helping him tie up loose ends. When Tycho learned from
Esge that the former pastor on Hven (who had left Denmark with
him in 1597 to avoid prosecution for religious offenses but had
subsequently fomented so much discord that Tycho had had to
dismiss him and give him expenses for the trip home)[39] had returned
to Hven and "requested that he might have the [workshop for the
artisans] that you erected there on the island, although it is falling
into disrepair, because the parsonage fell in disrepair" (some years
ago).[40] Tycho responded:

> I do not owe anything to the pastor on Hven, neither with respect
> to the smithy, which I had built with my own money, nor in other
> ways, for he deserves nothing of the sort from me. Were I to pay him
> as he deserves, I would deal quite differently with him, nor did he
> return there with my support and approval, though he should have
> sought it. If David Pedersen has done anything about this, it was not
> done with my knowledge. Morten Deyn can certainly tell you how
> this same pastor conducted himself toward me after he came there, if
> he is willing to come right out with it. The peasants themselves let the
> parsonage fall into ruin, and they stole away a large part of it despite
> my prohibition, as Sven Truissen well knows, whether or not he will
> admit the truth. I gave the pastor another place which I had bought
> myself and helped to build, and it would have been made more
> complete if I had not had reasons to depart from there. I do not know
> what to answer to the fact that the grange is falling into ruin and the
> castle is also in such bad repair besides what I have written before, that
> I am not of a mind to spend more money on them. I have already
> spent too much. If I still had that money, I would not use it on such an
> ill place again.[41]

Such an uncompromising stance was typical of Tycho, even in his
dealings with people who were his social equals. And his temper was
sufficiently quick that most of his friends must have heard his bark
more than once.[42] Holger Rosenkrantz's delay in getting Tycho's
carriage delivered to him at Wandesbeck and his reluctance to present
Tycho's books to King Christian both earned him harsh words from
Tycho. Axel Gyldenstierne, Tycho's older cousin and good friend,
came in for some recrimination when he was unable to meet quickly
Tycho's request for repayment of a loan.[43] One of Tycho's Oxe

---

[39] VIII, 12.   [40] See letter 1 of Appendix 6.
[41] See letter 6 of Appendix 6.
[42] For a sample of the cajolery Tycho must have had to use from time to time following such
outbursts, see the beginning of Tycho's letter to Niels Krag, dated 24 March 1600 (VIII,
272).
[43] See letter 5 of Appendix 6.

"relations" probably never did collect a debt allegedly owed to her, because Tycho simply insisted that it had already been covered in some previous transaction. Moreover, Tycho probably even left his Copenhagen house unsold at his death because he refused to consider at least two offers on the grounds that they were woefully less than the money he had spent in building and subsequently remodeling it.[44] The one instance in which his hard-nosed approach may have been fruitful was in reclaiming, after some exertion, the ten thousand dalers he had lent to the young dukes of Mecklenburg.[45]

All of the unfinished business and, no doubt, homesickness, brought Tycho to consider a trip back to Denmark. As of the fall of 1599, he was thinking about the following summer. By the spring of 1600, however, after having been told that the annual meeting of the Danish nobility that he wanted to attend probably would not be held that summer, and deciding that the emperor would be reluctant to give him leave so soon anyway, Tycho began to look toward the summer of 1601. For Esge's concern that there might be some kind of trouble (arrest?) if Tycho came, Tycho had nothing but bluster:

> What you mentioned to my son, Tyge, is of no danger whatever to me, for if it happened, though it has no right to, there would undoubtedly be one who would come to my defense and would not let me suffer any wrong, and if it struck again in another way, if they wanted to use force, and I take care to have such letters with me, if God wills that I come, they will have to think about what they do. If anybody has anything to say to me, let them say it to an unpartisan judge, and they will get their answer. I still intend to get back some of the money I have lent out, there in that country, and thought that I deserved other thanks and return for them than I got, but all things in their time, there is time for council and ill council.[46]

In contemplating his own trip to Denmark, Tycho assumed that the round trip with some time at his destination would take eight to ten weeks.[47] Given the financial and physical costs, Tycho could not really expect anyone to visit him at Benatky. But already in April 1600 a third cousin of his by the name of Frederick Rosenkrantz turned up unexpectedly. He was not exactly what one would call a visitor, because he was himself an exile. But it must have given Tycho a certain grim satisfaction to contemplate someone whose misfortune had been greater than his own.

[44] See letters 1 and 6 of Appendix 6, and also VIII, 91. For the disposition of the house in 1606, see XIV, 258.
[45] XIV, 213–14, 222–32. [46] See letters 1 and 6 of Appendix 6.
[47] See letter 6 of Appendix 6.

Frederick had started life in circumstances even more advantageous than Tycho's. His father, Holger, had been one of three *rigsraad* brothers (including Jørgen, the father of Tycho's correspondent, Holger the Learned) and was reputed to have been Frederick II's best friend. As befit a man whose namesake was a king, young Rosenkrantz had been raised in the lap of luxury. How well Tycho knew him is not clear, but Frederick had visited Hven in 1589[48] during his student travels. And the fact that those travels had been to foreign universities, under a preceptor who was to replace (and counter doctrinally) Niels Hemmingsen as Denmark's leading theologian, must have given Tycho and Rosenkrantz something to talk about.[49]

What they had most in common, whether or not they could comfortably discuss it, was Frederick's treatment by the regime in Denmark, which one Danish historian has compared with the "heavy hand" and "senseless brutality" wielded against Vedel and Tycho.[50] What had plunged Frederick into trouble and notoriety was an affair with a young lady-in-waiting named Rigborg Brockenhuus.[51] Frederick had fled the country when they learned she was pregnant, but it had not saved him. He had been brought back and sentenced to the loss of two fingers and his honor as a noble. In actuality, the sentence had subsequently been commuted to service in one of Christendom's campaigns against the Turks, and that had been what had brought him to Prague.

Frederick seems to have been with Tycho for some weeks before going to Vienna at the beginning of June, with a letter of introduction from Tycho to the emperor's brother, Archduke Matthias, who was the commander of the Austrian forces. Tycho's primary impression of Rosenkrantz was the alarming state of his health.[52] And indeed, Rosenkrantz soon died, but in the attempt to separate two comrades who were engaged in a duel.

Unbeknownst to either Tycho or Frederick, Rosenkrantz, in his short lifetime, had already been immortalized: While he had been on

---

[48] 9 June 1589: IX, 73.

[49] Tycho mentioned (VIII, 319) that Rosenkrantz knew Latin, Italian, French, Spanish, and German (nothing about English!). Rosenkrantz's tutor was Hans Poulsen Resen.

[50] Erik Arup, *Danmarks historie* (Copenhagen: H. Hagerup, 1925), vol. 1, pp. 645–8.

[51] For her part in their adventures, Rigborg was sentenced to house arrest for life. Rigborg's family was well known to Tycho. Her grandfathers, *rigsraads* Frands Brockenhuus and Peder Skram, had been Marshal of the Realm and Admiral of the Fleet, respectively, when Tycho was a young man. In 1571, Laurids Brockenhuus, Rigborg's father, had been fined 10,000 dalers for killing a woman in cold blood without provocation. A few months later he had married the sister of Tycho's Leipzig schoolmate, Knud Skram.

[52] VIII, 307–9 and 446–7. See Tycho's report to Holger Rosenkrantz (Frederick's cousin) and his fears for the latter's health in VIII, 331–2 and 368.

a tour of duty with the Danish legation in England in 1592, he and a slightly less remote cousin of Tycho's, Knud Gyldenstierne, somehow made an impression on the young William Shakespeare that was sufficient to get them "bit parts" in *Hamlet*.[53]

Sometime in May or June 1600, Tycho had another visitor, a writer on astronomy named David ben Solomon Gans. According to an enthusiastic account later published by Gans, Tycho's observatory consisted of thirteen rooms, all situated in a row (presumably along a south wall). Each room had a large and very imposing instrument mounted in it, and Tycho had twelve assistants to help him with his work.[54] From Gans's standpoint, clearly, Tycho had established his new Uraniborg.

The remodeled Benatky was surely larger, probably more comfortable, and, at least in some respects, objectively better than the old facility. Yet from Tycho's standpoint, it was still a mere shell of the establishment he needed to have for the final stage of his life's mission. The respect in which Gans would have found it most deficient, if he had had the perspective of the old Uraniborg, would have been the very one that seems to have impressed him most: the instruments.

Those instruments available at Benatky – "six larger ones and some smaller ones," as Tycho described them[55] – were there almost solely because their transport had posed minimal problems for Tycho. However imposing they may have looked to Gans, they were the culls of a negative selection process by which Tycho had first decided to leave his two large, permanently mounted quadrants and his armillary[56] in their emplacements at Hven and had then, a year and a half into his travels, stored the heaviest (and generally most effective) of the remaining ones at Magdeburg. Those that reached Benatky may have included the great globe, just because it was so expensive and spectacular that Tycho wanted to have it with

---

[53] Act 2, sc. ii. The Knud Gyldenstierne involved was probaby not the one who was the son of Axel Gyldenstierne but, rather, Knud *Henriksen* Gyldenstierne, the son of the man who had married (and been widowed by) Tycho's older sister. This Knud's mother (Henrik's second wife), Mette Rud, was a second cousin of Tycho's.

[54] *Norlind*, 304. That there was at least a modicum of faith in this report seems clear. Even if the cubicles were finished, Tycho's good instruments were still in Magdeburg.

[55] VIII, 370.

[56] Tycho left four instruments at Hven. The mural quadrant, the great steel quadrant, and the great armillary, the essence of his observing tools, were three of them. The fourth one was apparently the heavy steel semicirculus (XIII, 101). It is easy to see why Tycho took the revolving quadrant with him, because its wooden struts would have made it relatively light and (unlike the mural arch) reasonably impervious to damage. The only thing to be said about his having taken the large rulers with him is that their I-beam construction must have rendered them invulnerable to the rigors of moving.

him. They certainly included a couple of medium-sized instruments that had been built specifically to be portable and had been used in surveying activities. But it had probably been fifteen years since Tycho had handled any of the others with any serious intent. Although it is difficult to imagine a "new Uraniborg" that could not boast of the instruments that had been the focus of the old one, the fact is that by the time Tycho left Denmark, the observational phase of his work was essentially complete. Observations of unprecedented accuracy had been crucial to his research on the new star and comets, the solar and lunar theories, and his star catalogue. But with the possible exception of the appearance of another particularly spectacular ephemeral object (such as the supernova of 1604 that has come to be associated with the name of his successor, Johannes Kepler), no further observations were likely to affect the conclusions (and theories) that Tycho had already printed for the *Progymnasmata* and *De mundi*. Only for the planetary theories did any serious research remain to be done, and Tycho did take advantage of the visibility of all five planets in the spring of 1600 to make rough observations of them.[57] But in general, he was surely planning to be able to get most of the data he needed from the thirty years' worth of observations he already had.[58]

What Tycho needed most for the final aspect of his work was assistants who were skilled at astronomical computing. Many calculating hours would be required just to get the observational data transformed from the horizon or equatorial coordinates in which they had been recorded (by the big quadrants or armillary, respectively) to the ecliptic coordinates in which the planetary theories had to be expressed. Someone would have to decide which data were going to be transformed for trial, which theory they would be "tried" against, how far the selected data could depart from the selected theory without being deemed to impugn it, and what kind of changes should be made once it was decided that a theory needed rectification.

Tycho surely expected to preside over all of this work in at least a general way. But each aspect of it could be delegated to someone who was good with numbers, and Tycho, equally surely, planned to do exactly that. It was a scheme that had already worked well for the solar theory (even if Tycho had probably had to do more of the detail

---

[57] XIII, 204–35.
[58] Tycho had told Magini in 1598 that he still needed a few observations, because Saturn moved so slowly and Mercury and Venus stayed so close to the sun. But this was probably more of an excuse for having to tell Magini (in response to his inquiry) that he had not yet completed his planetary theories than a genuine assessment of his situation: VIII, 121.

work himself than he might have wished) and had also produced good – if still incomplete – results for the lunar theory. If the planetary theories had proved to be somewhat less tractable, there was no reason for Tycho to doubt that they too would yield to the ingenuity and labors of himself and a few good assistants. What he needed was the assistants.

Until mid-1600, assistance for Tycho meant Longomontanus. When he first traveled from Prague to Benatky, Tycho wrote to his long-time student to describe the enthusiastic reception he had been accorded in Bohemia.[59] And when he then sent the younger Tycho and Claus Mule home to deliver other reports of his success and fetch the four instruments still at Hven, he tried to reinvolve Longomontanus in the institutional enterprise, by soliciting his help in packing the valuable tools for a safe trip to Prague.[60] Only then did he come to his point: He wanted Longomontanus himself to return, too.

The tasks were not daunting: The quartos on the lunar theory would soon be printed, and if Christian (Longomontanus) would come, not only the *Progymnasmata* and *De mundi* but even the third volume (on all of the other comets) of the trilogy could be published. Moreover, there would be help with the work. Johannes Müller had already agreed to come if Tycho could get him released from his duties in Brandenburg for a year or two, and the emperor had already written the letter that should achieve that goal. David Fabricius, who had visited Tycho at Wandesbeck while Longomontanus was there might come too, to help with the observing. (Perhaps Tycho's theoretician was not fond of night air.) Jöstel had had inquiries from two candidates who were interested in coming down from Wittenberg, and even Rothmann had been heard from and might be induced to come.[61] For whatever reasons, this appeal worked. The reluctant disciple accompanied the younger Tycho (and a supply of salted fish from Hven) back to Prague[62] and, by mid-January 1600, was hard at work on the theory of Mars.[63]

Within a couple of weeks he was joined by the man whose results with the same tasks that Longomontanus had been pursuing (and eventually completed) would overshadow his so completely as to relegate Longomontanus to the footnotes of history.

[59] 20 August 1599: VIII, 158–9.   [60] VIII, 182.
[61] VIII, 182–4. Fabricius (1564–1617) had been parson at Resterhaave in Holland since 1584. He discovered the variable star Mira Ceti in 1596. On a visit to Tycho in Prague in 1601 he stayed with Rollenhagen, who reported to Tycho that it was a shame such a man would rather herd cows than do astronomy (VIII, 420). He was killed by an irate parishoner in 1617.
[62] VIII, 239–40.   [63] G. W. III, 109.

The path that took Johannes Kepler to Tycho's door was largely an economic one, hacked out of the thickets of political and religious intrigue that characterized the Counter-Reformation and culminated in the exercise in insanity called the Thirty Years' War. Given Tycho's general eagerness to recruit help, his reception of Kepler would have been guaranteed under any circumstances. But Kepler came with special credentials, credentials arising not so much from Tycho's perception that he might be an especially talented collaborator but from the happenstance of his having written the innocent letter that Ursus chose to print in his *De astronomicis hypothesibus*.

The guileless Kepler had no conception of the power that his capacity to repudiate his letter to Ursus gave him over Tycho and naturally even assumed that the letter was a liability in his dealings with Tycho. In fact, because both Kepler's self-introductory letter to Tycho in 1597 and Tycho's response to it in 1598 had traveled very round-about routes, Kepler did not even know that he had established any relation with Tycho until a year after he had sent his letter. And then he learned it from Mästlin, who had received a copy of Tycho's letter to Kepler but, of course, did not know that Kepler had not received the original. Therefore Mästlin only alluded to Tycho's being upset that Kepler's letter to Ursus (of which Kepler had not kept a copy) had been printed in a book (of which Kepler was not to see a copy until May 1599) in such a manner as to seem to put Kepler on Ursus's side in the priority dispute between him and Tycho.[64] All Kepler could do was explain to Mästlin that he had been urged to contact Ursus by a third party who thought the contact would be valuable:

> The theme [of the offending letter] was virtually the same as in my letter to Tycho [soliciting a reaction to the theory advanced in Kepler's *Mysterium cosmographicum*]. Since [Ursus] is a widely known man, from whose writings I may learn ..., I sought from him as an older man, his judgment about the matter of the five solids.... He heard from [me as] a pupil. Why is he happy? Had he called me a judge, perhaps he would have heard something different. Over a letter which I dash off, a big fuss is made by those whose pupil I acknowledge myself to be for the sake of honoring them.... He harms good men; he abuses the generosity of those who write courteously rather than sincerely, for the purpose of implementing his own madness. His misdeed is aggravated by his zeal to dishonor by means of my words a very great man whom he sees that I admire.... Would that I might see that monster, who carries that letter of mine around on his horns! I

[64] Mästlin to Kepler, 14 July 1598; G .W. XIII, 236. See a more detailed exposition of these events in *Rosen*, 106, 108 119, 132, 135, 147.

shall absolutely avoid the man, however, and when the opportunity presents itself, I shall clear myself with regard to Tycho, and I also ask you to do the same. Have a copy sent to me of my letter as printed. For even though I now no longer remember the words after such a long interval, nevertheless I shall readily recognize whether he violated what I wrote and, furthermore, whether perhaps he attributes to me words which are not mine. I cannot believe that I expended so much sweat in praising him that the jackass can justly strut around.[65]

Kepler's manifest mortification was a product of three different problems. First, he had wanted for some time to cultivate Tycho for purely professional reasons. The source of this desire was doubtless Mästlin, who expressed his view of Tycho's accomplishments in 1599 by saying that Tycho had scarcely left a shadow of what had hitherto been taken for astronomical science.[66] If Kepler's teacher and mentor, Mästlin, had corresponded with Tycho as Tycho had invited him to do, Kepler would have had ready access to – or at least information about – Tycho.

But in the spring of 1598, after having (as he assumed) failed to get either of two separately dispatched copies of his *Mysterium cosmographicum* delivered to Tycho, Kepler had had to give up and ask another correspondent for information about Tycho.[67] The basis of his interest was the judgment, expressed with Kepler's typical forthrightness, that Tycho had an abundance of wealth that he did not use properly (as was generally the way with rich people) and it would be good to extort his riches from him by getting him to publish his observations.[68]

Kepler's second motive sprang from the recent move to flush Protestantism out of Styria. As a result of it, he was almost certain to lose his job, and as of the beginning of 1599 he had still not found another post. So troubled was Kepler by his financial position that when he learned from Mästlin that Tycho had answered his letter (even though the response had not reached him in the intervening eight months) and that Tycho was on his way to Prague, he strongly considered a job-hunting trip to Wittenberg, where (as he had heard) Tycho had taken up winter quarters. Under the circumstances

[65] Kepler to Mästlin, 8 December 1598; *G. W.* XIII, 286 (as translated in *Rosen*, 134–5).
[66] *G. W.* XIII, 276. For Tycho's reciprocal evaluation of Mästlin, see Chapter 8, notes 52 and 53. For an extended discussion, see Richard Jarrell's "The Life and Work of the Tübingen Astronomer Michael Maestlin (1550–1631)" (Ph.D. diss., University of Toronto, 1971).
[67] *Rosen*, 104.
[68] Kepler wrote to Magini in 1601: "What influenced me most (to join Tycho) was the hope of completing my study of the harmony of the world – something that I have long contemplated and that I would be able to contemplate only if Tycho were to rebuild astronomy or if I could use his observations." *G. W.* XIV, 173.

that Ursus had created, the meeting would have had its unpleasant
aspects, but a perception that the air needed to be cleared may well
have been the primary reason Kepler wanted to see Tycho in
person.[69]

Kepler's third problem was that he had had more dealings with
Ursus since the initial letter. Ursus had initially found no reason to
answer the letter – even though he had published it – from an obscure
provincial mathematician. But after reading in the catalogue of the
Frankfurt book fair of the publication of the *Mysterium cosmographicum*, Ursus at least recognized Kepler's name, if only because he had
published his letter. He therefore unblushingly sent a copy of his
just-published *Chronotheatrum* "in expectation of your most thought-
ful judgment of it," along with a request for a copy of Kepler's
book.[70]

In retrospect, Kepler claimed to have been insulted by Ursus's
opportunistic response. He told Mästlin that he had sent the book to
Ursus (even though he was not a serious writer) because he was the
imperial mathematician and was capable of helping or harming
someone even as far away as Graz.

In fact, Kepler had sent two copies and had asked Ursus to
forward the second one to Tycho. He had even requested (galling as
it must have been to Ursus) that Ursus send, by collect mail, any
copies of Tycho's writings that he could lend him. But Kepler had
kept no copy of this letter either, and so he had no evidence to
mitigate his further traffic with Ursus and indeed no very clear
memory of what he had said.

Finally, when Ursus did not answer this second appeal for a judgment of ideas, Kepler had sent a third request – with the same effect
– at about the same time that he had put a copy of the *Mysterium cosmographicum* in the mail directly to Tycho in "Denmark."[71] Kepler
did not remember what he had said in that letter either and could
only guess how it might turn up to embarrass him.

These problems were compounded by the fact that Kepler did not
know what Ursus had said or done or exactly how Tycho was
reacting to the situation and Kepler's role in it. Tycho had in fact
"excused your error [in praising Ursus] because you would not
commit it except on the basis of hearsay and that publication which
he calls the *Foundation of Astronomy*." And although Tycho had
certainly recounted and rebutted each of Kepler's sins in "singing
[Ursus's] praises mightily [and to such excess!]," he had expressly
assumed that "you wrote these things through ignorance when you
were young. Maybe you had not thought that he would ever publish

⁶⁹ *Rosen*, 149.   ⁷⁰ Ibid., 90.   ⁷¹ Ibid., 90–2.

that letter, much less even misuse it to mock and insult others. That is why I am rather calm in this affair."[72]

But of course, Kepler had not actually received Tycho's letter. Mästlin had – as an enclosure without comment to a letter that (1) upbraided him for not having acknowledged receipt of the copy of *De mundi* or the cordial invitation to correspond that Tycho had sent ten years earlier, (2) reprimanded him for having (in his preface to Kepler's book) paid "too much attention to my plagiarists," by having "promiscuously attributed" to others the essence of Tycho's system of the world, (3) shared and solicited eclipse data, and (4) imparted news of a discovery he had made in regard to accounting for the "annular" appearances of solar eclipses.[73] And although that letter certainly referred to the "insolent chicaneries and shocking lies and insults with which [Ursus's] publication abounds to excess and beyond all shame,"[74] the book in question had not come into Mästlin's hands, either, and he seems to have assumed that the only problem was the priority issue.[75]

Perhaps embarrassed by Tycho's scolding, Mästlin rebuked Kepler for having written to Ursus – if he really had, which he doubted – a letter "in which you whitewash him with the most lavish praise":

> For you know my opinion of that man. What he published in his little book is not his own thinking. He does not even understand it, so that what is right in his book, he propounds with the wrong words. He takes a lot from Tycho and advertises it as his own. I showed you this in Tycho's book. Ursus's trigonometry contains nothing remarkable which is not written better elsewhere.... I frankly do not believe that you honored him with decorations of this kind.[76]

Mästlin's letter also went astray. Written in July, it reached Kepler (as already mentioned) only in December 1598. Kepler took Mästlin's advice to write to Tycho but seems to have required a couple of months to formulate his reply. After informing Tycho about the fate of his and Mästlin's letters "at the outset lest, from this delay of ten months, you should develop a greater suspicion of guilt with regard to me," Kepler hastened first to concede that Tycho had a right "in that most just quarrel of yours [to] demand that I shall explain my point of view to you at the very earliest time" and then to acknowledge that "most of the words which you quoted ... are mine."[77]

---

[72] VIII, 47; translated in *Rosen*, 110–13.
[73] VIII, 52–5. Tycho's "trick" for accounting for the later-to-be-discovered-and-named corona was to diminish arbitrarily ("sic me docuit experientia") the "book" diameter of the sun in solar eclipse situations. See Appendix 4.
[74] *Rosen*, 110.   [75] Ibid., 136.   [76] Ibid., 120–1.
[77] VIII, 141; translated in *Rosen*, 138.

Bolstered, however, by an apparent conflation of his second letter to Ursus (in which he had inquired eagerly after Tycho and his works) with his first, Kepler was righteously, but probably wrongly, suspicious that Ursus had done some knavish editing:

> God almighty, how great and how manifold are the injuries inflicted on me by the wild man! I was soliciting his friendship by sending the letter.... [Such things] ought to be confidential. He published it; he did not consult me; nor did he inform me that he would publish it, nor that he had published it; nor did he send me a copy.

After much more handwringing, Kepler converged on the issue. "The question is not how badly he behaved, but how well I behaved. There are indeed in that letter of mine very many things which should be defended, some excused, and others freed from the worst suspicion."[78]

Kepler proceeded to do this at considerable length. His letter reached Tycho at Wittenberg, before he was in a position to know how things would work out in Prague. And, of course, after Tycho got to Prague and then Benatky, there were other things to do. But Kepler had clearly made a good impression on him, for when Tycho finally got around to answering him on 9 December 1599, he was all flattery and courtesy:

> Even though I have not yet met you face to face, most learned sir, nevertheless I love you very dearly on account of the excellent qualities of your mind.... What you say in the beginning at great length for the purpose of exculpating yourself with regard to that misplaced Ursus would not indeed have required so many words and so elaborate an explanation, since I myself regard you as sufficiently excused on other grounds. Nor do I hold it against you at all that he inserted your letter without your knowledge or consent in that defamatory and foul-mouthed writing which he published. As I now learn from you, he chopped up your letter and distorted it for the purpose of making himself look good, for this is perhaps what he had to do.

Having come to the subject of Ursus, Tycho could not help giving vent to his anger:

> The issue concerning his misappropriation of my hypotheses and publicizing them as his own would not be so grave had he not with most outrageous and criminal lies shamelessly attacked with his scurrilous and misshapen pen in that disreputable book teeming with

---

[78] VIII, 141–2; translated in *Rosen*, 138–41, who continues with the entire letter.

falsehoods and insults my dignity and reputation, nay, my country, ancestry, and most honorable house, and defamed them as much as he can, and diffused such stuff in print to the public.[79]

In his initial letter to Kepler a year and a half earlier, Tycho had been very cautious. Only after responding politely and thoroughly to the professional concerns Kepler had expressed in sending the *Mysterium cosmographicum* to him had he broached the subject of his letter to Ursus. And then, as we have seen, he had given Kepler every chance to dissociate himself from the letter. But he had then asked him fairly bluntly to declare himself, "to deposit with me at the earliest possible moment a statement indicating whether you are satisfied with [Ursus's] behavior and what you think of his poisonous publication."[80] Even though Kepler had responded without actually seeing Tycho's letter, he had devoted many pages to expressing his anger and regret over Ursus's perfidy. At the same time, however, he had "added that it did not seem worthy of Tycho's exalted status to be aroused so passionately by this slander. Let him rather permit their works to speak for each of them with confidence in what scholars will decide."[81] Even though Kepler had no conception of the depth of Ursus's offense, his advice was too sound to require comment. If it had been offered to anyone but a person who had had his wife and daughter slandered in the way Tycho's had been, it would probably have cost Kepler his hopes for a job.

But, by December of 1599, Tycho clearly cherished his own hopes of using Kepler's testimony in his prosecution of Ursus. There were only two problems: the reservations that Kepler had expressed about the unseemliness of confronting Ursus, and the practical difficulty of getting Kepler from Graz to Prague for whatever activities might be required for the suit. Both could be solved by persuading Kepler to join the ranks of his assistants.

With the benefit of historical hindsight, it has generally been assumed that Tycho recruited Kepler on the basis of what he had shown in the *Mysterium cosmographicum*. It is certainly possible. But none of Kepler's other contemporaries saw in that work – or any of Kepler's later ones, for that matter – the promise of Kepler's ideas;[82] so what reason is there to assume that Tycho was so overwhelmed by either them or a few weeks of interaction with Kepler that he

---

[79] VIII, 203–4; translated in *Rosen*, 221–3.

[80] VIII, 47–8, translated in *Rosen*, 113.

[81] Postscript by Kepler (as he could remember it) to a lost letter to Tycho. Quoted in *Rosen*, 147, from a copy sent by Kepler to Mästlin.

[82] In a recent seminar paper, James Voelkel has pointed out that Mästlin, Praetorius, and Longomontanus all registered strong objections to Kepler's physical speculations. *G. W.* XIII, 111, 205–6, and XIV, 45.

would first overlook behavior that he would not have tolerated from anyone else, and then reward his temerity with an unprecedented status involving extra money, separate living quarters, and relief from observing duties? It seems much more likely that Kepler's special value for Tycho lay in the potential of his testimony against Ursus to reveal "the bear" for the desperate and despicable character he was. Thus, it can truly be said that to every bit the extent that Kepler was driven into Tycho's arms by the excesses of the Counter-reformation, Tycho was forced to embrace him by his hope of mitigating the excesses of Ursus.

Unable to content himself with Kepler's sincere flattery, then, and forgo his hopes for a piece of Ursus's hide, Tycho closed his second letter to Kepler by saying that he looked forward to discussing various aspects of planetary theory and cosmology

> and other topics with you at greater length with pleasure and much intensity, and I shall hand you more of my writings if you visit me some day as you promise. . . . Yet I would not want hard luck to force you to come to me, but rather your own judgment as well as your love and affection for the studies which we share. Nevertheless, whatever comes to pass, you will find in me not a follower of fortune, whatever that may be, but your friend who even under untoward circumstances will not fail you with his advice and help, but rather will advance you to everything that is best. And if you arrive soon, perhaps we shall find arrangements which will hereafter be more advantageous for you and your family than previously.[83]

Kepler did arrive soon, as we have seen: so soon, in fact, that his trip to Prague crossed Tycho's letter to him, and he arrived without getting it. Not surprisingly, the reason Kepler came was precisely the one Tycho had piously hoped would not bring him: misfortune. Perceiving that the only way in which his position in Graz could deteriorate further was to evaporate completely, Kepler journeyed to Prague at the beginning of 1600. Assured by the tone of Tycho's first letter to him – which had finally staggered in on 18 February 1599, the day after Kepler put his "response" (to Mästlin's report of the contents of Tycho's letter) in the mail to Tycho[84] – Kepler was apparently determined to seek out Tycho and propose some kind of arrangement.

Whether Kepler had a crisis of confidence once he got to Prague or whether he had trouble finding out where Tycho was (in the aftermath of the latter's temporary move to Girsitz), he managed to apprise Tycho of his arrival only on 25 January 1600, after he had

---

[83] VIII, 205; translated in *Rosen*, 229–31.   [84] *Rosen*, 147–8.

been in town for a couple of weeks (long enough, as it turned out, to have met Ursus at some kind of gathering and to have conversed with him without mentioning his own name).

Tycho immediately sent a cordial invitation lamenting the crossing of his letter with Kepler's movement, expressing dismay that Kepler had not simply traveled out to him at Benatky, and insisting that he now come out in Tycho's coach with Tengnagel and the younger Tycho when they returned from doing some business for Tycho in Prague.[85] On 12 July Kepler enthusiastically summed up the ensuing developments for a close correspondent:

> I ... finally reached Benatky on 5 February. From that day until 1 June, if you omit about three weeks, I lived uninterruptedly with Tycho and in his household.... My arrival was not only welcome but also anticipated.... [Tycho] wanted to reimburse me for my travel expenses, and he advanced me the money necessary for my return trip as far as Vienna. The gist of our agreement was that Tycho would obtain for me a letter from the Emperor by which I would be called to Bohemia for two years to supervise Tycho's publications, and yet I could hold on to my salary in Styria. He would supply the money needed to move my family. To these proposals presented by me, he added that he would see to it that I would receive at least a hundred thalers more in Bohemia.[86]

Actually things did not really go so smoothly as Kepler's report suggests. The terms of employment that Kepler described were actually negotiated only several weeks into Kepler's tenure and were not really settled even then. Kepler gave a far more candid assessment of his situation in a letter written to Mästlin after Tycho's death. Although he had been welcomed cordially enough, he said, he had been treated more like a hired hand than a colleague. The intensive conversation about learned things that Tycho had promised in his letters had occurred only when Kepler had been able to buttonhole Tycho on some subject at mealtime. Indeed, Kepler had been so far from having the status of an independent scholar that he had not even been given general access to Tycho's observations.[87]

How legitimate these complaints were is hard to assess. At the very least, Kepler can be said to have been warned of his coming status in the letter by which Tycho invited him to come from Prague out to Benatky. For in it, Tycho twice applied his habitual term,

[85] VIII, 246.
[86] G. W. XIV, 168 (translated in *Rosen*, 234–5). The correspondent was Herwart von Hohenburg, the chancellor of Bavaria.
[87] G. W. XIV, 130.

domesticus (meaning something between servant and employee), to "the noble and learned youth, Franz Tengnagel."[88] And when Kepler arrived at Benatky, Tycho surely did treat him as an assistant and probably not even as his first assistant. It could scarcely ever have occurred to Tycho that Kepler should have as high a status as Longomontanus did, who, after all, was a proven quantity. Nor is it obvious that any other astronomer before, say, Jeremiah Horrox in the 1630s, would have rated Kepler any higher than Longomontanus.

Moreover, as Kepler took stock of the situation during his first few weeks, he may well have perceived that he would be displaced to third in the pecking order when Johannes Müller arrived from Brandenburg. Already on 6 March, Tycho wrote Baron Hoffmann (governor of Stiermark and a member of Rudolph's council who, out of his interest in astronomy, had provided Kepler's transport to Prague and supported him in his quest for patronage) that he and Kepler would like to discuss with him the question of Kepler's position.[89] Before that meeting could be scheduled, however, Kepler boiled over. Beset by problems associated with the prospect of uprooting his wife from her home, family, and rather considerable property holdings in Graz and worried about the insecurity of his position and arrangements – if, indeed, there even were any with Tycho at that time – Kepler insisted on both an amelioration and a formal stipulation of the terms of his employment.

Presumably the contract Kepler envisioned included the terms mentioned as proposals by him in the letter just quoted (and in another memorandum composed at about the same time). There were other terms to which Tycho could agree but that he (or Kepler) would probably not have dignified with contract status; for example, Kepler would be able to go into Prague whenever he wished, provided that he did not stay there too long. (Kepler, in return, was prepared to guarantee strict silence, while he was there, on those aspects of their work that Tycho deemed worthy of secrecy.) There were some complaints that Tycho may have found instructive: When Kepler had to work into the night for Tycho, it was only fair that he should have compensatory time off for family affairs.[90] There was at least one recorded demand that Tycho definitely found insulting: When had he ever asked his assistants to work on Sundays or holidays? he indignantly responded.[91]

[88] VIII, 246. At the same time, however, Tycho effusively welcomed Kepler not so much as a guest, he said, but as a friend and companion in "observations of the heavens."
[89] VIII, 255.   [90] G. W. XIX, 40–2.
[91] G. W. XIX, 42. Kepler's view was that Tycho was full of promises and would keep them if it were up to him: G. W. XIV, 161.

Most of the issues under discussion had ramifications for the duties of Longomontanus and Müller. Tycho apparently agreed to relieve Kepler from the normal observing duties (Kepler pleaded that he had poor eyes and a frail constitution and was all thumbs, anyway); but he could scarcely have agreed to "share the work day" with Kepler by permitting him to switch over every day at noon to working on his own projects. On 5 April, he and Kepler met to discuss the demands in a conversation over which Jesensky had the thankless task of presiding as arbitrator.

In a written response that took up Kepler's complaints point by point, Tycho agreed to share Kepler's traveling expenses and to try to get him and his family separate quarters somewhere. But he did not see how he could guarantee Kepler's salary, whether from Graz or from the emperor, without getting an undertaking from the emperor for which he had applied but was still waiting.[92] Finding this position unacceptable, Kepler departed with Jesensky for Prague the next day.

Tycho's attitude to this point seems to have been one of quiet confidence that he would eventually succeed in working out some kind of satisfactory arrangement with the emperor, if he could just persevere in the task long enough. Thus, when Jesensky left with Kepler, Tycho sent along with him a letter to Baron Hoffmann,[93] asking him to intercede in the matter and straighten things out. At the same time, however, he told Jesensky that he would have to have a written apology from Kepler for his outbursts. Whether it was the latter requirement or what may have seemed to be Tycho's rather too casual attitude toward the whole matter, Kepler became even more upset the following day and sent out to Tycho a note that was sufficiently nasty that Tycho decided not to have one of his other assistants copy it to preserve it.

Thoroughly disgusted by what he felt was Kepler's most unseemly conduct, Tycho sent the letter back to Jesensky expressing his regret that he had ever had anything to do with Kepler and asking Jesensky to inquire discreetly as to whether Kepler had allied himself with Ursus, who was now back in Prague.[94] That Kepler may have made some kind of threat in regard to Ursus is not impossible. He was clearly very upset and must have perceived by then that Ursus was the one issue on which Tycho was vulnerable. Otherwise we can guess only that the letter did not descend to personal remarks in the style of Ursus, both because of the relative mildness of Tycho's reaction and the other information we have documenting Kepler's admirable character.

[92] VIII, 272.   [93] VIII, 297.   [94] VIII, 299.

But although Tycho may have professed doubts to Jesensky about being able to retain Kepler's services, he appears to have been able to resolve them when Kepler, under some combination of Hoffmann's advice and his own straightforward nature, wrote an apologetic letter to Tycho acknowledging that he had been treated with nothing but kindness and asking to be forgiven for his outburst, which had been the product of youth, mercurial temper, and ill health.[95]

Tycho, for his part, drove into Prague – something he had told Kepler in an earlier letter that he did not like to do – to reconcile with and fetch his errant assistant. Either at that time or sometime in the next month they hammered out the agreement that Kepler recounted so blithely in July. At the beginning of June[96] Kepler returned to Graz to settle his affairs and ask for a leave of absence from his post.

At the middle of 1600, when the summer had opened the shipping lanes to transport Tycho's instruments, and Kepler had set out (using the entourage of Frederick Rosenkrantz for this trip)[97] to fetch his family back to Benatky, Tycho must have thought that his struggles for the new Uraniborg were almost over. On 10 June, however, the emperor returned home from Pilsen[98] and summoned Tycho to Prague. In the nine months since the emperor had retreated into hiding, Tycho's contact with him had been purely literary. The separation of more than a hundred kilometers had doubtless complicated the tasks of getting Benatky remodeled and dealing with Kepler's demands, but it had probably also relieved Tycho of much of the burden of his office. It would therefore have been no more than enlightened self-interest for Tycho to have advised his patron – as Tycho asserted his enemies said he did – to stay away from the capital for so long a time.[99] Now, however, on his return to the world, Rudolph had a backlog of decisions of state that all required consultation with Tycho. And the time for sealed messages was past: Just as Tycho was finally getting Benatky altered to suit him, he was asked to move to Prague, suddenly, and, as it turned out, permanently.

Tycho and his company were initially put up in a hotel or rooming house called the Golden Griffin. There he was at the emperor's beck and call to such an extent that during one crisis at least, he was summoned to the palace twice in the same day. With an eye, no doubt, to escaping the emperor's flattering dependence by getting permission to retreat to Benatky, Tycho complained that his accommodations did not provide the peace and quiet he needed to do the

---

[95] VIII, 305–7.   [96] See Tycho's recommendation for him in VIII, 324.
[97] VIII, 332. This, no doubt, was the help with his travel expenses to which Kepler alluded.
[98] *Norlind*, 306.   [99] VIII, 273, 374.

special (astrological, presumably) projects the emperor had assigned to him. This plea got him moved to a private residence in the neighborhood of Baron Hoffmann's house.[100] The fact that it provided no better accommodation for Tycho's instruments than the Golden Griffin had was irrelevant, initially, because of the small number that Tycho actually had with him at Benatky.

The other instruments, however, at least those at Magdeburg, were finally on their way. After having somehow, during their year and a half of storage, become hostages to some combination of German politics and official avarice, so that even a letter from the emperor had failed to achieve their release,[101] Tycho's suggestion/ bribe that the city authorities might make a profit for themselves on the transport of his possessions, by using the return trip to import cheap Bohemian wine, had finally shaken them loose.[102] The instruments from Hven had proved to be almost as troublesome. Tycho had apparently been prepared for the possibility that he might have some difficulty even in getting them out of Denmark, for when he sent the younger Tycho and Claus Mule to fetch them he had armed them with a letter to his brother Axel[103] (who was fiefholder of Helsingborg) asking for whatever help (permission?) might be necessary. Things seem to have gone smoothly at that end.

But the instruments then got stuck in Lübeck over the winter and were held up by the authorities in Hamburg through the summer (despite another letter from the emperor), so that they did not reach Tycho until sometime in October 1600, a few weeks after those from Magdeburg did.[104] If Tycho tried to use the need for space for the instruments as an excuse to return to Benatky, he did not succeed. The emperor told him to have them set up on the second-story porch of a villa (the still-standing Belvedere) that had been built on one side of the palace grounds by Rudolph's grandfather (Ferdinand I).[105] By Christmas night Tycho was making observations.[106]

Whether Tycho accomplished anything at all during the second half of 1600 is doubtful. By all indications his summer must have been completely consumed by the tasks of uprooting his establishment at Benatky and satisfying the demands of his patron. Exactly what the latter were, Tycho did not specify. But he had done his best to limit the scope of his duties back in January, when someone, probably Barwitz, had requested forecasts on behalf of the emperor.

---

[100] VIII, 341. For a map of Tycho's environs in Prague, see Appendix 5.
[101] XIV, 163–4. Tycho's attempt is on XIV, 194–5, and his interpretation of the situation is in letter 6 of Appendix 6.
[102] XIV, 208–9, 211–12. [103] XIV, 160.
[104] XIV, 151–2, 153–5, 211–12, 216; VIII, 370, 387.
[105] VIII, 370, 387. [106] XIII, 241.

Tycho had responded that he would gladly comply with His Highness's wishes but that such predictions did not have his confidence because they were not as accurate as he could wish.[107] Perhaps when he did not have to commit himself in writing, Tycho was able to be a bit freer with his advice. But that did not make him any less apprehensive. In September he complained to Rollenhagen, an inveterate practitioner of the art, that Rudolph was consulting him even about such matters as the war against the Turks and that he was unwilling to trespass on the domains of others or to act as any kind of astrological prophet.[108]

Precisely what was bothering Tycho is hard to tell. Did Rudolph really press him for a forecast of the outcome of a specific military campaign? And was Tycho citing it as an example of Rudolph's naïveté in failing to envision the technical impossibility of summing up and weighing the collective fates of two armies, or of judging the uncertainties arising from the mass of individual volitions that Tycho would have insisted were involved? Was it a subtler form of astrological inquiry, such as the projected dates of death for Rudolph's generals, designed to suggest which ones augured best as leaders of his troops? Or was Tycho being called on for primarily logical, psychological, or political analysis, as he seems to be lamenting in the rest of his statement?

All that can be said is that Rudolph's problems were large and complex and that he tended to seek advice on them from people he could regard as neutrals in the highly charged atmosphere of Bohemian politics.[109] By the time Tycho arrived, the long-smoldering tensions in the multiethnic Bohemian society were being redirected along the two confessional battlelines of the Counter-Reformation. However threatening the Turks may have seemed at the time, it was these internal passions that broke out, after a generation of fanning, into the conflagration that brought the empire to its knees.

The leader of the Counter-Reformation offensive was Archduke Matthias. His brother, Rudolph, was, of course, Catholic too. But he was a firm believer in moderation and harmony and a dedicated opponent of the papacy's political aspirations, so he did what he could to keep the zealots of his faith at arm's length from the great Protestant majority.[110] Unfortunately, it was not always enough. In August 1600 an ecclesiastical commission went to Kepler's (predominantly Catholic) home province of Stiermark and proceeded to

---

[107] VIII, 240–1.   [108] VIII, 373–4.
[109] Evans, *Rudolf II and His World*, p. 63.
[110] Ibid., pp. 46–7. According to Evans, about 99 percent of the population was Protestant (p. 37).

purge the area of all Protestants. Despite what must have been frantic protests by the emperor's predominantly Protestant advisers, Rudolph was unwilling or unable to reverse the commission's action.

Perhaps in reaction, Rudolph banished a cloister of Capuchin monks that was domiciled near the palace and had been invited to Prague by Rudolph himself. That incident is of interest only because the monks attributed their ejection (which turned out to be only temporary) to the machinations of Tycho Brahe. In their view, the "alchemist" as they exclusively referred to their scapegoat, was frustrated in his various black magical practices – particularly gold making – by the aura of the prayers recited by the Capuchins, and Tycho therefore influenced Rudolph to get rid of them[111] Analysts with a smaller vested interest in the situation have noted that Rudolph was facing a general crisis of considerable proportions at the time: superstitions connected to the turn of the new century; fears that he might be assassinated before his fiftieth birthday, as his father had been (possibly by a mad monk, as Henry III of France had recently been); and above all, intense pressure from his Protestant advisers to stop vacillating over his choice of a mate and get on with the production of an heir to forestall a succession by Matthias.[112]

The expulsion of the Capuchins was only one consequence of what seems to have been a brief but drastic mental breakdown for Rudolph. Among other actions, he banished Rumpf and Trautson from court forever. Tycho emerged unscathed, except insofar as a vote by the Council to reimburse Ursus's widow three hundred guldens for the confiscation of her husband's books was a slap in Tycho's face that might not have been possible when Rudolph was sound.[113] As for Tycho's involvement in all of this drama, he said nothing about any aspect of it except (well after the fact) that he played no role in the expulsion of the Capuchins.[114] But we know that he had an extensive correspondence with Archbishop Wolf Dietrich, who approached him about influencing Rudolph to nominate his second brother Albrecht, instead of Matthias, as his successor.[115]

---

[111] *Norlind* (308–9) reviewed the Capuchin literature and arguments at some length. Evans, *Rudolf II and His World*, p. 90, stated that Rudolph "is reliably said to have been turned against them by Tycho Brahe and his [Rudolph's] scheming Calvinist servant Makofsky."

[112] Evans, *Rudolf II and His World*, pp. 47, 279.

[113] *Rosen*, 310.

[114] Tycho to Jöstel, 8 October 1600; VIII, 381: Tycho to the grand duke of Tuscany; VIII, 418–20.

[115] The three letters that have been preserved are in the Kepler papers. The intent of the correspondence, which included several other letters, emerges from correspondence between Archduke Albrecht and Dietrich in the state archives at Brussels: For discussion, see *Norlind*, 311–12. Wolf Dietrich was archbishop at Salzburg. The portrait of him hanging

Whatever Tycho did, he obviously managed to give sufficiently cogent and balanced advice to avoid the fates of Rumpf and Trautson. When in February of 1601, Tycho applied for citizenship and noble status for himself and his sons, it was Rudolph who sponsored his petition.[116] And when, two weeks after that, the Brahe household moved into the Kurtz summer house, it was because the emperor wanted to keep his adviser near him and had paid ten thousand dalers to ensure that he would be as comfortable as possible.[117]

It must have been a weary man who gathered his dependents, possessions, and professional equipment one last time to go through the motions of reestablishing his operation. The premises, built against the city wall, as his Copenhagen house had been, were those that Tycho had rejected eighteen months earlier on grounds that they were too small (and too accessible), and there is no reason to believe that his requirements had diminished in the meantime. To be sure, Tycho may have allowed his domestic staff to dwindle a bit, as his Danish servants gradually became homesick and he faced the problem of using shrinking – and uncertain – funds to hire alien-tongued locals who were ipso facto less congenial and less efficient. It is also true that young Tycho and Tengnagel were now mostly absent on various missions in their own and/or Tycho's interests. But Tycho had also acquired some assistants, two of them encumbered by families who had to be boarded. So, if his "household" at Rostock and Wandesbeck had been two dozen strong, it cannot have been much smaller in Prague. Moreover, if he had not already sold the extra horses he had had Esge Bille purchase for him and ship from Denmark, he had an embarrassing supply of them[118]

And of course, Tycho now had all of his instruments. In a letter written on 15 April, Tycho indicated that all of his instruments were now installed in the house.[119] That they were *in* the house we may believe – even if the house had not grown magically since Tycho had

in the fortress museum there includes an armillary sphere. He was too discreet to talk to Tycho directly but word of their communications reached Rollenhagen almost immediately (VIII, 420–1), and copies of some letters were available to Herwart von Hohenburg by the end of 1601. Not surprisingly, news of his activity reached Matthias, too. When he deposed Rudolph in 1611, he forced Dietrich to resign and live under house arrest at Salzburg. Politics was not Dietrich's only vice: He built the still-standing Schloss Mirabel (1606) for his mistress.

[116] XIV, 218–19.

[117] Tycho moved on 25 February: XIII, 265. The house had actually cost Kurtz more than twenty thousand dalers to build, Tycho told Rosenkrantz: VIII, 409. See also *Gassendi*, 321–2, and Appendix 5 for the location of the no-longer-extant premises.

[118] See letter 1 of Appendix 6.   [119] VIII, 416.

seen it eighteen months earlier. But that they were installed in any serious sense, particularly the large ones that provided Tycho's accuracy, seems doubtful. It is not difficult to imagine that a house built for a mathematician contained at least one north-south wall sufficiently true to allow perfect adjustment of the mural quadrant in the meridian. But if the "pit" instruments of Stjerneborg had anything approaching an unrestricted south horizon, it must have been from makeshift mountings on the city wall. Perhaps the most telling comment is the fact that Tycho was now even using the large rulers for his observing.

Fortunately, as we have seen, observing was no longer essential to Tycho's work. It was computing assistants that Tycho needed, and he had waged a continuous campaign to recruit them. The invitations to Rothmann and David Fabricius had produced nothing (although the latter was to visit for a few weeks in the summer of 1601 to take some of the load off Erikson).[120] Tycho had worked out some kind of arrangement with Jöstel whereby the latter had done some computations for him in Wittenberg during the summer of 1600. But the work had gone slowly because the professor was also preparing for the examination for his medical degree. After first wishing Jöstel good luck and success in his examination and then graciously declining (on October 9) an invitation by the university to attend the graduation ceremonies, Tycho seems to have decided that he could dispense with his services.[121]

Perhaps expense was a consideration: Negotiations with one of Jöstel's students, Johannes Oswald, seem to have foundered on Tycho's unwillingness to pay the extra expenses arising from the fact that the man was married. Still, Tycho was sufficiently pressed for computational help to agree to pay Oswald to work part time for him from his home in Wittenberg.[122] And when Ambrosius Rhodius turned up in Prague in November, delivering some computations done by Jöstel, Tycho enlisted him, too, for a few months.[123] At some unknown time Tycho also hired Matthias Seiffart, who stayed on after Tycho's death to serve Kepler.[124] Finally, in midsummer of 1601 he supervised the apprenticeship of Simon Marius (who later claimed to have made some of the first observations with the telescope) and acquired the services of a Dane by the name of Paul Jensen (from) Kolding.[125]

---

[120] Gassendi, 324.   [121] VIII, 278–9, 363, 381, 391–2.   [122] VIII, 278–9, 363.
[123] VIII, 391. Rhodius later became a professor at Wittenberg.
[124] Marius seems to have been sent by his patron, the Margrave of Ansbach; G. W. XIV, 168, 170, XVI, 394–5.
[125] Norlind, 315, and note 29 on pp. 408–9.

Exactly what Tycho had these people doing is not clear. Müller, who completed his fourteen-month stint with Tycho without leaving a trace of his activity, may have been conducting medical/ alchemical investigations. At least some of the work being done by mail was the computation of tables from (and perhaps for) the progressing work on the lunar theory, and much of the rest seems to have been similarly routine.

What was being done about the planetary theories was not recorded. But Tycho certainly knew by this time that (contrary to what he had told the *landgrave* in 1588) it was not merely a matter of observing the oppositions of the superior planets, greatest elongations of the inferior planets, and some intermediate positions for all of them, in order to (re)determine the apogee, eccentricities, and mutual proportions of the circles.[126] Already by 1590 he was sure not only that most of the orbital elements determined by Copernicus were at least slightly erroneous but also that there must be another inequality not noticed by him or Ptolemy that affected the motions of the planets. And although Tycho was confident that all of the planetary theories would have a similarity of form based on the fact that nature was known to rejoice in order and abhor complexity and confusion, he was sure that the new inequality would be much the most apparent in Mars, because of the similarity of its orbit to the sun's.

What Tycho was grappling with was the necessity of referring the theories of the planets to the *true* place of the sun instead of mean place. In the Ptolemaic system, achieving this feat would have entailed using eccentric epicycles. In the Copernican and Tychonic schemes, all that was required was to compute seriously the geometrical consequence of the sun's eccentric orbit. But Copernicus had taken the computational shortcut of making tables analogous to Ptolemy's, and everyone else down to Kepler, including Tycho, had followed him. In 1590 Tycho was as close to solving the problem as recognizing that his "other inequality" was an effect of the eccentricity of the sun,[127] and calling the sun the "king and commander" of the planets. Apparently, all of the oppositions he had charted to that point had been at least a degree or two away from where Copernican theory placed them. And Tycho had seen some correlation (whether just qualitative or actually quantitative, he did not say) with the fact that all of those oppositions had occurred outside the sun's line of apsides. Before the opposition of June 1591, anyway, Tycho told Magini he was looking forward to checking it because of the sun's position in apogee. And afterwards he happily reported that because Mars had

---

[126] VI, 131.   [127] VII, 291, 299.

been in the sun's perigee, the Copernican prediction had been almost perfect.[128]

That was as close as Tycho came, however. Longomontanus, who was at that time just starting his apprenticeship with Tycho, eventually managed to approximate the effects of the sun's eccentricity, by introducing some kind of variation into Mars's eccentricity (105', Tycho told Kepler in 1598).[129] If he or Tycho had conceived the best solution of the problem in 1591, so that Longomontanus could have applied his considerable ingenuity to coping with the other inadequacies of the theory, Tycho's redintegration of astronomy may well have been completed without the necessity for Kepler's spectacular war with Mars.

As it was, Tycho harbored illusions of completing it without Longomontanus in 1599 and when Longomontanus returned, he accounted for the longitudes of Mars within 2'.[130] But that was not until the first few months of 1600, and by that time either Tycho sensed an impasse in the work, or Longomontanus became frustrated with it, or the lunar theory suddenly seemed more important. Most likely, the forty-year-old assistant saw an open-ended task yawning in front of him (instead of the limited term he thought he had agreed to serve) and so told Tycho that he had to get back to Denmark to get on with his own career. Whatever the case, the long-suffering disciple did complete the revision of the lunar theory. And when Tycho was suddenly called into Prague, he stayed at Benatky long enough to supervise observations of the solar eclipse of 10 July and possibly also to preside over the closing down of the operation there.[131] But after a couple of weeks in Prague, and some ten years after he had first appeared at Tycho's doorstep on Hven, Longomontanus bade a final farewell to his long-time chief and set out on the road home with a gracious letter of recommendation dated 4 August:

> I, Tycho Brahe, lord of Knudstrup, citizen of Denmark, now resident at His Imperial Majesty's castle Benatky, in Bohemia, bear witness to one and all, of whatever station and whatever rank he might be, that the bearer of this my letter, the learned and distinguished gentleman, Master Kristen Sørensen Longomontanus, citizen of Jutland, has been my collaborator for eight years in Denmark and later also for some time in Germany and Bohemia. He has assisted me in my work with zeal and skill. But because circumstances now no longer permit him to

---

[128] VI, 239; VII, 306. [129] VIII, 45.

[130] G. W. III, 109. See also VIII, 64.

[131] XIII, 196–8. Tycho observed the eclipse in Prague with Baron Hoffmann: VIII, 353. Longomontanus probably left Benatky on 18 July, when he presented a manuscript to Erickson that he designated as a memento of his departure.

stay with me and his advancing age forces him to think of his own
situation and therefore to return to his homeland, I have not wished to
place obstacles in his path, but to let him happily and in all friendship
go his way. Noble and distinguished gentlemen, especially my friends
in Denmark, famed and learned gentlemen in other lands, I ask you
obligingly and urgently to consider the aforesaid Master Kristen and
not to fail to do your best for him.[132]

Longomontanus's departure cleared the stage for what has become
one of the celebrated episodes in the history of science. Once he left,
the likelihood that Tycho would manage to complete his planetary
theories in the way he wanted to do it (and the way Longomontanus
finally achieved it twenty years later) virtually vanished. As a result,
the final aspect of Tycho's redintegration of astronomy remained
uncompleted at his death and fell to the genius of the man who
would succeed him as imperial mathematician, Johannes Kepler.
Guided by the hindsight of Tycho's and Longomontanus's struggles
with the traditional models and inspired by the confidence that
"because divine goodness had provided such a careful observer as
Tycho . . ., it was only appropriate to recognize God's gift by taking
the pains necessary to search out the true form of the heavenly
motions,"[133] Kepler discovered the fatal flaw of the old theories and
pressed on to inaugurate a new astronomy.[134]

[132] VIII, 335.   [133] G. W. III, 178.
[134] The hindsight afforded by later developments in astronomy reveals at least six respects in
which the traditional theory had to be failing to represent Tycho's observational data.
Two of them – the perturbations by other planets of the basic elliptical orbits of the earth
(or sun) and the given planet – are so small that Tycho could scarcely have detected them
at their maximum, let alone unravel their several components, without the aid of either
physical theory or much more advanced observational and statistical techniques.
    Another two, arising from the difference between elliptical and circular orbits for the
planet in question and the earth (or sun, in Tycho's system) had been so effectively
neutralized by the near equivalence of the eccentric circle to the ellipse of low eccentricity
that their effects were detectable only in Kepler's celebrated octants of Mars.
    The third pair of potential problems stemmed from the fact that Tycho did not have his
planets moving in their orbits according to Kepler's second law; that is, describing arcs at a
rate proportional to the areas of their sectors in the eccentric. But even in that respect,
Ptolemy's bisection of the eccentricity and Copernicus's double epicycle approximated
Kepler's law of areas so miraculously that Mars was the only planet for which there was
any detectable error. (Mercury involved so many problems of both observation and
representation that it was destined to remain as Lalande described it, an embarrassment to
astronomers right into the twentieth century.)
    There was one planet, however, for which astronomers before Kepler did not use the
remarkable approximations invented for them: the earth. For although Ptolemaic astronomy had managed to simulate the motion of the earth in the planetary epicycles, it
provided no reason to treat those circular epicycles as preliminary approximations, subject
to correction in the way the eccentricities and motions of the primary planetary orbits had

While Longomontanus was laboring to complete the lunar theory, Kepler spent the summer in Graz, working on optics. His research had been inspired by Tycho – and before Tycho had met or even corresponded with Kepler. The problem had been conveyed to Kepler by Mästlin, who had relayed in January 1599 Tycho's report of a curious diminution of the apparent size of the moon during solar eclipses.[135] Kepler had probably never seen a solar eclipse. But he was sufficiently impressed by the direct quotation sent by Mästlin and Mästlin's very positive reaction to it – to give some very serious thought to this phenomenon that seemed to fly in the face of optical laws that had been unquestioned since Euclid.[136] In April 1599 Kepler unburdened himself on what he termed this "enigma" in a long letter to Herwart von Hohenburg, chancellor of Bavaria.[137] But that seems to have been the end of it at the time. During his months at Benatky, however, Kepler surely talked with Tycho about it and saw some kind of demonstration (perhaps for the first time) of the "canal" by means of which Tycho observed eclipses. One can scarcely doubt that a solution of this problem of the "camera obscura" was foremost among the projects for which Kepler negotiated with Tycho about having time for his own work. At Graz, of course, Kepler got it, and by the end of the summer he had the solution to the problem of the formation of optical images behind small apertures. Four years later it blossomed into the book that revolutionized the ancient science of optics for the seventeenth century, *Additions to Witelo, in Which Is Contained the Optical Part of Astronomy*.

As Kepler described his situation to Mästlin in September 1600, he had "left the negotiations hanging in the balance." What he had hoped to achieve through Tycho, Baron Hoffmann, and Corraduc was an imperial order securing his Styrian salary but reassigning him to duty with Tycho for the emperor. It was his assessment (to Mästlin) that "with the help of very excellent interventions [from Tycho and Hoffmann] I could have accepted a salary in the service of the Emperor. But I was deterred by the experiences of those awarded

been corrected. Logically, the situation should have changed with the advent of Copernican astronomy. But Copernicus not only failed to apply to the earth (or sun) the double-epicycle mechanism he used for the rest of the planets, but he even neglected to use the one epicycle actually present in his theory of the earth's motion, by referring all planetary computations to the center of the earth's orbit rather than to the sun itself. For the story of Kepler's struggle to reach this insight (and others), see Curtis A. Wilson, "From Kepler's Laws, So-called to Universal Gravitation: Empirical Factors," *Archive for History of Exact Sciences* 6 (1969–70): 89–170.

[135] *G. W.* XIII, 276–80. For Tycho's discovery, see Appendix 4.
[136] Stephen M. Straker, "Kepler's Optics: A Study in the Development of Seventeenth-Century Natural Philosophy" (Ph.D. diss., Indiana University, 1970).
[137] *G. W.* XIII, 305–6, 324.

magnificent salaries who extract barely half with immense difficulty."[138] Whether or not he would ever have discussed such a
personal matter with Kepler, Tycho was one of those "promisees,"
and that was doubtless why he had balked at guaranteeing Kepler's
salary, even while agreeing to fund certain fringe benefits out of his
own pocket. While Kepler was awaiting the outcome of the negotiations, however, his options were foreclosed in a way that conducted
him into Prague as inexorably as Tycho's instruments were being
brought there at about the same time: The ecclesiastical commission
authorized by Archduke Matthias came to Graz to conduct religious
tests of all Protestants.

> For all of the more than a thousand citizens and clergymen [who
> would not convert] including me, permanent exile was decreed,
> effective in forty five days. All personal goods have become very
> cheap; no money is exchanged among people except Hungarian small
> change; all property is minutely inventoried. You may imagine all the
> other evils. I reported this to Prague. Tycho answers that by word of
> mouth he recommended me to the Emperor and I was approved; only
> the letter concerning this matter is awaited. But because their plans are
> thrown into confusion by the termination of my Styrian salary (since I
> have given it up because I am about to go to another homeland and
> voluntarily [!] look for another employer), Brahe urges me to forget
> about Württemberg [where Kepler was hoping for a position – with
> Mästlin – at the University of Tübingen] and come to Bohemia as
> quickly as possible. I remain undecided. Yet I am planning to go to
> Linz with my family and, leaving them there [shades of Tycho], to
> proceed by myself to Prague. While I am there, I shall find out what
> salary will be awarded me, and what hope there will be of extracting
> it.[139]

[138] Kepler to Mästlin, 19 September 1600 (*G. W.* XIV, 175), translated in *Rosen*, 280–1.
[139] Ibid. The situation was graphically described by Kepler's biographer: "The shepherds had
      been driven away; now the herds had to be destroyed. Ever harder was the screw turned.
      After the expulsion of the preachers from Graz, the Protestant citizens proceeded to visit
      the divine service of their faith on the neighboring estates of the nobility and to receive the
      sacrament there. This was now made punishable and the people were forced to have their
      children baptized as Catholics and to be married according to Catholic ritual. Kepler
      himself received an order for a fine of more than 10 talers for evading the city clergy at the
      death of his little daughter. This fine was reduced by half at his request, but he had to pay
      the other before he could bury the child. Naturally, it was not very long before the
      expulsion of all Protestant clergymen still in the land was also ordered. All who received
      them were threatened in body and property. Attendance at other than Jesuit schools was
      forbidden; the installation of Catholic clergymen in the collegiate church was demanded.
      Anyone who sang hymns in the city, or read collections of homilies, or Luther's Bible,
      made himself liable to banishment from the city. Heretical books had to be rooted out and
      destroyed; barrels and chests, which held books had to be opened and examined in the

On 17 October 1600, Kepler wrote Tycho from a resting place somewhere on the road. In a short while he would arrive in Prague, exhausted from a two-week journey and ill with an intermittent fever that he did not shake for almost a year. He was wondering, he confessed, how to eke out from the money that remained after the frightful expenses of moving, a few weeks of living expenses in a capital where costs were four times what they had been in Graz. With his Styrian salary now gone, the agreement he had hoped Tycho would be able to work out with the emperor was no longer possible, but he would come, he told Tycho, because the emperor still represented his best hope. With no real alternatives in sight, he was at the mercy of whatever arrangement Tycho would be able to make for him.[140]

It does not appear that Tycho had been able to fund a formal contract for Kepler, from the negotiations that had been in progress for over six months. In fact, even the money the emperor had promised to pay Müller in lieu of his salary from Brandenburg was seriously overdue. Tycho had been paying the Müller family's living expenses for seven months and was eventually to pay his two-hundred-daler salary, too, out of his own pocket, after Müller postponed his departure two months past his contracted year and still had not been paid.[141] But two developments since Kepler's departure had now rendered his assistance more valuable to Tycho than ever. The first was that Tycho had lost Longomontanus. The other was that Ursus had died.

From his first look at Ursus's book in the spring of 1598 to his interrogation of Rollenhagen and Lange at Magdeburg in October of that year, Tycho had laid meticulous plans for the prosecution of his nemesis. The letter that arrived in March 1599, protesting Kepler's unawareness of and embarrassment over Ursus's publication of his flattering remarks, had been icing on the cake. Tycho had begun to investigate the channels of retribution shortly after his arrival in Prague. But when the imperial court had scattered because of the plague, he had had to defer his plans for almost a year. Only when first Rudolph and then Tycho himself had returned to Prague had he been able to initiate his legal action. And by that time Tycho already

---

presence of the archpresbyter. A watch was set at tolls and gates to keep heretical books out of the city." Max Caspar, *Kepler*, ed. and trans. C. D. Hellman (New York, Abelard Schuman, 1959), p. 97.

[140] VIII, 383–5.

[141] News sent by Barbara Kepler to her husband on 31 May 1601, while he was back in Graz on family business. *G. W.*, XIV 169–70. See also a letter from Johannes Eriksen to Kepler: *G. W.* XIV 168.

knew there was danger in delay, because Ursus was seriously ill.[142] In the hope of accelerating his cause, he applied through religious as well as secular channels. The archbishop of Prague had responded quickly, if somewhat inconclusively:

> With regard to the book of which you sent me a copy, and of which Nicholas Reymars Ursus says he is the author, I cannot find out who printed it in this city without my approval. But I shall take care to have the printer tracked down and, if he is found, to have others view him as an appropriate example, and to seize whatever copies he has. But since Ursus is under the jurisdiction of the Most Illustrious Court Marshal, you will be able to call on him and request him officially to demand that Ursus identify the printer and hand over whatever copies he has. In whatever way I can and should I shall not fail to assist in this matter.[143]

The civil proceedings for libel and defamation of character proceeded just as inconclusively. Tycho sent to Ursus on his deathbed

> two doctors of jurisprudence together with a public notary to ask whether he was willing to retract that malicious publication, chockful of insults. At the same time I prepared the main items of the insults to be read to him. He would not admit most of them, even though they were found spelled out on the pages indicated in his publication. Nevertheless he refused to recant. On the contrary, he submitted everything to the decision of the judges. Consequently, more than happy when I saw this, I most humbly obtained from the emperor the appointment of four commissioners to decide the case by law. Two of them were Barons, Christopher von Schleunitz and Ernfried von Minckwitz. The other two were doctors of jurisprudence from His Majesty's councillors. But the defendant died.[144]

Ursus died on 15 August, whether from "attack in the French camp [syphilis] to which he was accustomed, or by asthma and consumption, or rather by both,"[145] Tycho did not know. In the view of Jesensky, and doubtless also of Tycho, death had "struck that wild beast [Ursus, the bear] with special kindness and saved him from a thoroughly deserved punishment."[146] Tycho's understanding was that "had the author lived a while longer, he would have been sentenced, as I learned from the commissioners, to be branded in infamy, and beheaded or quartered according to Bohemian Law."[147]

---

[142] Tycho to Rosenkrantz, 3 August 1600: VIII, 334.
[143] 27 July 1600: XIII, 209. Translation from *Rosen*, 301.
[144] Tycho to Rollenhagen, 26 September: VIII, 371; translated by *Rosen*, 304–5.
[145] Ibid.
[146] Elegy at Tycho's funeral, 4 November 1601: XIV, 238–9.
[147] Letter to Rollenhagen, 26 September 1600 (VIII, 372), translated in *Rosen*, 307.

But if the pound of flesh was no longer (if it ever had been) an option, mitigation of damages still was. Ursus's infamous book, Tycho told Kepler and Rollenhagen, would speak forever in the place of its author: "The question was not so much destroying his person, which everyone knows was clownish and vainglorious, but rather his book, stuffed full of so many insults and lies, and restoring the glory and reputation of myself and my associates."[148] The political aspect of this task was soon accomplished. The archbishop was "charged by the emperor to notify the printer, have all the copies sought out, as many as are found in Prague, and have them consigned to the flames."[149] Corraduc "promised the promulgation of an imperial decree prohibiting the book throughout the entire Empire and declaring it banned as an infamous and scurrilous book."[150]

At the opposite extreme, Tycho recognized that the personal scars left from the insults and slanders put into circulation by Ursus simply could not be remedied. No matter how overwhelming and perhaps even excessive, Tycho's rage was, there was no way he could address the attacks on his family and his personal honor with any reasonable hope of ameliorating them.

There was, however, a middle intellectual ground on which Tycho wanted both to attack Ursus and to defend himself: the matter of priority in the development of the system of the world. In this restricted arena, Tycho could conduct a battle he might expect to win, while simultaneously venting his spleen on Ursus. Moreover, if he could destroy Ursus's scientific credibility, he might not only inhibit the circulation of Ursus's other statements but also discourage other people who Tycho thought had a predilection for expounding his system but no corresponding regard for giving full credit to its discoverer. The person Tycho had in mind was a Scotsman by the name of Duncan Liddell.

As with Ursus, Tycho's acquaintance with Liddell dated from a visit to Hven, this one in June 1587. Unlike Ursus, however, Liddell was apparently sufficiently well off to be living the life of a wandering scholar and was on the brink of a successful lecturing career of nearly twenty years. Whether because of or despite having been a student of Wittich, among others, Liddell got close enough to Tycho during his five-day visit[151] to be favored with some kind of partial glimpse of his new system.

So interested in the subject was Liddell that he wrote to Tycho to

---

[148] Ibid. Presumably Rothmann was the associate for whom Tycho was concerned.
[149] Ibid.
[150] Tycho to Kepler, 28 August 1600: VIII, 344, translated by *Rosen*, 302.
[151] IX, 55.

check his understanding of it, reappeared at Uraniborg the following summer,[152] and then delivered lectures on the system at the University of Rostock in 1588 or 1589.[153] After the lectures, however, Tycho was so far from being flattered or gratified by this interest in his system that in 1589 he issued a veiled complaint to Hayek about certain learned men in Germany who continually taught his ideas to others and "deck themselves in borrowed plumes like Aesop's crow."[154]

One of the auditors of Liddell's lectures was a German student named Daniel Cramer, who later visited Hven in 1592 as the perceptor of Holger Rosenkrantz.[155] On the way to the island Cramer refreshed his memory of the system – which he seems to have identified with Tycho – from his student lecture notes. But when he then sought to use his information for some name-dropping, he learned that Tycho was upset with Liddell for "blazing abroad information that he had received from Tycho in strict confidence." Only after Cramer produced his notes from Liddell's lectures and showed Tycho numerous references to himself, did Tycho reluctantly profess to be mollified.[156]

But the situation obviously continued to rankle. And it apparently did not improve when Liddell moved to Helmstadt in 1591 to take the chair of higher mathematics. Sometime before 1598, whether because he had heard tales, received a copy of a circular printed at Helmstadt in 1595, or simply learned that someone he met or knew either was going to pass through or was already living at Helmstadt, Tycho asked for and got a private inquiry as to whether Liddell was crediting him with the invention of his system. The word that came back was that Liddell not only had acknowledged that Tycho had published the system, but had publicly expressed disgust with Ursus as one who had "drained dry the cup of impudence."[157]

Tycho was still not satisfied. Whether there were lingering doubts from the reports he had heard or whether he now had a copy of Liddell's circular of 1595, which referred to the third system of the world not as "Tychonic" but as "the one that Tycho Brahe mentions in his book on the comet of 1577,"[158] Tycho was as agitated as ever.

[152] The letter, and Tycho's response to it on 14 October 1587, were preserved with Kepler's papers because they were given to him by Tycho. They have been printed as *Bilaga* 2 by Norlind, 366–8.
[153] VIII, 39.   [154] VII, 201.
[155] VIII, 42. Although Cramer recalls the year as 1591, Tycho's log (IX, 112) shows them as having been there from 6 to 8 July 1592.
[156] Cramer to Tycho: VIII, 42.
[157] Tycho to Hayek: VIII, 56–8.
[158] Tycho to Cramer: VIII, 185.

The origin of such possible hairsplitting and of Tycho's anger lies in the two very cordial letters that he and Liddell exchanged in 1587.

Although the portion of Liddell's letter that contained some geometrical models of planetary theories was not deemed important enough to save, it is clear from the context that Liddell was a technician in the tradition of Wittich, who liked to juggle circles just for the sport of considering alternative possibilities. Most likely, he presented these models to his classes in essentially the context in which he related the new Tychonic system to the old Ptolemaic and Copernican ones. Under such circumstances, it is likely that a few of his students might have lost track of what Liddell was and was not claiming as his, no matter how scrupulous he may have been. But at the same time, Tycho's letter provides reasons for wondering whether Liddell was always as scrupulous with the distinction as he might have been. Tycho's response congratulates Liddell for having correctly inferred the principles of his system ("hit the nail on the head, as the saying goes")[159] from his conversation with Tycho three months earlier. This locution strongly suggests that the once-bitten twice-shy Tycho had not actually shown his system to Liddell. If so, there may have been a real point to the circumlocution by which Liddell referred to the system not simply as "Tycho's" but as one mentioned or (in a later variant) "described" by Tycho.

For whatever reason, Liddell's transgressions came up in the conversation when Holger Rosenkrantz visited Tycho at Wandesbeck in February 1598. Not surprisingly, word of the discussion quickly reached Rosenkrantz's former preceptor. With that communication, apparently, went some kind of statement from Liddell, possibly in the aforementioned 1595 printing from Helmstadt, to the effect that "the learned Cramer, pastor at Stettin," had passed on to Tycho "the hypothesis that I proposed seven years ago at Rostock."

Poor Cramer was clearly unnerved by the accusation and its implications. Puzzled by the difference between the Duncan of Rostock who had been so scrupulous about crediting Tycho and the one at Helmstadt whom he could now only label a plagiarist, he admitted that Liddell had been known, after a particularly stimulating presentation of the system, to mention to an interested student or two that he had a letter from Tycho congratulating him for having "hit the nail on the head."[160]

In the spring of 1599, Tycho was visited at Wittenberg by a remote kinsman, Steen Bille, and his preceptor, Barthold Mule, who was a

---

[159] *Norlind*, 367.   [160] Cramer to Tycho: VIII, 38–41.

brother of Else and Claus Mule. During one of their conversations, Liddell's name came up, apparently because Bille had been a student of Liddell's. When Tycho responded with the ill humor that seems to have been absolutely predictable in connection with any reference to his system and produced Cramer's letter to back up his remarks, Mule had the letter copied (furtively, Tycho asserted: He would not want to have been accused of willfully circulating uncomplimentary reports). When the two students returned to Helmstedt, Mule confronted Liddell with the letter and reported his reaction to Tycho. As far as Bille was concerned, Liddell probably passed the test (doubtless designed by Tycho, who related only selected parts of the report).

Apart from cursing Cramer as a "miserable ass," Liddell hunted out the university's catalogue of lectures and showed Bille the announcement that his course treated "the planetary theories according to the hypotheses of Ptolemy and Copernicus, and that other hypothesis of the system of the world that Tycho Brahe describes in Volume II of *De mundi*." For Tycho, however, this was the perfectly consistent behavior of a "sly fox." Unwilling either to stake a claim or to renounce credit, Tycho told Cramer, Liddell split hairs by refusing to acknowledge Tycho's priority explicitly: For the mere fact that an author had described or expounded an hypothesis in a book did not make it his.[161]

It seems safe to say that almost everyone knows someone who would have responded to Ursus's singularly offensive babblings at least as strongly as Tycho did. In fact, given Tycho's temper and resources and the generally violent atmosphere of the times, it seems remarkable that Tycho did not take the law into his own hands. In regard to Wittich, who certainly transgressed considerably, if not so provocatively, Tycho's suspicions and mutterings are understandable, even if not admirable: The most remarkable thing about them is the double standard under which Tycho laid – or at least magnified – his claim to a share of the credit for work whose creative component must have been due entirely to Wittich.

Tycho's reaction to Liddell, however, is hard to characterize as normal. The man was promulgating from the lectern a system that Tycho had published and he had not. Most people would regard such a person as a disciple, no matter how he was mentioning their name. How Tycho could have imagined that Liddell's petty reminiscences could diminish his stature defies conception. But he clearly did. One way of dealing with Liddell's impertinence was to persuade Cramer to take legal action against him, and Tycho exhorted Cramer to do so. Another was to do something about it himself. But that would have to wait until after he (or, rather, Kepler) had dealt with Ursus.

Tycho's original expectations of Kepler may have been reasonably minimal. Whether because of the advice imparted in Kepler's letter or because of a perception that he was personally reluctant to enter the dispute, Tycho approached him on the issue only after Kepler had been at Benatky for a month. And even then, it was through Tengnagel, who professed a burning desire to have spelled out for him the technical issues of Ursus's claim that he had merely "extracted the hypotheses from the same source as Tycho, since they are available explicitly in Copernicus and Ptolemy" (through their presentation of ideas developed by Apollonius).

In March 1600, Kepler obligingly produced a scholarly two-page document headed "Quarrel between Tycho and Ursus over Hypothesis."[162] An associated statement written by Tycho shows clearly that he intended this account only for purposes of his lawsuit. (However, it is interesting that one of the terms Kepler sought to establish at their confrontation a month later was that if Tycho wanted to publish anything under Kepler's name, it would have to be at his own cost and subject to Kepler's agreement.) But with the death of Ursus and the consequent cancellation of the tribunal through which Tycho had been expecting to be able to pillory his enemy, Tycho decided that he had to do something else to vindicate his honor. In a postscript to the letter of 28 August in which he informed Kepler of Ursus's death, Tycho related his plans for pursuing Ursus. He would

> publish the entire proceeding, together with the commissioners' decision and His Imperial Majesty's decree [banning the book], in a special book which will expand to a moderate size. I shall, God willing, do this with His Most Clement Majesty's consent and Corraduc's approval already vouchsafed me. In the second part of the book I shall reply to the questions which are mathematical and concern the hypotheses. That is why I should also like to have your views of those matters soon. Or if (as I hope) you come here as quickly as possible, the matter can wait until you arrive.[163]

Presumably Kepler at least had the psychological comfort of having received this guarantee of his welcome from Tycho, when he loaded his worldly possessions onto two carts and left Graz on 30 September 1600. When he arrived in Prague at Baron Hoffmann's house three weeks later, he desperately played out the various slim

---

[162] The entire document is translated in *Rosen*, 290–6.
[163] VIII, 344, as translated in *Rosen*, 302.

hopes he still held for obtaining some kind of regular professional appointment.

But from the best possible intentions everywhere came nothing anywhere. By December, all that remained was the least attractive alternative for both him and Tycho: Tycho's ability and willingness to support Kepler out of his own funds. Under terms that seem not to have been written and may not even have been fixed, Kepler, his wife, and her young daughter joined Tycho's household. There he was put to work

> to rebut even more clearly and more fully than you have [done] pre-
> viously Ursus's distorted and dishonest objections to my invention
> of the new hypothesis, namely, that it was derived from Copernicus
> or Apollonius of Perga, and not to think it disagreeable to ascribe the
> new hypothesis to me, as is right, just as you did before with dem-
> onstrable reasoning.[164]

In the preface to his "Defense of Tycho Brahe against Ursus," Kepler said he had "volunteered to transfer this task to myself from Tycho Brahe's shoulders." Later, he told Mästlin that he "wrote against Ursus at the command of Tycho." Whatever the case, he was still "doing nothing but writing against Ursus"[165] in February 1601 and presumably continued to do so until the spring thaws, when he journeyed back to Graz "in a fruitless effort of four or five months to salvage his wife's . . . holdings" in Styria.[166]

Kepler probably never resumed his work on the manuscript. When he returned to Prague in August, Tycho took him to meet Rudolph. In what was probably an attempt to secure a separate salary for Kepler, Tycho negotiated a deal whereby he would publish his planetary theories under the name of the *Rudolphine Tables* and Kepler would help him with the work as his official assistant. It turned out to be a happy arrangement all the way around. When Tycho died a few weeks later, Kepler inherited the job and the post, although at a salary that was one-sixth the amount that Tycho's had been and every bit as uncertain.[167]

Kepler then declined Fabricius's urgings to publish his "Apologia Tychonis contra Ursum" on the grounds that he needed to do more historical research on "hypotheses," and that, because Ursus was a "predecessor in the office of imperial mathematicians, any publica-

---

[164] VIII, 343, as translated by *Rosen*, 299. Kepler's plight is described in some detail by Caspar in *Kepler*, pp. 108–18.

[165] *G. W.* XV, 139, and XIV, 125.   [166] *Rosen*, 322.

[167] Caspar, *Kepler*, pp. 121–3. Tycho assigned Kepler to the task of determining "the quantity of the sun's second eccentricity": *G. W.* XV, 139 (letter to Mästlin).

tion should be postponed until it would involve less ill will than now,"[168] But Kepler found ample opportunity to honor Tycho's memory in other ways. In particular, when he finally completed the *Rudolphine Tables* in 1627 (long after Rudolph's death in 1612), he listed himself on the title page as assisting author to "that Phoenix of Astronomers Tycho Brahe."

By the time the Brahe entourage was established in the Kurtz house, Tycho was, or at least perceived himself to be, in visible decline. Perhaps it was the news that his younger brother Jørgen Brahe had died on 4 February that made him especially aware of his own mortality.[169] At any rate, he was still up to promoting an expedition to Alexandria (for the younger Tycho), writing a good technical response to a letter from Magini and defending himself against the charge of having influenced Rudolph against the Capuchins[170] (when the younger Tycho was made persona non grata in Italy on that account), but that was about all. What little observing he did in 1601 was directed almost exclusively toward reorienting his instruments, and the books that had been almost completed for many years remained unpublished at his death.

Although Tycho's professional working life had stalled – and indeed had never really gotten restarted in exile – the other issues that had taken him out of his homeland seem to have been reasonably satisfactorily resolved. His financial affairs in Denmark were still messy, and as for his situation in Bohemia all that can be said is that no one ever has *enough* money. But as we have seen, he was probably living at least as comfortably as he would have been in Denmark and would soon have his loan back from Mecklenburg with which to buy property that would be his estate.

What made the whole business worthwhile was that his wife and children now enjoyed a status that would permit them to inherit that estate. If Kirsten was not the social asset that most wives were, no one was labeling her his mistress. Even after his death, she would be able to buy a country estate to maintain herself and his other dependents,[171] and when she died, she would be encrypted next to Tycho in the great cathedral of Prague, the Teyn Church. And if the younger Tycho still had to be called his "natural son,"[172] Tycho nevertheless had ambitions for him that he could not have entertained in Denmark.

[168] G. W. XIV. 334. (letter to David Fabricius).
[169] In the published eulogy of Tycho, Jesensky reported that Tycho sometimes broke up a cheerful conversation among friends to talk of death: XIV, 240.
[170] VIII, 243, 394–8, 418–20.   [171] See letter 1 of the Epilogue.   [172] VIII, 398, 418.

Exactly how much money Tycho was able to leave his family to support their enhanced status is debatable. Just a few months before his death he was obviously stretched to the limit when he paid a twenty-daler stipend to Kepler in ten- and six-daler installments with a plea that he did not have much ready cash at the time.[173] This embarrassment may have been the temporary result of what was apparently a rather lavish wedding that Tycho was planning for his daughter Elizabeth. But even assuming that Tycho had managed to retain the incomes and produce from Benatky, when he had moved to Prague the year before, the higher prices and more expensive lifestyles of the capital and his failure to collect any more of his pay makes it unlikely that there was much left for the family when he died.

In fact, because Tycho still owed money to the widow Rantzov, the family might well have been impoverished if Tycho's determined efforts to recover his ten-thousand-daler loan to the young dukes of Mecklenburg had not finally produced payment a month before his death.[174] As it was, a loan repayment by Axel Gyldenstierne probably more than covered the debt to Rantzov's widow so that the ten thousand dalers was all available to the family.[175] Nor was that the full extent of Tycho's estate. Nearly three thousand guldens of back pay were due, and Rudolph soon offered the heirs twenty thousand guldens (from the government, as opposed to his own exchequer) for Tycho's instruments and observations.[176] About half of the back pay was collected, but only a few thousand of the rest ever was.[177] Still, however, it was a truly princely sum and must have

---

[173] XIV, 231. Johannes Eriksen to Kepler; *G. W.* XIV, 185.

[174] XIV, 174, 200, 231. For evidence that the Rantzov debt was still unpaid a year and a half after Tycho left Wandesbeck, see XIV, 169, 174; letters 1 and 6 of Appendix 6; and XIV, 198, and 200. Tycho's gradually escalating efforts to collect the debt from Mecklenburg are related in XIV, 213–14, 219, 222–32.

[175] See letters 5 and 6 of Appendix 6. Axel's reputation suggests that he indeed repaid the loan. But it is not obvious that the money from Mecklenburg would have come back if Tycho had died before collecting it.

[176] The initial computation of Tycho's back pay was 2,333 guldens, later corrected to 2,966 guldens (XIV, 232, 242). On the purchase of Tycho's instruments and observations (for all of which the heirs originally asked 100,000), see XIV, 246, and the letters in the Epilogue.

[177] The heirs received four thousand guldens in 1604 and another one thousand in 1608. In 1643 some of the heirs signed their interest in the debt over to two monasteries. Another heir traded his rights for 300 flasks of wine: XIV, 251–306, 318. Kepler retained the observation logs, first for use in his work, then as security for back pay owed him by the emperor. Eventually, when the debt on the back pay reached 12,694 gulden, Kepler's son pragmatically sold the logs to Frederick III of Denmark for 600 dalers. After World War I, when the modern state of Czechoslovakia was formed, its government sued for recovery of the observation logs on the grounds that Tycho's heirs had deeded them to the

contributed considerably – even on paper – to the Brahes' general eligibility.

The younger Tycho managed to marry a noble widow of good family and sire five children, and his sister, Sidsel, married a Swedish baron.[178] And because good estates could be purchased for 2,500 to 3,000 dalers,[179] there should have been enough patrimony to support a decent, if modest, living for all, at least to the age of majority. Best of all, Tycho got to witness some proof of his family's acceptance before he died, in the form of Elizabeth's wedding to the Westphalian noble, Franz Tengnagel van Kamp.

By the spring of 1601, Tengnagel had been with Tycho for over six years. During that time he seems to have acquired some pretensions to expertise in things astronomical, because after Tycho's death he was to contest Kepler's assumption of the role of Tycho's literary executor.[180] But as far as Tycho was concerned, politics was Tengnagel's métier. Indeed, Tycho's judgment in using Tengnagel as his ambassador for trips through most of Europe during 1598 and 1599 was later ratified by a career that took him right up to the emperor's privy council.

Exactly when and how the arrangements for the match were made is not known, but by the time the wedding was announced in April – for sometime between Easter and Pentecost[181] (10 June) – the matter was out of Tycho's hands. Because of the couple's anticipation of the formalities, Tycho became a grandfather (28 September) before he died.[182] Whether for that reason or others, there was friction between Tycho and the bridegroom, so that the festivities had to be postponed at least once and did not actually take place until 17 June.[183]

Whatever his private reaction, Tycho was enthusiastic publicly. There could be no more conclusive vindication of his decision to emigrate from Denmark than nuptials between his daughter and a nobleman, to be held in the palace built by the former vice-chancellor of the empire and now occupied by Tycho. Tycho, understandably with great delight, invited all of his friends and many

---

emperor. The Danish government refused to consider the suit on the basis that the emperor had never completed payment for the logs. See Harald Mortensen, "Fra Tycho Brahes Bogsamling," *Fund og Forskning* 2, 25–32.

[178] XIV, 324–5. His mother was a Swedish Brahe.   [179] See letter 1 of the Epilogue.

[180] Kepler likened Tengnagel's efforts to take control of Tycho's observations to the barking of a dog in a manger: It prevents other creatures from eating the hay but cannot benefit from the hay itself. See Caspar, *Kepler*, pp. 140–1.

[181] VIII, 409.

[182] See the news and the infant daughter's horoscope, sent by Johannes Eriksen to Kepler: *G. W.* XIV, 190–2.

[183] *G. W.* XIV, 170, 185.

of his noble acquaintances in Denmark, not because he expected them to come, he told Rosenkrantz, but just so they would know.[184] By the time Tycho invited Duke Ulrich of Mecklenburg in mid-April, he had included invitations to "the emperor and his most distinguished advisers, as well as to various German princes, both political and ecclesiastic."[185] Whether any of them came is not recorded. But no matter how many of them might have appeared, Tycho would have been disappointed, for there was no one at all from Denmark.

Although Tycho had probably not expected even Holger, let alone any of his less affluent friends, to come all the way to Prague for a wedding, he had expected his sister Sophie. It was not that he would have missed her so much at the wedding per se (although it would have been nice to have one member of the family there to carry an account back home) but that he had been anticipating her arrival for some time.

Sophie had indeed started out for Prague in the fall of 1599. Her plan had been to accompany her son south on his tour of foreign universities until he reached the point in his journey that was nearest to where her perambulating fiancé, Erik Lange, happened to be at the time. She would then spend some time with Lange and eventually move on to Prague to visit Tycho. Unfortunately, however, she got no further than Holstein before deciding, for some reason, that she had to return to Denmark.[186] Son Tage continued south to Heidelberg, from where, displaying already the tact and savoir faire that eventually made him one of the richest men in Denmark, he wrote Tycho a very nice greeting and received, in turn, some cordial avuncular advice about his plans to go to Basel.[187]

What called Sophie back to Denmark was probably some kind of financial problem, because her life at the time seems to have consisted largely of financial problems. Erik's share of his ancestral estate had long since been sold to his sister and her husband (Sophie's brother Knud), so that Lange had apparently been depending on whatever money Sophie could send him to augment the living gained by his wits. Sophie seems to have sent him enough money over the years to necessitate finally the mortgaging of her son's ancestral Thott estate to the family entrepreneur, brother Steen.

After being gone from Denmark just long enough to have missed Tycho, Jr.'s, brief return at the end of 1599, Sophie tried again in the summer of 1600. This time she got as far as Lübeck before deciding that there was no point continuing the journey, because Lange had let their communications lapse and she did not know where he

[184] VIII, 409.   [185] XIV, 221.   [186] XIV, 157.   [187] XIV, 170-2.

was.[188] By the spring of 1601 she had apparently told Tycho she would try again, for Tycho seemed confident that she would be with him for the wedding. Unhappily, however, she got only across the Baltic to Rostock before falling sick. If Sophie really had left early enough to get to the wedding (perhaps after Jørgen's burial in mid-March), she must have become sick, indeed, because she was still recuperating when word came of Tycho's death.

Given that the turnout for Elizabeth's wedding must have been almost entirely local, it would be interesting to know – as a gauge of Tycho's acceptance by Bohemian society – how large an event it was. A year earlier, attendance would surely have been small, for at that time the Brahes had been living at rural Benatky, where Tycho, as he told Rosenkrantz, was not interrupted or disturbed by anyone.[189] As it turned out, most of the success of this attempt to emulate the seclusion of Hven was actually due to Rudolph's flight from the plague. For with the consequent disbanding of court and the minimization of contacts with anyone who might carry the plague Rudolph also minimized his exposure to advice that might have been contrary to Tycho's and so saved Tycho the messy business of defending his own or rebutting competing views.

What Tycho's isolation at Benatky spared him were the tiresome social/political exercises of court and capital: dodging the thrusts of frustrated competitors for imperial favor, fending off opportunists who aspired to hitch their own careers to his coattails, and plotting with like-minded advisers to achieve mutual goals. Relief from this latter activity was a mixed blessing, because in such political alliances lay whatever hope Tycho might have had of extracting his salary from the courtiers who controlled the administration of funds for the Bohemian bureaucracy.

By March 1600, a few weeks from the end of his salary year, when Tycho had yet to see the first of the two thousand guldens due him from the civil exchequer, he realized that he had a serious problem. At the very time he was telling Jesensky that he did not willingly become involved in the affairs of others but devoted time only to his own researches, and bragging to Krag that his favor with the emperor increased daily,[190] he expressed doubts to Corraduc about being able to maintain his position. Although he had had the support and protection of friends in high places in Denmark, as he told the emperor's chamberlain, it had not prevented his being assaulted by envious courtiers who despised his studies and cast aspersions on his long-pursued research and finally forced him and his family to leave

---

[188] VIII, 369.   [189] VIII, 168.   [190] VIII, 282, 273.

the country. How much more vulnerable would he not be to such intrigues in a foreign land, he asked his correspondent, if he were not protected by such powerful men as Corraduc?[191] Tycho echoed this theme two weeks later, as the context of his just-cited remarks to Jesensky: "I am an immigrant here. I am not supported, as I was previously in my home country, by the prestige of the nobility, and kinship and connections with the most outstanding men."[192] Whether this manifest financial insecurity was leading Tycho to doubt the wisdom of being isolated at Benatky is hard to say. And whether Tycho could ever have established himself as a serious factor in Bohemian politics is even more doubtful. Within three months, however, the issue was taken out of his hands by Rudolph's call to Prague. Within days of Tycho's arrival there, he suddenly received his long-delayed salary.[193] He doubtless also entered the circles of the capital elite. But during the fifteen months remaining before his death, he did not manage to collect another year's salary.

From his first and very brief appearance at Prague in August 1599, Tycho must have been a subject of considerable curiosity. On his return eleven months later, his confirmed status as the latest and (by all indications) greatest focus of Rudolph's hopes and enthusiasms must have made him a very celebrated person, indeed. One can thus scarcely doubt that Tycho had the opportunity to participate in the activities of the Bohemian social scene. A significant component of this scene was the circle of foreign diplomats. One of Tycho's daughters eventually married into that circle, and Tycho himself apparently became fairly close to a distant Swedish cousin of his, Count Erik Brahe.[194] Tycho also seems to have seen quite a bit of Ernfried von Minckwicz, one of the emperor's council of advisers.[195]

Most of the people with whom Tycho would have wished to associate, however, must have shared at least some interest in things intellectual and must therefore have belonged to one of Prague's literary circles. Tycho was probably introduced by Hayek to the cream of Prague's intellectuals when he first arrived in Bohemia. But because of the alacrity with which Tycho retreated to Benatky and

---

[191] VIII, 258. [192] VIII, 282. [193] XIV, 208.

[194] Four generations and about a hundred years before Tycho's birth, Johanna Brahe, a granddaughter of the "Halland boy" (see Chapter 2) married a Swedish aristocrat who elected to adopt the Brahe family name. By Tycho's day her descendants (Tycho's fifth or sixth cousins) were probably even more distinguished in Sweden than the Danish Brahes were in Denmark. Erik Brahe's father was Per Brahe, the elder, who was a nephew of Gustavus Vasa and a historical writer of some note.

[195] Minckwicz was apparently a neighbor of Tycho's at Benatky and had been on the panel that was to have judged Ursus; VIII, 371.

the unlikelihood that seventy-five-year-old Hayek was up to anything very strenuous, the introduction cannot have been complete. Tycho thus had to finish it on his own, in the summer of 1600, for Hayek was now so weak that he had had to have his wife write to Tycho in the spring when they needed help with the collection of a debt in Hamburg,[196] and he died on 1 September.

In an age when public libraries were unknown, the basis of membership in literary circles, and indeed the hallmark of anyone with pretensions to the status of intellectual, was the private collection of books and manuscripts. Almost any accumulation of them involved considerable expense, and one of Tycho's acquaintances, Peter Vok Rozmberk, is supposed to have owned ten thousand.[197] Whatever element of conspicuous consumption may have entered into a library of this size, one must assume that its owner was also a genuine lover of both books and ideas and that this characteristic was a significant part of Tycho's attraction to him.

Indeed, until Tycho finally pried his own library out of Magdeburg and got it moved (twice?) into the Kurtz house, it would have been convenient for him to have access to someone else's collection. In this respect, however, Baron Hoffmann's library of four thousand volumes may have been more useful.[198] His interest in astronomy was sufficiently strong to have induced him to have an artisan copy one of the quadrants Tycho had depicted in the *Mechanica*, and he and Tycho observed the solar eclipse of 10 July with it when Tycho was newly displaced from Benatky to Prague.[199]

A substantial part of a library was the collecting itself, particularly for a person who was as far away from civilization's beaten paths as Tycho had been at Hven. After an early lesson from having a messenger abscond with books that he had been paid to purchase,[200] Tycho restricted himself to having only friends or associates buy books for him.

But just getting information about books that had been published was an uncertain undertaking. For example "Hipparchus's Commentary on the Poem of Aratus," published in Florence in 1567, remained unknown to Tycho for thirty years, until one of his students saw it during a sojourn in Holland. Tycho was not always so fortunate as to have the owner of a book (Scaliger in this case) send it

---

[196] XIV, 198–9.
[197] Evans, *Rudolf II and His World*, p. 41. A substantial part of Rozmberk's library eventually wound up in Swedish libraries as booty from the Thirty Years' War.
[198] Ibid., 53.
[199] VIII, 353. This instrument may be extant. L. Weinek described *Die Tychonische Instrumente auf der Prager Sternwarte* (Prague, 1901).
[200] VII, 15, 19.

to him as a gift.[201] But a considerable degree of accommodation in such matters was routine among collectors, and Tycho benefited from as much of it while he was in Bohemia as he had when he was at Uraniborg. His principal patron in this regard was Herwart von Hohenburg, chancellor to the Duke of Bavaria and, like Scaliger, a man Tycho never met in person. Herwart's interest in Tycho seems to have stemmed from the news of Tycho's arrival in Prague. Whether because Tycho's expertise had not earlier commended itself to Herwart or because he had not known how to get mail to Tycho before then, Herwart now excitedly posed to Tycho a question (he had already asked of Kepler) having to do with a supposed sudden diminution in the brightness of Venus in antiquity.[202]

After a couple of letters on other semi-technical astronomical matters (Tycho thought the ancient account of Venus's "eclipse" was poetic license and doubted either that the seventh sister of the Pleiades had once been brighter or that the pole star had lost its splendor after the fall of Constantinople)[203] in the course of which Tycho sent Herwart tables of his solar and lunar theories, Herwart began to scout out books for Tycho, both from his own and the duke's libraries.[204] When Tycho mentioned a manuscript by Leovitius, for instance, Herwart even traveled from Munich to Augsburg to borrow it from the Fugger library for Tycho.[205]

By all indications, then, Tycho had enough kindred spirits and well-wishers in Bohemia to conduct as busy a social schedule as he wished to. And as he gradually lost his Danish-speaking collaborators, students, and servants, he may well have come to welcome the diversions afforded by residence in Prague. But whether because of Bohemian society's conception of him as an immigrant or (more likely) because of his own perception of himself as an emigrant, Tycho appears to have been unable to achieve a sense of "belonging" in his new circumstances. In the last recorded episode of his life, on 13 October 1601,

> [he] accompanied Councillor Minckwicz to dinner at the home of Peter Vok Ursinus [Rozmberk]. Holding his urine longer than he was accustomed to doing, Brahe remained seated. Although he drank a bit overgenerously and felt pressure on his bladder [he told Kepler], he had less concern for the state of his health than for [the breach of]

---

[201] VIII, 102, 151. See also Kepler's reference to the book in a letter of January 1603: *G. W.* XVII, 354.

[202] VIII, 157.   [203] VIII, 160, 189.

[204] VIII, 293, 302 305, 313 385, 393, 394.

[205] VIII, 313. The manuscript may not have been returned. It somehow got into the National Library in Vienna: *Gassendi*, 211.

etiquette [involved in excusing himself from the table]. By the time
Brahe returned home, he could no longer urinate.[206]

Physician though he essentially was, Tycho could not heal himself. Through five days and nights of sleepless agony, he pondered
the irony of paying so great a price for, as he thought, having
committed so trivial an offense.

> Finally, with the most excruciating pain he barely passed some urine.
> But, yet it was blocked. Uninterrupted insomnia followed; intestinal
> fever; and little by little, delirium.... On 24 October, when his de-
> lirium had subsided for a few hours, amid the prayers, tears, and
> efforts of his family to console him, his strength failed and he passed
> away very peacefully.
>
> At this time, then, his series of celestial observations was inter-
> rupted, and the observations of thirty-eight years came to an end.[207]

Interment was on 4 November 1601, not in a local parish church
such as Vidskøfle, Tøstrup, or Kågerød, where his *rigsraad* great-
grandfather, grandfather, and father lay buried, but in the great Teyn
Cathedral of Prague. The procession to the cathedral was sufficiently
imposing that someone preserved a detailed account of the event:

> The casket was draped with black cloth and decorated in gold with the
> Brahe coat of arms. In front of the casket were carried candlesticks,
> likewise adorned with his arms, and a black damask banner displaying
> his titles and arms, in gold. Behind the casket was led his riding horse,
> followed by a black taffeta banner and then another horse draped in
> black cloth. [Following the horses were several men walking single
> file and carrying Tycho's arms and armor.] The casket was borne by
> twelve imperial officials, all noblemen. Behind the casket walked
> Tycho's younger son, between the Swedish count Erik Brahe and
> Baron Ernfried von Minckwicz, in long mourning dress. They were
> followed by other Imperial councillors, barons, and noblemen,
> Tycho's assistants and servants, then Tycho's wife, guided by two
> distinguished old royal judges, and finally his three daughters, one
> after the other, each escorted by two noble gentlemen. [Tycho junior
> was traveling in Italy, and Elizabeth and Tengnagel were on their way

---

[206] Entry in Kepler's hand at the end of Tycho's observation log (XIII, 283), translated in *Rosen*, 313.

[207] Ibid. A young (26) physician named James Wittich who heard of Tycho's death attributed the cause to kidney stone. But modern medical opinion seems to favor uremia due to hypertrophy of the prostate. No kidney stones were found at the exhumation, even though they were explicitly sought. Edvard Gøtfredsen, "Tyge Brahes sidste sygdom og død," *Fund og Farskning* 2 (1955): 32–5.

to a diplomatic assignment in England.] Then proceeded many stately women and girls, and after them the most distinguished citizens. The chairs in the church in which the family sat all were draped with black English cloth. The streets were so full of people that those in the procession walked as if between two walls, and the church was so crowded with both nobles and commoners that one could scarcely find room in it. When the sermon [Jesensky's eulogy, presumably][208] was over, [Tycho's] banners, helmet, shields, and other arms were hung over the crypt.[209]

The Teyn Church still stands. And despite the wholesale removal of Protestant graves during the Thirty Years' War, the renovation of the church floor in the eighteenth century,[210] and the disinterment and autopsy of Tycho's remains in 1901,[211] Tycho still rests there, too. An impressive, life-sized effigy of Tycho in armor, carved in relief on red marble, marks the crypt in the front of the church. More fitting is an epitaph[212] above it that proclaims Tycho's renunciation of the old world – "To be rather than to be perceived" – and provides a brief list of the accomplishments that made him an inspiration to generations of successors.

[208] XIV, 234–40. The eulogy was reprinted in 1610.
[209] XIV, 233.   [210] Norlind, 318–19.
[211] Heinrich Matiegka, *Bericht über die Untersuchung der Gebeine Tycho Brahes* (Prague, 1901).
[212] Norlind, 318.

Epilogue

(translated by J. R. Christianson)

30 JULY 1602, MAGDALENE BRAHE TO ESGE BILLE

My most friendly greetings now and always sent with Our Lord. Dear Esge Bille, trustworthy, especially good friend,

My most friendly and great thanks for all the manifold goodness you have shown my dear mother, brothers and sisters, and me, for which we are obligated to you all of our days and will be found right willing and full ready with our small fortune in all ways to do whatever we can to your honor and service.

This is to inform you most cordially that I could not burden you with this my humble letter but that my dear mother, my brothers and sister, and I long so much to know how you and your dear wife are. God in heaven grant that all things go fortunately and well for you always, to the glory of God, and to your own and our gladness. We wish you this with all our hearts.

My dear mother, brothers and sisters, praise be to God, are all healthy and sound, may God in Heaven continue to help with His spirit and grace. But that our greatest joy and pleasure after God is now gone from this earth, in that God in Heaven has called home our dearest father, now blessed in God, places us in the hands of God, who rules over us all. God grant us to be found again in the realm of God, in the eternal joy which never ends. We were most reluctant to lose him, but we must now be satisfied with the divine will of God. God in Heaven protect my poor, grieving mother and we poor children, who survive in such sorrow.

We pray you most humbly and earnestly, for the sake of all the good will my late father showed towards you, and which you also always showed towards him, that you will not abandon my dear, bereaved mother and us poor, fatherless children, for after God in Heaven we put our trust primarily in you. God shall reward you for it, and we with our small fortune will always be found entirely thankful and willing to serve you all of our days.

Dear Esge Bille, concerning our situation here at court, I cannot refrain from telling you in confidence that we have still not come to any real conclusion concerning my late father's instruments. The reason is that His Imperial Majesty's Council wants us to acccept their assurances for them, saying that they will give us strong enough assurances and pay interest each year on the money until the principal sum is paid, which is twenty thousand dalers, the amount that they have offered us for them. We have accepted this amount on the condition that we either receive immediate cash payment or else a landed estate for it, here in this country, to the value of the aforementioned sum of money. But to give away our late father's heritage of learned skill in return for assurances – that we neither can nor ever will do, God help us. For experience shows well enough, to others as to us, what gain one has here at court if one does not quickly seize it in one's hands. Now they have held us up so long, as is unfortunately the custom here, that we neither can nor by any means will keep it up any longer with them, but will now have a final reply and be paid right away for the same instruments and our late father's observational books, and the other things they received, or else we will keep our own.

Concerning this, we have a short time ago sent His Imperial Majesty various supplications, to His Majesty's own hands, and in them, we humbly requested that His Majesty would simply be a gracious emperor and lord to us, even though we kept our own which we are forced to do because we could not be paid for it,[1] for it was no longer possible to live here in such an extraordinarily expensive city and make our solicitations without results, especially now that we have no income to live by. Then His Majesty replied to us that he had again commanded the Lord of Lichtenstein, His Majesty's Most High Steward and the Bohemian President, that they should pay in good order and satisfy our claims. But it still is such that they want to come together and decide in the Council how they best can pay us for it, and this is supposed to happen any day so we will soon learn what they will decide about it. But we will by no means be satisfied without cash money or good landed property for it.

If it does not happen soon, then we will immediately take back our own. We have already begun to have some of the instruments brought from the Kurtz house where they stood to the house in the old city where we now live, because if we did not do this it would be even longer before we saw an end to it all. For otherwise they would think that we did not seriously intend to keep them ourselves. That they should get to have them for nothing, however, shall never

---

[1] This strange locution seems to be an expression of hope that the emperor will forgive the heirs if it should prove necessary for them to keep Tycho's instruments because the council refuses to pay.

be: with the help of God, we will find lords who will buy them. The emperor does not want to lose them, but he wants the Treasury to pay us, and the lords want the emperor to pay out of his own treasure, and that is what is delaying us so long.... It would also be just in and of itself that they gave us some compensation for the great expense our late father had on the journey to His Imperial Majesty, and also for what he spent of his own in His Majesty's service. Moreover, His Majesty promised that he would provide for us after demise of our late father, but God forgive them for the little they hold to it, and we will soon learn whether or not they will do it. His Majesty is compassionate and has asked about it often enough.

Around next St. Michael's day [Sepetember 29], God willing, my dear mother plans to buy a good grange, here in Bohemia, in the direction of Dressen, for 2,500 or 3,000 dalers, where she can have an income to live on and maintain her household. Landed property is rather good here in this country, and easy to buy for money, and it is impossible for her with her large household to live here any longer in such an expensive city. As soon as God wills that she get money for the instruments, either from His Imperial Majesty or from another lord, from wheresoever God will send it to us, she intends with the help of God to put it out at interest in this country to her own and our gain. I hope that we, with the help of God, can get something for them someplace. I wish to God that our fortune were such that we did not need to sell them; then we would keep them ourselves in honor of our late father, but because that is not so, we will undoubtedly have to part with them.

Dear Esge Bille, trustworthy commissioner, true and good friend, my dear mother and all of us most humbly and dearly pray you to do well and take the trouble to let us know, as soon as possible, how things stand with that which fell to us after our late father, both with respect to livestock and the like on Hven, and our house in Copenhagen, and whatever else you know of that was his in Denmark, and also with respect to any debts. Dear Esge Bille, we pray earnestly that you will not blame us for writing so boldly to you about this. We know of no other single person there in that country who would know all aspects of this as well as you. We commend ourselves fully and entirely to you and do not doubt in the least that you, as our true defender and friend, will arrange all things as far as it is possible to do so, to our gain and benefit. God in heaven will reward you for helping us poor, fatherless children, to a good outcome. Therefore, as long as we live, we will remain grateful and willing to serve you with our modest fortune, and all of our days will busy ourselves wholeheartedly to do so.

Dear Esge Bille, we pray you most earnestly that you will do well and let us know what you think of the German letter that we sent you last winter, which Count Erik [Brahe of Sweden] and Lord Ernfried Minckwicz with a sworn notary and His Imperial Majesty's doctor gave to us. We confidently commend ourselves to you for your final opinion on this. We also hope that you are the one who will help us do our best in this and in other ways, for after God, we have nobody on this earth besides you to whom we entrust ourselves, and we truly hope that you will not fail us. God knows that you are most charitable in this, because our situation is now most piteous. If God in heaven had wanted to spare our dear father to be with us, he would undoubtedly have arranged our affairs otherwise for us, but now we must commend ourselves so ardently to you, that you as our father will not fail us, but help us to what is best, as far as it is at all possible for you. We will always pray God in heaven that he reward you for it, which we with our modest fortune cannot do. We have twice written to you about out situation, and I certainly hope you have received those letters.

We have also written to all of our father's brothers, our father's mother, and our father's sisters, but have not received any reply from any of them. We truly hope, however, that they will not simply forget and abandon us after our late father's demise, especially now that we are so miserable, here in a foreign land, and we hope even more that they do not wish the worst of us or that we should suffer need, and they also, praise be to God, rather easily could do that which would be no loss to them and great assistance and benefit to us. We hope now that they consider that which is right and proper, and also that they do a great, good, and pious deed thereby towards us poor, fatherless children, who have nobody but them, our kinsmen, to depend upon. We also trust absolutely in you, that you as our true and loyal friend will commend us in all ways to our dear father's mother and father's brothers. I hope that my dear grandmother will never abandon us, but always remain good to us, and you could help plead our cause with her, which we also turn over completely to you and cordially ask you to do in any and all ways.

Dear Esge Bille, we pray you most earnestly not to blame us for so audaciously burdening you in all ways, for God knows that in all things we put our trust in you, and without your help and advice, we do not know how we can best pursue our case. We await your good reply about all things at the very earliest possibility. We are so anxious for your reply that we pray you most earnestly to hire a courier at our expense so that we can get report from you about everything you know that we should know, and what your advice is to us, how we should pursue our case, for we have a great desire to know your advice in all ways.

I will not keep you longer with this my lengthy letter, but will now and always commend you, with your dear wife, to the eternal, omnipotent, gracious God, in long-lasting health, fortune, and benediction of body and soul. My dear mother, brothers and sister, my brother-in-law, Franz Tengnagel and I send both of you our best regards and wish you many, many good nights. My brother, Tyge, is not home now, or else he would have liked to write to you. He has departed for Breslau in Silesia after our late father's arrears in salary, of which he has now received 1,300 guldens. But over 1000 guldens is still unpaid and we cannot get them until sometime around next Christmas.

Iterum vale. Dated at Prague, the 30th of July, stylo novo Anno 1602.[2]

Magdalene Brahe Tygesdatter, own hand.

28 FEBRUARY 1604, TYGE TYGESEN BRAHE TO ESGE BILLE

My most friendly greetings sent now and always with Our Lord.

Dear Esge Bille, trustworthy, especially good friend, my great and repeated thanks for all the goodness you have shown, for which I shall remain indebted to you all of my days, willing and true to honor and serve you and yours with all of my fortune in whatever I can do.

I want to inform you most cordially that I have received your kind letter, which was most heartwarming and pleasing to me, especially because I perceived from it that you and yours are healthy and well. When you, dear Esge, report in your letter that the house which we have in Copenhagen is falling into ruin, I can well imagine it, for it is not properly looked after. With respect to the title, which my late father supposedly did not have entirely completely clear,[3] I do not really know anything about that, but when, God willing, I come to Prague, I will look through his letters to see if I can find anything that pertains to it. Regardless of how that turns out, I think we should be able to see that which is ours. I have also received with your letter the copies of the receipts and the last business that you and my father dealt with, and if possible, I will with the help of God try to come to you in Denmark, and then I will speak personally and deal further with you about everything. Though to say the truth, I

---

[2] XIV, 247–51.
[3] This is probably a reference to the requirement that Tycho provide space and/or accommodation for a dyer; see the postscript to letter 1 of Appendix 6.

am not particularly eager to come there at any time, for where my late father, who had done no little good for the realm, was not very well liked or respected, there I have no desire to come, and the good which I hope for there (except for you and yours, for we expect much good of you) is very slight. Meanwhile, I put my hope in God, that he will not abandon us here in a foreign land, even if we be abandoned by others in other places.

If that woman in our house asks for the money our late father supposedly owes her, I do not know anything about it, nor can I acknowledge it, for I know well that our late father did not do so either.

Dear Esge, I cannot refrain from telling you cordially that I am betrothed to a widow, here in this country, who is of the house of Vitzthum and was previously married to a Kinsky, and I intend with the help of God to hold my wedding with her in fourteen days. God knows how much I would have liked to see you here, but because I know very well that it cannot happen, I must accept it. I will not have a particularly large wedding, but many good lords of her family and others will come, for although she is not particularly rich, she comes of a distinguished family, and many of her family and kin live in this country as well as in Meissen, as you undoubtedly are aware. It was the will of God that I marry so soon, for I could have bided my time, but what will be, will be. I intend with the help of God to enter right into the service of the emperor, for she has some cousins who are in the Council, and I will look into that as soon as the wedding is over. May God only give his good fortune and benediction, so that everything may go successfully and well. I could not refrain from writing to you about it because I know well your good disposition towards me and all of us. If God were only to grant that I could serve you in some manner, I would do it gladly, but even if I cannot do it in deed, my will shall always be ready to do so.

Now I will not occupy you any longer, dear Esge, with this my letter, especially since Jørgen Daa, who has been here with us for awhile, and is now traveling to Denmark, can give you a more detailed report about everything. I commend you now and always to God almighty in body and soul, with many good nights, which my dear mother, brother-in-law, brothers and sisters wish you. I look forward to a letter from you again at the earliest occasion. Datum Lindaw, the 28th of February, Anno 1604.[4]

Your obliging good friend always,
Thyge Brahe T. S.
own hand.

[4] XIV, 253-4.

Besides health and thanks, dear Master Christen Lomborg, I give
you cordially to know that I received your letter this past summer, in
which you told me about your situation, that you are now a
professor in Copenhagen and maintain your own household, and I
was pleased to hear that things are going well for you. As for me and
my dear brothers and sisters here in a foreign land, with time we
have become rather used to things here in this country, so that we
have as well as forgotten how things are in Denmark. We have now
suffered both bad and good in this land, insofar as we suffered great,
heartfelt bereavement when God almighty so soon after by his divine
will took both of our dear parents away. For it is now in the seventh
year since God called our dear late father, now with God, and in the
fourth year since God in heaven called our dear late mother to
heaven, and they both lie buried together in one grave in the most
distinguished church in the old town, called Teyn Church. May God
in heaven grant them and all true Christians eternal joy and resurrec-
tion and grant us to meet again in the kingdom of God.

My brother, Tyge, is in the fourth year of his marriage and has
taken a widow from Meissen, of the family of Vitzthum, who had
previously been married to a Bohemian nobleman in this country, of
the family of Rathiski and lives less than a Danish mile from the
estate which our dear mother purchased after our late father had
died, where she lived for a time with us until Our Lord called her.[5]
She died of *vatersoet* and was always very melancholy after Our Lord
called our late father, and had suffered much illness before Our Lord
came and released her, to her great joy but to our great sorrow. But
we must surrender all things to the divine will of God.

My brother Tycho now has two children with his wife, a son and a
daughter. The daughter, who was named after my late mother, is
named Christine Barbara, and the son, Otte Tyge Brahe, is named
after my late father. They begin to look a lot like my brother
and us, and will be big, tall people. May Our Lord preserve them
that they may grow up in the fear of God.

My sister, Elisabet, now has three children and will soon come
into the childbed with the fourth. God has called away one of her
children, a daughter, who was two years old and named Vendele.
She has two beautiful children remaining, a daughter named Ide
Katherine and a son named Rudolf Tycho after His Imperial Majesty

---

[5] As a result of the family's move into rural Bohemia, Blaeu could not obtain information
about any of them when he inquired in Prague sometime before 1616: *Gassendi*, 182.

and our late father. He looks a great deal like my late father, plump, with a thick head of hair, so that he looks exactly like him. May God spare them and grant that they may grow up in the fear of God, to the joy of us all.

Concerning my other brothers and sisters, my brother, Jørgen, is also here in Prague, and my sister, Sophie, and my sister, Cecilia, and I, since our brother-in-law returned here to Prague from England, have lived with him and our sister, Elisabet. We now plan soon, with the help of God, to celebrate the wedding of our sister, Cecilia, who gets a handsome hereditary baron. They have been betrothed to each other for some years, and we, praise be to God, are quite pleased with it. God grant continued good fortune and benediction.

Our dear brother-in-law also has a good position with His Imperial Majesty, is his Councillor and has 1,500 dalers [per annum] from His Imperial Majesty, who has given him, since he recently returned again from England, 12,000 dalers.

Concerning our late father's instruments and observational books, we have long since sold them to His Imperial Majesty for 20,000 dalers and have several thousand, but the remainder is drawing interest, so that the principal and interest together total 26,000 dalers, which we have received for our late father's instruments. The greater part of this sum is still owed to us, and the total sum is to be paid to us between Easter and Pentecost. The reason it has taken so long is that our brother-in-law has been so long in England, and our own negligence in not being persistent enough, because everything moves slowly at this court. But now we will be paid within a few weeks, praise be to God, and when we have settled it between us, each of us will have enough to live on and can comfortably maintain a style of life as it should be. Those of whom Our Lord demands more, to them he also provides sufficient patience to endure. Our Lord grant that we so use our time in this world, wheresoever we may be or stay, that we might at last be found with our dear parents and all of God's chosen Christians amidst the joys of heaven.

Dear Master Christen, I do not know what else to write to you at this time. I would have written to you long ago, but I could not get hold of a dependable courier until one who belongs to Duke Hans of Holstein now came before me with a notarized message. He says that he will come to Copenhagen himself and has promised to deliver your letter to your own hand and take my letter from you again. There, I pray you earnestly to be certain to write to me again and send it with this courier, or wherever else you can find a dependable courier, and let me know how you and other good friends and relatives are doing, and what other news there is from that country:

how the king is, what he is engaged in doing, and all news of that kind I would like to know to pass my time.

I know of nothing else in particular to write about, but that there is a disagreement between His Imperial Majesty and his brother, Duke Mathias of Austria which one does not dare to write about, but we hope, with God's gracious help, that everything will be arranged for the best. They take on secret troops on both sides. Austria and Hungary are supporting Duke Mathias of Austria, and the emperor has a great support on his side from Bohemia, Silesia, and his other surrounding lands, and the princes in those realms and other potentates will support the emperor, so a terrible war will come of it if it is allowed to develop, but may God graciously avert it. If war were to reach here into Bohemia, we brothers and sisters are agreed that we will remove ourselves to Meissen, for it would not be safe to be here in such a time.[6]

Dear Master Christen, now I will not keep you any longer this time. Write to me soon, about how things are in Denmark, and how our relatives and friends are doing. Greet Master Cort [Axelson]; I have heard that he is married. You should also get married because you need a wife to manage your house for you.[7] Be so kind as to greet all of our good friends there in that country, who ever inquire about us.

Datum Prague, the 3rd day of Easter 1608.[8]
Magdalene Brahe Tygesdater
own hand.

---

[6] Matthias did, indeed, depose Rudolph in 1611, shortly before his death in 1612. The expected war, however, did not begin until 1618. It was as terrible as Magdalene feared it would be, but the family seems to have been able to avoid the worst of it by moving into Protestant territory.

[7] Longomontanus married Dorothea Bartholin, whose brother Caspar sired the Bartholinus family that dominated Danish intellectual life in the seventeenth century.

[8] XIV, 261–3.

Appendix 1: Abbreviations for Frequently Cited Sources

All citations of the form VII, 192, refer to the volume and page number of Tycho's collected works, Tychonis Brahe Dani Opera Omnia. Edited by J. L. E. Dreyer, 15 vols., Copenhagen Libraria Gyldendaliana, 1913–29.

Cassiopeia	*Cassiopeia: Tycho Brahe – Sällskapets Årsbok.* Lund, Sweden: Hakan Ohlssona Boktryckeri, 1939–48.
DAA	*Danmarks Adels Aarbog.* Copenhagen, 1884–1919
DBL	Dansk Biografisk Leksicon. Edited by J. H. Schultz. 27 vols. Copenhagen: Forlag, 1933–44.
Dreyer	*Tycho Brahe: A Picture of Scientific Life and Work in the Sixteenth Century.* Edited by J. L. E. Dreyer. Edinburgh: Adam & Charles Black, 1890; 2nd ed., New York: Dover, 1963.
DSB	*Dictionary of Scientific Biography.* Edited by C. C. Gillispie. 16 vols. New York: Scribner, 1970–80.
Friis	*Tyge Brahe: En Historisk Fremstelling.* Edited by F. R. Friis. Copenhagen, 1871.
Gade	*The Life and Times of Tycho Brahe.* Edited by John Allyne Gade. Princeton, N.J.: Princeton University Press, 1947.
Gassendi	*Tychonis Brahei Vita, Accessit Nicolai Copernici, Georgii Peurbachii et Joannis Regiomontani Vita.* By Pierre Gassendi. Paris, 1654. Because of the rareness of the original edition, I have cited the Swedish translation, *Tycho Brahe: Mannen och Verket,* Efter Gassendi översatt med kommentar av Wilhelm Norlind. Lund, Sweden: C. W. K. Gleerup, 1951.
G. W.	*Johannes Kepler Gesammelte Werke.* Edited by Walter Van Dyck, Max Caspar, and Franz Hammer. Munich: C. H. Beck, 1939–.
Norlind	*Tycho Brahe: En Levnadsteckning,* med nya bidrag belysande hans liv och verk Wilhelm Norlind. Lund, Sweden: C. W. K. Gleerup, 1970.
Raeder	*Tycho Brahe's Description of His Instruments and Scientific Work.* Translated from *Astronomiae instauratae mechanica* by Hans Raeder, Elis Strömgren, and Bengt Strömgren. Copenhagen: Ejnar Munksgaard, 1946.

Appendix 1

Rosen	*Three Imperial Mathematicians: Kepler Trapped Between Tycho Brahe and Ursus.* By Edward Rosen. New York: Abaris Books, 1986.
Vistas	*Vistas in Astronomy.* Edited by Arthur Beer. Oxford, England: Pergamon Press, 1955–.
Westman	*The Copernican Achievement.* Edited by Robert S. Westman. Berkeley and Los Angeles: University of California Press, 1975.
Zinner	*Deutsche und Niederländische Astronomische Instrumente des 11.–18. Jahrhunderts.* By Ernst Zinner. Zweite ergänzte Auflage. Munich: C. H. Beck, 1967.

Appendix 2: Glossary of Technical Terms

anomaly: the angular distance of a body from some reference point, apogee, in traditional astronomy. The mean anomaly is the distance a body would be from apogee after a given time if it were traveling with its average (mean) orbital speed.

apogee: the point in an eccentric or epicycle that is farthest from the earth, or the point in an epicycle that is farthest from the center of the deferent: marked *A* in Figure A.2.1.

concentric: a circle on which a point moves with uniform speed and therefore with uniform angular velocity, as seen from the center of the circle.

conjunction: the (essentially invisible) position in a planet's orbit at which its longitude coincides with the sun's.

coordinate system: the means by which the locations of celestial objects are specified. They rest on the common (Greek) conception of the heavens as a hollow shell of very large radius, centered on the earth and called the celestial sphere (see Figure A.2.2). Any position (*S*) on the sphere can be completely described by reference to its angular distance (ϕ) above or below a plane (*LMN*) bisecting the sphere, and the angular distance (θ) in that plane between the meridian through *S* and a given "zero" meridian. Note the resemblance to terrestrial coordinates. The Greeks developed not only the geographical system but also all systems of celestial coordinates. The latter are denominated by their respective fundamental planes, and their components are named as shown in the table A.2.1.

deferent: a concentric circle (solid in Figure A.2.1) carrying a second circle (epicycle) moving uniformly.

eccentric: the path of a concentric, as seen from a point (excenter) other than its center. Also, the geometrical resultant of a body moving on an epicycle revolving on a deferent, as illustrated by the dashed circle in Figure A.2.1.

eccentricity: the ratio between the displacement of an eccentric point and the radius of its concentric ($\frac{EC}{BC}$ in Figure A.2.1), or the ratio between the radius of an epicycle and the radius of its deferent (r/R).

ecliptic: the path of the sun in the celestial sphere.

epicycle: see *deferent.*

equant: circular motion in which the motion is uniform with respect to some point other than the center of the circle.

inclination: the angle between one orbital plane and another.

line of apsides: the line connecting apogee and perigee (*AB* in Figure A.2.1) or that connecting the center and eccenter of an eccentric (*EC*).

line of nodes: the line of intersection between two orbital planes.

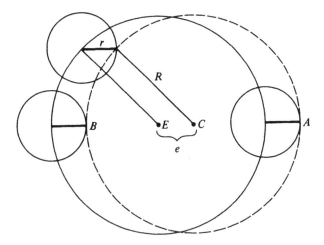

Figure A.2.1. Equivalence of epicycle and eccentric mechanisms.

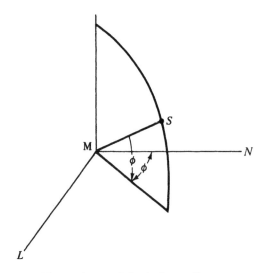

Figure A.2.2. Spherical coordinator.

opposition: the position in a planet's orbit (analogous to full moon) at which its longitude is 180° away from the sun's.

parallax: an apparent displacement of an object due to a motion of the observer (see Figure 10.2).

perigee: the point in an eccentric or epicycle that is closest to the earth: *P* in Figure A.2.1.

Table A.2.1. *Astronomical coordinate system*

System	Equator	Horizon	Ecliptic
fundamental plane	plane of the diurnal rotation of the celestial sphere (earth)	instantaneous horizon of the observer	plane of the sun's (earth's) orbit
poles	north/south celestial poles	observer's zenith or nadir	north/south ecliptic poles
name of ϕ component	declination	altitude	celestial latitude
name of component	right ascension	azimuth	celestial longitude
reference of θ component, measured eastward from	vernal equinox	south point on the horizon	vernal equinox

Appendix 3: The Tychonic Lunar Theory

Until its final revision by Longomontanus in 1600, Tycho's lunar theory was fundamentally Copernican: Tycho (or Longomontanus) had been able to represent the Variation by simply putting an extra circle, AB in Figure A.3.1, into the center of Copernicus's theory. By making the center of the deferent (B) revolve around the earth (A) at twice the synodic velocity of the moon, he produced displacements that mounted to $\pm 45'20''$ when the moon was in the octants and yet vanished in syzygy and quadrature. But the Variation was not the only extra circle grafted onto the traditional theory. The center of the large epicycle also moved on a new circle, CD, producing a periodic displacement up to $\pm 11'$.

This correction appears without a prior hint of any kind in either Tycho's log notes or his letters and would therefore be unknown if Tycho had not allowed Jöstel to "prepublish" his theory in 1599.[1] What it represented, although somewhat imperfectly, was a phenomenon called the *annual equation*. When it finally attained recognition as a legitimate lunar phenomenon and was explained by Newton a hundred years later, it was seen as an annual variation of the speed of the moon in its orbit, owing to the annual fluctuation of the sun's distance from (and hence influence on) the earth–moon system.

That Tycho also associated the phenomenon with the sun in some sense is obvious from the fact that his period for it was a year. He even established the "right" maximum value for it: With a circle of radius 320 parts (on a unit deferent, BC, taken as 100,000 parts), he generated exactly the $11'$ that is deduced from modern theory. But Tycho had its period referred not to the solar apogee but to the sun's angular distance from the beginning of Leo, $24\frac{1}{2}°$ ahead of apogee.[2]

This may seem like a relatively minor error, particularly because Tycho can scarcely be faulted for having had no physical conception of the phenomenon. Nobody else – except Kepler – did either, until Newton. But actually, this small displacement of the line of maximum effect introduces

---

[1] See Chapter 12. Although Jöstel's tract was listed by Lalande, it was apparently treated as a ghost by Dreyer , so that it was reprinted only as an addendum to Tycho's works, in 1972. See Victor E. Thoren, "An 'Unpublished' Version of Tycho Brahe's Lunar Theory," *Centaurus* 16 (1972): 203–30. Actually, the diagram survived in Kepler's correspondence (*G. W.* 14, 16–18), in a letter from Herwart von Hohenburg. But its value was not recognized by the editors, perhaps because they wished too strongly to award Kepler priority for the annual equation (*G. W.* 13, 409: 14, 466).

[2] Thoren, "An 'Unpublished' Version," 222, line 17.

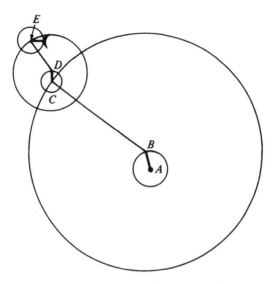

Figure A.3.1. Tycho's first Lunar Theory.

errors of nearly 5′ everywhere the phenomenon is appreciable[3] and makes it truly remarkable that Tycho was able to discover and generalize it at all. Exactly when he did it, that is, how long he had known about it before Longomontanus performed his "restitution of 1597,"[4] can only be a matter of speculation.

The essence of, and almost certainly the rationale for, Longomontanus's subsequent remodeling of the lunar theory in mid-1600 was a rectification of the representation of the moon's distances. In March 1600 Tycho wrote Jöstel that he would "soon see another theory of the motions and diameters of the moon through which everything agrees with the appearances more accurately than before."[5] In the final exposition of the theory for the *Progymnasmata* Tycho (or Longomontanus) stressed that although the Copernican model was justly celebrated for having reduced the range of the moon's distances to nearly half of the Ptolemaic range, the observed range was still much less.[6]

Whether the problem of distances was a general one that Tycho had seen for some time or a specific one that he thought was responsible for the erroneous eclipse prediction of 1599 is impossible to say. But to further re-

---

[3] The errors would have been 0.′99 sin *m* + 4.′56 cos *m*. They would therefore have mounted above 4′ for all mean anomalies within 40° of the line of apsides.

[4] XII, 10, 115.

[5] VIII, 280. See also VIII, 345, for a statement in August 1600 that the moon's parallaxes still require better representation.

[6] II, 131.

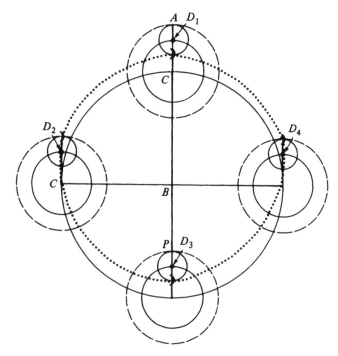

Figure A.3.2. The representation of the first inequality in Tycho's final Lunar Theory.

duce the range of distances implied by the model, Longomontanus discarded the large single epicycle (dashed in four positions in Figure A3.2) previously used to represent the first inequality and substituted a double epicycle very similar to the mechanism already adopted by Tycho (from Copernicus) for the planets.

In this mechanism, the first (larger) epicycle was conceived to operate exactly as the single epicycle always had. Its center (C) moved counterclockwise from apogee (A) in the anomalistic period of the moon, and point D on its periphery moved clockwise in the same period, starting from position D_1, in apogee. On this epicycle, Longomontanus mounted a second one, precisely half its size, in which the moon was conceived to travel counterclockwise around the moving point D at twice the angular rate of the first two motions. As in the planetary theory, it was referenced so as to be at perigee (the point closest to C) of the small epicycle when the large epicycle was in the line of apsides (A or P) and at apogee of the small epicycle when the large one was in the mean distances (D_2 and D_4). The result was angular corrections that were nearly equal to those entailed by the single large epicycle[7] but radius vectors that varied by only one-third of the traditional amount.[8] Once the representation of the first inequality had been altered in this way, the rest of the model was virtually determined. Having preempted both Copernican epicycles for the first inequality, Longomontanus was more or less forced to push the second one into the center and thence to move the third – the Variation – back out to the deferent where the fourth – the annual equation – had been. The result of this shuffling, Figure A.3.3, is what the world has known since then as Tycho Brahe's lunar theory.[9] Schematically it bears a strong resemblance to Tycho's earlier version. The only difference is that whereas the Variation had involved both additions and subtractions, its replacement in the center circle – the second inequality – was always absolutely additive to the first.[10] Accordingly, the

---

[7] The correction (Figure A.4.7 in Appendix 4) due to the simple (dashed line) epicycle is found by the construction of perpendicular L_1F: $\tan L_1BF = \dfrac{L_1F}{BC + CF}$; setting $L_1C = e$; and $BC = r$; $\tan L_1BF = \dfrac{e \sin m}{r + e \cos m}$. The correction of the compound epicycle appears from the construction of perpendicular L_2F_2 and parallel (to CB) DG. Now, $\tan L_2BF_2 = \dfrac{L_2F_2}{BC + CF_2}$. But because $CDL_2 = 2m$ and $GDC = m$, $L_2DG = m$. And because $CD = \dfrac{2e}{3}$ and $DL_2 = \dfrac{e}{3}$, $\tan L_2BF_2 = \dfrac{e \sin m}{r + \frac{e}{3} \cos m}$.

[8] The distances resulting from Longomontanus's epicycles have been misunderstood by both *Dreyer* and *Herz*. From Figure A.4.7 in Appendix 4. it is easy to see that the square of the distance (r) of point L_2 from a point $\dfrac{e}{3}$ above B will be $\overline{BC}^2 + \left(\dfrac{2}{3}\overline{L_2F_2}\right)^2$. Thus, for a deferent of radius R, r^2 at any given time will be $R^2 + \left(\dfrac{2}{3}e \sin m\right)^2$. Then $r \doteq R\left[1 + \dfrac{2}{9}\left(\dfrac{e \sin m}{R}\right)^2\right]$. Because $\dfrac{e}{R}$ is only 0.087, the orbit will not differ greatly from the standard eccentric circle. Everywhere except in the line of apsides, when $\sin m = 0$, r will be slightly greater than R, reaching a maximum of $1.0017\,R$ when $m = \pm 90°$. In the line of apsides, $r = R$, and so the distances from B will be $R + \dfrac{e}{3}$ and $R - \dfrac{e}{3}$, as is shown by Tycho's tables (II, 116). Yet *Dreyer* (p. 338) stated that "at apogee the moon was 0.029 $\left[\dfrac{e}{3}\right]$ outside the deferent, at perigee 0.058 + 0.029 = 0.087 inside it." He repeated this in Latin for the *Opera omnia* (I, LII). Norbert Herz ("Allgemeine Einleitung in die Astronomie," in *Handworterbuch der Astronomie*, vol. I, p. 69) stated – truly, but misleadingly – that "durch diese Annahme wird überdies die Enfernung (❱ B) stets kleiner bleiben als $(R + e)$ und stets grösser als $(R - e)$."

[9] Tycho's diagram and "explicatio hypotheseos" are in II, 100–1.

[10] Although the modern form of the second inequality, $1°20' \sin(2\alpha - m)$, is algebraically additive, it is associated with a first inequality of $6°20' \sin m$. If the traditional first inequality, $5° \sin m$, is split off from the latter, the remaining $1°20' \sin m$ can be combined with the $1°20' \sin(2\alpha - m)$ to obtain the identity $2°40' \sin \alpha \cos(\alpha - m)$, which is the

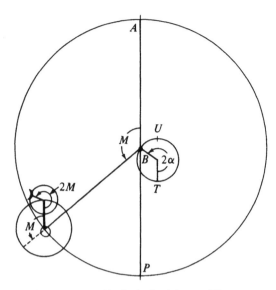

Figure A.3.3. Tycho's final Lunar Theory.

earth (*T*) had to be placed *on* what had fomerly been the "variation" circle
instead of at its center. The center of the deferent (*B*) was still constrained to
revolve counterclockwise in the circle twice each synodic month but was
at *T*, where it produced no correction at all whenever the moon was in
syzygy, and at *U*, where it provided as much as $2\frac{1}{2}°$ correction in quadrature.
How much of the available correction was used at any given time depended,
of course, on the moon's anomalistic position, just as in the Ptolemaic and
Copernican versions of the same inequality.[11]

Given the different functions of the respective circles in the two theories,
their resemblance could be only superficial. All the dimensions had to be
altered to correspond to the new functions. The two epicycles (of Figure
A.3.3) had formerly had radii of 11,000 and 2,200, providing first and second inequalities represented by the ratios of 8,800 and 4,400 to 100,000. The
first inequality Longomontanus lowered to 8,700, presumably at least partly

traditional absolutely additive form of the second inequality. The trigonometric equivalence of the two methods is demonstrated in V. E. Thoren, Tycho Brahe on the Lunar
Theory (Ph.D. diss., Indiana University 1965), p. 151.

[11] The correction for the second inequality, *E*, as a function of the diameter of the center
circle, *d*, is shown in Figure A.4.8 in Appendix 4. by $r' \sin E = BT \sin θ$. But $θ + (m + Q)$
$+ BTU = 180°$, where *Q* is the correction of the first inequality, and $BTU = \frac{1}{2}(180° -
BUT) = \frac{1}{2}(180° - 2α) = 90° - α$. So $θ = 180° - (m + Q) - (90° - α)$, and $\sin θ =
\sin [90° + α - m - Q]$. Finally, because $BT = d \sin α$, $r' \sin E = d \sin α \cos (α - m - Q)$.
Thus *E* differs from those of Ptolemy and Copernicus only insofar as the models provide
slightly different values for *r'* and *Q*.

to make it divisible by 3 so that he could assign two parts (5,800) to the large epicycle and one part (2,900) to the small one. The second he likewise lowerd so that it would be half the value of the first inequality, just as it had been in the old theory.

The variation and annual equation were treated more radically. From $45\frac{1}{2}'$, the former was decreased to $40\frac{1}{2}'$, only about 1 minute above its modern value. In addition, it was reduced to a libratory motion and thus, in effect, to the modern status of a pure equation; that is, it was removed from the geometrical reckoning of distances altogether, on the grounds that "any other dimension than the arc of $40'30''$ that it subtends ... would be insensible."[12] The fourth inequality, having literally no place to fit into the model, was simply omitted. Only the fact that it was partially reinserted by means of a peculiar version of the equation of time has allowed a posterity ignorant of Tycho's first lunar theory to associate Tycho's name with even a partial discovery of the annual equation.[13]

Although the fact that a sundial will not keep precisely uniform time throughout the year was probably realized (theoretically) long before Ptolemy, it is only in the *Almagest* that the first discussion of what has come to be known as the equation of time is preserved.[14] From then until Tycho's day it remained a standard feature of astronomical thought. Already in 1580, however, Tycho seems to have been exposed by Wittich to some kind of doubts on the nature, validity, or necessity of this traditional nicety.[15]

Exactly how Wittich might have influenced Tycho is impossible to say because neither man's views are available, but by 1587 Tycho was confiding his doubts on the subject to Rothmann:

> On the equations of days, which you mention among other things, you should know that I hold ideas on this somewhat different from the opinons of the ancients and think it not to be so great as is assumed by Copernicus himself, having been taught not only by a priori principles but also by experience itself.[16]

In 1589, when Tycho printed up his solar theory for the *Progymnasmata*,

---

[12] "Denique circelli ad F semidiametrum, non est necesse aliter dimetiri, quam arcu $40'30''$ quem subtendit, cum differentia in tanta exilitate sit insensibilis," II, 101. But concerning this claim, see note 20.

[13] Actually, this is a rather charitable view. Anyone who looked at Jöstel's tract would have seen it displayed there. Moreover, in a letter to Herwart von Hohenburg (VIII, 345) Tycho stated expressly: that "Verum ex quo in neoterica reformatione revolutionum Lunarium haec annua variatio, quae per se exigua est, et sextantem gradus modice excedit, aliis quibusdam mediis absorbetur." Dreyer (VIII, 449) remarked on the closeness of the value and cited similar statements by Kepler.

[14] For an exposition of Ptolemy's treatment of the Equation of Time, see O. Neugebauer, *History of Ancient Mathematical Astronomy* (Berlin: 1975), vol. 1, pp. 61–8.

[15] This story was related by Longomontanus in his *Astronomia Danica* (1622), p. 681.

[16] VI, 147.

he also printed an equation of days, reduced in magnitude from the traditional one: Delambre's analysis revealed that a constant amounting to about 7′36″ had been subtracted from each entry in the table.[17] It seems unlikely that anyone will ever reconstruct the rationale for Tycho's alteration. It was surely theoretical and virtually immune to empirical reproach, as the motion of most bodies during such a time interval could not be detected and the interval was doubtless compensated in Tycho's epochs anyway.

There was one body, however, whose motions during times of such magnitude were appreciable: the moon. Already by 1598, Tycho had noticed that the effect of corrections for the equation of time on the motion of the moon, from $-12′$ to $+4′$, was often quite comparable in magnitude to the effect of his annual equation ($-11′$ to $+11′$). In particular, one of the two components of the former had precisely the same argument and, conveniently, the opposite sign. Thus, the part that depended on the mean anomaly of the sun, $-4.5′$ sin M, could be canceled against part of his annual equation, $11′$ sin M.

Apparently Tycho then decided that the remainder of his annual equation simply was not worth the trouble it took to deal with it. This seems most uncharacteristic of Tycho, because the neglected portion would surpass 3′ over two-thirds of the time. Yet, several independent pieces of evidence – statements in Jöstel's tract, Tycho's summary for the emperor, and two eclipse computations for 1598[18] – all suggest this very conclusion. And by no means the least significant argument for it is the fact that this is what Longomontanus did in 1600.

Faced with a theory so complicated that it was the first one in the history of science that could not be put entirely into tables, to spare practitioners the labor of complicated trigonometrical computations,[19] Longomontanus had already taken a shortcut that neglected effects ranging up to 10′, through his elimination of the radial effects of the circle representing the Variation.[20] It

---

[17] J.-B. J. Delambre, *Histoire de l'Astronomie Moderne* (Paris, 1821), vol. 1, p. 159. Tycho's table is on p. 97.

[18] Thoren, "An 'Unpublished' Version," p. 221; and V, 165–89. The eclipse computations were not published by Dreyer for some reason but are in Tycho's logs at the Royal Library in Copenhagen. See GKS 317, 117r, and 129v.

[19] See Tycho's cookbook procedure in II, 117–21 or the gloss on it in Thoren, "Tycho Brahe on the Lunar Theory," pp. 158–61. What the procedure boils down to is the solution of the formula $\dfrac{\tan \frac{1}{2}(A+B)}{\tan \frac{1}{2}(A-B)} = \dfrac{a + b}{a - b}$.

[20] At quadrature and mean distances, where the correction with a librating variation was tan $(E + Q)$ = 13,048/100,000, the radial effect of a circular Variation would have entailed a denominator of 100,000 either plus or minus 1,320, either reducing or increasing the correction by nearly 6′. The corresponding effect in syzygy and mean distance (where the center of the epicycle would have to have been at the opposite end of the little circle) would have been nearly 4′. Tycho defended his shortcut by saying that Copernicus often used it (II, 101) and the difference was insensible anyway.

cannot be surprising, therefore, that he should have seen no good alternative to representing the annual equation in this strange way. Whether because he could not conveniently represent it on the model or because he had genuine doubts about the nature of the phenomenon, Longomontanus never used the word *inequality* in connection with it.[21]

Interestingly enough, Kepler was doing equally remarkable things with the equation of time and the annual equation during this period. One of his obligations as provincial mathematician of Styria was to issue an annual almanac, which naturally included predictions of such significant phenomena as eclipses. Unlike Tycho's eclipse predictions for 1598, Kepler's results did not fare well. In fact, they were so wide of the mark that Kepler felt obliged to say something about them when compiling his prognostications for 1599. Noticing that the eclipses in February had run late and that the one in August had occurred an hour early, Kepler began twenty-five years of personal speculation on the subject, by suggesting in his next almanac that the synodic month might be longer in winter than in summer.[22]

Unfortunately, Kepler's thoughts seem to have been more inspired than informed, for the primary source of error in his eclipse predictions was actually Copernicus's solar theory.[23] But Kepler's fertile mind continued to work on the issue and eventually led him to conclude that it was not the rate of the moon's revolution but the rate of the earth's rotation that varied annually with the distance of the earth–moon system from the sun.[24]

Perhaps Kepler convinced Longomontanus on the subject before he accepted it himself. At any rate, all Longomontanus said by way of justifying a special equation of time for the moon was that extensive experience had

[21] Tycho terminated his discussion of the Variation (the third inequality) with "atque haec est motuum Lunarium circularis ex ipsis apparentiis deprehensa compositio, "II, 101. Longomontanus, in his own exposition of the theory (p. 243) introduced the Variation as "ultimam denique in motu Lunari inaequalitatem." According to Kepler, Tycho "often said to Christian, 'I cannot stand this.' But Christian responded, 'Then you find another theory that agrees with your observed places.' And because Tycho could not at this time treat these studies with very much attention, he, forced by necessity, left unchanged this well-concealed evil." *G.W.* XV, 343.

[22] The calendars of 1598 and 1599 are where Kepler registered his speculation in a "Bericht an den Gunstigen Leser": *Astronomi Johannis Kepleri Opera Omnia*, ed. Christian Frisch (Frankfurt, A. M. 1858), pp. 396–7. Kepler elaborated his ideas a bit for Mästlin and Herwart von Hohenburg in *G. W.* XIII, 253, 282–5.

[23] In the preface to some ephemerides generated for Rudolph II, Tycho mentioned explicitly that his solar theory perpetually anticipated Copernicus's, by a maximum of 37' on 10 February each year, to a minimum of 7' in mid-August. The 30' "acceleration" of the Copernican sun in the interval was most of the effect Kepler was seeing, even though Kepler expicitly ruled out that source.

[24] For Tycho, of course, such an alternative would have required conceiving of an annual irregularity in the diurnal motion of the heavens. On Kepler's ruminations, see C. Anschütz, "Ueber die Entdeckung der Variation und der jahrlichen Gleichung des Mondes," *Zeitschrift für Mathematik und Physik*, 31–2 (Leipzig, 1886–7), pp. 161–71, 201–19, 1–15.

proved it necessary to eliminate the part depending on the difference between the sun's mean and true motion and to use only the part constructed from the right ascensions of the ecliptic. Years later, in Longomontanus's own *Astronomia danica*, he merely was to use many more words to say essentially the same thing and to provide only the one-component table to be used for all applications of the equation of time.[25] Through the combination of Tycho's, Longomontanus's, and Kepler's expressions of doubt about the equation of time and the annual equation, both corrections languished under a cloud of uncertainty for most of the seventeenth century.

Longomontanus's rectification of the theory of longitudes marked the first serious attempt in history to account rigorously for the distances of a celestial body. As might be expected, it was a curious compound of success and failure. Formally, the ratio of the distance extremes implied by the theory was 1.156: 107,248 units in apogee/quadrature, to 92,752 in perigee/quadrature. To atttach these numbers to the phenomena, Tycho adopted a value of $56\frac{1}{2}$ earth radii[26] for his mean distance (100,000 units) and deduced proportionate extreme distances of $52\frac{2}{5}$ and $60\frac{3}{5}$ earth radii. This range of distances finally brought the theory into some kind of proximity to the 8 *minus* $\frac{1}{8}$ earth radii of reality,[27] at least superficially.

Unfortunately, however, although the size of the range was reasonable enough, the nature of its variation was erroneous to an extent such that Tycho must have known that the theory still did not represent the "real" orbit of the moon. There were two ways to monitor this variation: parallaxes and angular diameters. Unlike the situation for the planets, Tycho must surely have conducted some hard tests of the moon's parallaxes. And those done after he had his new theory of latitudes might have been moderately good, although it is worth noting in this context that Tycho altered the constants of his theory of latitudes several times between 1595 and 1601.[28] Yet the inextricable association of parallax with both latitudes and refractions rendered impossible a serious erɔpirical analysis of any single component of the problem. Lurking in the background, moreover, were Tycho's solar parallaxes. Because of their insidious effect (through the values used for the obliquity and refractions) on the reduction of Tycho's observations, Tycho placed the moon nearer the earth than it truly is, so that he reduced Copernicus's fortunate stab at the mean distance, $60\frac{1}{2}$ earth radii, to $56\frac{1}{2}$.[29]

---

[25] *Astronomia Danica* (1622), p. 181.
[26] II, 131. This ratio, 1.5649, is almost precisely equal to that of the theoretical extremes (107, 248 to 92, 752), 1.5629.
[27] 63.836 earth radii to 55.943 e.r. gives a ratio of 1.411 and a range of 7.89 e.r.
[28] See Thoren, "Tycho Brahe," pp. 111–12, 115–16, 162–3.
[29] Delambre's description (*Histoire de l'Astronomie Moderne*, vol. 1, p. 174) of Tycho's parallaxes is badly garbled. Most significantly, he read as the parallax at apogee the value tabulated (II, 132–4) for altitude 3° instead of for the horizon, which gave him a minimum

The moon's angular diameters, hower, were not compounded with such complexities. If the angles involved were smaller and sufficiently difficult to measure to rule out any "plotting" of the shape of the moon's orbit, they could, and clearly did, nevertheless show that Longomontanus's shape was far from correct.

In the (perturbed) orbit of the moon, the variations in distance are caused chiefly by the elliptical shape of the orbit. These variations had always been simulated tolerably well by the large epicycle used to represent the first inequality. What had distorted the moon's distances in previous lunar theories was the representation of the second inequality. Unfortunately, Longomontanus solved the radius vector problems not by rectifying the offending second inequality but by distorting the representation of the first inequality to compensate for the distortions of the second. As might be expected, the result was less than perfect.

Although the ratio between Longomontanus's extreme distances (and, inversely, diameters) was close enough – 1.156 to 1.141 – these extremes, produced by the center circle representing the second inequality, occurred in quadrature, The relative extremes available in syzygy – where radius vectors were most important, because of eclipse computations – were significantly less (102,900 to 97,100, for a ratio of 1.060), and Tycho knew it. Converted to an average value of 34' (for 100,000 units), they would have implied a variation of almost exactly ±1'. Such a small range could be falsified through observation by a straightforward measurement of angular diameters, and was certain to produce unacceptable consequences for eclipse calculations, as well. Accordingly, Tycho just used ±2',[30] a ratio (1.125) that accorded reasonably well with both observation (1.141) and the absolute extremes of his theory (1.156, for quadrature). This solved his problems for full moon and therefore for lunar eclipses.

For solar eclipses, Tycho found it necessary to make one last adjustment. Contrary to Ptolemy, whose semidiameters for the sun and moon theoretically ruled out the possibility of an annular eclipse,[31] Tycho found that the

---

parallax of 48'43" instead of 56'21". With an upper limit of 66'6", this gave him (1) a very broad range of parallax and (2) an average value of 57'24" (instead of the 60'51" quoted by Tycho, II, 131). As a result, Delambre summed up Tycho's efforts as providing a good mean value but an erroneous range, instead of vice versa. To compound the problem, Delambre overlooked the fact that the tabulated extremes he was wanting to read were actually outside the range of Tycho's parallaxes, as Tycho tabulated his columns by integral values of earth radii (pp. 52–61), even though his theory actually permitted extremes of only $52\frac{2}{5}$ and $60\frac{3}{5}$. Thus his 66'6" to 48'43" range should have been the 65'36" to 56'44" quoted by Tycho in I, 131. As final proof that Delambre can scarcely even have read this section, let alone given any thought to it, he quoted Tycho's minimum semidiameter of the moon as $14\frac{2}{5}$ instead of the 16' – or $12\frac{4}{5}$, that is, $\frac{4}{5}$ of 16' – really given by Tycho (II, 148).

[30] II, 148.
[31] Ptolemy had used $33\frac{1}{2}$ ± 2' for the diameters of the moon and a constant $31\frac{1}{2}$' for the sun: *Almagest* VI, 5.

solar eclipse of February 1598 was much more noticeably annular than it should have been. Reflection on his experiences with previous solar eclipses seems to have convinced him that this phenomenon (which later astronomers would dub the solar corona) was fairly general,[32] that indeed all central eclipses would probably appear annular.

On the other hand, Tycho had apparently made enough measurements of the diameter of the new moon to feel confident that the effect was only apparent, some kind of optical illusion rather than an actual recession of the moon in distance. Of course, it still had to be taken into account if astronomers were to be able to predict its appearances. But that could be done artificially, without any real modification of the theory. In tabulating the sizes of the moon in conjunction, therefore, Tycho departed in a second respect from the strict geometrical consequences of his model. In addition to eliminating the variation of angular diameters arising from synodic motion, and doubling the anomalistic variation in syzygy, he arbitrarily diminished the moon's diameter by one-fifth for all solar eclipse computations.[33]

After Tycho's death, the long-postponed preface to the *Progymnasmata* had to be written by his heirs, and the printing of it and the lunar quartos fell to Kepler, who decided that it was advisable to add an appendix explaining the long (and obvious) delay in the publication of the *Progymnasmata*, and the presence of an elaborate, entirely new lunar theory.[34] (By this time, Tycho's discoveries had so expanded the subject matter that six and a half quartos were required for its exposition. Because this had entailed the addition of twenty-eight separately numbered pages to those originally available in the three-quarto gap, the discontinuity between Chapters I and II was more noticeable than ever.[35]) Under the circumstances, it seemed justifiable to dignify the entire section with its own title: "Appendix concerning the restored motion of the moon," thereby indicating that Tycho had finally achieved his long-sought restitution of the motion of the moon.

---

[32] Tycho discussed his finding in a letter to Mästlin (VIII, 155) and in the preface to his ephemerides (V, 183). In the latter he mentioned a dimination of one-fourth. Even as late as 1842, the solar corona generated this kind of impression.

[33] II, 148. Delambre (see n. 29) seems to have missed this point completely and in fact even read one column upside down to extract the minimum diameter he quoted (14′24″ instead of 12′48″).

[34] III, 320–1.

[35] The twenty-eight pages come between pages 112 and 113. In most copies the numbering runs from 01 to 129, with 021 missing. *Dreyer's* (p. 187) report that the insert contains thirty-two pages originates in the fact that in some copies the numbering continues to 032, with the result that pages 113 through 115 have been omitted.

Appendix 4: Figures for Footnotes

Figure A.4.1, to note 29 of Chapter 7.

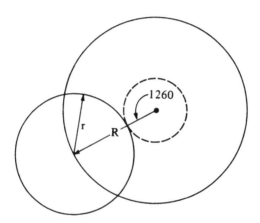

Figure A.4.2, to note 20 of Chapter 8.

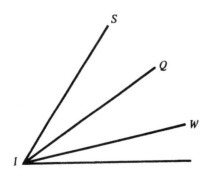

Figure A.4.3, to note 106 of Chapter 9.

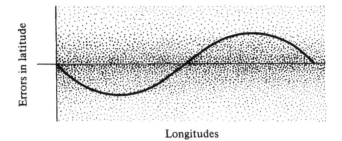

Longitudes

Figure A.4.4, to note 109 of Chapter 9.

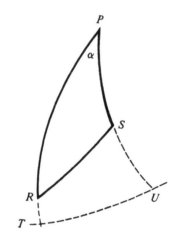

Figure A.4.5, to note 120 of Chapter 9.

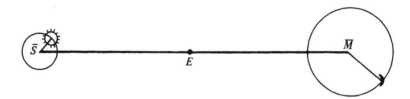

Figure A.4.6, to note 47 of Chapter 10.

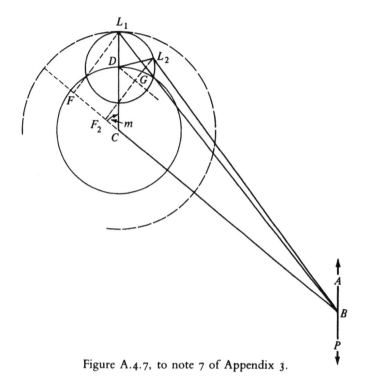

Figure A.4.7, to note 7 of Appendix 3.

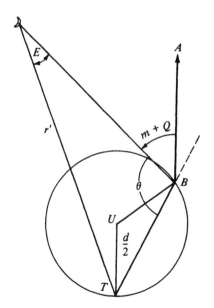

Figure A.4.8, to note 11 of Appendix 3.

Appendix 5: Tycho's Dwellings in Exile

Czechoslovakia preserves a considerable number of mementos of Tycho's residence there. Both Brandeis and Benatky are reached via Highway 10, some thirty kilometers north of Prague. The former, although now in a sad state of disrepair, was once a truly magnificent estate. If it had been built to even an approximation of its final grandeur when Tycho declined it, he must have been very concerned, indeed, about depriving the emperor of it. A minute portion of it is now used as an office for the Czech forestry service, and access to the entire building is forbidden. Farther up the highway, Benatky becomes visible from the road, high up on a hill that offers as commanding a view as one can imagine. It has been considerably enlarged since Tycho's day, and there is no sign of the wooden observing cubicles that Tycho was given permission to build. But even the portion available to him (and made available to me through the kind offices of Dr. Pavel Štifter) involved three floors of nearly five hundred square meters each and therefore provided quite a bit of space. It is currently undergoing renovation and will, in time, house a small Tycho museum.

For most of the time he was in Prague, Tycho lived in the summer home (1, in the map on page 501) built by Jacob Kurtz a few minutes' walk west of the entrance to the still preserved Imperial Castle. The site, on the crest of a hill, is definitely an astronomer's choice, and it is now graced (since 1984) by large statues of Tycho and Kepler. Remnants of the building are supposed to have been discovered when the site was cleared for the construction of a school at the beginning of the twentieth century. But if these remnants are the foundation preserved in front of the school, that 7-×-16-meter outline can scarcely have supported very much of the 20,000-gulden edifice described by Tycho.

The "Golden Griffin," in which Tycho lived temporarily when he was recalled to Prague in mid-1600, still stands. Located on "New World" Street (3) adjacent to the castle grounds, it was marked in 1901 with a plaque (in Czech) noting that Tycho resided there briefly. The Capuchin monastery a few meters around the corner (4) still stands, too, and seems now to be a military facility of some kind. Tycho stayed only a few weeks at the Golden Griffin before moving to another temporary house "near Baron Hoffmann's." Where that house might have been, I have been unable to discover. However, a street on the other side of the castle has come to be called "Tychonova" (5), possibly completely gratuitously or perhaps because it runs into the Imperial summer house, Belvedere (6), where Tycho's instruments were set up during the winter of 1600–1. On the other hand, if

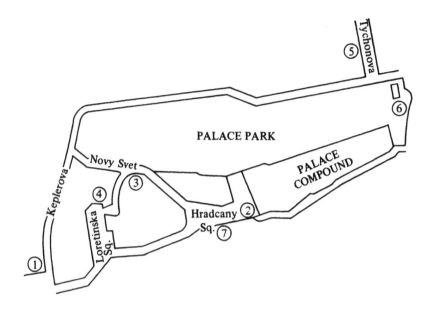

the rental house were on that street, it would have been a very good reason for letting Tycho use Belvedere for his observing.

Tycho's "last meal" was at the so-called Schwarzenberg palace (7), which Rozmberk acquired by a trade with Rudolph in 1600. It is currently open to the public as a museum of medieval military history and a repository of political propaganda.

Finally, even Vienna, a city that Tycho seems never to have visited, claims to have a souvenir of Tycho – the clock in the tower of the Amalienburg. The building itself was built by Rudolph, apparently as a repository for his various collections of curiosities. Rudolph also had access to Tycho's instruments – including his clocks, presumably – through purchase from the heirs after Tycho's death. Although the instruments apparently languished until nobody knew (or cared) what they were, the clocks would certainly have been recognized as valuable objects. It is, therefore, by no means impossible that the largest of these may have been preserved with Rudolph's curios until the clock tower was added in 1784. Unfortunately, the administrative underlings at the Hofburg refused to permit inquiries directed at investigating the authenticity of the clock described in its official guidebook. So the strong doubts of professional horologists that any such clock of Tycho's – which, in any case, would scarcely have been made in his shop – exists, must represent the best estimate of the situation.

Appendix 6: Letters, 1599–1601

(translated by John R. Christiansen)

LETTER 1: ESGE JØRGENSEN BILLE TO TYCHO BRAHE, 11 NOVEMBER (NEW STYLE) 1599

Dear Tycho Brahe, my dear brother,

Now and always, I send you my most friendly greetings in the name of Our Lord. I send you my most friendly thanks for all the goodness you have done for me, which you will always find me willing to repay with any service I can render to your honor and welfare.

Dear brother, your letter came into my hands here in Copenhagen through your son, Tyge, and in it you let me know about the good circumstances you have come into at the court of the Holy Roman Emperor, which I (as well as many of your good friends, who asked about you at this Diet of the Nobility) was most pleased to hear. May God continue to give you good fortune.

Concerning the fifty dalers that you asked me to pay to Christian Longomontanus, your son Tyge will undoubtedly tell you about that, because I note that he is coming with that money, otherwise he would certainly have gotten it [from me], regardless of the circumstances.

Concerning your instruments, I have given your son my thoughts and advice on what I consider to be the best way to get them out, which would be by way of Lübeck. I cannot stay here in the city until Tyge is through having the instruments packed and doing the other things he has to do, for I am in His Royal Majesty's command and must leave in a few days in order to go to the place where these commands are to be executed.

I note that the 700 dalers which I sent to you some time ago had still not arrived when your son departed from you, but Tyge informs me that he has gotten some information about that money during his trip. I still have not settled Lady Cecilia Urne's account[1] because I cannot come to an agreement with her. She wants the interest as well as the principal sum, and I am still offering her the principal, and if she will not accept that, I am afraid that she will sue you for the principal and interest. When you write that I should serve her a summons at the Diet of the Nobility, you surely know that it is not worth it in a matter like this until the case has been tried before the Hundred Court and then the Provincial Court, and maybe it could go all the

---

[1] Cecelia Urne apppears to have been one of the coheirs of Inger Oxe; see also letter 6.

way to the Chancellor of the Realm if she were summoned according to the letter of Lady Johanne Oxe, and you can well imagine how much harm and difficulty it would cause you if it went against you. So that you can see more clearly that the letter of Lady Johanne Oxe cannot carry much weight, I am sending you a copy of a judgment of the Diet of the Nobility which contains the original contract, and you can see for yourself. I have also shown your son a letter, and which you have signed with your own hand, in which you promise to repay the principal sum and a reasonable rate of interest. When you wrote that you would come here for the next Diet of the Nobility, you probably believed that there would not be a session in the summer, since this present session is so late in the year. I have discussed this with your son, Tyge, who will bring you a verbal report.

As for the 300 dalers in interest money which were to be sent to Holstein, they will be sent there, whether I get them from Axel Gyldenstierne or elsewhere. And I note that you want the principal sum repaid, which is 5,000 dalers, so that you can get back your note, which is to the amount of 10,000 dalers,[2] but I do not know whether I need to give him three or six month's notice. Therefore, please send me a copy of the contract, so that I can give him due notice, in another year, that you are recalling the loan, for I doubt that it can be done by the next annual money market. A person cannot raise that sum of money in a short time, unless one knew about it around Midsummer or by St. Lawrence's day [August 10] at the latest.

Regarding the stable of horses you had on Hven, I was not able to sell them without sending them over to Ellinge, where the steward has now sold them, some to peasants and some to clergymen, and I will not know how much he got for them until I come home and can find out. Regarding the horses that you wrote you wanted me to buy for you, I have gotten five from Oluf Bille, three of them were his carriage horses. The oldest is over seven, and I also bought a young foal from him, in its fourth year, which will make a good carriage horse together with the other three. The fifth I have bought to be your personal riding horse, a four year old, and without doubt he is certainly a good and noble horse, and if he is cared for well and given enough oats, he will undoubtedly be the kind of a horse that a prince would be pleased to ride. If I had not bought him for you, I would certainly not have sold him again for 150 dalers. I had to give him 300 dalers for all five horses and pay him by Christmas of this year. There are two foals on Hven that belong to you, one with red spots and the other with a brown star, which cannot be your favorites. I wanted to sell them for forty dalers but was offered no more than thirty for the pair, so I have advised your son

---

[2] The most promising interpretation of this confusing passage is that Tycho still owed 5,000 dalers to the widow Rantzov (from an original loan of 10,000) and that the 300 dalers is the 6 percent interest that Tycho mentioned (in letter 6) as the going rate. Axel Gyldenstierne clearly owed Tycho money (letter 5), and Esge's reference to giving "him" notice must refer to recalling that loan.

to take them along with him: let the boy ride on one and lead the other in order to spare it. I have not yet been able to get the sixth horse for you, and the reason is that everybody expects a mustering, so that everybody has need of his horses.

I have hired a coachman for you, and he has promised to accompany your son abroad, and I have promised him that when he no longer wants to serve you, that he should get his pay and a letter of recommendation, so that he can return here to Denmark. He has also promised to comport himself well in your service. I have also hired a stableboy for you, who gave me his hand and promised to serve you for one year in return for shoes and shirts, six dalers, English clothes, and a pair of knee boots. I have promised him that when his year is up, if he does not want to serve you any longer, then he will receive his wages and a letter of recommendation, so that he can return to Denmark, and if he had good service and chose to remain, that I could get him released from any obligations to the place where he had previously served.

Regarding the fish that you wanted to have salted down, your son Tyge has been there himself and taken care of it.

Now I have answered everything in your letter, and I have also spoken with your son, as he can report verbally to you. I would like to have David Pedersen's accounts in here because of a couple of matters.

Moreover, I cannot settle with Dirick Farver's wife[3] unless I have her account again, which I have sent to you, and I am afraid that she will take her supplication to His Royal Majesty or to the Lord Steward, and then I will not be able to make any progress with her until I have her account again, which I sent to you, and from which one can see that the account is not entirely clear. But it seems to me from the slips you have issued and the evidence she has from merchants and shopkeepers that you must owe her some money.

The pastor on Hven has requested that he might have the smithy that you erected there on the island, although it is falling into disrepair, because the parsonage fell into disrepair in your day, but I did not want to let him have it until I knew what you wanted to do about this. The grange is falling into disrepair, so that I am afraid that the cowbarn will have to be torn down this summer or else it will fall down on its own.[4]

I cannot think of anything else to write about to you at this time, but will

---

[3] This appears to concern the dyer who lived in or adjacent to Tycho's house in Copenhagen on Farversgade ("dyers" street). See the postscript to this letter, and letter 6.

[4] This litany of decay seems to describe the fate of everything on Hven except Stjerneborg. Perhaps Tycho's master builder (possibly himself?) was simply incompetent. Possibly the peasants sabotaged the original construction or (more likely) cannibalized the abandoned buildings for their own homes. At any rate, in a few years Uraniborg was virtually nothing but the hole in the ground that has been preserved there since. Plans are currently afoot to erect at least an outline of the former house/observatory, rebuild the wall, and replant the gardens. The masonry pits of Stjerneborg fared better, and have been protected since the 1930s by a concrete shell that simulates the outline of the original wooden turrets.

commend you now and always to God Almighty with many good nights, which my father and mother and Esge Brok and Jørgen Brockenhuus and I wish you and yours and those you wish well.

From Copenhagen in great haste,
Esge Bille

Dear Tycho,

Since this letter was written, I have returned here to the city and have spoken with your son and learned that he has received a letter asking him to acquire a total of as many as fourteen horses to bring along. I do not think it is possible to get hold of another seven horses in such a short time, because I have had to do a great deal of trading in the four weeks that Tyge has been here in Denmark before I was able to put together the deal for those I have gotten. Today I bought two horses, one from Axel Aagesen for 45 dalers, which is described to me as a good, fast, strong little stallion, suitable for you personally to ride, or for your friend if you give it away; you can certainly find a good use for it. I bought the other one from Baldzer Fox for 40 dalers, and it should also turn out to be a good horse, but it is still not broken. When it comes on fodder, you can also make good use of this one, for he should have a good ride and a handsome trot. Neither of these horses is older than five years. For these seven horses I have gotten for your son, I have paid out a total of 385 dalers. God grant that they may come to you without accident. I would not advise you to part with the one horse that I bought from Oluf Bille. Even if I wanted to pay 200 dalers for a horse for you or another friend, I do not know where I could get another horse like that.

Your son and I have negotiated with your old coachman today, and he has promised to accompany your son, and I have promised him wages of 14 dalers a year and a long coat, and I have also promised that when he does not want to stay down there any longer, then you will be willing to let him go on good terms, so that he can return to Denmark again.

Concerning your house here in town, Sophie Paxis has offered me 1,500 dalers for it, but she wants two years to pay it. Lady Margrete Rosenkrantz has offered me 1,400 dalers for it and will pay in one year. But I have not made an agreement with either of them because your sister, Sophie, is not in the country, and I understand that she has to be informed if the house is to be sold. Moreover, whoever buys this house has to promise you to maintain a dyer in the house, which I do not believe that either of these ladies is willing to do.

Fare well, and keep for yourself the horse that I got from Oluf Bille. Your son will talk more with you about that. I commend you to the care of God almighty, and likewise to all of you.

Copenhagen[5]

---

[5] XIV, 160–3.

LETTER 2: BEATE BILLE TO TYCHO BRAHE, 26 NOVEMBER (OLD STYLE) 1599

Dear son Tycho,
   May God almighty be with you and yours now and always and graciously preserve you from all evil. Most friendly thanks for the kind letter you wrote to me, dear son, to let me know how things are now. All is still well with me, thanks be to God. Dear son, I was most pleased to hear that you, thank God, have now come to settle down, and I was pleased to learn from your letter that all is well with you and yours, thanks be to God. For God will not forget those who fear him. And your son has also reported to me that your affairs are going fortunately and well, thanks be to God. Therefore, I will not hold you longer, dear son, with this my short letter.
   All my days, you will find me true to you, and I now commend you and yours to God almighty, and wish you all good fortune and welfare. Billesgaard.[6]

Beate Bille

LETTER 3: TYCHO BRAHE TO BEATE BILLE, 21 MARCH 1600

Dear Mother,
   Now and always, I send you my friendly filial greetings in the name of Our Lord. May God almighty be with you now and always, and graciously preserve you from all harm to soul or life. Many thanks for all the goodness you have shown me, for which you will always find me willing to serve you in all ways, whenever I can do anything to your honor and good, as I remain obliged to do.
   I want to let you know that your kind letter was brought to me by my son, Tyge, who returned here to me shortly after New Year's day of this year. I was most pleased to learn that you are well, thanks be to God. May the same gracious God always grant me the same tidings from you. And I thank you as well for all the kindness you showed my son when he was with you, and for the money you gave him. He used it to buy a fine golden bracelet which he will keep and wear for the rest of his life to remind him of you.
   Dear Mother, I want you to know that my household and I are doing well by the grace of God, and the same gracious God has saved us from the great peril which was here last summer and a good part of the winter until around Christmas time, with much pestilence and pox throughout the whole Kingdom of Bohemia and surrounding lands, worse than in any of

[6] XIV, 167.

the previous years, so that the Holy Roman Emperor had to leave Prague, where His Majesty preferred to be, and was forced to go to another city by the name of Pilsen, ten Danish miles from there, where His Majesty is still residing, and he will probably not return to Prague until after Easter. The same dangerous plagues have also raged here in the region which belongs to Benatky Castle, where I reside with my household, so that around two thousand have died, both here in the town, which lies on a large hill near the castle, and in the surrounding villages which belong to the estate. And when His Majesty learned that there was such danger here, he has several times sent gracious inquiries about my condition and that of my household, for His Majesty has a strong personal concern for me and them. And he twice graciously invited me to withdraw to another place in His Majesty's lands where it is safer than here, even to Vienna, which is His Majesty's principal capital in Austria, where the previous Holy Roman Emperors usually resided, and offered to find a suitable place for me and pay my expenses. But I declined His Majesty's offers, saying that I would rather remain here and commend everything to God, which His Majesty graciously accepted. And he graciously requested me to send him some of the medicine which I used against this same illness, and a recipe for preparing it, which I immediately did. This pleased His Majesty well, because His Majesty has otherwise shown me all possible imperial favor and grace, so that I cannot thank Our Lord enough for making my circumstances different and better than they previously were in my fatherland. I will not write you more about this now because when my son was there, he told you a good deal about it.

All things are still the same, and I hope with the help of God that they can become even better in the future, if the good Lord will give fortune and blessings. I commend you to the same gracious God and pray that he will grant you continued good health, fortune, and welfare.

Dated at the imperial castle of Benatky.[7]

LETTER 4: TYCHO BRAHE TO SOPHIE BRAHE, 21 MARCH 1600

Dear sister Sophie,

May God almighty be with you now and always and graciously preserve you from all harm. Many thanks for all the goodness you have shown me, for which you will always find me your true brother, who will gladly do anything that can serve to your honor and welfare.

I received your letter with my son Tyge, some time after New Year's day, though you had not yet received the long letter I sent to you with him,

---

[7] XIV, 175–6.

for you were not in Denmark then. I assume that you have since received it
and learned from it about my situation, so I do not consider it necessary to
repeat anything that was in it. I learned from your letter that you had
traveled through Schleswig-Holstein to Hamburg with your son, Tage, in
order to send him on to Heidelberg or Basel, and I was glad to hear that he
was coming out to try himself. He wrote to me just before Christmas from
Heidelberg and let me know that he liked it there and that he intended to go
with his preceptor to Basel in the spring. I wrote back to him to give him
my views and recommend him to the most famous men of learning with
whom I was acquainted. He also let me know how far you accompanied
him, and who came to meet you, and where you then traveled, but he did
not really know how long your travels lasted. You told me that and more in
your own letter to me.

Among other things, you reported in the same letter about the conversation you had with *Mercurius* [Chancellor Christian Friis] in Landskrona, and
that, among other things, he said something about the emperor that did not
please you. I would like to know what that was, because I know of nothing
he could truthfully say about His Imperial Majesty which would displease
you for my sake. I thank the good Lord, who has provided me with such a
good and pious and mild lord in him, so that I am well satisfied. His
Majesty not only shows me all goodness and does not allow me to lack
anything but has also graciously shown a personal, fatherly concern for me
. . . and lets me have whatever I wish, and sends me game from time to
time, harts, hinds, and wild boars, and also various living fresh fish to the
kitchen.

Recently, he has also graciously turned over to me a suitable grange,
near the castle here, where I can have meadowland and sow as much as I can
of all kinds, and there are also some fine fishponds there, and he will also
give me a fine garden and vineyard here by the castle to use and enjoy
myself in. In addition, His Majesty has given me more on his own than I
would ever dare to request, although His Majesty still has not seen very
much of what I can do because my finest instruments are still back in
Germany, since they are not able to come up the Elbe until winter is past.
His Majesty has recently written to Magdeburg about them, and I now
expect them at the earliest opportunity. Then His Majesty intends to come
here to the castle to see them and my other work, both distillation and other
things, for His Majesty has a strong interest and pleasure in all such things.
Some weeks ago, His Majesty asked his first secretary, Mr. Barwitz, who is
his factotum, with whom he prefers to talk, whether he had anything from
me that he could let him see, and then he said that I had recently sent him
one of my books, which was still not completely printed, and that I wanted
to know whether I might humbly dedicate it to His Majesty. Then he had to
get immediately the book entitled *Astronomiae Instauratae progymnasmata*,
and the emperor kept it with him for more than fourteen days, and read

often and at length in it, and took great pleasure in it, and commanded that I should be told that he definitely did not want me to dedicate it to anyone else than His Majesty himself, and that I should put in it, before all of the other statements of privileges and copyrights, a letter to His Majesty himself, and that I should set below the title that it was begun in Denmark but completed and published here at Benatky in his kingdom of Bohemia. And he also sent me verses which His Majesty wants to be carved in marble over the doors of the two portals of the buildings that His Majesty is having built for me here, the one a special observatory in which all of my instruments can be set up properly and in order, so that each will have its enclosed room to stand in, and the other a special laboratory, so that His Majesty can go down on the one side from the house which is reserved for His Majesty to be in when he comes here, and go through it all and see whatever His Majesty desires. And from the other house, on the other side, the whole of which I occupy, I can also go down through the laboratory and then on to all the instruments. The clerk who is handling the accounts has calculated that the work that has already been done has cost close to two thousand dalers, but it is not even near to being finished because of the winter, which is much harder than usual here in this country, so that it is still not possible to do masonry work outside.

I report all of this so that you can know that things are not at all as Mercurius wanted you to believe, and there is much more to report if the pen could stand it and it would not become too lengthy. When he said that two of my best friends at the imperial court had been dismissed and fallen from favor, that is just as true as the rest, for the emperor has not dismissed any members of his council, nor is it so easy to do so, even if they deserve it, so that I cannot understand who or what he meant by that, other than to say something that might upset you in order to produce the effect he desired.

I am pleased to hear that when you write about Titan [Erik Lange], that he was able to come out right with that which he had undertaken, and I would like to get an accurate account of how it all fit together. I have not yet been able to discover whether Hans Ernst actually wrote anything to the emperor about it, nor would it be wise to do so until it had been thoroughly tried to see that the art was true, and that there was an advantage in it, because there have been so many who have made such claims before the emperor without having anything, so he does not now believe it readily, nor could his Privy Council, who read all the letters sent to him, allow him to indulge in things like that, since such huge expenses have been incurred in vain with such attempts in the past.[8] One or two of them have the power to take up the letters sent to him and, if they contain something displeasing to them, they remain silent and do not let him know about it. It happens in greater matters than this. Therefore, if such a matter

---

[8] Tycho probably was referring here to alchemy or some kindred enterprise.

were certain, and there was no mistaking it, then someone whom His Majesty likes to talk with and in whom he trusts should say it to His Majesty by word of mouth, so that nobody knew of it, or it should be written to His Majesty on a slip in a sealed letter, and then the others will not learn what the slip contains, even if he later chooses to let them see the letter. That is what I usually do when I have something to tell His Majesty which is of vital concern either to His Majesty or myself, for His Majesty wants it that way, and nobody else may open my letter when there is written on the outside, *For the hands of his Imperial Majesty himself.* Therefore, I would know how to assure that His Majesty were properly informed of it, so that nobody else would know it, if I were certain in advance that there were no shortcoming in the matter, for otherwise I would not dare to present it at all, either by word of mouth or in writing, for His Majesty does not believe anymore in those who have once failed him, for he is firm and steady in his dealings.

They know very well that I write to His Majesty in this way at times, and that His Majesty prefers it like that. Therefore, I had to listen to quite a bit when I was in Prague recently, from both high and low, and I was to have written to His Majesty to advise him not to come to Prague until after Easter because there was danger. It did not help that I said no to this, for it did not actually happen. I personally wanted His Majesty to come there soon. They thought that I would not persist, for he had forbidden me to let others know such things, and some thought that it was right to warn him if I knew. Some who have profit and gain from his being in Prague resented it and blamed me for informing the court that nothing of the kind had occurred, for the sake of the other councillors who would rather be in Prague than stay so long in Pilsen, where the emperor is, for it is not so convenient for them there. But folks must speak, for flocks have not learned the art.

It was not prudent of Titan to offer in writing to give the art to the king in that country, without first getting another opinion on it. I knew very well that he himself and those he trusts most would only mock and ridicule it, and that Titan would receive no answer from him, because when a person is suspect and hated, then whatever he proposes is suspect and ill received, even if it were proven to be right and true. I would like you to send me a copy of Titan's supplication, because I want to know what he said to the king in it. If I were like him, I would not bother them any more, and think so little of them as they of me. I have not received a letter from him for a long time, although I have written to him two or three times, and I do not know for certain where he is, but you will undoubtedly write to me about everything that you know about him and his affairs, for I long to know about it. Also write to me everything you know that might pertain to me, and what is said there in that country, both by friends and enemies, now that they have learned that my situation and circumstances are other than they might perhaps have thought earlier. Let me know, too, how things are going with our friends and acquaintances, and if anybody wants to write anything to me, let them know that you are now my representative/ messenger. You can also tell that sometime to our sister, Margrethe, if she wants to write to me in reply to my last letter, and let me know how things are going with her youngest sons since they have come to court, and also about our mother's sister's daughter's sons, the young Lunges, whom I have heard are also at court. God preserve them from all harm. Do not forget to mention anything else that you know I would like to know. You usually do not do so, for you know well that I would rather read a long letter than play cards. Also because I know that you enjoy it I will write the more lengthily to you, for the Lord does not send a message every day. I will not keep you any longer now, however, but commend you now and always to God almighty, that He may protect you and yours graciously and well from all evil. Please wish our mother's brother's sons and daughters many good nights on my behalf. Claus Steensen Bille was here with me for a quarter of a year and then left for France, accompanied by Master Niels Hammer, who arrived here at the same time and knows the language and some of the people in that country. He promised me that he would take good care of Claus and assure his gain and welfare wherever possible. I have not heard anything from them since, and I have written a letter to his mother which you can help to forward to her. Wish Maid Dorete many good nights as well on my behalf. *Iterum et saepius quam felicissime vale.* Dated at the imperial castle of Benatky.[9]

LETTER 5: TYCHO BRAHE TO AXEL GYLDENSTIERNE, 21 MARCH 1600

Dear Axel Gyldenstierne, my dear mother's sister's son,
My most friendly greeting now and always in Our Lord. May God almighty preserve you graciously and well from all evil. I thank you for all goodness shown to me, for which I am ever indebted to you with my fortune and in whatever way I can do anything to serve your honor and welfare.

Dear Axel, I received your letter with my son, Tyge, who returned home to me shortly after New Year's day. Thanks for the good will you expressed in your pleasure at my situation and circumstances in this land. All things are still, praise be to God, just the same as I described them when I wrote to you, and you have undoubtedly learned further from Esge Bille that I hope they can become even better in the future, if God grant them benediction and good fortune.

Dear Axel, you report in your letter concerning the money between us

9 XIV, 178–82.

that you could not put it out until the last annual money market in Holstein, and you therefore request that it be allowed to remain until after the coming market. I had certainly counted on its being available after the last market on the basis of your own letter, but since that cannot happen I shall count on its definitely being paid at the next annual market, which you in this your second letter definitely promise. This is very important to me, as Esge Bille knows.

I am calling in the money I have loaned out in Mecklenburg because I want to have it here, since I intend to use it, with the help of God, to buy a landed estate, here in this country, which is available at a very good price and gives a good profit. When I am accepted as a resident of this realm by means of the emperor's authority and support, and with the approval of the order of knights and nobility, which can easily take place at the next noble diet here in this country, then I will be able to buy and possess as a hereditary estate all the landed property I can get for money, just as if I were born in this country, and enjoy all the privileges the knights of this land have, which are not negligible. If this had occurred already at the last diet, I would have wanted my money on time for that purpose. I am in trouble on your account, therefore, dear Axel, when you do not uphold our agreement any longer, because I can seek my gain and advantage with it in other ways. You can let Esge Bille have it if you do not know of other means to handle it, because he knows about my affairs in both Holstein and Mecklenburg.

Dear Axel, I have no other news in particular to write you, except that they say the Turk is arming himself strongly, and our emperor goes and does what is possible for his. The German princes will not give any support because of the Spaniards, who threaten Germany. The Tartars did great damage in Lower Hungary and Mehrland during a short time last summer, and carried off many people. You have undoubtedly heard how the emperor recaptured Transylvania. God grant that it might remain ever in his hands.

Dear Axel, I will not keep you any longer with this letter, but now and always commend you to God almighty with all that you will. Dated at the imperial castle of Benatky,[10]

LETTER 6: TYCHO BRAHE TO ESGE BILLE, 22 MARCH 1600

Dear Esge Bille and dear brother,
    Now and always, I send you my most friendly greetings in Our Lord. ...[I] wrote to you from [one of his Majesty's houses, named Girsitz, where I was for six or seven weeks] in November, which letter of mine I

---

[10] XIV, 182–3.

expect that you have since received. In it, I believe that I answered some of the matters your recent letter raises.

I perceive that my son, Tyge, has gotten all of my instruments out of the country [Denmark], and that they are now in Lübeck or Hamburg, and I hope at the first opportunity to get them up here to me via the Elbe River, insofar as His Majesty graciously wrote last year to the burgomaster and council in Hamburg about it. Most of the others that I had with me are still in Magdeburg with a great pile of the rest of my goods. The burgomaster and council of Magdeburg found a legal loophole so that they did not have to bring all of it here by this winter at their own expense, and His Majesty wrote to them twice about it. Now they would gladly do it if they could, but His Majesty took such offense at their earlier refusal that now he does not want it done by them. He has written to the prelates of Magdeburg chapter about it, and they would be pleased to comply with His Majesty's wishes and had even offered to do so before they received the letter. Therefore, I hope to have it here very soon at their expense.

The seven hundred dalers which you sent to me did not arrive until just before last Christmas. They were on the way among merchants that long.

See if you can satisfy Lady Cecilia Urne with respect to the principal sum. You should not pay her any interest because I want to take that to court, for I still believe that she should pay me the five hundred dalers that Maid Johanne owed me. I have no other evidence of this than the letter she gave me, which you have. Had I known that Cecilia Urne would not be satisfied with that, I would certainly have gotten other satisfaction from Maid Johanne for it, nor did I previously have any knowledge of the Hundred Court judgment Cecilia has obtained and will not be satisfied with it. I do not think it concerns me as much as it does the coheirs jointly, who thereby had granted to me and consented unknowingly, nor can one find in the contract of the council adequate treatment of the special circumstances. But if she can produce her receipt for the money that was granted to me, I will be satisfied with that. If that had been, then Maid Johanne would undoubtedly have let me know about it while she was still alive and would have gotten Cecilia's manuscript back from me, and in other ways satisfied my claims, so that I could have paid Cecilia the five hundred dalers from that, but because both she and Frederick remained silent about that, and did not require them [the dalers] of me while Frederick was alive, I did not know what else to believe than that they were satisfied to quit the same sum against Cecilia's manuscript. I pray you will speak at length with Axel Guildenstern about this, and get his good advice, so that I am not short-changed in any way. If I cannot get any credit for the money that I was owed by the late Albert, could I not claim it from Maid Johanne's heirs, so that I could receive adequate compensation for the loss? If you thought it to be good, one could take friends on both sides, who divide Cecilia and me in this. As for the letter which you report that you have let my son see, which I supposedly wrote to Cecilia about this in my own hand, I would like to know how it reads and when it was dated; maybe it was written before I received her letter from Maid Johanne. In that case, it would be good to answer that I know well that I wanted to have the money from Maid Johanne in order to pay Cecilia, for she urged me, sometime after God had called Albert Oxe, and then I undoubtedly wrote such a letter to her. But after I received the other one from Maid Johanne to set against it, she would hardly have received such a message from me, but I would undoubtedly have written the same to her which I now stand on, if indeed she would allow such letters to come forth. Please send me a copy of the letter of mine that you report on, and then I will send you a better account of it, as far as I can recall.

I had intended to come there and attend the Diet of the Nobility this year, but I learn from your letter that it is not certain whether there will be a diet this year, since the last one was so late in the year. Moreover, I cannot get leave of His Roman Imperial Majesty to travel so far from here in this first year, nor are my affairs properly arranged yet, so that I will be busy with that most of the summer. If all goes well, I plan, with the help of God, to come and attend the following Diet of the Nobility next year, to speak with my family and friends and conduct my business as necessary. I imagine that I can get a leave by His Majesty of some eight or ten weeks to do it....

You can use whatever you can get for my house in Copenhagen and whatever else you know is coming to me, and if there is anything left over, which cannot be sent right away, I want to loan out in Holstein at six percent interest, which is now the customary rate there. You can undoubtedly get some reliable guarantors to whom I will give a contract.

You wrote that you have had my stable of horses brought over to Skaane and still did not know how they could be sold. I will let you decide this and like matters as you will, for I do not doubt that you will handle them gainfully and well.

You deserve thanks for the horses you acquired for me, although they were expensive (one can make a better buy on horses here in this country), but I am satisfied with what you did. I am quite pleased with the one you bought from Oluf Bille for me personally. He did not look too good when he first arrived here, and he had lost his trot on the way, but he has now recovered both his weight and his trot. I will keep it to ride myself in Prague and elsewhere, and nobody will easily get it from me, as you also advise. Many thanks also for the handsome little coach you gave to me. As for any foals still on Hven that belong to me, you can do with them as you please, but I would prefer that you sold them to others than to peasants there on the island. Concerning the fish on Hven which I wanted to have salted down, Tyge tells me that he had fourteen barrels of them salted down. I had expected to receive many more, but you can see to it, both with this and other matters, that all goes sufferably and well, and that some carp and smaller fish remain in the largest fishponds, and whatever else needs to be done.

David Pedersen's accounts and that which concerns them I cannot now obtain for you because they are with those of my goods which are stored in Magdeburg. But you know what to do with respect to him because you know his affairs inside and out.

The accounts of Diderich Dyer's woman are also with the others, so that I cannot send them to you either. You know well how they were scraped off and falsified, from which it is easy to imagine how good the claim is when she brings up some wagonloads of rye and barley which she has had in her care for some years, and which were always short by four or five measures out of every thirty-two when they came to Hven. That is my reply to her fictitious bills. She can show her receipt for the same rye and barley. She knows well that when her husband was alive, I found no small shortage in his, and that this was in all fairness, but I have not received any compensation since that time. . . .

Dear Esge, you report in your postscript, after you had returned to Copenhagen, concerning my house there in the city, that you cannot get more than fifteen hundred dalers for it from the ladies. It cost me over two thousand to build it in the beginning, besides the lot and the other buildings that stood there previously, and I had it remodelled and improved, not long before I left that country, for at least another thousand dalers, and Michel Wibe knows most of the details. I should like you to take down right away the bell hanging in the little spire which I had built over the bay and send it to Lübeck to Jochum Buch, because it belongs to the clockwork which is in Lübeck, and I do not want it to remain there when I have good use for it here. In addition, I have learned that one of my clocks, which shows minutes, also remained behind on Hven, and Tyge reports that you have loaned it to Jørgen Brockenhus to make a copy of it. I would also like to have it back because I have need of it here in my observation, even though I had several like it. You can have a clockmaker pack it in a little square chest and also send it to Lübeck to Jochum Buch, because I imagine that Jørgen Brockenhus has probably now used it as much as he needs to. However, if he cannot get another made there and wants to keep it, then you may let him have it, because he is my good friend and dares to speak the truth for my sake on occasion. I would like to see him have some instruments made and observe with them, there in that country, as he told Tyge he wants to do, and when I learn that he actually intends to do so, I will write him an instruction concerning what in particular to observe each year. If others who do not have any understanding of such things may mock, he should pay no attention to them, nor should he reply anything to them. If he were able sometime to come down here to me, then I would certainly direct him in a short time so that he could feel comfortable with it and work with it more easily and with greater success thereafter. You can say this to him on my behalf when you talk with him. I have also made such arrangements various places in Germany with those who like to do it. There are also some of the distinguished lords in Prague who have already had some instruments made after those I brought with me, and want to observe with them, and His Imperial Majesty himself has some in stock and intends to have more made after some of mine, when they all arrive, which His Majesty will have at the castle in Prague for himself and whomever in particular he wants to use for that, for His Majesty has great interest and pleasure in such things, and right after Easter, he intends to send to me his most noted clockmaker, who is a rare artist, and from me, he will learn everything involved, and then make some like them for His Majesty.

Dear Esge and dear brother, do not blame me for burdening you so often with my business. I have nobody else there in Denmark who is willing to go to so much trouble for me. If I can repay your services in any way, you will find me more than willing to do so. Now and always, I commend you to God almighty. May he preserve you from all that may be harmful to either body or soul. Please wish your dear father and mother and all of our other good friends many good nights on my behalf.

Dated in the land of Bohemia at the imperial castle of Benatky.[11]

---

[11] XIV, 185–9. The old walled city of Copenhagen is commemorated today in streets that bear the name *Voldgade*, or "Rampart Street." In Tycho's day, the west rampart street (Vestervoldgade) ran along the inside of the city wall to Langangstraade, where it ended at a great watermill powered by a millstream flowing from the east. There the wall angled east to enclose the millstream, but continued only as a wooden palisade.

In 1559–60, King Frederick II purchased two properties along that palisade for the purpose of establishing a dye works and fulling mill, the first factory in Copenhagen. The properties included the watermill to provide power for the fulling of cloth.

Tycho inherited from his father the property in front of the dye works on the street that had acquired (and retains) the name "Dyer street" (Farvergade). It was about a hundred meters from what is now called H. C. Andersen's square.

Perhaps thinking already of transforming his property into a kind of urban Uraniborg, with its observatory on the watermill tower, in 1589 Tycho purchased the dye works and fulling mill. As part of the purchase agreement, he promised to build a good new dwelling and workshop closer to the mill pond for the master dyer.

With three lots and the tower bastion to build on, Tycho erected what must have been a splendid townhouse. It had a bay in front and a spire with a clock and bell, which was quite unusual in that day. Perhaps the best gauge of its grandeur is the fact that he spent on it four times the 500 dalers that Frederick originally gave him to build Uraniborg. And since he claims to have spent an additional 1000 dalers to improve the site, he may have planned to move there (permanently) for some time before he actually went there (briefly) in April of 1597.

When King Christian IV eventually acquired the complex in 1605, he transformed it into a poorhouse. By the end of the year, it had some fifty to seventy-five residents, which gives some idea of the size of the complex. In the nineteenth century, the house was torn down and replaced by the Vartov hospital. See O. Nielsen, *Kjøbenhavns historie og beskrivelse*, Copenhagen, G.E.C. (Gad, 1881), pp. 361 et seq.

LETTER 7: TYCHO BRAHE TO HOLGER ROSENCRANTZ, 25 MARCH/4 APRIL 1601

My most friendly greetings now and always in Our Lord. Dear Holger Rosencrantz, kinsman by birth, brother-in-law,[12] and especially good friend, may God almighty now and always be with you and preserve you long and well from all that might be harmful either to soul or life. Many thanks for all the good you have shown me, for which in return with my fortune I will always be found willing and full ready.

This is to inform you most cordially that, inasmuch as I, with the help of God, intend at a time between the coming Easter and Pentecost, here in Prague in the Kurtz house, where I now reside, to hold my dear daughter, Honorable and Wellborn Maid Elisabet Brahe's wedding with a nobleman, born in Westphalia, who serves here at the court of the Roman Imperial Majesty, by name, Franz Tengnagel von Kampe. My most friendly request to you is that you will kindly come here at the aforementioned time, and be lusty and glad with other distinguished lords and good people of His Imperial Majesty's Council, and others in this lawful kingdom of Bohemia, which I hope to assemble here at my residence at that time. Dear Holger, my dear brother-in-law, please do in this which I commend to you whatever it is possible for you to do. You will always find me willing to serve you and yours in return. And now I will not keep you longer with this my letter, but now and always commend you to God almighty, both in life and soul. Dated in Prague at the Kurtz house, at the edge of town, where I now live.[13]

---

[12] Holger was married to Tycho's niece, not his sister. See a similar shortcut in Tycho's salutation to Axel Gyldenstierne (letter 5), who was a son of Tycho's mother's aunt, rather than Tycho's mother's sister.

[13] XIV, 218.

Author Index

See also Appendix 2. All references are to footnotes on page cited.

Aaboe, 55
Abell, 87
Ackerman, 108
Aiton, 254
Andersen, 8, 40
Argelander, 191
Arhnung, 142

Baade, 61
Baldwin, 61
Bastholm, 45, 75
Bech, 5, 27, 46
Beckett, 108
Bialas, 191, 210, 222
Bowden, 214, 219, 301
Braunmühl, 237
Brønstad, 205

Caspar, 452, 460
Christensen, 108
Christianson, x, 10, 48, 80, 101, 118, 127,
    129, 193, 214, 216, 276, 339, 346, 425
Clark, 61, 62

Delambre, 156, 223, 234, 249, 260, 299, 318,
    492, 494
Dobrzycki, 248, 253, 274
Donahue, 254
Dreyer, vii, 89, 282, 288, 290, 295, 299, 317,
    364, 382

Edge, 61
Erslev, 39
Evans, James, 174, 299
Evans, Robert, 212, 215, 389, 444, 445

Figala, 213
Friis, vii, 353, 364
Frisch, 495

Garstein, 202
Gingerich, ix, 86, 91, 222, 236, 238, 248, 250,
    281, 409
Glarbo, 14
Goldstein, 62, 156, 245
Grant, 88, 89, 90, 275
Guerlac, 264

Halley, 294
Hammer, 297
Hannaway, 135

Hartfelder, 11
Hartner, 245
Heath, 227
van Helden, 304
Helk, 54
Hellman, 69, 73, 127, 138, 248, 264
Hellmann, 214
Henderson, 310
Hon, 317

Jansson, 198
Jardine, 80, 393, 394
Jensen, 108
Jern, 109, 134
Johnson, 57, 90
Jones, 239, 256

Klitgaard, 47
Kolb, 100
Krabbe, 44
Kunitzsch, 297

Lejeune, 227
Lengertz, 364
Lindroth, 25

Maeyama, ix, 230–4, 299
Matiegka, 26
Moesgaard, x, 128, 256, 286, 289, 298, 303
Møllerup, 24, 354
Moran, 93, 236
Mortensen, 187, 339, 366, 463
Mukhopadhyaya, 327

Nelleman, 360
Neugebauer, 491
Newton, 155, 174, 299
Norlind, viii, 366, 386
North, x, 89

Overmann, 275

Pannekoek, 289, 299
Parsons, 62
Peng-Yoke, 62
Peters, 190
Plaskett, 222
Price, 155, 205, 233, 253

Raeder, 4
Reedy, 82

Repsold, 151
Riddell, 88
Roberts, 318
Roche, 149
Rordam, 42, 132
Rosen, ix, 90, 276, 391, 459

Schofield, (*see* Jones)
Secher, 352
Skov, 45, 75
Skrubbeltang, 112
Stephenson, 61, 62
Stolterfoht, 23
Straker, 451
Strömgren, 481
Studnicka, 282
Swerdlow, 89, 254, 289, 290, 300, 310

Thoren, 151, 238, 281, 316, 320, 321, 322,
    325, 407

Toomer, 172
Tupman, 191

Vieth, 25
Vogt, 55, 242, 299

Wad, 74
Wanscher, 109
Warburg, 189
Warner, 298
Wegener, 17
Wesley, 191
Westman, ix, 68, 90, 236, 238, 243, 248, 253,
    254, 260, 276, 281, 409
Wilson, x, 233, 293, 451
Wittkower, 108
Woldstedt, 190

Zimmermann, 25

Subject Index

Aalborg, 22, 315, 387, 401
alchemy, 98, 206, 210–13, 445
Alfonso, 277n
Alhazen, 227–8
annual equation, 486
Apian, 10, 30, 249
astrology, 12, 60–1, 69–70, 81–4, 120–2, 141–2, 213–19, 443–4
Augustus (Duke, Elector), 5, 100, 118
Axelsen, Cort, 80n, 129n, 202–3, 479

Bacmeister, 22, 376
Barwitz, 410, 413, 416, 443
Bertelsen, Ivar, 127, 373
Bille: family, 3, 341; Anders, 351; Beate (mother), 3, 119, 340, 424; Esge, 388, 425, 447, 471; Steen (uncle), 49n, 50, 102, 119, 267
Blaeu, 200, 205, 422n, 477n
Bølle, 341, 371
Brahe: family, 1–9, 21, 38, 48, 341; Swedish family, 466, 474; Axel, (great-uncle) 1, 341, (brother) 8, 342, 371–2, 387, 424, 443; Beate (mother), see Beate Bille; Elizabeth (daughter), 141, 357, 463, 477; Jørgen, (uncle) 1, 4–5, 20, 337, (brother) 355, 368, 461, (son) 192; Kirsten (wife), 45–8, 75, 104, 115; Knud (brother), 52, 141, 351, 355, 464; Lange (cousin), 2, 25; Lisbet (sister), 3, 21; Magdalene (daughter), 78, 356, 359–60, 409, 471–5, 477–9; Margarethe (sister), 388, 424, 511; Otte (father), 1, 20, 36–7, 115–16; Sidsel and Sophie (daughters), 141, 463, 478; Sophie (sister), 75, 140, 205, 213, 335, 357–9, 403, 424, 464–5; Steen (brother), 8, 119, 141, 342, 348, 355, 424, 464; Tyge, or Tycho, Jr. (son), 192, 384, 405, 419, 431, 461, 463, 475–7
Brahe, Tycho, books of: De nova stella (1573), 55–73, 75; De mundi ... (1588), 123–7, 135–8, 248–50, 259–64, 272–3, 312, 420–21; Epistolae astronomicae (1596), 314, 362–3, 366; Mechanica (1598), 150–80, 367, 381–2, 414; Astronomiae instauratae progymnasmata (1602), 283–5, 313, 362, 382, 421
Brahe, Tycho: education of, (Univ. of Copenhagen) 9–12, (Leipzig) 14–19,

(Rostock [duel]) 22–4; estate and rank (of children), 45–7, 350–5, 461–3; exile, (at Rostock) 376–80, (at Wandesbeck) 381–403, (at Magdeburg) 403–4, (at Dresden) 404–5, (at Wittenberg) 405–8, (at Prague) 410–15, 442–70, (at Brandeis) 414, (at Benatky) 414; fiefs and incomes, 103–5, 122, 132–3, 142–3, 188–9, 337–40, 344, 367–70, 418–19, 422, 425, 430, 452, 462, 473, 477; grand tour, 92–101; lawsuits, (against Pedersen) 346–8, (against Gallius) 358–61, (against Ursus) 390, 396, 403–4, 420, 453–5, 459–61; lectures at Copenhagen, 78–92; loans, (from Queen Sophie) 340, 349n, (to Mecklenburg) 378–9, 427, 462, (from Rantzov) 388, 462, 512; portraits, 365–6; students of, 192–205; travels, (Augsburg) 30–5, 97, (Basel) 30, 96–7, (Cassel) 93–6, (Herrevad) 47–52, 74, (Venice) 97
Braun, 208
Brockenhuus family, 428, 515
Brucaeus, 24, 117, 139, 178, 211–12, 215, 251, 274, 277, 313, 315, 376
Brunswick (Duke), 335–403, 411
Buchanan, 117, 277
Bürgi, 271, 283, 394
Buridan, 89, 117

calendar, 313, 362
Calvin, 373, 100–1, 118–19
Camerarius, 99, 315
canonry, see Roskilde
Capuchins, 445
Catalogue of Stars, 288–9, 294–300, 383
Chapel of the Wise Men, 338–9, 348–9, 357
Christian (Ripensis), see Ripensis
Christian (Sorensen), see Longomontanus
Christian IV, 348–9, 377–82, 384–6, 400
Chytraeus, 376, 423
clocks, 157–9
comet: 1577, 123–32, 136–8, 249; others, 265
Copernicus, 85–6, 90–2, 238–47; see also Earth, motion of
copyrights, 315–16
Corraduc, 396, 409, 420, 455, 465
Council of the Realm, see Rigsraad
Craig, 312, 364, 408, 420

Cramer, 456–8
Crol, 180, 303

Daney, 42, 62, 79, 84, 102, 118, 315
Dasypodious, 91, 219, 238
Dee, 57, 98
Diary, meteorological, 194
Dybvad, 100, 128–30, 132

Earth, motion of, 87, 250–1, 254–5, 261,
  274–80, 337
eclipses, 11–12, 75, 123, 127, 141, 210, 316,
  322, 407, 449
Elegy to Denmark, 381, 385, 399–400
Elegy to Urania, 71, 76
equant, 88, 91, 483
equation of time, 491–4
Eriksen, Johannes, from Hamburg, 99n, 397,
  447, 463n
Ernst, Elector of Cologne, 396, 401

Fabricius, 431, 447, 460
Flemløse, 92n, 134, 193, 199n, 207, 210–15,
  269, 283
Frauenburg, 194–6
Frederick II, 41, 43, 54, 118, 336–40
Friis: Christian, 367–71, 376, 381, 508;
  Johann, 41, 309n
Frisius, Gemma, 18, 174, 249
Frobenius, 204–5, 406n

Galen, 24, 53, 75
Ganz, 429
Gassendi, vii, viii, 4, 160n, 278
Gellius, 193, 212, 356–62, 372, 405
Gemperlin, 99, 108, 125, 365
geodetic work, 208, 210
globe, large, 159–62
Golden Griffin, 442, 500
Goye: Falk, 355; Mogens, 6, 351
Gyldenstierne family, 341; Axel, 212, 345n,
  426, 462; Henrik, 429; Peder, 348;
  Knud, 429

Hainzel, Paul, 31, 92, 97
Hardenberg: Anne, 353; Erik, 343
Hayek, 98, 117, 178, 211, 247, 257n, 267–70,
  273, 300n, 314–16, 327, 364, 374–5, 382,
  389–90, 395, 410, 418, 466–7
Hegelund: Jacob, 193; Peder; 14, 19, 193
Hemmingsen, 10, 42, 79–84, 100, 118–19,
  373
Hipparchus, 66, 81, 292, 467
Hoffman, Baron, 440, 451, 459, 466
von Hohenburg, 439n, 451, 468, 486
Holck, 142, 378
Huitfeld, 342, 348

Instruments: first purchases, 14–18;
  construction of, 31–3, 75–8, 149–52,
  159–71, 172–80, 196; sights and
  divisions; 152–7; accuracy of, 190–1;
  removal to Prague, 429, 443

Jachinow, 205
James I, 334–5
Jeppe, 193, 194n
Jessenius, 405–6, 454, 465
Johannes Franciscus (Ripensis), 13, 43, 70,
  85, 114, 186
Jöstel (Joestelius), 406–8, 431, 447, 486, 492

Kaas, Niels, 43, 101, 126, 342, 344, 348, 367
Kepler, 72, 258, 276, 383, 419, 421, 423, 430,
  432–42, 450–3, 459, 469n, 493–4
Krag: Anders, 358; Niels, 207, 356, 358, 374,
  402, 424
Kronborg, 103, 113, 122, 135
Kurtz, 273, 316, 390, 396, 411, 413

Laetus, 42, 78
Landgrave, see Wilhelm
Lange: Erik, 186, 206, 211, 255, 355, 399,
  403; Niels, 351
van Langren, 422n
latitude: of Frauenburg, 194–5; of Hven,
  224–6
Laurentius, 117, 141n, 408
Leovitius, 30, 219
Liddel, 455–8
longitudes (of stars), 271–2, 290–4
Longomontanus (Christian Sorensen), 99n,
  199, 255, 297, 369n, 383, 397–8, 477,
  486–95
Lunar Theory, Chapter 10, 397, 404, 407,
  419, 421, 431, 449–50, App. 3
Luther, 8, 81, 276

Magini, 254, 273, 405–6, 448
Major, 31, 117, 313
Mästlin, 73, 139, 253n, 254, 257, 276, 313,
  422, 432–5, 451
Matthias, Archduke, 428, 444–5
Mecklenburg, Duke of, 217, 266, 377, 388,
  396, 412, 427, 462
Melanchthon, 8, 11, 41, 81, 86, 276, 405
Mercator, 18, 422
Monau, 408–9
Mühlstein, 416–18, 420
Mule: Claus, 356, 398; Else, 355, 360; family,
  355
Müller, 398, 419, 431, 440–1, 453

New Star of 1572, 55–70, 307
Nuñez, 156–7

Obliquity (of ecliptic), 195, 224, 226, 231–2
observations: accuracy, 190–1, 285–6; frequency, 201, 220, 296; procedure, 201
Offusius, 99, 303
Olsen, 194, 207, 210
Osiander, 90, 275–6
Oxe: Inger, 4, 20, 102, 340–1, 423; Peder, 5, 26–8, 41, 101–2

Palladio, 106–8
paper, 266, 314–15, 362–3, 366
Paracelsus, 24, 52–3, 59, 79, 82
parallax: comets, 124–5; Mars, 250–1, 255–8; moon, 56, 300, 327; new star, 57–8, 67–8; sun, 227–34; stars, 88, 279–80, 303–6
Parsberg, 22, 140, 343
Pedersen, Rasmus, 346–8
Peucer, 17, 86, 91n, 100, 218, 238, 259, 312n, 405
Plato, 109, 264n, 277
Pontanus, 203, 205, 316
Praetorius, 100, 254
Pratensis, 43, 62–5, 69–72, 102–4, 113, 186, 211
prosthaphaereses, 237–8, 280–3, 407
Ptolemy, 81, 88, 227, 292

Ramus, Petrus, 33–5, 42
Rantzov, Heinrich, 140, 186, 208, 255, 266, 314, 341, 343, 364, 370, 380–5, 396, 462
refraction, 94, 226, 235
regency, 344, 354
Regiomontanus, 10, 137, 248, 263, 318n
Reinhold, 86, 99
Reymers, *see* Ursus
Rheticus, 86, 237
Rigsraad, 1, 340–4, 354
Ripensis, Christian, 198, 211, 401n
Rollenhagen, 260, 262, 273, 393, 403
Rosenkrantz: family, 341–2, 410, 424–26, 457, 464; Frederick, 351n, 427–9; Holger, 343, 387–8, 400–2; Jørgen, 101, 119, 343–4, 349, 360, 369
Roskilde, 29, 142–3, 337–9, 348–9
Rothmann, 204, 236, 257, 271, 276–80, 283, 290–3, 301, 336–7, 363–7, 391–3, 422, 423n, 431, 447
Rozmberk, 467
Rud, 1, 20, 37
Rudolph II, 97, 255, 315, 382, 389, 404, 409–19, 442–6, 460–1
Rumpf, 411–2, 445

Saxo Grammaticus, 41, 51, 206
Scaliger, 401, 418, 467
Scavenius, 9–10, 79
Schissler, 30, 36, 117, 159
Schultz, 17, 217n, 248–50, 280, 374, 421
Severinus, *see* Peder Sørensen
Shakespeare, 113
sizes & distances (of heavenly bodies), 302–8
Skeel: Albert, 356; Christen, 342; Hans, 47
Skram, 14, 428n
Snel, 422
Sophie (Queen), 54, 119, 340, 349n, 354
Sørensen: Christian, *see* Longomontanus; Peder, 43, 75n, 79, 98, 113–14, 116, 127, 212n
spheres, 254–62, 273–4, 306–8
stars: catalogue, 294–300; proper motion, 294, 306–7; scintillation, 307
van Steenwinckel, 133, 187, 606
Stephanius, 193, 203
Stjerneborg, 180–5
Stygge, 386–7, 401
supernova, *see* New Star of 1572

Tables: Alfonsine (Ptolemaic), 16, 91, 326; Prutenic (Copernican), 16, 85
Tengnagel, 396–8, 401, 404–5, 439, 459, 463, 478
theology, 275–7, *see also* Calvin
Thott, 2, 140

Ursus, 255, 260–1, 283, 314, 390–6, 399, 404, 408, 420, 432–9, 441, 453, 458–60

Valkendorf, 342, 344, 368, 370–2, 381, 386
Variation, 324–7, 486
Vedel, 10, 24, 51, 70, 206–7, 314, 374, 388–9
Venus: comet's tail pointed towards, 249; intermediary for absolute longitudes, 287–8
Venusin, 356, 358–9, 373

Wachtmeister, xi, 115n
Walther: Bernard, 227, 288; Michael, 399, 403
Wilhelm, Landgrave of Hesse, 57, 93–6, 102, 224n, 228, 256, 262–3, 266–72, 287–9, 336, 344–6, 363, 367
Witelo, 227–8, 451
Wittich, 236–49, 267–71, 280–3, 408–9, 422, 458
Wolf, 31, 92, 97, 139, 216n
Wolf-Dietrich, 445, 446n